The
Colossal
Book of
Mathematics

Also by Martin Gardner

Also by Martin Gardner

Martin Gardner

The
COLOSSAL
Book of
MATHEMATICS
Classic Puzzles, Paradoxes, and Problems

Number Theory

Algebra

Geometry

Probability

Topology

Game Theory

Infinity

and other Topics
of Recreational
Mathematics

W. W. Norton & Company
New York · London

Copyright © 2001 by Martin Gardner

The text of this book is composed in Melior, with the display set in Goudy Sans Italic with Augustea Open.
Composition by Allentown Digital Services
Manufacturing by The Maple-Vail Book Manufacturing Group
Book design by Charlotte Staub
Production manager: Andrew Marasia

Library of Congress Cataloging-in-Publication Data
Gardner, Martin, 1914–
    The colossal book of mathematics: classic puzzles, paradoxes, and problems: number theory, algebra, geometry, probability, topology, game theory, infinity, and other topics of recreational mathematics / Martin Gardner.
        p. cm.
    Includes bibliographical references and index.
    ISBN 0-393-02023-1
    1. Mathematical recreations. I. Title.
QA95.G245 2001
793.7'4—dc21   2001030341

W. W. Norton & Company, Inc., 500 Fifth Avenue, New York, N.Y. 10110
www.wwnorton.com

W. W. Norton & Company Ltd., Castle House, 75/76 Wells Street, London W1T 3QT

1 2 3 4 5 6 7 8 9 0

For Gerard Piel

Who made all this possible

Contents

Preface

My long and happy relationship with *Scientific American,* back
in the days when Gerard Piel was publisher and Dennis Flanagan was
the editor, began in 1952 when I sold the magazine an article on the history of logic machines. These were curious devices, invented in pre-
computer centuries, for solving problems in formal logic. The article
included a heavy paper insert from which one could cut a set of window cards I had devised for solving syllogisms. I later expanded the article to *Logic Machines and Diagrams,* a book published in 1959.

My second sale to *Scientific American* was an article on hexaflexagons, reprinted here as Chapter 29. As I explain in that chapter's
addendum, it prompted Piel to suggest a regular department devoted to
recreational mathematics. Titled "Mathematical Games" (that M. G. are
also my initials was a coincidence), the column ran for a quarter of a
century. As these years went by I learned more and more math. There
is no better way to teach oneself a topic than to write about it.

Fifteen anthologies of my *Scientific American* columns have been
published, starting with the *Scientific American Book of Mathematical
Puzzles and Diversions* (1959) and ending with *Last Recreations* (1997).
To my surprise and delight, Robert Weil, my editor at W. W. Norton,
suggested that I select 50 of what I consider my "best" columns, mainly
in the sense of arousing the greatest reader response, to make this hefty,
and, in terms of my career, definitive book you now hold. I have not included any of my whimsical interviews with the famous numerologist
Dr. Irving Joshua Matrix because all those columns have been gathered
in *The Magic Numbers of Dr. Matrix* (1985).

To each chapter I have added an addendum, often lengthy, to update
the material. I also have provided selected bibliographies for further
reading.

Martin Gardner
Hendersonville, NC

I
Arithmetic and Algebra

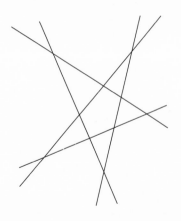

Chapter 1

The Monkey and the Coconuts

In the October 9, 1926, issue of *The Saturday Evening Post* appeared a short story by Ben Ames Williams entitled "Coconuts." The story concerned a building contractor who was anxious to prevent a competitor from getting an important contract. A shrewd employee of the contractor, knowing the competitor's passion for recreational mathematics, presented him with a problem so exasperating that while he was preoccupied with solving it he forgot to enter his bid before the deadline.

Here is the problem exactly as the clerk in Williams's story phrased it:

Five men and a monkey were shipwrecked on a desert island, and they spent the first day gathering coconuts for food. Piled them all up together and then went to sleep for the night.

But when they were all asleep one man woke up, and he thought there might be a row about dividing the coconuts in the morning, so he decided to take his share. So he divided the coconuts into five piles. He had one coconut left over, and he gave that to the monkey, and he hid his pile and put the rest all back together.

By and by the next man woke up and did the same thing. And he had one left over, and he gave it to the monkey. And all five of the men did the same thing, one after the other; each one taking a fifth of the coconuts in the pile when he woke up, and each one having one left over for the monkey. And in the morning they divided what coconuts were left, and they came out in five equal shares. Of course each one must have known there were coconuts missing; but each one was guilty as the others, so they didn't say anything. How many coconuts were there in the beginning?

Williams neglected to include the answer in his story. It is said that the offices of *The Saturday Evening Post* were showered with some

2,000 letters during the first week after the issue appeared. George Horace Lorimer, then editor-in-chief, sent Williams the following historic wire:

For 20 years Williams continued to receive letters requesting the answer or proposing new solutions. Today the problem of the coconuts is probably the most worked on and least often solved of all the Diophantine brainteasers. (The term Diophantine is descended from Diophantus of Alexandria, a Greek algebraist who was the first to analyze extensively equations calling for solutions in rational numbers.)

Williams did not invent the coconut problem. He merely altered a much older problem to make it more confusing. The older version is the same except that in the morning, when the final division is made, there is again an extra coconut for the monkey; in Williams's version the final division comes out even. Some Diophantine equations have only one answer (e.g., $x^2 + 2 = y^3$); some have a finite number of answers; some (e.g., $x^3 + y^3 = z^3$) have no answer. Both Williams's version of the coconut problem and its predecessor have an infinite number of answers in whole numbers. Our task is to find the smallest positive number.

The older version can be expressed by the following six indeterminate equations which represent the six successive divisions of the coconuts into fifths. N is the original number; F, the number each sailor received on the final division. The 1's on the right are the coconuts tossed to the monkey. Each letter stands for an unknown positive integer:

$$N = 5A + 1,$$
$$4A = 5B + 1,$$
$$4B = 5C + 1,$$
$$4C = 5D + 1,$$
$$4D = 5E + 1,$$
$$4E = 5F + 1.$$

It is not difficult to reduce these equations by familiar algebraic methods to the following single Diophantine equation with two unknowns:

$$1{,}024N = 15{,}625F + 11{,}529.$$

This equation is much too difficult to solve by trial and error, and although there is a standard procedure for solving it by an ingenious use

ARITHMETIC AND ALGEBRA

of continued fractions, the method is long and tedious. Here we shall be concerned only with an uncanny but beautifully simple solution involving the concept of *negative* coconuts. This solution is sometimes attributed to the University of Cambridge physicist P.A.M. Dirac (1902–1984), but in reply to my query Professor Dirac wrote that he obtained the solution from J.H.C. Whitehead, professor of mathematics (and nephew of the famous philosopher). Professor Whitehead, answering a similar query, said that he got it from someone else, and I have not pursued the matter further.

Whoever first thought of negative coconuts may have reasoned something like this. Since N is divided six times into five piles, it is clear that 5^6 (or 15,625) can be added to any answer to give the next highest answer. In fact any multiple of 5^6 can be added, and similarly any multiple can be subtracted. Subtracting multiples of 5^6 will of course eventually give us an infinite number of answers in negative numbers. These will satisfy the original equation, though not the original problem, which calls for a solution that is a positive integer.

Obviously there is no small positive value for N which meets the conditions, but possibly there is a simple answer in negative terms. It takes only a bit of trial and error to discover the astonishing fact that there is indeed such a solution: –4. Let us see how neatly this works out.

The first sailor approaches the pile of –4 coconuts, tosses a positive coconut to the monkey (it does not matter whether the monkey is given his coconut before or after the division into fifths), thus leaving five negative coconuts. These he divides into five piles, a negative coconut in each. After he has hidden one pile, four negative coconuts remain— exactly the same number that was there at the start! The other sailors go through the same ghostly ritual, the entire procedure ending with each sailor in possession of two negative coconuts, and the monkey, who fares best in this inverted operation, scurrying off happily with six positive coconuts. To find the answer that is the lowest positive integer, we now have only to add 15,625 to –4 to obtain 15,621, the solution we are seeking.

This approach to the problem provides us immediately with a general solution for n sailors, each of whom takes one nth of the coconuts at each division into $n$ths. If there are four sailors, we begin with three negative coconuts and add 4^5. If there are six sailors, we begin with five negative coconuts and add 6^7, and so on for other values of n. More formally, the original number of coconuts is equal to $k(n^{n+1}) - m(n - 1)$, where n is the number of men, m is the number of coconuts given to the monkey at each division, and k is an arbitrary integer called the parameter. When n is 5 and m is 1, we obtain the lowest positive solution by using a parameter of 1.

Unfortunately, this diverting procedure will not apply to Williams's modification, in which the monkey is deprived of a coconut on the last division. I leave it to the interested reader to work out the solution to the Williams version. It can of course be found by standard Diophantine techniques, but there is a quick shortcut if you take advantage of information gained from the version just explained. For those who find this too difficult, here is a very simple coconut problem free of all Diophantine difficulties.

Three sailors come upon a pile of coconuts. The first sailor takes half of them plus half a coconut. The second sailor takes half of what is left plus half a coconut. The third sailor also takes half of what remains plus half a coconut. Left over is exactly one coconut which they toss to the monkey. How many coconuts were there in the original pile? If you will arm yourself with 20 matches, you will have ample material for a trial and error solution.

Addendum

If the use of negative coconuts for solving the earlier version of Ben Ames Williams's problem seems not quite legitimate, essentially the same trick can be carried out by painting four coconuts blue. Norman Anning, now retired from the mathematics department of the University of Michigan, hit on this colorful device as early as 1912 when he published a solution (*School Science and Mathematics,* June 1912, p. 520) to a problem about three men and a supply of apples. Anning's application of this device to the coconut problem is as follows.

We start with 5^6 coconuts. This is the smallest number that can be divided evenly into fifths, have one-fifth removed, and the process repeated six times, with no coconuts going to the monkey. Four of the 5^6 coconuts are now painted blue and placed aside. When the remaining supply of coconuts is divided into fifths, there will of course be one left over to give the monkey.

After the first sailor has taken his share, and the monkey has his coconut, we put the four blue coconuts back with the others to make a pile of 5^5 coconuts. This clearly can be evenly divided by 5. Before making this next division, however, we again put the four blue coconuts aside so that the division will leave an extra coconut for the monkey.

This procedure—borrowing the blue coconuts only long enough to see that an even division into fifths can be made, then putting them aside again—is repeated at each division. After the sixth and last division, the blue coconuts remain on the side, the property of no one. They play no essential role in the operation, serving only to make things clearer to us as we go along.

A good recent reference on Diophantine equations and how to solve them is *Diophantus and Diophantine Equations* by Isabella Bashmakova (The Mathematical Association of America, 1997).

There are all sorts of other ways to tackle the coconut problem. John M. Danskin, then at the Institute for Advanced Study, Princeton, NJ, as well as several other readers, sent ingenious methods of cracking the problem by using a number system based on 5. Scores of readers wrote to explain other unusual approaches, but all are a bit too involved to explain here.

Answers

The number of coconuts in Ben Ames Williams's version of the problem is 3,121. We know from the analysis of the older version that $5^5 - 4$, or 3,121, is the smallest number that will permit five even divisions of the coconuts with one going to the monkey at each division. After these five divisions have been made, there will be 1,020 coconuts left. This number happens to be evenly divisible by 5, which permits the sixth division in which no coconut goes to the monkey.

In this version of the problem a more general solution takes the form of two Diophantine equations. When n, the number of men, is odd, the equation is

$$\text{Number of coconuts} = (1 + nk)n^n - (n - 1).$$

When n is even,

$$\text{Number of coconuts} = (n - 1 + nk)n^n - (n - 1).$$

In both equations k is the parameter that can be any integer. In Williams's problem the number of men is 5, an odd number, so 5 is substituted for n in the first equation, and k is taken as 0 to obtain the lowest positive answer.

A letter from Dr. J. Walter Wilson, a Los Angeles dermatologist, reported an amusing coincidence involving this answer:

> Sirs:
>
> I read Ben Ames Williams's story about the coconut problem in 1926, spent a sleepless night working on the puzzle without success, then learned from a professor of mathematics how to use the Diophantine equation to obtain the smallest answer, 3,121.
>
> In 1939 I suddenly realized that the home on West 80th Street, Inglewood, California, in which my family and I had been living for several months, bore the street number 3121. Accordingly, we entertained all of our most erudite friends one evening by a circuit of games and puzzles, each arranged in a different room, and visited by groups of four in rotation.
>
> The coconut puzzle was presented on the front porch, with the table placed directly under the lighted house number blazingly giving the secret away, but no one caught on!

The simpler problem of the three sailors, at the end of the chapter, has the answer: 15 coconuts. If you tried to solve this by breaking matches in half to represent halves of coconuts, you may have concluded that the problem was unanswerable. Of course no coconuts need be split at all in order to perform the required operations.

Ben Ames Williams's story was reprinted in Clifton Fadiman's anthology, *The Mathematical Magpie* (1962), reissued in paperback by Copernicus in 1997. David Singmaster, in his unpublished history of famous mathematical puzzles, traces similar problems back to the Middle Ages. Versions appear in numerous puzzle books, as well as in textbooks that discuss Diophantine problems. My bibliography is limited to periodicals in English.

Bibliography

N. Anning, "Solution to a Problem," *School Science and Mathematics,* June 1912, p. 520.

N. Anning, "Monkeys and Coconuts," *The Mathematics Teacher,* Vol. 54, December 1951, pp. 560–62.

J. Bowden, "The Problem of the Dishonest Men, the Monkeys, and the Coconuts," *Special Topics in Theoretical Arithmetic,* pp. 203–12. Privately printed for the author by Lancaster Press, Inc., Lancaster, PA, 1936.

P. W. Brashear, "Five Sailors and a Monkey," *The Mathematics Teacher*, October 1967, pp. 597–99.

S. King, "More Coconuts," *The College Mathematics Journal*, September 1998, pp. 312–13.

R. B. Kirchner, "The Generalized Coconut Problem," *The American Mathematical Monthly*, Vol. 67, June–July 1960, pp. 516–19.

R. E. Moritz, "Solution to Problem 3,242," *The American Mathematical Monthly*, Vol. 35, January 1928, pp. 47–48.

T. Shin and G. Salvatore, "On Coconuts and Integrity," *Crux Mathematicorum*, Vol. 4, August/September 1978, pp. 182–85.

S. Singh and D. Bhattacharya, "On Dividing Coconuts: A Linear Diophantine Problem," *The College Mathematics Journal*, May 1997, pp. 203–4.

Chapter 2

The Calculus of Finite Differences

*T*he calculus of finite differences, a branch of mathematics that is not too well known but is at times highly useful, occupies a halfway house on the road from algebra to calculus. W. W. Sawyer, a mathematician at Wesleyan University, likes to introduce it to students by performing the following mathematical mind-reading trick.

Instead of asking someone to "think of a number" you ask him to "think of a formula." To make the trick easy, it should be a quadratic formula (a formula containing no powers of x greater than x^2). Suppose he thinks of $5x^2 + 3x - 7$. While your back is turned so that you cannot see his calculations, ask him to substitute 0, 1, and 2 for x, then tell you the three values that result for the entire expression. The values he gives you are −7, 1, 19. After a bit of scribbling (with practice you can do it in your head) you tell him the original formula!

The method is simple. Jot down in a row the values given to you. In a row beneath write the differences between adjacent pairs of numbers, always subtracting the number on the left from its neighbor on the right. In a third row put the difference between the numbers above it. The chart will look like this

$$-7 \quad 1 \quad 19$$
$$8 \quad 18$$
$$10$$

The coefficient of x^2, in the thought-of formula, is always half the bottom number of the chart. The coefficient of x is obtained by taking half the bottom number from the first number of the middle row. And the constant in the formula is simply the first number of the top row.

What you have done is something analogous to integration in calculus. If y is the value of the formula, then the formula expresses a function of y with respect to x. When x is given values in a simple arithmetic progression $(0, 1, 2, \ldots)$, then y assumes a series of values $(-7, 1, 19, \ldots)$. The calculus of finite differences is the study of such series. In this case, by applying a simple technique to three terms of a series, you were able to deduce the quadratic function that generated the three terms.

The calculus of finite differences had its origin in *Methodus Incrementorum,* a treatise published by the English mathematician Brook Taylor (who discovered the "Taylor theorem" of calculus) between 1715 and 1717. The first important work in English on the subject (after it had been developed by Leonhard Euler and others) was published in 1860 by George Boole, of symbolic-logic fame. Nineteenth-century algebra textbooks often included a smattering of the calculus, then it dropped out of favor except for its continued use by actuaries in checking annuity tables and its occasional use by scientists for finding formulas and interpolating values. Today, as a valuable tool in statistics and the social sciences, it is back in fashion once more.

For the student of recreational mathematics there are elementary procedures in the calculus of finite differences that can be enormously useful. Let us see how such a procedure can be applied to the old problem of slicing a pancake. What is the maximum number of pieces into which a pancake can be cut by n straight cuts, each of which crosses each of the others? The number is clearly a function of n. If the function is not too complex, the method of differences may help us to find it by empirical techniques.

No cut at all leaves one piece, one cut produces two pieces, two cuts yield four pieces, and so on. It is not difficult to find by trial and error that the series begins: 1, 2, 4, 7, 11, ... (*see* Figure 2.1). Make a chart as before, forming rows, each representing the differences of adjacent terms in the row above:

NUMBER OF CUTS	0	1	2	3	4
Number of pieces	1	2	4	7	11
First differences		1	2	3	4
Second differences			1	1	1

If the original series is generated by a linear function, the numbers in the row of first differences will be all alike. If the function is a quadratic, identical numbers appear in the row of second differences. A

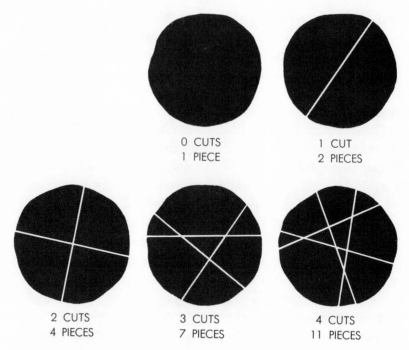

0 CUTS	1 CUT
1 PIECE	2 PIECES

2 CUTS	3 CUTS	4 CUTS
4 PIECES	7 PIECES	11 PIECES

Figure 2.1. The pancake problem

cubic formula (no powers higher than x^3) will have identical numbers in the row of third differences, and so on. In other words, the number of rows of differences is the order of the formula. If the chart required 10 rows of differences before the numbers in a row became the same, you would know that the generating function contained powers as high as x^{10}.

Here there are only two rows, so the function must be a quadratic. Because it is a quadratic, we can obtain it quickly by the simple method used in the mind-reading trick.

The pancake-cutting problem has a double interpretation. We can view it as an abstract problem in pure geometry (an ideal circle cut by ideal straight lines) or as a problem in applied geometry (a real pancake cut by a real knife). Physics is riddled with situations of this sort that can be viewed in both ways and that involve formulas obtainable from empirical results by the calculus of finite differences. A famous example of a quadratic formula is the formula for the maximum number of electrons that can occupy each "shell" of an atom. Going outward from the nucleus, the series runs 0, 2, 8, 18, 32, 50. . . . The first row of differences is 2, 6, 10, 14, 18. . . . The second row is 4, 4, 4, 4. . . . Apply-

Arithmetic and Algebra

ing the key to the mind-reading trick, we obtain the simple formula $2n^2$ for the maximum number of electrons in the nth shell.

What do we do if the function is of a higher order? We can make use of a remarkable formula discovered by Isaac Newton. It applies in all cases, regardless of the number of tiers in the chart.

Newton's formula assumes that the series begins with the value of the function when n is 0. We call this number a. The first number of the first row of differences is b, the first number of the next row is c, and so on. The formula for the nth number of the series is

$$a + bn + \frac{cn(n-1)}{2} + \frac{dn(n-1)(n-2)}{2 \cdot 3}$$
$$+ \frac{en(n-1)(n-2)(n-3) \ldots}{2 \cdot 3 \cdot 4}$$

The formula is used only up to the point at which all further additions would be zero. For example, if applied to the pancake-cutting chart, the values of 1, 1, 1 are substituted for a, b, c in the formula. (The rest of the formula is ignored because all lower rows of the chart consist of zeros; $d, e, f, \ldots$ therefore have values of zero, consequently the entire portion of the formula containing these terms adds up to zero.) In this way we obtain the quadratic function $\frac{1}{2}n^2 + \frac{1}{2}n + 1$.

Does this mean that we have now found the formula for the maximum number of pieces produced by n slices of a pancake? Unfortunately the most that can be said at this point is "Probably." Why the uncertainty? Because for any finite series of numbers there is an infinity of functions that will generate them. (This is the same as saying that for any finite number of points on a graph, an infinity of curves can be drawn through those points.) Consider the series 0, 1, 2, 3. . . . What is the next term? A good guess is 4. In fact, if we apply the technique just explained, the row of first differences will be 1's, and Newton's formula will tell us that the nth term of the series is simply n. But the formula

$$n + \frac{1}{24}n(n-1)(n-2)(n-3)$$

also generates a series that begins 0, 1, 2, 3. . . . In this case the series continues, not 4, 5, 6, . . . but 5, 10, 21. . . .

There is a striking analogy here with the way laws are discovered in science. In fact, the method of differences can often be applied to physical phenomena for the purpose of guessing a natural law. Suppose, for example, that a physicist is investigating for the first time the way in which bodies fall. He observes that after one second a stone drops 16 feet, after two seconds 64 feet, and so on. He charts his observations like this:

$$0 \quad\quad 16 \quad\quad 64 \quad\quad 144 \quad\quad 256$$
$$16 \quad\quad 48 \quad\quad 80 \quad\quad 112$$
$$32 \quad\quad 32 \quad\quad 32$$

Actual measurements would not, of course, be exact, but the numbers in the last row would not vary much from 32, so the physicist assumes that the next row of differences consists of zeros. Applying Newton's formula, he concludes that the total distance a stone falls in n seconds is $16n^2$. But there is nothing certain about this law. It represents no more than the simplest function that accounts for a finite series of observations: the lowest order of curve that can be drawn through a finite series of points on a graph. True, the law is confirmed to a greater degree as more observations are made, but there is never certainty that more observations will not require modification of the law.

With respect to pancake cutting, even though a pure mathematical structure is being investigated rather than the behavior of nature, the situation is surprisingly similar. For all we now know, a fifth slice may not produce the sixteen pieces predicted by the formula. A single failure of this sort will explode the formula, whereas no finite number of successes, however large, can positively establish it. "Nature," as George Pólya has put it, "may answer Yes or No, but it whispers one answer and thunders the other. Its Yes is provisional, its No is definitive." Pólya is speaking of the world, not abstract mathematical structure, but it is curious that his point applies equally well to the guessing of functions by the method of differences. Mathematicians do a great deal of guessing, along lines that are often similar to methods of induction in science, and Pólya has written a fascinating work, *Mathematics and Plausible Reasoning*, about how they do it.

Some trial and error testing, with pencil and paper, shows that five cuts of a pancake do in fact produce a maximum of sixteen pieces. This successful prediction by the formula adds to the probability that the formula is correct. But until it is rigorously *proved* (in this case it is not hard to do) it stands only as a good bet. Why the simplest formula is so often the best bet, both in mathematical and scientific guessing, is one of the lively controversial questions in contemporary philosophy of science. For one thing, no one is sure just what is meant by "simplest formula."

Here are a few problems that are closely related to pancake cutting and that are all approachable by way of the calculus of finite differences. First you find the best guess for a formula, then you try to prove the formula by deductive methods. What is the maximum number of pieces that can be produced by n simultaneous straight cuts of a flat figure shaped like a crescent moon? How many pieces of cheesecake can be produced by n simultaneous plane cuts of a cylindrical cake? Into how many parts can the plane be divided by intersecting circles of the same size? Into how many regions can space be divided by intersecting spheres?

Recreational problems involving permutations and combinations often contain low-order formulas that can be correctly guessed by the method of finite differences and later (one hopes) proved. With an unlimited supply of toothpicks of n different colors, how many different triangles can be formed on a flat surface, using three toothpicks for the three sides of each triangle? (Reflections are considered different, but not rotations.) How many different squares? How many different tetrahedrons can be produced by coloring each face a solid color and using n different colors? (Two tetrahedrons are the same if they can be turned and placed side by side so that corresponding sides match in color.) How many cubes with n colors?

Of course, if a series is generated by a function other than a polynomial involving powers of the variable, then other techniques in the method of differences are called for. For example, the exponential function 2^n produces the series 1, 2, 4, 8, 16. . . . The row of first differences is also 1, 2, 4, 8, 16, . . . , so the procedure explained earlier will get us nowhere. Sometimes a seemingly simple situation will involve a series that evades all efforts to find a general formula. An annoying example is the necklace problem posed in one of Henry Ernest Dudeney's puzzle books. A circular necklace contains n beads. Each bead is black or white. How many different necklaces can be made with n beads? Starting with no beads, the series is 0, 2, 3, 4, 6, 8, 13, 18, 30. . . . (Figure 2.2 shows the 18 different varieties of necklace when $n = 7$.) I suspect that two formulas are interlocked here, one for odd n, one for even, but whether the method of differences will produce the formulas, I do not know. "A general solution . . . is difficult, if not impossible," writes Dudeney. The problem is equivalent to the following one in information theory: What is the number of different binary code words of a given length, ruling out as identical all those words that have the same cyclic order of digits, taking them either right to left or left to right?

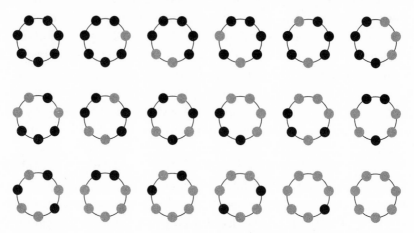

Figure 2.2. Eighteen different seven-beaded necklaces can be formed with beads of two colors.

A much easier problem on which readers may enjoy testing their skill was sent to me by Charles B. Schorpp and Dennis T. O'Brien, of the Novitiate of St. Isaac Jogues in Wernersville, PA: What is the maximum number of triangles that can be made with n straight lines? Figure 2.3 shows how 10 triangles can be formed with five lines. How many can be made with six lines and what is the general formula? The formula can first be found by the method of differences; then, with the proper insight, it is easy to show that the formula is correct.

Addendum

In applying Newton's formula to empirically obtained data, one sometimes comes up against an anomaly for the zero case. For instance, *The Scientific American Book of Mathematical Puzzles & Diversions,* page 149, gives the formula for the maximum number of pieces that can be produced by n simultaneous plane cuts through a doughnut. The formula is a cubic,

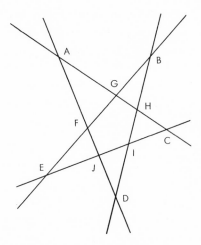

Figure 2.3. Five lines make ten triangles.

$$\frac{n^3 + 3n^2 + 8n}{6},$$

that can be obtained by applying Newton's formula to results obtained empirically, but it does not seem to apply to the zero case. When a doughnut is not cut at all, clearly there is one piece, whereas the formula says there should be no pieces. To make the formula applicable, we must define "piece" as part of a doughnut *produced by cutting*. Where there is ambiguity about the zero case, one must extrapolate backward in the chart of differences and assume for the zero case a value that produces the desired first number in the last row of differences.

To prove that the formula given for the maximum number of regions into which a pancake (circle) can be divided by n straight cuts, consider first the fact that each nth line crosses $n - 1$ lines. The $n - 1$ lines divide the plane into n regions. When the nth line crosses these n regions, it cuts each region into two parts, therefore every nth line adds n regions to the total. At the beginning there is one piece. The first cut adds one more piece, the second cut adds two more pieces, the third cut adds three more, and so on up to the nth cut which adds n pieces. Therefore the total number of regions is $1 + 1 + 2 + 3 + \cdots + n$. The sum of $1 + 2 + 3 + \cdots + n$ is $\frac{1}{2}n(n - 1)$. To this we must add 1 to obtain the final formula.

The bead problem was given by Dudeney as problem 275 in his *Puz-*

zles and Curious Problems. John Riordan mentions the problem on page 162, problem 37, of his *Introduction to Combinatorial Analysis* (Wiley, 1958; now out of print), indicating the solution without giving actual formulas. (He had earlier discussed the problem in "The Combinatorial Significance of a Theorem of Pólya," *Journal of the Society for Industrial and Applied Mathematics,* Vol. 5, No. 4, December 1957, pp. 232–34.) The problem was later treated in considerable detail, with some surprising applications to music theory and switching theory, by Edgar N. Gilbert and John Riordan, in "Symmetry Types of Periodic Sequences," *Illinois Journal of Mathematics,* Vol. 5, No. 4, December 1961, pages 657–65. The authors give the following table for the number of different types of necklaces, with beads of two colors, for necklaces of 1 through 20 beads:

NUMBER OF BEADS	NUMBER OF NECKLACES
1	2
2	3
3	4
4	6
5	8
6	13
7	18
8	30
9	46
10	78
11	126
12	224
13	380
14	687
15	1,224
16	2,250
17	4,112
18	7,685
19	14,310
20	27,012

The formulas for the necklace problem do not mean, by the way, that Dudeney was necessarily wrong in saying that a solution was not possible, since he may have meant only that it was not possible to find a polynomial expression for the number of necklaces as a function of *n* so that the number could be calculated directly from the formula without requiring a tabulation of prime factors. Because the formulas include Euler's phi function, the number of necklaces has to be calculated recursively. Dudeney's language is not precise, but it is possible that he would not have considered recursive formulas a "solution." At any rate, the calculus of finite differences is not in any way applicable to the problem, and only the recursive formulas are known.

Several dozen readers (too many for a listing of names) sent correct solutions to the problem before Golomb's formulas were printed, some of them deriving it from Riordan, others working it out entirely for themselves. Many pointed out that when the number of beads is a prime (other than 2), the formula for the number of different necklaces becomes very simple:

$$\frac{2^{n-1} - 1}{n} + 2^{\frac{n-1}{2}} + 1.$$

The following letter from John F. Gummere, headmaster of William Penn Charter School, Philadelphia, appeared in the letters department of *Scientific American* in October 1961:

Sirs:

I read with great interest your article on the calculus of finite differences. It occurs to me that one of the most interesting applications of Newton's formula is one I discovered for myself long before I had reached the calculus. This is simply applying the method of finite differences to series of powers. In experimenting with figures, I noticed that if you wrote a series of squares such as 4, 9, 16, 25, 36, 49 and subtracted them from each other as you went along, you got a series that you could similarly subtract once again and come up with a finite difference.

So then I tried cubes and fourth powers and evolved a formula to the effect that if *n* is the power, you must subtract *n* times, and your constant difference will be factorial *n*. I asked my father about this (he was for many years director of the Strawbridge Memorial Observatory at Haverford College and teacher of mathematics). In good Quaker language he said: "Why, John, thee has discovered the calculus of finite differences."

Donald Knuth called my attention to the earliest known solution of Dudeney's bead problem. Percy A. MacMahon solved the problem as early as 1892. This and the problem are discussed in Section 4.9 of *Concrete Mathematics* (1994), by Ronald Graham, Donald Knuth, and Oren Patashnik.

The Calculus of Finite Differences

Answers

How many different triangles can be formed with n straight lines? It takes at least three lines to make one triangle, four lines will make four triangles, and five lines will make 10 triangles. Applying the calculus of finite differences, one draws up the table in Figure 2.4.

NUMBER OF LINES	0	1	2	3	4	5
NUMBER OF TRIANGLES	0	0	0	1	4	10
FIRST DIFFERENCES		0	0	1	3	6
SECOND DIFFERENCES			0	1	2	3
THIRD DIFFERENCES				1	1	1

Figure 2.4. The answer to the triangle problem

The three rows of differences indicate a cubic function. Using Newton's formula, the function is found to be: $\frac{1}{6}n(n - 1)(n - 2)$. This will generate the series 0, 0, 0, 1, 4, 10, . . . and therefore has a good chance of being the formula for the maximum number of triangles that can be made with n lines. But it is still just a guess, based on a small number of pencil and paper tests. It can be verified by the following reasoning.

The lines must be drawn so that no two are parallel and no more than two intersect at the same point. Each line is then sure to intersect every other line, and every set of three lines must form one triangle. It is not possible for the same three lines to form more than one triangle, so the number of triangles formed in this way is the maximum. The problem is equivalent, therefore, to the question: In how many different ways can n lines be taken three at a time? Elementary combinatorial theory supplies the answer: the same as the formula obtained empirically.

Solomon W. Golomb, was kind enough to send me his solution to the necklace problem. The problem was to find a formula for the number of different necklaces that can be formed with n beads, assuming that each bead can be one of two colors and not counting rotations and reflections of a necklace as being different. The formula proves to be far beyond the power of the simple method of differences.

Let the divisors of n (including 1 and n) be represented by d_1, d_2, d_3. . . . For each divisor we find what is called Euler's phi function for that divisor, symbolized $\Phi(d)$. This function is the number of positive integers, not greater than d, that have no common divisor with d. It is assumed that 1 is such an integer, but not d. Thus $\Phi(8)$ is 4, because 8 has the following four integers that are prime to it: 1, 3, 5, 7. By convention, $\Phi(1)$ is taken to be 1. Euler's phi functions for 2, 3, 4, 5, 6, 7 are 1, 2, 2, 4, 2, 6, in the same order. Let a stand for the number of different colors each bead can be. For necklaces with an odd number of beads the formula for the number of different necklaces with n beads is the one given at the top of Figure 2.5. When n is even, the formula is the one at the bottom of the illustration.

$$\frac{1}{2n}[\phi(d_1) \bullet a^{\frac{n}{d_1}} + \phi(d_2) \bullet a^{\frac{n}{d_2}} \ldots + n \bullet a^{\frac{n+1}{2}}]$$

$$\frac{1}{2n}\left[\phi(d_1) \bullet a^{\frac{n}{d_1}} + \phi(d_2) \bullet a^{\frac{n}{d_2}} \ldots + \frac{n}{2} \bullet (1 + a) \bullet a^{\frac{n}{2}}\right]$$

Figure 2.5. Equations for the solution of the necklace problem

The single dots are symbols for multiplication. Golomb expressed these formulas in a more compressed, technical form, but I think the above forms will be clearer to most readers. They are more general than the formulas asked for because they apply to beads that may have any specified number of colors.

The formulas answering the other questions in the chapter are:

1. Regions of a crescent moon produced by n straight cuts:

$$\frac{n^2 + 3n}{2} + 1.$$

2. Pieces of cheesecake produced by n plane cuts:

$$\frac{n^3 + 5n}{6} + 1.$$

3. Regions of the plane produced by n intersecting circles:

$$n^2 - n + 2.$$

4. Regions of the plane produced by n intersecting ellipses:

$$2n^2 - 2n + 2.$$

5. Regions of space produced by n intersecting spheres:

$$\frac{n(n^2 - 3n + 8)}{3}.$$

The Calculus of Finite Differences

6. Triangles formed by toothpicks of n colors:

$$\frac{n^3 + 2n}{3}.$$

7. Squares formed with toothpicks of n colors:

$$\frac{n^4 + n^2 + 2n}{4}.$$

8. Tetrahedrons formed with sides of n colors:

$$\frac{n^4 + 11n^2}{12}.$$

9. Cubes formed with sides of n colors:

$$\frac{n^6 + 3n^4 + 12n^3 + 8n^2}{24}.$$

Bibliography

E. L. Baker, "The Method of Differences in the Determination of Formulas," *School Science and Mathematics,* April 1967, pp. 309–15.

C. Jordan, *The Calculus of Finite Differences,* Chelsea, 1965.

K. S. Miller, *An Introduction to the Calculus of Finite Differences and Difference Equations,* Henry Holt, 1960.

W. E. Milne, *Numerical Calculus,* Princeton University Press, 1949.

L. M. Milne-Thomson, *The Calculus of Finite Differences,* Macmillan, 1951.

R. C. Read, "Pólya's Theorem and its Progeny," in *Mathematics Magazine,* Vol. 60, December 1987, pp. 275–82. Eighteen references are listed.

Chapter 3

Palindromes: Words and Numbers

A man, a plan, a canal—Suez!

—Ethel Merperson, a "near miss" palindromist,
in *Son of Giant Sea Tortoise*, edited by
Mary Ann Madden (Viking, 1975)

A palindrome is usually defined as a word, sentence, or set of sentences that spell the same backward as forward. The term is also applied to integers that are unchanged when they are reversed. Both types of palindrome have long interested those who amuse themselves with number and word play, perhaps because of a deep, half-unconscious aesthetic pleasure in the kind of symmetry palindromes possess. Palindromes have their analogues in other fields: melodies that are the same backward, paintings and designs with mirror-reflection symmetry, the bilateral symmetry of animals and man (see Figure 3.1) and so on. In this chapter we shall restrict our attention to number and language palindromes and consider some entertaining new developments in both fields.

Figure 3.1. Flying seagull: a visual palindrome

An old palindrome conjecture of unknown origin (there are references to it in publications of the 1930s) is as follows. Start with any positive integer. Reverse it and add the two numbers. This procedure is repeated with the sum to obtain a second sum, and the process continues until a palindromic sum is obtained. The conjecture is that a palindrome always results after a finite number of additions. For example, 68 generates a palindrome in three steps:

$$
\begin{array}{r}
68 \\
+\ 86 \\
\hline
154 \\
+451 \\
\hline
605 \\
+506 \\
\hline
1,111
\end{array}
$$

For all two-digit numbers it is obvious that if the sum of their digits is less than 10, the first step gives a two-digit palindrome. If their digits add to 10, 11, 12, 13, 14, 15, 16, or 18, a palindrome results after 2, 1, 2, 2, 3, 4, 6, 6 steps, respectively. As Angela Dunn points out in *Mathematical Bafflers* (McGraw-Hill, 1964; Dover, 1980), the exceptions are numbers whose two digits add to 17. Only 89 (or its reversal, 98) meets this proviso. Starting with either number does not produce a palindrome until the 24th operation results in 8,813,200,023,188.

The conjecture was widely regarded as being true until 1967, although no one had succeeded in proving it. Charles W. Trigg, a California mathematician well known for his work on recreational problems, examined the conjecture more carefully in his 1967 article "Palindromes by Addition." He found 249 integers smaller than 10,000 that failed to generate a palindrome after 100 steps. The smallest such number, 196, was carried to 237,310 steps in 1975 by Harry J. Saal, at the Israel Scientific Center. No palindromic sum appeared. Trigg believed the conjecture to be false. (The number 196 is the square of 14, but this is probably an irrelevant fact.) Aside from the 249 exceptions, all integers less than 10,000, except 89 and its reversal, produce a palindrome in fewer than 24 steps. The largest palindrome, 16,668,488,486,661, is generated by 6,999 (or its reversal) and 7,998 (or its reversal) in 20 steps.

The conjecture has not been established for any number system and has been proved false only in number notations with bases that are powers of 2. (See the paper by Heiko Harborth listed in the bibliography.) The smallest binary counterexample is 10110 (or 22 in the decimal system). After four steps the sum is 10110100, after eight steps it is 1011101000, after 12 steps it is 101111010000. Every fourth step increases by one digit each of the two sequences of underlined digits. Brother Alfred Brousseau, in "Palindromes by Addition in Base Two," proved that this asymmetric pattern repeats indefinitely. He also found other repeating asymmetric patterns for larger binary numbers.

There is a small but growing literature on the properties of palindromic prime numbers and conjectures about them. Apparently there are infinitely many such primes, although so far as I know this has not been proved. It is not hard to show, however, that a palindromic prime, with the exception of 11, must have an odd number of digits. Can the reader do this before reading the simple proof in the answer section? Norman T. Gridgeman conjectured that there is an infinity of prime pairs of the form 30,103–30,203 and 9,931,399–9,932,399 in which all digits are alike except the middle digits, which differ by one. But Gridgeman's guess is far from proved.

Gustavus J. Simmons wrote two papers on palindromic powers. After showing that the probability of a randomly selected integer being palindromic approaches zero as the number of digits in the integer increases, Simmons examined square numbers and found them much richer than randomly chosen integers in palindromes. There are infinitely many palindromic squares, most of which, it seems, have square roots that also are palindromes. (The smallest nonpalindromic root is 26). Cubes too are unusually rich in palindromes. A computer check on all cubes less than 2.8×10^{14} turned up a truly astonishing fact. The only palindromic cube with a nonpalindromic cube root, among the cubes examined by Simmons, is 10,662,526,601. Its cube root, 2,201, had been noticed earlier by Trigg, who reported in 1961 that it was the only nonpalindrome with a palindromic cube less than 1,953,125,000,000. It is not yet known if 2,201 is the only integer with this property.

Simmons' computer search of palindromic fourth powers, to the same limit as his search of cubes, failed to uncover a single palindromic fourth power whose fourth root was not a palindrome of the general form 10 . . . 01. For powers 5 through 10 the computer found no palindromes at all except the trivial case of 1. Simmons conjectured that there are no palindromes of the form X^k where k is greater than 4.

"Repunits," numbers consisting entirely of 1's, produce palindromic squares when the number of units is one through nine, but 10 or more units give squares that are not palindromic. It has been erroneously stated that only primes have palindromic cubes, but this is disproved by an infinity of integers, the smallest of which is repunit 111. It is divisible by 3, yet its cube, 1,367,631, is a palindrome. The number 836 is also of special interest. It is the largest three-digit integer whose square, 698,896, is palindromic, and 698,896 is the smallest palindromic square with an even number of digits. (Note also that the number remains palindromic when turned upside down.) Such palindromic squares are extremely rare. The next-larger one with an even number of digits is 637,832,238,736, the square of 798,644.

Turning to language palindromes, we first note that no common English words of more than seven letters are palindromic. Examples of seven-letter palindromes are *reviver, repaper, deified,* and *rotator.* The word "radar" (for radio detecting and ranging) is notable because it was coined to symbolize the reflection of radio waves. Dmitri Borgmann, whose files contain thousands of sentence palindromes in all major languages, asserts in his book *Language on Vacation* that the largest nonhyphenated word palindrome is *saippuakauppias,* a Finnish word for a soap dealer.

Among proper names in English, according to Borgmann, none is longer than Wassamassaw, a swamp north of Charleston, SC. Legend has it, he writes, that it is an Indian word meaning "the worst place ever seen." Yreka Bakery has long been in business on West Miner Street in Yreka, CA. Lon Nol, the former Cambodian premier, has a palindromic name, as does U Nu, once premier of Burma. Revilo P. Oliver, a classics professor at the University of Illinois, has the same first name as his father and grandfather. It was originally devised to make the name palindromic. If there is anyone with a longer palindromic name I do not know of it, although Borgmann suggests such possibilities as Norah Sara Sharon, Edna Lala Lalande, Duane Rollo Renaud, and many others.

There are thousands of excellent sentence palindromes in English, a few of which were discussed in a chapter on word play in my *Sixth Book of Mathematical Games from Scientific American.* The interested reader will find good collections in the Borgmann book cited above and in the book by Howard Bergerson. Composing palindromes at night is one way for an insomniac to pass the dark hours, as Roger Angell so amusingly details in his article "Ainmosni" ("Insomnia" backward) in *The New Yorker.* I limit myself to one palindrome that is not well known, yet is remarkable for both its length and naturalness: "Doc note, I dissent. A fast never prevents a fatness. I diet on cod." It won a prize for James Michie in a palindrome contest sponsored by the *New Statesman* in England; results were published in the issue for May 5, 1967. Many of the winning palindromes are much longer than Michie's, but, as is usually the case, the longer palindromes are invariably difficult to understand.

Palindromists have employed various devices to make the unintelligibility of long palindromes more plausible: presenting them as telegrams, as one side only of a telephone conversation, and so on. Leigh Mercer, a leading British palindromist (he is the inventor of the famous "A man, a plan, a canal—Panama!"), has suggested a way of writing a palindrome as long as one wishes. The sentence has the form, " '———,' sides reversed, is '———.' " The first blank can be any sequence of letters, however long, which is repeated in reverse order in the second blank.

Good palindromes involving the names of U.S. presidents are exceptionally rare. Borgmann cites the crisp "Taft: fat!" as one of the shortest and best. Richard Nixon's name lends itself to "No 'x' in 'Mr. R. M. Nixon'?" although the sentence is a bit too contrived. A shorter, capitalized version of this palindrome, NO X IN NIXON, is also invertible.

The fact that "God" is "dog" backward has played a role in many sentence palindromes, as well as in orthodox psychoanalysis. In *Freud's Contribution to Psychiatry* A. A. Brill cites a rather farfetched analysis by Carl Jung and others of a patient suffering from a ticlike upward movement of his arms. The analysts decided that the tic had its origin in an unpleasant early visual experience involving dogs. Because of the "dog-god" reversal, and the man's religious convictions, his unconscious had developed the gesture to symbolize a warding off of the evil "dog-god." Edgar Allan Poe's frequent use of the reversal words "dim" and "mid" is pointed out by Humbert Humbert, the narrator of Vladimir Nabokov's novel *Lolita*. In the second canto of *Pale Fire*, in Nabokov's novel of the same title, the poet John Shade speaks of his dead daughter's propensity for word reversals:

> . . . *She twisted words: pot, top,*
> *Spider, redips. And "powder" was "red wop."*

Such word reversals, as well as sentences that are different sentences when they are spelled backward, are obviously close cousins of palindromes, but the topic is too large to go into here.

Palindrome sentences in which words, not letters, are the units have been a specialty of another British expert on word play, J. A. Lindon. Two splendid examples, from scores that he has composed, are

"You can cage a swallow, can't you, but you can't swallow a cage, can you?"

"Girl, bathing on Bikini, eyeing boy, finds boy eyeing bikini on bathing girl."

Many attempts have been made to write letter-unit palindrome poems, some quite long, but without exception they are obscure, rhymeless, and lacking in other poetic values. Somewhat better poems can be achieved by making each line a separate palindrome rather than the entire poem or by using the word as the unit. Lindon has written many poems of both types. A third type of palindrome poem, invented by Lindon, employs lines as units. The poem is unchanged when its lines are read forward but taken in reverse order. One is allowed, of course, to punctuate duplicate lines differently. The following example is one of Lindon's best:

> As I was passing near the jail
> I met a man, but hurried by.
> His face was ghastly, grimly pale.
> He had a gun. I wondered why
> He had. A gun? I wondered . . . why,
> His face was ghastly! Grimly pale,
> I met a man, but hurried by,
> As I was passing near the jail.

This longer one is also by Lindon. Both poems appear in Howard W. Bergerson's *Palindromes and Anagrams* (Dover, 1973).

DOPPELGÄNGER

> Entering the lonely house with my wife,
>   I saw him for the first time
> Peering furtively from behind a bush—
>   Blackness that moved,
>   A shape amid the shadows,
> A momentary glimpse of gleaming eyes
>   Revealed in the ragged moon.
> A closer look (he seemed to turn) might have
> Put him to flight forever—
>   I dared not
> (For reasons that I failed to understand),
>   Though I knew I should act at once.
>
> I puzzled over it, hiding alone,
> Watching the woman as she neared the gate.
>   He came, and I saw him crouching

Night after night.
Night after night
He came, and I saw him crouching,
Watching the woman as she neared the gate.

I puzzled over it, hiding alone—
Though I knew I should act at once,
For reasons that I failed to understand
I dared not
Put him to flight forever.

A closer look (he seemed to turn) might have
Revealed in the ragged moon
A momentary glimpse of gleaming eyes,
A shape amid the shadows,
Blackness that moved.

Peering furtively from behind a bush,
I saw him, for the first time,
Entering the lonely house with my wife.

Lindon holds the record for the longest word ever worked into a letter-unit palindrome. To understand the palindrome you must know that Beryl has a husband who enjoys running around his yard without any clothes on. Ned has asked him if he does this to annoy his wife. He answers: "Named undenominationally rebel, I rile Beryl? La, no! I tan. I'm, O Ned, nude, man!"

Addendum

A. Ross Eckler, editor and publisher of *Word Ways,* a quarterly journal on word play that has featured dozens of articles on palindromes of all types, wrote to say that the "palindromic gap" between English and other languages is perhaps not as wide as I suggested. The word "semitime" can be pluralized to make a 9-letter palindrome and "kinnikinnik" is an 11-letter palindrome. Dmitri Borgmann pointed out in *Word Ways,* said Eckler, that an examination of foreign dictionaries failed to substantiate such long palindromic words as the Finnish soap dealer, suggesting that they are artificially created words.

Among palindromic towns and cities in the United States, Borgmann found the 7-letter Okonoko (in West Virginia). If a state (in full or abbreviated form) is part of the palindrome, Borgmann offers Apollo, PA, and Adaven, Nevada. Some U.S. towns, Eckler continued, are intentional reversal pairs, such as Orestod and Dotsero, in Eagle County, Colorado, and Colver and Revloc, in Cambria County, Pennsylvania. Nova and Avon, he added, are Ohio towns that are an unintentional reversal pair.

George L. Hart III sent the following letter, which was published in *Scientific American,* November 1970:

Sirs:

Apropos of your discussion of palindromes, I would like to offer an example of what I believe to be the most complex and exquisite type of palindrome ever invented. It was devised by the Sanskrit aestheticians, who termed it *sarvatobhadra,* that is, "perfect in every direction." The most famous example of it is found in the epic poem entitled *Sisupālavadha.*

sa - kā - ra - nā - nā - ra - kā - sa -
kā - ya - sā - da - da- sā - ya - kā
ra - sā - ha - vā vā - ha - sā - ra -
nā - da - vā - da - da - vā - da - nā.
(nā da vā da da vā da nā
ra sā ha vā vā ha sā ra
kā ya sā da da sā ya kā
sa kā ra nā nā ra kā sa)

Here hyphens indicate that the next syllable belongs to the same word. The last four lines, which are an inversion of the first four, are not part of the verse but are supplied so that its properties can be seen more easily. The verse is a description of an army and may be translated as follows: "[That army], which relished battle [rasāhavā], contained allies who brought low the bodes and gaits of their various striving enemies [sakāranānarakāsakāyasādadasāyakā], and in it the cries of the best of mounts contended with musical instruments [vāhasāranādavādadavādanā]."

Two readers, D. M. Gunn and Rosina Wilson, conveyed the sad news that the Yreka Bakery no longer existed. However, in 1970 its premises were occupied by the Yrella Gallery, and Ms. Wilson sent a Polaroid picture of the gallery's sign to prove it. Whether the gallery is still there, I do not know.

Lee Sallows repaired the "near miss" palindrome in this chapter's epigraph by adding a word: "Zeus! A man, a plan, a canal—Suez!"

Answers

Readers were asked to prove that no prime except 11 can be a palindrome if it has an even number of digits. The proof exploits a well-known test of divisibility by 11 (which will not be proved here): If the difference between the sum of all digits in even positions and the sum of all digits in odd positions is zero or a multiple of 11, the number is a multiple of 11. When a palindrome has an even number of digits, the digits in odd positions necessarily duplicate the digits in even positions; therefore the difference between the sums of the two sets must be zero. The palindrome, because it has 11 as a factor, cannot be prime.

The same divisibility test applies in all number systems when the factor to be tested is the system's base plus one. This proves that no palindrome with an even number of digits, in any number system, can be prime, with the possible exception of 11. The number 11 is prime if the system's base is one less than a prime, as it is in the decimal system.

Bibliography

The palindrome number conjecture

D. Ahl, "Follow-up on Palindromes," *Creative Computing,* May 1975, p. 18.

A. Brousseau, "Palindromes by Addition in Base Two," *Mathematics Magazine,* Vol. 42, November 1969, pp. 254–56.

F. Gruenberger, "How to Handle Numbers With Thousands of Digits, and Why One Might Not Want To," *Scientific American,* Vol. 205, April 1984, pp. 19–26.

H. Harborth, "On Palindromes," *Mathematics Magazine,* Vol. 46, March 1973, pages 96–99.

G. Johns and J. Wiegold, "The Palindromic Problem in Base 2," *Mathematical Gazette,* Vol. 78, November 1994, p. 312.

W. Koetke, "Palindromes: For Those Who Like to Start at the Beginning." *Creative Computing,* January 1975, pp. 10–12.

G. Kröber, "On Non-Palindromic Patterns in Palindromic Processes," *Mathematical Gazette,* Vol. 80, November 1996, pp. 577–79.

D. J. Lanska, "The 196 Problem," *Journal of Recreational Mathematics,* Vol. 18, No. 2, 1985–86, pp. 152–53.

D. J. Lanska, "More on the 196 Problem," *Journal of Recreational Mathematics,* Vol. 18, No. 4, 1985–86, pp. 305–6.

R. Sprague, *Recreation in Mathematics,* Dover, 1963, Problem 5.

C. W. Trigg, "Palindromes by Addition," *Mathematics Magazine,* Vol. 40, January 1967, pp. 26–28.

C. W. Trigg, "More on Palindromes by Reversal-Addition," *Mathematics Magazine,* Vol. 45, September 1972, pp. 184–86.

C. W. Trigg, "Versum Sequences in the Binary System." *Pacific Journal of Mathematics,* Vol. 47, 1973, pp. 263–75.

"The 196 Problem," *Popular Computing,* Vol. 3, September 1975, pp. 6–9.

Palindromic primes

L. Card, "Patterns in Primes," *Journal of Recreational Mathematics,* Vol. 1, April 1968, pp. 93–99.

H. Gabai and D. Coogan, "On Palindromes and Palindromic Primes," *Mathematics Magazine,* Vol. 42, November 1969, pp. 252–54.

C. W. Trigg, "Special Palindromic Primes," *Journal of Recreational Mathematics,* Vol. 4, July 1971, pages 169–70.

Palindromic Powers

G. J. Simmons, "Palindromic Powers," *Journal of Recreational Mathematics,* Vol. 3, April 1970, pp. 93–98.

G. J. Simmons, "On Palindromic Squares of Non-Palindromic Numbers," *Journal of Recreational Mathematics,* Vol. 5, January 1972, pp. 11–19.

C. W. Trigg, "Palindromic Cubes," *Mathematics Magazine,* Vol. 34, March 1961, p. 214.

Word Palindromes

R. Angell, "Ainmosni," *The New Yorker,* May 31, 1969, pp. 32–34.

H. W. Bergerson, *Palindromes and Anagrams,* Dover, 1973.

C. C. Bombaugh, *Oddities and Curiosities of Words and Literature,* M. Gardner (ed.), Dover, 1961.

D. Borgmann, *Language on Vacation,* Scribner's, 1965.

D. Borgmann, *Beyond Language,* Scribner's, 1967.

S. J. Chism, *From A to Zotamorf: The Dictionary of Palindromes,* Word Ways, 1992.

M. Donner, *I Love Me, Vol. I: S. Wordrow's Palindrome Encyclopedia,* Algonquin Books, 1996.

M. Gardner, "The Oulipo," *Penrose Tiles To Trapdoor Ciphers,* Chapters 6 and 7, W. H. Freeman, 1989.

S. W. Golomb, "I Call on Professor Osseforp," *Harvard Bulletin,* March 1972, p. 45.

W. Irvine, "Madam, I'm Adam and Other Palindromes," *Scribner's,* 1987.

J. Kuhn and M. Kuhn, *Rats Live on No Evil Star,* Everett House, 1981.

R. Lederer, "Palindromes: The Art of Reverse English," *Verbatim,* Vol. 11, Winter 1985, pp. 14–15.

L. Levine. *Dr. Award and Olson in Oslo: A Palindromic Novel,* By author, 1980.

J. Pool, *Lid Off a Daffodil: A Book of Palindromes,* Holt, Rinehart, and Winston, 1985.

R. Stuart, *Too Hot To Hoot,* McKay, 1977.

Word Ways: The Journal of Recreational Linguistics. Vol. 1, February 1968.

"Weekend Competition," *New Statesman,* April 14, 1967, p. 521; May 5, 1967, p. 630; December 22, 1972, p. 955; March 9, 1973, p. 355.

"Results of Competition," *New York Magazine,* December 22, 1969, p. 82.

"The Contest," *Maclean's Magazine,* December 1969, pp. 68–69.

II

Plane Geometry

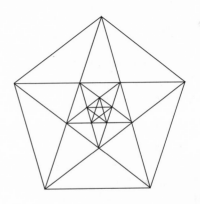

Chapter 4

Curves of Constant Width

*I*f an enormously heavy object has to be moved from one spot to another, it may not be practical to move it on wheels. Axles might buckle or snap under the load. Instead the object is placed on a flat platform that in turn rests on cylindrical rollers. As the platform is pushed forward, the rollers left behind are picked up and put down again in front.

An object moved in this manner over a flat, horizontal surface obviously does not bob up and down as it rolls along. The reason is simply that the cylindrical rollers have a circular cross section, and a circle is a closed curve possessing what mathematicians call "constant width." If a closed convex curve is placed between two parallel lines and the lines are moved together until they touch the curve, the distance between the parallel lines is the curve's "width" in one direction. An ellipse clearly does not have the same width in all directions. A platform riding on elliptical rollers would wobble up and down as it rolled over them. Because a circle has the same width in all directions, it can be rotated between two parallel lines without altering the distance between the lines.

Is the circle the only closed curve of constant width? Most people would say yes, thus providing a sterling example of how far one's mathematical intuition can go astray. Actually there is an infinity of such curves. Any one of them can be the cross section of a roller that will roll a platform as smoothly as a circular cylinder! The failure to recognize such curves can have and has had disastrous consequences in industry. To give one example, it might be thought that the cylindrical hull of a half-built submarine could be tested for circularity by just measuring maximum widths in all directions. As will soon be made clear, such a hull can be monstrously lopsided and still pass such a test. It is precisely for this reason that the circularity of a submarine hull is always tested by applying curved templates.

The simplest noncircular curve of constant width has been named the Reuleaux triangle after Franz Reuleaux (1829–1905), an engineer and mathematician who taught at the Royal Technical High School in Berlin. The curve itself was known to earlier mathematicians, but Reuleaux was the first to demonstrate its constant-width properties. It is easy to construct. First draw an equilateral triangle, *ABC* (see Figure 4.1). With the point of a compass at *A*, draw an arc, *BC*. In a similar manner draw the other two arcs. It is obvious that the "curved triangle" (as Reuleaux called it) must have a constant width equal to the side of the interior triangle.

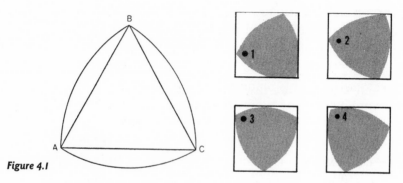

Figure 4.1

Construction of Reuleaux triangle. Reuleaux triangle rotating in square.

If a curve of constant width is bounded by two pairs of parallel lines at right angles to each other, the bounding lines necessarily form a square. Like the circle or any other curve of constant width, the Reuleaux triangle will rotate snugly within a square, maintaining contact at all times with all four sides of the square (see Figure 4.2). If the reader cuts a Reuleaux triangle out of cardboard and rotates it inside a square hole of the proper dimensions cut in another piece of cardboard, he will see that this is indeed the case.

As the Reuleaux triangle turns within a square, each corner traces a path that is almost a square; the only deviation is at the corners, where there is a slight rounding. The Reuleaux triangle has many mechanical uses, but none is so bizarre as the use that derives from this property. In 1914 Harry James Watts, an English engineer then living in Turtle Creek, PA, invented a rotary drill based on the Reuleaux triangle and

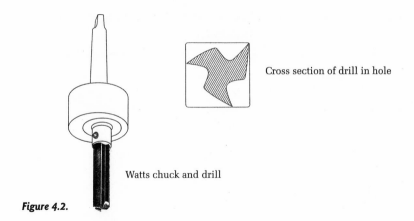

Cross section of drill in hole

Watts chuck and drill

Figure 4.2.

capable of drilling square holes! Beginning in 1916 these curious drills were manufactured by the Watts Brothers Tool Works in Wilmerding, PA. "We have all heard about left-handed monkey wrenches, fur-lined bathtubs, cast-iron bananas," reads one of their descriptive leaflets. "We have all classed these things with the ridiculous and refused to believe that anything like that could ever happen, and right then along comes a tool that drills square holes."

The Watts square-hole drill is shown in Figure 4.2. At right is a cross section of the drill as it rotates inside the hole it is boring. A metal guide plate with a square opening is first placed over the material to be drilled. As the drill spins within the guide plate, the corners of the drill cut the square hole through the material. As you can see, the drill is simply a Reuleaux triangle made concave in three spots to provide for cutting edges and outlets for shavings. Because the center of the drill wobbles as the drill turns, it is necessary to allow for this eccentric motion in the chuck that holds the drill. A patented "full floating chuck," as the company calls it, does the trick. (Readers who would like more information on the drill and the chuck can check United States patents 1,241,175; 1,241,176; and 1,241,177; all dated September 25, 1917.)

The Reuleaux triangle is the curve of constant width that has the smallest area for a given width (the area is $1/2\ (\pi - \sqrt{3})w^2$, where w is the width). The corners are angles of 120 degrees, the sharpest possible on such a curve. These corners can be rounded off by extending each side of an equilateral triangle a uniform distance at each end (see Figure 4.3). With the point of a compass at A draw arc DI; then widen the compass and draw arc *FG*. Do the same at the other corners. The resulting curve has a width, in all directions, that is the sum of the same two radii. This of course makes it a curve of constant width. Other symmetrical curves of constant width result if you start with a regular pentagon (or any regular polygon with an odd number of sides) and follow similar procedures.

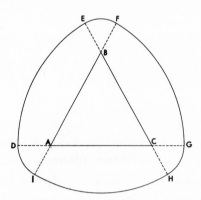

Figure 4.3. Symmetrical rounded-corner curve of constant width

There are ways to draw unsymmetrical curves of constant width. One method is to start with an irregular star polygon (it will necessarily have an odd number of points) such as the seven-point star shown in black in Figure 4.4. All of these line segments must be the same length. Place the compass point at each corner of the star and connect the two opposite corners with an arc. Because these arcs all have the same radius, the resulting curve (shown in gray) will have constant width. Its corners can be rounded off by the method used before. Extend the sides of the star a uniform distance at all points (shown with broken lines) and then join the ends of the extended sides by arcs drawn with the compass point at each corner of the star. The rounded-corner curve, which is shown in black, will be another curve of constant width.

Figure 4.5 demonstrates another method. Draw as many straight lines as you please, all mutually intersecting. Each arc is drawn with the compass point at the intersection of the two lines that bound the arc. Start with any arc, then proceed around the curve, connecting each arc to the preceding one. If you do it carefully, the curve will close and will have constant width. (Proving that the curve must close and have con-

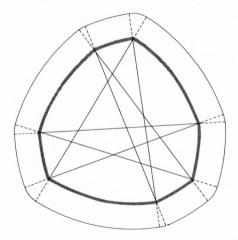

Figure 4.4. Star-polygon method of drawing a curve of constant width

stant width is an interesting and not difficult exercise.) The preceding curves were made up of arcs of no more than two different circles, but curves drawn in this way may have arcs of as many different circles as you wish.

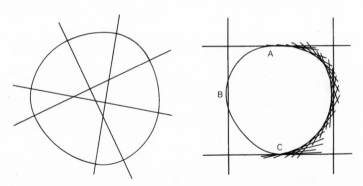

Figure 4.5.   Crossed-lines method   Random curve and tangents

A curve of constant width need not consist of circular arcs. In fact, you can draw a highly arbitrary convex curve from the top to the bottom of a square and touching its left side (arc *ABC* in Figure 4.5), and this curve will be the left side of a uniquely determined curve of constant width. To find the missing part, rule a large number of lines, each parallel to a tangent of arc *ABC* and separated from the tangent by a distance equal to the side of the square. This can be done quickly by using both sides of a ruler. The original square must have a side equal to the ruler's width. Place one edge of the ruler so that it is tangent to arc

ABC at one of its points, then use the ruler's opposite edge to draw a parallel line. Do this at many points, from one end of arc *ABC* to the other. The missing part of the curve is the envelope of these lines. In this way you can obtain rough outlines of an endless variety of lopsided curves of constant width.

It should be mentioned that the arc *ABC* cannot be completely arbitrary. Roughly speaking, its curvature must not at any point be less than the curvature of a circle with a radius equal to the side of the square. It cannot, for example, include straight-line segments. For a more precise statement on this, as well as detailed proofs of many elementary theorems involving curves of constant width, the reader is referred to the excellent chapter on such curves in *The Enjoyment of Mathematics,* by Hans Rademacher and Otto Toeplitz.

If you have the tools and skills for woodworking, you might enjoy making a number of wooden rollers with cross sections that are various curves of the same constant width. Most people are nonplused by the sight of a large book rolling horizontally across such lopsided rollers without bobbing up and down. A simpler way to demonstrate such curves is to cut from cardboard two curves of constant width and nail them to opposite ends of a wooden rod about six inches long. The curves need not be of the same shape, and it does not matter exactly where you put each nail as long as it is fairly close to what you guess to be the curve's "center." Hold a large, light-weight empty box by its ends, rest it horizontally on the attached curves and roll the box back and forth. The rod wobbles up and down at both ends, but the box rides as smoothly as it would on circular rollers!

The properties of curves of constant width have been extensively investigated. One startling property, not easy to prove, is that the perimeters of all curves with constant width n have the same length. Since a circle is such a curve, the perimeter of any curve of constant width n must of course be πn, the same as the circumference of a circle with diameter n.

The three-dimensional analogue of a curve of constant width is the solid of constant width. A sphere is not the only such solid that will rotate within a cube, at all times touching all six sides of the cube; this property is shared by all solids of constant width. The simplest example of a nonspherical solid of this type is generated by rotating the Reuleaux triangle around one of its axes of symmetry (see Figure 4.6, *left*). There is an infinite number of others. The solids of constant width that have the smallest volumes are derived from the regular tetrahedron in somewhat the same way the Reuleaux triangle is derived from the equilateral triangle. Spherical caps are first placed on each face of the tetrahedron, then it is necessary to alter three of the edges slightly. These altered edges may either form a triangle or radiate from one corner. The solid at the right of Figure 4.6 is an example of a curved tetrahedron of constant width.

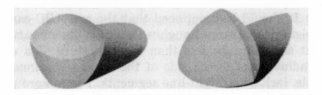

Figure 4.6. Two solids of constant width

Since all curves of the same constant width have the same perimeter, it might be supposed that all solids of the same constant width have the same surface area. This is not the case. It was proved, however, by Hermann Minkowski (the Polish mathematician who made such great contributions to relativity theory) that all *shadows* of solids of constant width (when the projecting rays are parallel and the shadow falls on a plane perpendicular to the rays) are curves of the same constant width. All such shadows have equal perimenters (π times the width).

Michael Goldberg, an engineer with the Bureau of Naval Weapons in Washington, has written many papers on curves and solids of constant width and is recognized as being this country's leading expert on the subject. He has introduced the term "rotor" for any convex figure that can be rotated inside a polygon or polyhedron while at all times touching every side or face.

The Reuleaux triangle is, as we have seen, the rotor of least area in a square. The least-area rotor for the equilateral triangle is shown at the left of Figure 4.7. This lens-shaped figure (it is not, of course, a curve of constant width) is formed with two 60-degree arcs of a circle having a radius equal to the triangle's altitude. Note that as it rotates its corners trace the entire boundary of the triangle, with no rounding of corners. Mechanical reasons make it difficult to rotate a drill based on this figure, but Watts Brothers makes other drills, based on rotors for higher-order regular polygons, that drill sharp-cornered holes in the shape of pentagons, hexagons, and even octagons. In three-space, Goldberg has shown, there are nonspherical rotors for the regular tetrahedron and octahedron, as well as the cube, but none for the regular dodecahedron and icosahedron. Almost no work has been done on rotors in dimensions higher than three.

Figure 4.7.

Least-area rotor in equilateral triangle   Line rotated in deltoid curve

Closely related to the theory of rotors is a famous problem named the Kakeya needle problem after the Japanese mathematician Sôichi Kakeya, who first posed it in 1917. The problem is as follows: What is the plane figure of least area in which a line segment of length 1 can be rotated 360 degrees? The rotation obviously can be made inside a circle of unit diameter, but that is far from the smallest area.

For many years mathematicians believed the answer was the deltoid curve shown at the right of Figure 4.7, which has an area exactly half that of a unit circle. (The deltoid is the curve traced by a point on the circumference of a circle as it rolls around the inside of a larger circle, when the diameter of the small circle is either one third or two thirds that of the larger one.) If you break a toothpick to the size of the line segment shown, you will find by experiment that it can be rotated inside the deltoid as a kind of one-dimensional rotor. Note how its end points remain at all times on the deltoid's perimeter.

In 1927, ten years after Kakeya popped his question, the Russian mathematician Abram Samoilovitch Besicovitch (then living in Copenhagen) dropped a bombshell. He proved that the problem had no answer. More accurately, he showed that the answer to Kakeya's question is that there is *no* minimum area. The area can be made as small as one wants. Imagine a line segment that stretches from the earth to the moon. We can rotate it 360 degrees within an area as small as the area of a postage stamp. If that is too large, we can reduce it to the area of Lincoln's nose on a postage stamp.

Besicovitch's proof is too complicated to give here (see references in bibliography), and besides, his domain of rotation is not what topologists call simply connected. For readers who would like to work on a much easier problem: What is the smallest *convex* area in which a line segment of length 1 can be rotated 360 degrees? (A convex figure is one in which a straight line, joining any two of its points, lies entirely on the figure. Squares and circles are convex; Greek crosses and crescent moons are not.)

Addendum

Although Watts was the first to acquire patents on the process of drilling square holes with Reuleaux-triangle drills, the procedure was apparently known earlier. Derek Beck, in London, wrote that he had met a man who recalled having used such a drill for boring square holes when he was an apprentice machinist in 1902 and that the practice then seemed to be standard. I have not, however, been able to learn anything about the history of the technique prior to Watts's 1917 patents.

In 1969 England introduced a 50-pence coin with seven slightly curved sides that form a circle of constant width, surely the first seven-sided coin ever minted. The invariant width allows the coin to roll smoothly down coin-operated machines.

Answers

What is the smallest convex area in which a line segment of length 1 can be rotated 360 degrees? The answer: An equilateral triangle with an altitude of 1. (The area is one third the square root of 3.)

Any figure in which the line segment can be rotated obviously must have a width at least equal to 1. Of all convex figures with a width of 1, the equilateral triangle of altitude 1 has the smallest area. (For a proof of this the reader is referred to *Convex Figures,* by I. M. Yaglom and V. G. Boltyanskii, pp. 221–22.) It is easy to see that a line segment of length 1 can in fact be rotated in such a triangle (see Figure 4.8).

The deltoid curve was believed to be the smallest simply connected area solving the problem until 1963 when a smaller area was discovered independently by Melvin Bloom and I. J. Schoenberg.

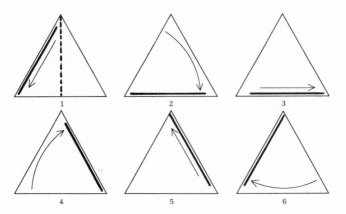

Figure 4.8. Answer to the needle-turning problem

Bibliography

W. Blaschke, *Kreis und Kugel,* Leipzig, 1916; Berlin: W. de Gruyter, 1956.

J. H. Cadwell, *Topics in Recreational Mathematics,* Cambridge, England: Cambridge University Press, 1966, Chapter 15.

J. Casey, "Perfect and Not-So-Perfect Rollers," *The Mathematics Teacher,* Vol. 91, January 1998, pp. 12–20.

G. D. Chakerian and H. Groemer, "Convex Bodies of Constant Width," *Convexity and Its Applications,* P. M. Gruber and J. M. Wills (eds.), Boston: Birkhauser, 1983.

J. A. Dossey, "What?—A Roller with Corners?," *Mathematics Teacher,* December 1972, pp. 720–24.

J. C. Fisher, "Curves of Constant Width from a Linear Viewpoint," *Mathematics Magazine,* Vol. 60, June 1987, pp. 131–40.

J. A. Flaten, "Curves of Constant Width," *The Physics Teacher,* Vol. 37, October 1999, pp. 418–19.

M. Goldberg, "Trammel Rotors in Regular Polygons," *American Mathematical Monthly,* Vol. 64, February 1957, pp. 71–78.

M. Goldberg, "Rotors in Polygons and Polyhedra," *Mathematical Tables and Other Aids to Computation,* Vol. 14, July 1960, pp. 229–39.

M. Goldberg, "N-Gon Rotors Making N + 1 Contacts with Fixed Simple Curves," *American Mathematical Monthly,* Vol. 69, June/July 1962, pp. 486–91.

M. Goldberg, "Two-lobed Rotors with Three-Lobed Stators," *Journal of Mechanisms,* Vol. 3, 1968, pp. 55–60.

C. G. Gray, "Solids of Constant Breadth," *Mathematical Gazette,* December 1972, pp. 289–92.

R. Honsberger, *Mathematical Gems,* Washington, D.C., 1973, Chapter 5.

H. Rademacher and O. Toeplitz, *The Enjoyment of Mathematics,* Princeton, NJ: Princeton University Press, 1957; Dover, 1990. See pp. 163–77, 203.

F. Reuleaux, *The Kinematics of Machinery,* New York: Macmillan, 1876; Dover Publications, 1964, pp. 129–46.

S. G. Smith, "Drilling Square Holes," *The Mathematics Teacher,* Vol. 86, October 1993, pp. 579–83.

I. M. Yaglom and V. G. Boltyanskii, *Convex Figures,* New York: Holt, Rinehart & Winston, 1961, Chapters 7 and 8.

On Kakeya's needle problem:

A. S. Besicovitch, "The Kakeya Problem," *American Mathematical Monthly,* Vol. 70, August/September 1963, pp. 697–706.

A. A. Blank, "A Remark on the Kakeya Problem," *American Mathematical Monthly,* Vol. 70, August/September 1963, pp. 706–11.

J. H. Cadwell, *Topics in Recreational Mathematics,* Cambridge, England: Cambridge University Press, 1966. See pp. 96–99.

F. Cunningham, Jr., "The Kakeya Problem for Simply Connected and Star-shaped Sets," *American Mathematical Monthly,* Vol. 78, February 1971, pp. 114–29.

I. M. Yaglom and V. G. Boltyanskii, *Convex Figures,* New York: Holt, Rinehart & Winston, 1961, pp. 61–62, 226–27.

Only three regular polygons—the equilateral triangle, the square, and the regular hexagon—can be used for tiling a floor in such a way that identical shapes are endlessly repeated to cover the plane. But there is an infinite number of irregular polygons that can provide this kind of tiling. For example, a triangle of any shape whatever will do the trick. So will any four-sided figure. The reader can try the following test. Draw an irregular quadrilateral (it need not even be convex, which is to say that it need not have interior angles that are all less than 180 degrees) and cut 20 or so copies from cardboard. It is a pleasant task to fit them all together snugly, like a jigsaw puzzle, to cover a plane.

There is an unusual and less familiar way to tile a plane. Note that each trapezoid at the top of Figure 5.1 has been divided into four smaller trapezoids that are exact replicas of the original. The four replicas can, of course, be divided in the same way into four still smaller replicas, and this can be continued to infinity. To use such a figure for tiling we have only to proceed to infinity in the opposite direction: we put together four figures to form a larger model, four of which will in turn fit together to make a still larger one. The British mathematician Augustus De Morgan summed up this sort of situation admirably in the following jingle, the first four lines of which paraphrase an earlier jingle by Jonathan Swift:

> *Great fleas have little fleas*
>   *Upon their backs to bite 'em,*
> *And little fleas have lesser fleas,*
>   *And so ad infinitum.*
> *The great fleas themselves, in turn,*
>   *Have greater fleas to go on;*

While these again have greater still,
And greater still, and so on.

Until 1962 not much was known about polygons that have this curious property of making larger and smaller copies of themselves. In
1962 Solomon W. Golomb, who was then on the staff of the Jet Propulsion Laboratory of the California Institute of Technology and is now a
professor at the University of Southern California, turned his attention
to these "replicating figures"—or "rep-tiles," as he calls them. The result was three privately issued papers that lay the groundwork for a
general theory of polygon "replication." These papers, from which almost all that follows is extracted, contain a wealth of material of great
interest to the recreational mathematician.

In Golomb's terminology a replicating polygon of order k is one that
can be divided into k replicas congruent to one another and similar to
the original. Each of the three trapezoids in Figure 5.1, for example, has
a replicating order of 4, abbreviated as rep-4. Polygons of rep-k exist for
any k, but they seem to be scarcest when k is a prime and to be most
abundant when k is a square number.

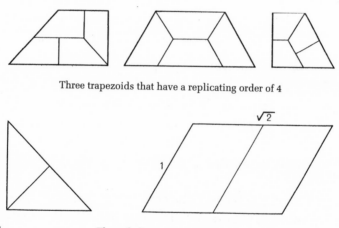

Three trapezoids that have a replicating order of 4

Figure 5.1.   The only known rep-2 polygons

Only two rep-2 polygons are known: the isosceles right triangle and
the parallelogram with sides in the ratio of 1 to the square root of 2 (see
bottom of Figure 5.1). Golomb found simple proofs that these are the
only possible rep-2 triangles and quadrilaterals, and there are no other
convex rep-2 polygons. The existence of concave rep-2 polygons appears unlikely, but so far their nonexistence has not been proved.

Rep-Tiles

The interior angles of the parallelogram can vary without affecting its rep-2 property. In its rectangular form the rep-2 parallelogram is almost as famous in the history of art as the "golden rectangle," discussed in the *Second Scientific American Book of Mathematical Puzzles and Diversions.* Many medieval and Renaissance artists (Albrecht Dürer, for instance) consciously used it for outlining rectangular pictures. A trick playing card that is sometimes sold by street-corner pitchmen exploits this rectangle to make the ace of diamonds seem to diminish in size three times (see Figure 5.2). Under cover of a hand movement the card is secretly folded in half and turned over to show a card exactly half the size of the preceding one. If each of the three smaller aces is a rectangle similar to the original, it is easy to show that only a 1-by-$\sqrt{2}$ rectangle can be used for the card. The rep-2 rectangle also has less frivolous uses. Printers who wish to standardize the shape of the pages in books of various sizes find that in folio, quarto, or octavo form it produces pages that are all similar rectangles. European writing paper also has a similar shape.

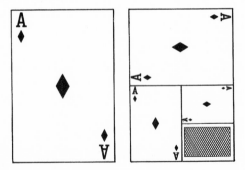

Figure 5.2. A trick diminishing card based on the rep-2 rectangle

The rep-2 rectangle belongs to the family of parallelograms shown in the top illustration of Figure 5.3. The fact that a parallelogram with sides of 1 and $\sqrt{k}$ is always rep-k proves that a rep-k polygon exists for any k. It is the only known example, Golomb asserts, of a family of figures that exhibit all the replicating orders. When k is 7 (or any prime greater than 3 that has the form $4n - 1$), a parallelogram of this family is the only known example. Rep-3 and rep-5 triangles exist. Can the reader construct them?

A great number of rep-4 figures are known. Every triangle is rep-4 and can be divided as shown in the second illustration from the top of

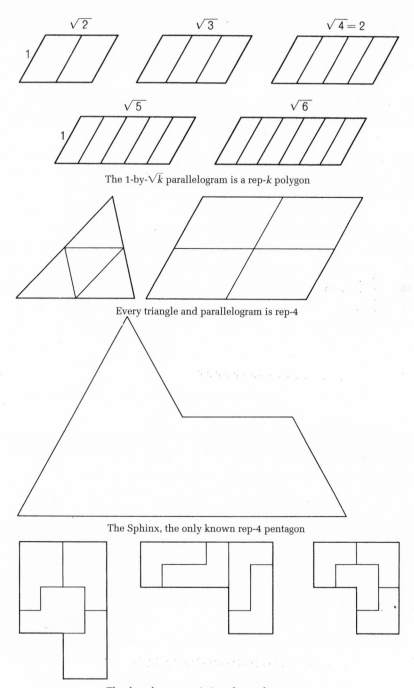

The 1-by-$\sqrt{k}$ parallelogram is a rep-k polygon

Every triangle and parallelogram is rep-4

The Sphinx, the only known rep-4 pentagon

The three known varieties of rep-4 hexagons

Figure 5.3.

Rep-Tiles

Figure 5.3. Among the quadrilaterals, any parallelogram is rep-4, as shown in the same illustration. The three trapezoids in the top illustration of Figure 5.1 are the only other examples of rep-4 quadrilaterals so far discovered.

Only one rep-4 pentagon is known: the sphinx-shaped figure in the third illustration from the top of Figure 5.3. Golomb was the first to discover its rep-4 property. Only the outline of the sphinx is given so that the reader can have the pleasure of seeing how quickly he can dissect it into four smaller sphinxes. (The name "sphinx" was given to this figure by T. H. O'Beirne of Glasgow.)

There are three known varieties of rep-4 hexagons. If any rectangle is divided into four quadrants and one quadrant is thrown away, the remaining figure is a rep-4 hexagon. The hexagon at the right at the bottom of Figure 5.3 shows the dissection (familiar to puzzlists) when the rectangle is a square. The other two examples of rep-4 hexagons (each of which can be dissected in more than one way) are shown at the middle and left in the same illustration.

No other example of a standard polygon with a rep-4 property is known. There are, however, "stellated" rep-4 polygons (a stellated polygon consists of two or more polygons joined at single points), two examples of which, provided by Golomb, are shown at the top of Figure 5.4. In the first example a pair of identical rectangles can be substituted for the squares. In addition, Golomb has found three nonpolygonal figures that are rep-4, although none is constructible in a finite number of steps. Each of these figures, shown at the left in the bottom illustration of Figure 5.4, is formed by adding to an equilateral triangle an endless series of smaller triangles, each one-fourth the size of its predecessor. In each case four of these figures will fit together to make a larger replica, as shown at the right in the same illustration. (There is a gap in each replica because the original cannot be drawn with an infinitely long series of triangles.)

It is a curious fact that every known rep-4 polygon of a standard type is also rep-9. The rep-4 Nevada-shaped trapezoid of Figure 5.5 can be dissected into nine replicas in many ways, only one of which is shown. (Can the reader dissect each of the other rep-4 polygons, not counting the stellated and infinite forms, into nine replicas?) The converse is also true: All known standard rep-9 polygons are also rep-4.

Three interesting examples of stellated rep-9 polygons, discovered and named by Golomb, are shown in Figure 5.6. None of these polygons is rep-4.

Two stellated rep-4 polygons

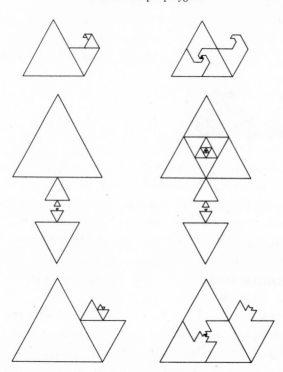

Figure 5.4.   Three examples of rep-4 nonpolygons

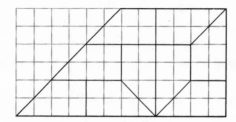

Figure 5.5. Every rep-4 polygon is also rep-9

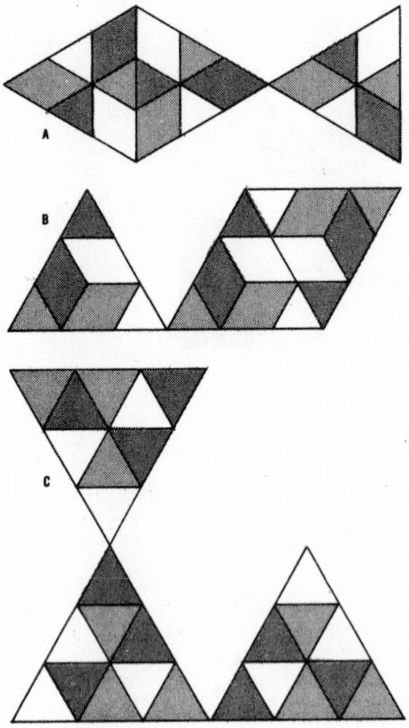

Figure 5.6. Stellated rep-9 polygons:

The Fish (A),

The Bird (B), and

The Ampersand (C)

Any method of dividing a 4 × 4 checkerboard along grid lines into four congruent parts provides a figure that is rep-16. It is only necessary to put four of the squares together to make a replica of one of the parts as in Figure 5.7. In a similar fashion, a 6 × 6 checkerboard can be quartered in many ways to provide rep-36 figures, and an equilateral triangle can be divided along triangular grid lines into rep-36 polygons (see Figure 5.8). All of these examples illustrate a simple theorem, which Golomb explains as follows:

Consider a figure P that can be divided into two or more congruent figures, not necessarily replicas of P. Call the smaller figure Q. The number of such figures is the "multiplicity" with which Q divides P. For example, in Figure 5.8 the three hexagons divide the triangle with a multiplicity of 3 and small equilateral triangles will divide each hexagon with a multiplicity of 12. The product of these two multiplicities (3 × 12) gives a replicating order for both the hexagon and the equilateral triangle: 36 of the hexagonal figures will form a larger figure of similar shape, and 36 equilateral triangles will form a larger equilateral triangle. In more formal language: If P and Q are two shapes such that P divides Q with a multiplicity of s and Q divides P with a multiplicity of t, then P and Q are both replicating figures of order st ($s \times t$). Of course, each figure can have lower replicating orders as well. In the ex-

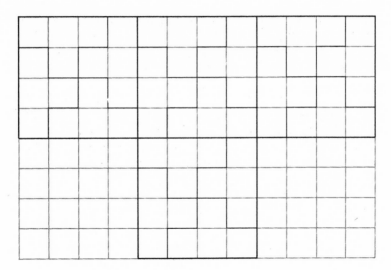

Figure 5.7. A rep-16 octagon

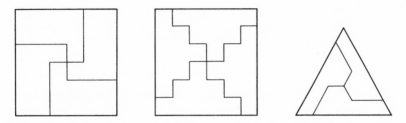

Figure 5.8. Three rep-36 polygons

ample given, the equilateral triangle, in addition to being rep-36, is also rep-4, rep-9, rep-16, and rep-25.

When P and Q are similar figures, it follows from the above theorem that if the figure has a replicating order of k, it will also be rep-k^2, rep-k^3, rep-k^4, and so on for all powers of k. Similarly, if a figure is both rep-s and rep-t, it will also be rep-st.

The principle underlying all of these theorems can be extended as follows. If P divides Q with a multiplicity of s, and Q divides R with a multiplicity of t, and R divides P with a multiplicity of u, then P and Q and R are each rep-stu. For instance, each of the polygons in Figure 5.9 will divide a 3 × 4 rectangle with a multiplicity of 2. The 3 × 4 rectangle in turn divides a square with a multiplicity of 12, and the square divides any one of the three original shapes with a multiplicity of 6. Consequently, the replicating order of each polygon is 2 × 12 × 6, or

144. It is conjectured that none of the three has a lower replicating order.

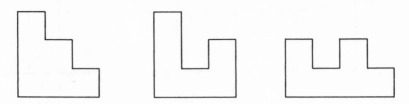

Figure 5.9. Three rep-144 polygons

Golomb has noted that every known polygon of rep-4, including the stellated polygons, will divide a parallelogram with a multiplicity of 2. In other words, if any known rep-4 polygon is replicated, the pair can be fitted together to form a parallelogram! It is conjectured, but not yet proved, that this is true of all rep-4 polygons.

An obvious extension of Golomb's pioneer work on replication theory (of which only the most elementary aspects have been detailed here) is into three or even higher dimensions. A trivial example of a replicating solid figure is the cube: it obviously is rep-8, rep-27, and so on for any order that is a cubical number. Other trivial examples result from giving plane replicating figures a finite thickness, then forming layers of larger replicas to make a model of the original solid. Less trivial examples certainly exist; a study of them might lead to significant results.

In addition to the problems already posed, here are two unusual dissection puzzles closely related to what we have been considering (see Figure 5.10). First the easier one: Can the reader divide the hexagon *(left)* into two congruent stellated polygons? More difficult: Divide the pentagon *(right)* into four congruent stellated polygons. In neither case are the polygons similar to the original figure.

Addendum

I gave the conjecture that none of the three polygons shown in Figure 5.9 has a replicating order lower than 144. Wrong for all three! Mark A. Mandel, then 14, wrote to show how the middle polygon could be cut into 36 replicas (see Figure 5.11). Robert Reid, writing from Peru, found a way for 121 copies of the first hexomino to replicate and 64

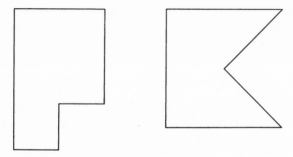

Figure 5.10. Two dissection problems

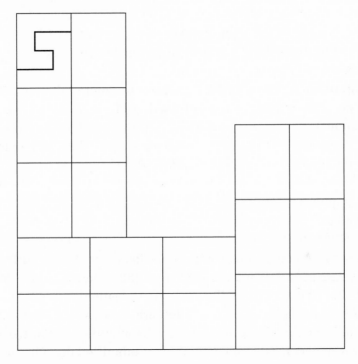

Figure 5.11. Hexomino: rep-tile of order 36

Figure 5.12. A Rep-81 polygon

copies of the third polygon to do the same. Reid also proved that the hexomino shown in Figure 5.12 is a rep-tile of order 81.

Rep-Tiles 55

Ralph H. Hinrichs, Phoenixville, PA, discovered that if the middle hexagon at the bottom of Figure 5.3 is dissected in a slightly different way (the pattern within each rectangle is mirror-reflected), the entire figure can undergo an infinite number of affine transformations (the 90-degree exterior angle taking any acute or obtuse value) to provide an infinity of rep-4 hexagons. Only when the angle is 90 degrees is the figure also rep-9, thus disproving an early guess that all rep-4 standard polygons are rep-9 and vice versa.

The three nonpolygon rep-tiles shown in Figure 5.4 are now called "self-similar fractals." Such fractals are being extensively studied. I have not listed recent references in my bibliography, but interested readers can consult Christoph Bandt's "Self-Similar Sets 5," in the *Proceedings of the American Mathematical Society,* Vol. 112, June 1991, pages 349–62, and his list of references.

The rep-tile triangle second from the top in Figure 5.12—incidently it is not accurately drawn—is mentioned in Plato's *Timaeus.* Timaeus points out that it is half of an equilateral triangle and that he considers it the "most beautiful" of all scalene triangles. (See the Random House edition of Plato, edited by Benjamin Jowett, Vol. 2, p. 34.)

Sol Golomb is best known for his work on polyominoes—shapes formed by joining *n* unit squares along their edges. Golomb's classic study, *Polyominoes,* was reissued by Princeton University Press in 1994.

The most spectacular constructions of what Golomb calls "infin-tiles (rep-tiles with infinitely many sides) are in the papers by Jack Giles, Jr., cited in the bibliography. Giles calls them "superfigures." Many of Golomb's infin-tiles, and those of Giles, are early examples of fractals. Golomb tells me that Giles was a parking lot attendant in Florida when he sent his papers to Golomb, who in turn submitted them to the *Journal of Combinatorial Theory.*

Answers

The problem of dissecting the sphinx is shown in Figure 5.13, top. The next two illustrations show how to construct rep-3 and rep-5 triangles. The bottom illustration gives the solution to the two dissection problems involving stellated polygons. The first of these can be varied in an infinite number of ways; the solution shown here is one of the simplest.

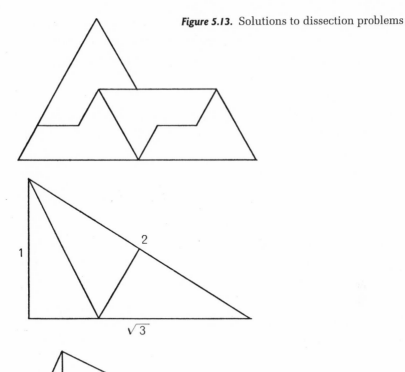

Figure 5.13. Solutions to dissection problems

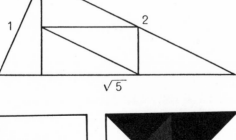

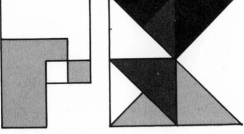

The second solution is an old-timer. Sam Loyd, in his puzzle column in *Woman's Home Companion* (October 1905) points out that the figure is similar to the one shown here in the lower right corner of Figure 5.3 in that one-fourth of a square is missing from each figure. He writes that he spent a year trying to cut the mitre shape into four congruent parts, each simply connected, but was unable to do better than the solution reproduced here. It can be found in many old puzzle books antedating Loyd's time.

Bibliography

R. O. Davies, "Replicating Boots," *Mathematical Gazette,* Vol. 50, May 1966, p. 157.

F. M. Dekking, "Replicating Superfigures and Endomorphisms of Free Groups," *Journal of Combinatorial Theory,* Vol. 32A, 1982, pp. 315–20.

J. Doyen and M. Lenduyt, "Dissection of Polygons," *Annals of Discrete Mathematics,* Vol. 18, 1983, pp. 315–18.

J. Giles, Jr., "Infinite-level Replication Dissections of Plane Figures," "Construction of Replicating Superfigures," and "Superfigures Replicating with Polar Symmetry," *Journal of Combinatorial Theory,* 26A, 1979, pp. 319–27, 328–34, 335–37.

J. Giles, Jr., "The Gypsy Method of Superfigure Construction," *Journal of Recreational Mathematics,* Vol. 13, 1980/81, pp. 97–101.

M. Goldberg and B. M. Stewart, "A Dissection Problem for Sets of Polygons," *American Mathematical Monthly,* Vol. 71, December 1964, pp. 1077–95.

S. W. Golomb, "Replicating Figures in the Plane," *Mathematical Gazette,* Vol. 48, December 1964, pp. 403–12.

H. D. Grossman, "Fun with Lattice Points," *Scripta Mathematica,* Vol. 14, June 1948, pp. 157–59.

C. D. Langford, "Uses of a Geometrical Puzzle," *Mathematical Gazette,* Vol. 24, July 1940, pp. 209–11.

R. Sibson, "Comments on Note 1464," *Mathematical Gazette,* Vol. 24, December 1940, p. 343.

G. Valette and T. Zamfirescu, "Les Partages d'un Polygone Convexe en 4 Polygones Semblables au Premier," *Journal of Combinatorial Theory,* Ser. B, Vol. 16, 1974, pp. 1–16.

Chapter 6

Piet Hein's Superellipse

There is one art, no more, no less: to do all things with artlessness.　　　—Piet Hein

Civilized man is surrounded on all sides, indoors and out, by a subtle, seldom-noticed conflict between two ancient ways of shaping things: the orthogonal and the round. Cars on circular wheels, guided by hands on circular steering wheels, move along streets that intersect like the lines of a rectangular lattice. Buildings and houses are made up mostly of right angles, relieved occasionally by circular domes and windows. At rectangular or circular tables, with rectangular napkins on our laps, we eat from circular plates and drink from glasses with circular cross sections. We light cylindrical cigarettes with matches from rectangular packs, and we pay the rectangular bill with rectangular credit cards, checks, or dollar bills and circular coins.

Even our games combine the orthogonal and the round. Most outdoor sports are played with spherical balls on rectangular fields. Indoor games, from pool to checkers, are similar combinations of the round and the rectangular. Rectangular playing cards are held in a fanlike circular array. The very letters on this rectangular page are patchworks of right angles and circular arcs. Wherever one looks the scene swarms with squares and circles and their affinely stretched forms: rectangles and ellipses. (In a sense the ellipse is more common than the circle, because every circle appears elliptical when seen from an angle.) In op paintings and textile designs, squares, circles, rectangles, and ellipses jangle against one another as violently as they do in daily life.

The Danish writer and inventor Piet Hein asked himself a fascinating question: What is the simplest and most pleasing closed curve that mediates fairly between these two clashing tendencies? Originally a scientist, Piet Hein (he is always spoken of by both names) was well known throughout Scandinavia and English-speaking countries for his enormously popular volumes of gracefully aphoristic poems (which critics have likened to the epigrams of Martial) and for his writings on scientific and humanistic topics. To recreational mathematicians he is best known as the inventor of the game Hex, of the Soma cube, and of other remarkable games and puzzles. He was a friend of Norbert Wiener, whose last book, *God and Golem, Inc.,* is dedicated to him.

The question Piet Hein asked himself had been suggested by a knotty city-planning problem that first arose in 1959 in Sweden. Many years earlier Stockholm had decided to raze and rebuild a congested section of old houses and narrow streets in the heart of the city, and after World War II this enormous and costly program got under way. Two broad new traffic arteries running north–south and east–west were cut through the center of the city. At the intersection of these avenues a large rectangular space (now called Sergel's Square) was laid out. At its center is an oval basin with a fountain surrounded by an oval pool containing several hundred smaller fountains. Daylight filters through the pool's translucent bottom into an oval self-service restaurant, below street level, surrounded by oval rings of pillars and shops. Below that there are two more oval floors for dining and dancing, cloakrooms, and kitchen.

In planning the exact shape of this center the Swedish architects ran into unexpected snags. The ellipse had to be rejected because its pointed ends would interfere with smooth traffic flow around it; moreover, it did not fit harmoniously into the rectangular space. The city planners next tried a curve made up of eight circular arcs, but it had a patched-together look with ugly "jumps" of curvature in eight places. In addition, plans called for nesting different sizes of the oval shape, and the eight-arc curve refused to nest in a pleasing way.

At this stage the architectural team in charge of the project consulted Piet Hein. It was just the kind of problem that appealed to his combined mathematical and artistic imagination, his sense of humor, and his knack of thinking creatively in unexpected directions. What kind of curve, less pointed than the ellipse, could he discover that would nest pleasingly and fit harmoniously into the rectangular open space at the heart of Stockholm?

To understand Piet Hein's novel answer we must first consider the ellipse, as he did, as a special case of a more general family of curves with the following formula in Cartesian coordinates:

$$\left|\frac{x}{a}\right|^n + \left|\frac{y}{b}\right|^n = 1,$$

where *a* and *b* are unequal parameters (arbitrary constants) that represent the two semiaxes of the curve and *n* is any positive real number. The vertical brackets indicate that each fraction is to be taken with respect to its absolute value; that is, its value without regard to sign.

When *n* =2, the real values of *x* and *y* that satisfy the equation (its "solution set") determine the points on the graph that lie on an ellipse with its center at the origin of the two coordinates. As *n* decreases from 2 to 1, the oval becomes more pointed at its ends ("subellipses," Piet Hein called them). When *n* = 1, the figure is a parallelogram. When *n* is less than 1, the four sides are concave curves that become increasingly concave as *n* approaches 0. At *n* = 0 they degenerate into two crossed straight lines.

If *n* is allowed to increase above 2, the oval develops flatter and flatter sides, becoming more and more like a rectangle; indeed, the rectangle is its limit as *n* approaches infinity. At what point is such a curve most pleasing to the eye? Piet Hein settled on *n* = 2½. With the help of a computer, 400 coordinate pairs were calculated to 15 decimal places and larger, precise curves were drawn in many different sizes, all with the same height-width ratios (to conform with the proportions of the open space at the center of Stockholm). The curves proved to be strangely satisfying, neither too rounded nor too orthogonal, a happy blend of elliptical and rectangular beauty. Moreover, such curves could be nested, as shown in Figures 6.1 and 6.2, to give a strong feeling of harmony and parallelism between the concentric ovals. Piet Hein called all such curves with exponents above 2 "superellipses." Stockholm immediately accepted the 2 1/2-exponent superellipse as the basic motif of its new center. Already the large superelliptical pool has conferred upon Stockholm an unusual mathematical flavor, like the big catenary curve of St. Louis's Gateway Arch, which dominates the local skyline.

Meanwhile Piet Hein's superellipse has been enthusiastically adopted by Bruno Mathsson, a well-known Swedish furniture designer. He first produced a variety of superelliptical desks, now in the offices of many Swedish executives, and has since followed with superelliptical tables, chairs, and beds. (Who needs the corners?) Industries in Denmark, Sweden, Norway, and Finland turned to Piet Hein for solutions to various orthogonal-versus-circular problems, and he worked on superelliptical furniture, dishes, coasters, lamps, silverware, textile patterns, and so on. The tables, chairs, and beds embodied another Piet

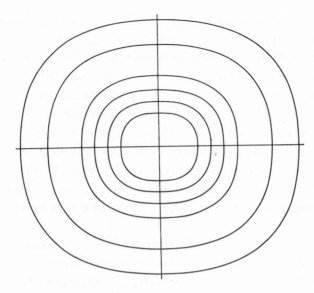

Figure 6.1. Concentric superellipses

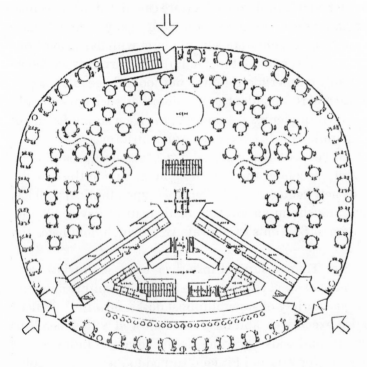

Figure 6.2. Plan of Stockholm's underground restaurants and the pools above them

Hein invention: unusual self-clamping legs that can be removed and attached with great ease.

"The superellipse has the same convincing unity as the circle and ellipse, but it is less obvious and less banal," Piet Hein wrote in a leading Danish magazine devoted to applied arts and industrial design. (The magazine's white cover for that issue bore only the stark black line of a superellipse, captioned with the formula of the curve.)

"The superellipse is more than just a new fad," Piet Hein continued; "it is a relief from the straitjacket of the simpler curves of first and second powers, the straight line and the conic sections." Incidentally, one must not confuse the Piet Hein superellipse with the superficially similar potato-shaped curves one often sees, particularly on the face of television sets. These are seldom more than oval patchworks of different kinds of arc, and they lack any simple formula that gives aesthetic unity to the curve.

When the axes of an ellipse are equal, it is of course a circle. If in the circle's formula, $x^2 + y^2 = 1$, the exponent 2 is replaced by a higher number, the graphed curve becomes what Piet Hein called the "supercircle." At 2½ it is a genuine "squared circle" in the sense that it is artistically midway between the two extremes. The changing shapes of curves with the general formula $x^n + y^n = 1$, as n varies from 0 to infinity, are graphed in Figure 6.3. If the graph could be stretched uniformly along one axis (one of the affine transformations), it would depict the family of curves of which the ellipse, subellipses, and superellipses are members.

Figure 6.3. Silver superegg, stable on either end

In the same way, one can raise the exponent in the corresponding Cartesian formulas for spheres and ellipsoids to obtain what Piet Hein called "superspheres" and "superellipsoids." If the exponent is 2½, such solids can be regarded as spheres and ellipsoids that are halfway along the road to being cubes and bricks.

The true ellipsoid, with three unequal axes, has the formula

$$\frac{x^2}{a^2} + \frac{y^2}{b^2} + \frac{z^2}{c^2} = 1,$$

where a, b, and c are unequal parameters representing half the length of each axis. When the three parameters are equal, the figure is a sphere. When only two are equal, the surface is called an "ellipsoid of rotation" or a spheroid. It is produced by rotating an ellipse on either of its axes. If the rotation is on the longer axis, the result is a prolate spheroid—a kind of egg shape with circular cross sections perpendicular to the axis.

It turns out that a solid model of a prolate spheroid, with homogeneous density, will no more balance upright on either end than a chicken egg will, unless one applies to the egg a stratagem usually credited to Columbus. Columbus returned to Spain in 1493 after having discovered America, thinking that the new land was India and that he had proved the earth to be round. At Barcelona a banquet was given in his honor. This is how Girolamo Benzoni, in his *History of the New World* (Venice, 1565), tells the story (I quote from an early English translation):

> Columbus, being at a party with many noble Spaniards . . . one of them undertook to say: "Mr. Christopher, even if you had not found the Indies, we should not have been devoid of a man who would have attempted the same thing that you did, here in our own country of Spain, as it is full of great men clever in cosmography and literature." Columbus said nothing in answer to these words, but having desired an egg to be brought to him, he placed it on the table saying: "Gentlemen, I will lay a wager with any of you, that you will not make this egg stand up as I will, naked and without anything at all." They all tried, and no one succeeded in making it stand up. When the egg came round to the hands of Columbus, by beating it down on the table he fixed it, having thus crushed a little of one end; wherefore all remained confused, understanding what he would have said: That after the deed is done, everybody knows how to do it.

The story may be true, but a suspiciously similar story had been told 15 years earlier by Giorgio Vasari in his celebrated *Lives of the Most Eminent Painters, Sculptors and Architects* (Florence, 1550). Young Filippo Brunelleschi, the Italian architect, had designed an unusually large and heavy dome for Santa Maria del Fiore, the cathedral of Florence. City officials had asked to see his model, but he refused, "proposing instead . . . that whosoever could make an egg stand upright on a flat piece of marble should build the cupola, since thus each man's intellect would be discerned. Taking an egg, therefore, all those Masters sought to make it stand upright, but not one could find a way. Whereupon Filippo, being told to make it stand, took it graciously, and, giving one end of it a blow on the flat piece of marble, made it stand upright. The craftsmen protested that they could have done the same; but Filippo answered, laughing, that they could also have raised the cupola, if they had seen the model or the design. And so it was resolved that he should be commissioned to carry out this work."

The story has a topper. When the great dome was finally completed (many years later, but decades before Columbus's first voyage), it had the shape of half an egg, flattened at the end.

What does all this have to do with supereggs? Well, Piet Hein (my source, by the way, for the references on Columbus and Brunelleschi) discovered that a solid model of a 2½-exponent superegg—indeed, a superegg of *any* exponent—if not too tall for its width, balances immediately on either end without any sort of skulduggery! Indeed, dozens of chubby wooden and silver supereggs are now standing politely and permanently on their ends all over Scandinavia.

Consider the silver superegg shown in Figure 6.3, which has an exponent of 2½ and a height–width ratio of 4 : 3. It *looks* as if it should topple over, but it does not. This spooky stability of the superegg (on both ends) can be taken as symbolic of the superelliptical balance between the orthogonal and the round, which is in turn a pleasant symbol for the balanced mind of individuals such as Piet Hein who mediated so successfully between C. P. Snow's "two cultures."

Addendum I

The family of plane curves expressed by the formula $|x/a|^n + |y/b|^n = 1$ was first recognized and studied by Gabriel Lamé, a 19th-century French physicist, who wrote about them in 1818. In France they are called *courbes de Lamé;* in Germany, *Lamesche kurven.* The curves are algebraic when *n* is rational, transcendent when *n* is irrational.

When *n* = 2/3 and *a* = *b* (see Figure 6.4) the curve is an astroid. This is the curve generated by a point on a circle that is one-fourth or three-fourths the radius of a larger circle, when the smaller circle is rolled around the inside of the larger one. Solomon W. Golomb called attention to the fact that if *n* is odd, and the absolute value signs are dropped in the formula for Lamé curves, you get a family of curves of which the famous Witch of Agnesi (The curve studied by Maria Gaetana Agnesi) is a member. (The witch results when *n* = 3.) William Hogan wrote to say that parkway arches, designed by himself and other engineers, often are Lamé curves of exponent 2.2. In the thirties, he said, they were called "2.2 ellipses."

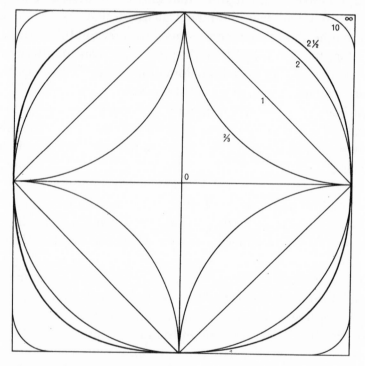

Figure 6.4. Supercircle and related curves

When a superellipse (a Lamé curve with exponent greater than 2) is applied to a physical object, its exponent and parameters *a* and *b* can, of course, be varied to suit circumstances and taste. For the Stockholm center, Piet Hein used the parameters *n* = 2½ and *a*/*b* = 6/5. A few years later Gerald Robinson, a Toronto architect, applied the superellipse to a parking garage in a shopping center in Peterborough, a Toronto suburb. The length and width were required to be in the ratio $a/b = 9/7$. Given this ratio, a survey indicated that an exponent slightly greater than 2.7 produced a superellipse that seemed the most pleasing to those polled. This suggested e as an exponent (since $e = 2.718\ldots$).

Readers suggested other parameters. J. D. Turner proposed mediating between the extremes of circle and square (or rectangle and ellipse) by picking the exponent that would give an area exactly halfway between the two extreme areas. D. C. Mandeville found that the exponent mediating the areas of a circle and square is so close to pi that he wondered if it actually *is* pi. Unfortunately it is not. Norton Black, using a computer, determined that the value is a trifle greater than 3.17. Turner also proposed mediating between ellipse and rectangle by choosing an exponent that sends the curve through the midpoint of a line joining the rectangle's corner to the corresponding point on the ellipse.

Turner and Black each suggested that the superellipse be combined with the aesthetically pleasing "golden rectangle" by making a/b the golden ratio. Turner's vote for the most pleasing superellipse went to the oval with parameters $a/b =$ golden ratio and $n = e$. Michel L. Balinski and Philetus H. Holt III, in a letter published in *The New York Times* in December 1968 (I failed to record the day of the month) recommended a golden superellipse with $n = 2\frac{1}{2}$ as the best shape for the negotiating table in Paris. At that time the diplomats preparing to negotiate a Vietnam peace were quarreling over the shape of their table. If no table can be agreed upon, Balinski and Holt wrote, the diplomats should be put inside a hollow superegg and shaken until they are in "superelliptic agreement."

The superegg is a special case of the more general solid shape which one can call a superellipsoid. The superellipsoid's formula is

$$\left|\frac{x}{a}\right|^n + \left|\frac{y}{b}\right|^n + \left|\frac{z}{c}\right|^n = 1.$$

When $a = b = c$, the solid is a supersphere, its shape varying from sphere to cube as the exponent varies. When $a = b$, the solid is supercircular in cross section, with the formula

$$\left|\frac{x}{a}\right|^n + \left|\frac{y}{a}\right|^n + \left|\frac{z}{b}\right|^n = 1.$$

Supereggs, with circular cross-sections, have the formula

$$\left|\frac{\sqrt{x^2 + y^2}}{a}\right|^n + \left|\frac{z}{b}\right|^n = 1.$$

When I wrote my column on the superellipse, I believed that any solid superegg based on an exponent greater than 2 and less than infinity would balance on its end provided its height did not exceed its width by too great a ratio. A solid superegg with an exponent of infinity would, of course, be a right circular cylinder that would, in principle, stand on its flat end regardless of how much higher it was than wide. But short of infinity it seemed intuitively clear that for each exponent there was a critical ratio beyond which the egg would be unstable. Indeed, I even published the following proof that this is the case:

If the center of gravity, *CG*, of an egg is below the center of curvature, *CC*, of the egg's base at the central point of the base, the egg will balance. It balances because any tipping of the egg will raise the *CG*. If the *CG* is above the *CC*, the egg is unstable because the slightest tipping lowers the *CG*. To make this clear, consider first the sphere shown at the left in Figure 6.5. Inside the sphere the *CG* and *CC* are the same point: the center of the sphere. For any supersphere with an exponent greater than 2, as shown second from left in the illustration, the *CC* is above the *CG* because the base is less convex. The higher the exponent, the less convex the base and the higher the *CC*.

Now suppose the supersphere is stretched uniformly upward along its vertical coordinates, transforming it into a superellipsoid of rotation, or what Piet Hein calls a superegg. As it stretches, the *CC* falls and the *CG* rises. Clearly there must be a point *X* where the *CC* and the *CG* coincide. Before this crucial point is reached the superegg is stable, as shown third from left in Figure 6.5. Beyond that point the superegg is unstable *(right)*.

C. E. Gremer, a retired U.S. Navy commander, was the first of many readers to inform me that the proof is faulty. Contrary to intuition, at the base point of *all* supereggs, the center of curvature is infinitely high! If we increase the height of a superegg while its width remains constant, the curvature at the base point remains "flat." German mathematicians call it a *flachpunkt*. The superellipse has a similar *flachpunkt* at its ends. In other words, all supereggs, regardless of their height–width ratio, are theoretically stable! As a superegg becomes taller and thinner, there is of course a critical ratio at which the degree of tilt needed to

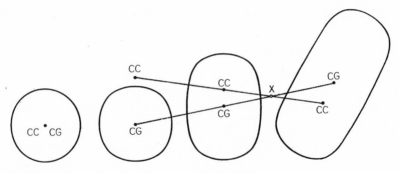

Figure 6.5. Diagrams for a false proof of superegg instability

topple it comes so close to zero that such factors as inhomogeneity of the material, surface irregularity, vibrations, air currents, and so on make it practically unstable. But in a mathematically ideal sense there *is* no critical height–width ratio. As Piet Hein put it, in theory one can balance any number of supereggs, each an inch wide and as tall as the Empire State Building, on top of one another, end to end, and they will not fall. Calculating precise "topple angles" at which a given superegg will not regain balance is a tricky problem in calculus. Many readers tackled it and sent their results.

Speaking of egg balancing, the reader may not know that almost any chicken egg can be balanced on its broad end, on a smooth surface, if one is patient and steady-handed. Nothing is gained by shaking the egg first in an attempt to break the yolk. Even more puzzling as a parlor trick is the following method of balancing an egg on its *pointed* end. Secretly put a tiny amount of salt on the table, balance the egg on it, then gently blow away the excess grains before you call in viewers. The few remaining grains which hold the egg are invisible, especially on a white surface. For some curious reason, balancing chicken eggs legitimately on their broad ends became a craze in China in 1945—at least, so said *Life* in its picture story of April 9, 1945.

The world's largest superegg, made of steel and aluminum and weighing almost a ton, was set up outside Kelvin Hall in Glasgow, October 1971, to honor Piet Hein's appearance there as a speaker during an exhibition of modern homes. The superellipse has twice appeared on Danish postage stamps: In 1970 on a blue two-kroner honoring Bertel Thorvaldsen and in 1972 on a Christmas seal bearing portraits of the queen and the prince consort.

Supereggs, in a variety of sizes and materials, are on sale throughout the world in stores that specialize in unusual gifts. Small, solid-steel supereggs are marketed as an "executive's toy." The best trick with one of them is to stand it on end, give it a gentle push, and try to make it turn one, two, or more somersaults before coming to rest again on one end. Hollow supereggs, filled with a special chemical, are sold as drink coolers. Larger supereggs are designed to hold cigarettes. More expensive supereggs, intended solely as art objects, are also available.

Figure 6.6 shows two stamps honoring Piet Hein's superellipse that were issued by Denmark. When he died in 1996, *Politiken,* Denmark's leading newspaper, devoted a full page to his obituary. It was a privilege to have known him as a friend.

Figure 6.6. The superellipse on two Danish stamps

Addendum II

Piet Hein's mathematical recreations were the topics of many of my *Scientific American* columns, most of which have been reprinted in books. For his game of Hex, see *The Scientific American Book of Mathematical Puzzles and Diversions* (Simon & Schuster, 1959), chapter 8. The nim-like game of Tac-Tix, which later became known as Nimbi, is covered in chapter 15 of the same volume. Piet Hein's famous Soma cube was the topic of chapter 6 in *The Second Scientific American Book of Mathematical Puzzles and Diversions* (Simon & Schuster, 1961; University of Chicago Press, 1987) and chapter 3 of *Knotted Doughnuts and Other Mathematical Entertainments* (W.H. Freeman, 1986).

In 1972 Hubley Toys, a division of Gabriel Industries, an American firm, marketed five unusual mechanical puzzles invented by Piet Hein. They are no longer on the market, but here is how I described them in my February 1973 *Scientific American* column:

1. Nimbi. This is a 12-counter version of Piet Hein's nim-type game. The counters are locked-in, sliding pegs on a reversible circular board so that after a game is played by pushing the pegs down and turning the board over it is set for another game.
2. Anagog. Here we have a spherical cousin of Piet Hein's Soma cube. Six pieces of joined unit spheres are to be formed into a 20-sphere tetrahedron or two 10-sphere tetrahedrons or other solid and flat figures.
3. Crux. A solid cross of six projecting arms is so designed that each arm rotates separately. One of several problems is to bring three spots of different colors together at each intersection.
4. Twitchit. A dodecahedron has rotating faces and the problem is to turn them until three different symbols are together at each corner.
5. Bloxbox. W. W. Rouse Ball, discussing the standard 14–15 sliding-block puzzle in his *Mathematical Recreations and Essays,* wrote in 1892: "We can conceive also of a similar cubical puzzle, but we could not work it practically except by sections." Eighty-one years later, Piet Hein found an ingenious practical solution. Seven identical unit cubes are inside a transparent plastic order-2 cube. When the cube is tilted properly, gravity slides a cube (with a pleasant click) into the hole. Each cube has three black and three white sides. Problems include forming an order-2 cube (minus one corner) with all sides one color, or all sides checkered, or all striped, and so on.

   Does the parity principle involved in flat versions apply to the three-dimensional version? And what are the minimum required moves to get from one pattern to another? Bloxbox opens a Pandora's box of questions.

Scantion International, a Danish management and consulting company, adopted the superegg as its logo. In 1982 it moved its world headquarters to Princeton, NJ, where Scantion-Princeton, as it is called, built a luxurious hotel and conference center hidden within the 25 acres of Princeton's Forrestal Center. An enormous stone superegg stands on the plaza in front of the hotel. The Schweppes Building, in Stamford, CT, just south of Exit 25 on the Merritt Parkway, has the shape of a superellipse.

Hermann Zapf designed a typeface whose "bowls" are based on the superellipse. He called it "Melior" because the curve meliorates between an ellipse and a rectangle. You'll find a picture of the upper- and lower-case letters on page 284 of Douglas Hofstadter's *Metamagical Themas* (Basic Books, 1985), with comments on page 291.

In 1959 Royal Copenhagen produced a series of small ceramic plaques in the shape of a superellipse, each with one of Piet Hein's Grooks together with one of the author's drawings to illustrate it. In 1988 Donald Knuth, Stanford University's distinguished computer scientist, and his wife commissioned David Kindersley's Workshop in Cambridge, England, to cut one of their favorite Grooks in slate, using the superellipse as a boundary (see Figure 6.7).

Figure 6.7. One of Piet Hein's Grooks, cut in slate and bounded by a superellipse.

Bibliography

The superellipse

J. Allard, "Note on Squares and Cubes," *Mathematics Magazine,* Vol. 37, September 1964, p. 210–14.

N. T. Gridgeman, "Lamé Ovals," *Mathematical Gazette,* Vol. 54, February 1970, pp. 31–37.

Piet Hein

A. Chamberlin, 'King of Supershape," *Esquire,* January 1967, p. 112ff.

J. Hicks, "Piet Hein, Bestrides Art and Science," *Life,* October 14, 1966, pp. 55–66.

Chapter 7

Penrose Tiles

$\boldsymbol{A}$t the end of a 1975 *Scientific American* column on tiling the plane periodically with congruent convex polygons (reprinted in my *Time Travel and Other Mathematical Bewilderments*) I promised a later column on nonperiodic tiling. This chapter reprints my fulfillment of that promise—a 1977 column that reported for the first time a remarkable nonperiodic tiling discovered by Roger Penrose, the noted British mathematical physicist and cosmologist. First, let me give some definitions and background.

A periodic tiling is one on which you can outline a region that tiles the plane by translation, that is, by shifting the position of the region without rotating or reflecting it. M. C. Escher, the Dutch artist, was famous for his many pictures of periodic tilings with shapes that resemble living things. Figure 7.1 is typical. An adjacent black and white bird constitute a fundamental region that tiles by translation. Think of the plane as being covered with transparent paper on which each tile is outlined. Only if the tiling is periodic can you shift the paper, without rotation, to a new position where all outlines again exactly fit.

An infinity of shapes—for instance the regular hexagon—tile only periodically. An infinity of other shapes tile both periodically and nonperiodically. A checkerboard is easily converted to a nonperiodic tiling by identical isosceles right triangles or by quadrilaterals. Simply bisect each square as shown in Figure 7.2A, *left*, altering the orientations to prevent periodicity. It is also easy to tile nonperiodically with dominoes.

Isoceles triangles also tile in the radial fashion shown in the center of Figure 7.2(A). Although the tiling is highly ordered, it is obviously not periodic. As Michael Goldberg pointed out in a 1955 paper titled "Central Tessellations," such a tiling can be sliced in half, and then the

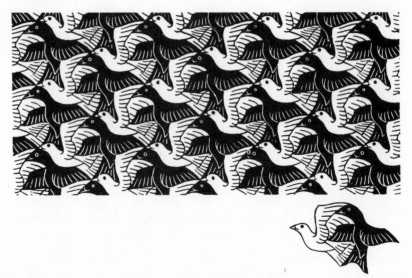

Figure 7.1. A periodic tessellation by M. C. Escher (1949)

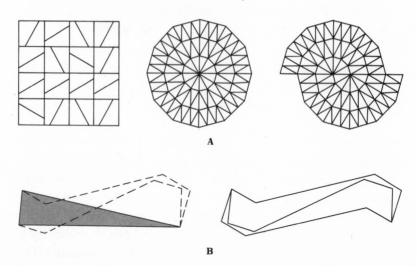

A

B

Figure 7.2. (A) Nonperiodic tiling with congruent shapes. (B) An enneagon (*dotted at left*) and a pair of enneagons (*right*) forming an octagon that tiles periodically.

half planes can be shifted one step or more to make a spiral form of nonperiodic tiling, as shown in Figure 7.2*(A), right.* The triangle can be distorted in an infinity of ways by replacing its two equal sides with congruent lines, as shown at the left in Figure 7.2*(B).* If the new sides have straight edges, the result is a polygon of 5, 7, 9, 11, . . . edges that tiles spirally. Figure 7.3 shows a striking pattern obtained in this way from a nine-sided polygon. It was first found by Heinz Voderberg in a complicated procedure. Goldberg's method of obtaining it makes it almost trivial.

Figure 7.3. A spiral tiling by Heinz Voderberg

In all known cases of nonperiodic tiling by congruent figures the figure also tiles periodically. Figure 7.2*(B), right,* shows how two of the Voderberg enneagons go together to make an octagon that tiles periodically in an obvious way.

Another kind of nonperiodic tiling is obtained by tiles that group together to form larger replicas of themselves. Solomon W. Golomb calls them "reptiles." Figure 7.4 shows how a shape called the "sphinx" tiles nonperiodically by giving rise to ever larger sphinxes. Again, two sphinxes (with one sphinx rotated 180 degrees) tile periodically in an obvious way.

Are there sets of tiles that tile only nonperiodically? By "only" we mean that neither a single shape or subset nor the entire set tiles periodically but that by using all of them a nonperiodic tiling is possible. Rotating and reflecting tiles are allowed.

For many decades experts believed no such set exists, but the sup-

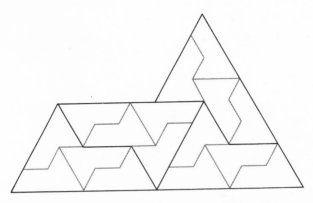

Figure 7.4. Three generations of sphinxes in a nonperiodic tiling

position proved to be untrue. In 1961 Hao Wang became interested in tiling the plane with sets of unit squares whose edges were colored in various ways. They are called Wang dominoes, and Wang wrote a splendid article about them for *Scientific American* in 1965. Wang's problem was to find a procedure for deciding whether any given set of dominoes will tile by placing them so that abutting edges are the same color. Rotations and reflections are not allowed. The problem is important because it relates to decision questions in symbolic logic. Wang conjectured that any set of tiles which can tile the plane can tile it periodically and showed that if this is the case, there is a decision procedure for such tiling.

In 1964 Robert Berger, in his thesis for a doctorate in applied mathematics from Harvard University, showed that Wang's conjecture is false. There is no general procedure. Therefore there is a set of Wang dominoes that tiles only nonperiodically. Berger constructed such a set, using more than 20,000 dominoes. Later he found a much smaller set of 104, and Donald Knuth was able to reduce the number to 92.

It is easy to change such a set of Wang dominoes into polygonal tiles that tile only nonperiodically. You simply put projections and slots on the edges to make jigsaw pieces that fit in the manner formerly prescribed by colors. An edge formerly one color fits only another formerly the same color, and a similar relation obtains for the other colors. By allowing such tiles to rotate and reflect Robinson constructed six tiles (see Figure 7.5) that force nonperiodicity in the sense explained above. In 1977 Robert Ammann found a different set of six tiles that also force nonperiodicity. Whether tiles of this square type can be re-

duced to less than six is not known, though there are strong grounds for
believing six to be the minimum.

Figure 7.5. Raphael M. Robinson's six tiles that force a nonperiodic tiling

At the University of Oxford, where he is Rouse Ball Professor of
Mathematics, Penrose found small sets of tiles, not of the square type,
that force nonperiodicity. Although most of his work is in relativity theory and quantum mechanics, he continues the active interest in recreational mathematics he shared with his geneticist father, the late L. S.
Penrose. (They are the inventors of the famous "Penrose staircase" that
goes round and round without getting higher; Escher depicted it in his
lithograph "Ascending and Descending.") In 1973 Penrose found a set
of six tiles that force nonperiodicity. In 1974 he found a way to reduce
them to four. Soon afterward he lowered them to two.

Because the tiles lend themselves to commercial puzzles, Penrose
was reluctant to disclose them until he had applied for patents in the
United Kingdom, the United States, and Japan. The patents are now in
force. I am equally indebted to John Horton Conway for many of the results of his study of the Penrose tiles.

The shapes of a pair of Penrose tiles can vary, but the most interesting pair have shapes that Conway calls "darts" and "kites." Figure
7.6(A) shows how they are derived from a rhombus with angles of 72
and 108 degrees. Divide the long diagonal in the familiar golden ratio

of $(1 + \sqrt{5})/2 = 1.61803398\ldots$, then join the point to the obtuse corners. That is all. Let phi stand for the golden ratio. Each line segment is either 1 or phi as indicated. The smallest angle is 36 degrees, and the other angles are multiples of it.

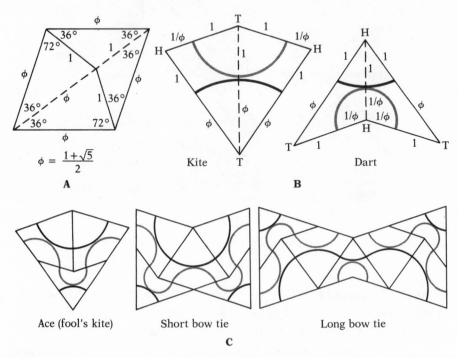

Figure 7.6. (**A**) Construction of dart and kite. (**B**) A coloring *(black and gray)* of dart and kite to force nonperiodicity. (**C**) Aces and bow ties that speed constructions.

The rhombus of course tiles periodically, but we are not allowed to join the pieces in this manner. Forbidden ways of joining sides of equal length can be enforced by bumps and dents, but there are simpler ways. For example, we can label the corners H and T (heads and tails) as is shown in Figure 7.6(*B*), and then give the rule that in fitting edges only corners of the same letter may meet. Dots of two colors could be placed in the corners to aid in conforming to this rule, but a prettier method, proposed by Conway, is to draw circular arcs of two colors on each tile, shown in the illustration as black and gray. Each arc cuts the sides as well as the axis of symmetry in the golden ratio. Our rule is that abutting edges must join arcs of the same color.

To appreciate the full beauty and mystery of Penrose tiling one should make at least 100 kites and 60 darts. The pieces need be colored on one side only. The number of pieces of the two shapes are (like their areas) in the golden ratio. You might suppose you need more of the smaller darts, but it is the other way around. You need 1.618 . . . as many kites as darts. In an infinite tiling this proportion is exact. The irrationality of the ratio underlies a proof by Penrose that the tiling is nonperiodic because if it were periodic, the ratio clearly would have to be rational.

A good plan is to draw as many darts and kites as you can on one sheet, with a ratio of about five kites to three darts, using a thin line for the curves. The sheet can be photocopied many times. The curves can then be colored, say, red and green. Conway has found that it speeds constructions and keeps patterns stabler if you make many copies of the three larger shapes as is shown in Figure 7.6*(C)*. As you expand a pattern, you can continually replace darts and kites with aces and bow ties. Actually an infinity of arbitrarily large *pairs* of shapes, made up of darts and kites, will serve for tiling any infinite pattern.

A Penrose pattern is made by starting with darts and kites around one vertex and then expanding radially. Each time you add a piece to an edge, you must choose between a dart and a kite. Sometimes the choice is forced, sometimes it is not. Sometimes either piece fits, but later you may encounter a contradiction (a spot where no piece can be legally added) and be forced to go back and make the other choice. It is a good plan to go around a boundary, placing all the forced pieces first. They cannot lead to a contradiction. You can then experiment with unforced pieces. It is always possible to continue forever. The more you play with the pieces, the more you will become aware of "forcing rules" that increase efficiency. For example, a dart forces two kites in its concavity, creating the ubiquitous ace.

There are many ways to prove that the number of Penrose tilings is uncountable, just as the number of points on a line is. These proofs rest on a surprising phenomenon discovered by Penrose. Conway calls it "inflation" and "deflation." Figure 7.7 shows the beginning of inflation. Imagine that every dart is cut in half and then all short edges of the original pieces are glued together. The result: a new tiling (shown in heavy black lines) by larger darts and kites.

Inflation can be continued to infinity, with each new "generation" of pieces larger than the last. Note that the second-generation kite, although it is the same size and shape as a first-generation ace, is formed

Figure 7.7. How a pattern is inflated

differently. For this reason the ace is also called a fool's kite. It should never be mistaken for a second-generation kite. Deflation is the same process carried the other way. On every Penrose tiling we can draw smaller and smaller generations of darts and kites. This pattern too goes to infinity, creating a structure that is a fractal.

Conway's proof of the uncountability of Penrose patterns (Penrose had earlier proved it in a different way) can be outlined as follows. On the kite label one side of the axis of symmetry L and the other R (for left and right). Do the same on the dart, using l and r. Now pick a random point on the tiling. Record the letter that gives its location on the tile. Inflate the pattern one step, note the location of the same point in a second-generation tile and again record the letter. Continuing through higher inflations, you generate an infinite sequence of symbols that is a unique labeling of the original pattern seen, so to speak, from the selected point.

Pick another point on the original pattern. The procedure may give a sequence that starts differently, but it will reach a letter beyond which it agrees to infinity with the former sequence. If there is no such agreement beyond a certain point, the two sequences label distinct patterns.

Not all possible sequences of the four symbols can be produced this way, but those that label different patterns can be shown to correspond in number with the number of points on a line.

We have omitted the colored curves on our pictures of tilings because they make it difficult to see the tiles. If you work with colored tiles, however, you will be struck by the beautiful designs created by these curves. Penrose and Conway independently proved that whenever a curve closes, it has a pentagonal symmetry, and the entire region within the curve has a fivefold symmetry. At the most a pattern can have two curves of each color that do not close. In most patterns all curves close.

Although it is possible to construct Penrose patterns with a high degree of symmetry (an infinity of patterns have bilateral symmetry), most patterns, like the universe, are a mystifying mixture of order and unexpected deviations from order. As the patterns expand, they seem to be always striving to repeat themselves but never quite managing it. G. K. Chesterton once suggested that an extraterrestrial being, observing how many features of a human body are duplicated on the left and the right, would reasonably deduce that we have a heart on each side. The world, he said, "looks just a little more mathematical and regular than it is; its exactitude is obvious, but its inexactitude is hidden; its wildness lies in wait." Everywhere there is a "silent swerving from accuracy by an inch that is the uncanny element in everything . . . a sort of secret treason in the universe." The passage is a nice description of Penrose's planar worlds.

There is something even more surprising about Penrose universes. In a curious finite sense, given by the "local isomorphism theorem," all Penrose patterns are alike. Penrose was able to show that every finite region in any pattern is contained somewhere inside every other pattern. Moreover, it appears infinitely many times in every pattern.

To understand how crazy this situation is, imagine you are living on an infinite plane tessellated by one tiling of the uncountable infinity of Penrose tilings. You can examine your pattern, piece by piece, in ever expanding areas. No matter how much of it you explore you can never determine which tiling you are on. It is no help to travel far out and examine disconnected regions, because all the regions belong to one large finite region that is exactly duplicated infinitely many times on all patterns. Of course, this is trivially true of any periodic tessellation, but

Penrose universes are not periodic. They differ from one another in infinitely many ways, and yet it is only at the unobtainable limit that one can be distinguished from another.

Suppose you have explored a circular region of diameter d. Call it the "town" where you live. Suddenly you are transported to a randomly chosen parallel Penrose world. How far are you from a circular region that exactly matches the streets of your home town? Conway answers with a truly remarkable theorem. The distance from the perimeter of the home town to the perimeter of the duplicate town is never more than d times half of the cube of the golden ratio, or 2.11 + times d. (This is an upper bound, not an average.) If you walk in the right direction, you need not go more than that distance to find yourself inside an exact copy of your home town. The theorem also applies to the universe in which you live. Every large circular pattern (there is an infinity of different ones) can be reached by walking a distance in some direction that is certainly less than about twice the diameter of the pattern and more likely about the same distance as the diameter.

The theorem is quite unexpected. Consider an analogous isomorphism exhibited by a sequence of unpatterned digits such as pi. If you pick a finite sequence of 10 digits and then start from a random spot in pi, you are pretty sure to encounter the same sequence if you move far enough along pi, but the distance you must go has no known upper bound, and the expected distance is enormously longer than 10 digits. The longer the finite sequence is, the farther you can expect to walk to find it again. On a Penrose pattern you are always very close to a duplicate of home.

There are just seven ways that darts and kites will fit around a vertex. Let us consider first, using Conway's nomenclature, the two ways with pentagonal symmetry.

The sun (shown in white in Figure 7.8) does not force the placing of any other piece around it. If you add pieces so that pentagonal symmetry is always preserved, however, you will be forced to construct the beautiful pattern shown. It is uniquely determined to infinity.

The star, shown in white in Figure 7.9, forces the 10 light gray kites around it. Enlarge this pattern, always preserving the fivefold symmetry, and you will create another flowery design that is infinite and unique. The star and sun patterns are the only Penrose universes with perfect pentagonal symmetry, and there is a lovely sense in which they are equivalent. Inflate or deflate either of the patterns and you get the other.

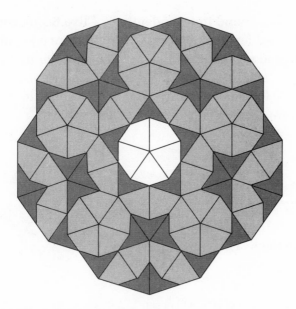

Figure 7.8. The infinite sun pattern

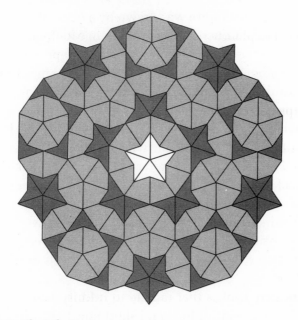

Figure 7.9. The infinite star pattern

The ace is a third way to tile around a vertex. It forces no more pieces. The deuce, the jack, and the queen are shown in white in Figure 7.10, surrounded by the tiles they immediately force. As Penrose discovered (it was later found independently by Clive Bach), some of the seven vertex figures force the placing of tiles that are not joined to the immediately forced region. Plate 1 shows in deep color the central portion of the king's "empire." (The king is the dark gray area.) All the deep colored tiles are forced by the king. (Two aces, just outside the left and right borders, are also forced but are not shown.)

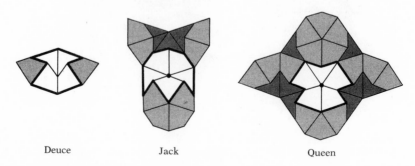

Deuce   Jack   Queen

Figure 7.10. The "empires" of deuce, jack, and queen

This picture of the king's empire was drawn by a computer program written by Eric Regener of Concordia University in Montreal. His program deflates any Penrose pattern any number of steps. The heavy black lines show the domain immediately forced by the king. The thin black lines are a third-generation deflation in which the king and almost all of his empire are replicated.

The most extraordinary of all Penrose universes, essential for understanding the tiles, is the infinite cartwheel pattern, the center of which is shown in Figure 7.11. The regular decagon at the center, outlined in heavy black (each side is a pair of long and short edges), is what Conway calls a "cartwheel." Every point on any pattern is inside a cartwheel exactly like this one. By one-step inflation we see that every point will be inside a larger cartwheel. Similarly, every point is inside a cartwheel of every generation, although the wheels need not be concentric.

Note the 10 light gray spokes that radiate to infinity. Conway calls them "worms." They are made of long and short bow ties, the number of long ones being in the golden ratio to the number of short ones. Every Penrose universe contains an infinite number of arbitrarily long worms. Inflate or deflate a worm and you get another worm along the same axis. Observe that two full worms extend across the central cartwheel in the infinite cartwheel pattern. (Inside it they are not gray.) The

Figure 7.11. The cartwheel pattern surrounding Batman

remaining spokes are half-infinite worms. Aside from spokes and the interior of the central cartwheel, the pattern has perfect tenfold symmetry. Between any two spokes we see an alternating display of increasingly large portions of the sun and star patterns.

Any spoke of the infinite cartwheel pattern can be turned side to side (or, what amounts to the same thing, each of its bow ties can be rotated end for end), and the spoke will still fit all surrounding tiles except for those inside the central cartwheel. There are 10 spokes; thus there are $2^{10} = 1,024$ combinations of states. After eliminating rotations and reflections, however, there are only 62 distinct combinations. Each combination leaves inside the cartwheel a region that Conway has named a "decapod."

Decapods are made up of 10 identical isosceles triangles with the shapes of enlarged half darts. The decapods with maximum symmetry are the buzzsaw and the starfish shown in Figure 7.12. Like a worm, each triangle can be turned. As before, ignoring rotations and reflections, we get 62 decapods. Imagine the convex vertexes on the perimeter of each decapod to be labeled T and the concave vertexes to be labeled *H*. To continue tiling, these *H*'s and *T*'s must be matched to the heads and tails of the tiles in the usual manner.

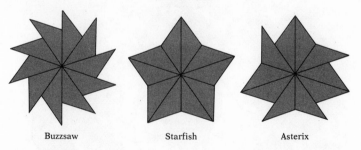

Buzzsaw   Starfish   Asterix

Figure 7.12. Three decapods

When the spokes are arranged the way they are in the infinite cartwheel pattern shown, a decapod called Batman is formed at the center. Batman (shown in dark gray) is the only decapod that can legally be tiled. (No finite region can have more than one legal tiling.) Batman does not, however, force the infinite cartwheel pattern. It merely allows it. Indeed, no finite portion of a legal tiling can force an entire pattern, because the finite portion is contained in *every* tiling.

Note that the infinite cartwheel pattern is bilaterally symmetrical, its axis of symmetry going vertically through Batman. Inflate the pattern and it remains unchanged except for mirror reflection in a line perpendicular to the symmetry axis. The five darts in Batman and its two central kites are the only tiles in any Penrose universe that are not inside a region of fivefold symmetry. All other pieces in this pattern or any other one are in infinitely many regions of fivefold symmetry.

The other 61 decapods are produced inside the central cartwheel by the other 61 combinations of worm turns in the spokes. All 61 are "holes" in the following sense. A hole is any finite empty region, surrounded by an infinite tiling, that cannot be legally tiled. You might suppose each decapod is the center of infinitely many tilings, but here Penrose's universes play another joke on us. Surprisingly, 60 decapods force a unique tiling that differs from the one shown only in the composition of the spokes. Only Batman and one other decapod, called Asterix after a once-popular French cartoon character, do not. Like Batman, Asterix allows an infinite cartwheel pattern, but it also allows patterns of other kinds.

Now for a startling conjecture. Conway believes, although he has not completed the proof, that every possible hole, of whatever size or shape, is equivalent to a decapod hole in the following sense. By rearranging tiles around the hole, taking away or adding a finite number of pieces if necessary, you can transform every hole into a decapod. If this is true, any finite number of holes in a pattern can also be reduced to one decapod. We have only to remove enough tiles to join the holes into one big hole, then reduce the big hole until an untileable decapod results.

Think of a decapod as being a solid tile. Except for Batman and Asterix, each of the 62 decapods is like an imperfection that solidifies a crystal. It forces a unique infinite cartwheel pattern, spokes and all, that goes on forever. If Conway's conjecture holds, any "foreign piece" (Penrose's term) that forces a unique tiling, no matter how large the piece is, has an outline that transforms into one of 60 decapod holes.

Kites and darts can be changed to other shapes by the same technique described earlier for changing isosceles triangles into spiral-tiling polygons. It is the same technique that Escher employed for transforming polygonal tiles into animal shapes. Figure 7.13 shows how Penrose changed his darts and kites into chickens that tile only nonperiodically. Note that although the chickens are asymmetrical, it is never necessary to turn any of them over to tile the plane. Alas, Escher died before he could know of Penrose's tiles. How he would have reveled in their possibilities!

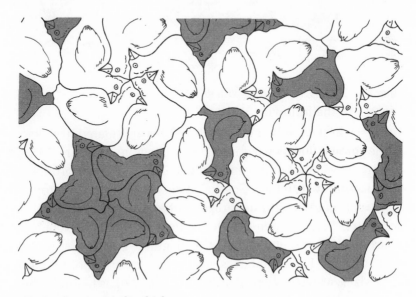

Figure 7.13. Penrose's nonperiodic chickens

By dissecting darts and kites into smaller pieces and putting them together in other ways you can make other pairs of tiles with properties similar to those of darts and kites. Penrose found an unusually simple pair: the two rhombuses in the sample pattern of Figure 7.14. All edges are the same length. The larger piece has angles of 72 and 108 degrees and the smaller one has angles of 36 and 144 degrees. As before, both the areas and the number of pieces needed for each type are in the golden ratio. Tiling patterns inflate and deflate and tile the plane in an uncountable infinity of nonperiodic ways. The nonperiodicity can be forced by bumps and dents or by a coloring such as the one suggested by Penrose and shown in the illustration by the light and dark gray areas.

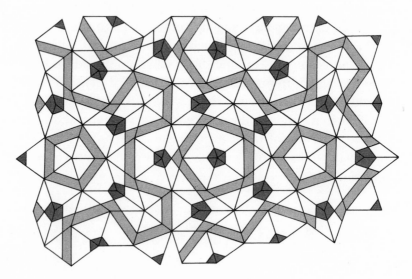

Figure 7.14. A nonperiodic tiling with Roger Penrose's rhombuses

We see how closely the two sets of tiles are related to each other and to the golden ratio by examining the pentagram in Figure 7.15. This was the mystic symbol of the ancient Greek Pythagorean brotherhood and the diagram with which Goethe's Faust trapped Mephistopheles. The construction can continue forever, outward and inward, and every line segment is in the golden ratio to the next smaller one. Note how all four Penrose tiles are embedded in the diagram. The kite is *ABCD,* and the dart is *AECB.* The rhombuses, although they are not in the proper relative sizes, are *AECD* and *ABCF.* As Conway likes to put it, the two sets of tiles are based on the same underlying "golden stuff." Any theorem about kites and darts can be translated into a theorem about the Penrose rhombuses or any other pair of Penrose tiles and vice versa. Conway prefers to work with darts and kites, but other mathematicians prefer working with the simpler rhombuses. Robert Ammann has found a bewildering variety of other sets of nonperiodic tiles. One set, consisting of two convex pentagons and a convex hexagon, forces nonperiodicity without any edge markings. He found several pairs, each a hexagon with five interior angles of 90 degrees and one of 270 degrees.

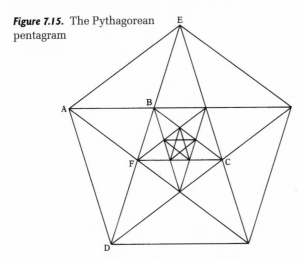

Figure 7.15. The Pythagorean pentagram

Are there pairs of tiles not related to the golden ratio that force nonperiodicity? Is there a pair of *similar* tiles that force nonperiodicity? Is there a pair of convex tiles that will force nonperiodicity without edge markings?

Of course, the major unsolved problem is whether there is a *single* shape that will tile the plane only nonperiodically. Most experts think not, but no one is anywhere near proving it. It has not even been shown that if such a tile exists, it must be nonconvex.

Addendum

For much more on Penrose tiles, see Chapter 2 in my *Penrose Tiles to Trapdoor Ciphers* (Mathematical Association of America 1989; paperback, 1997) in which I discuss solid forms of the tiles and their astonishing application to what are called quasicrystals. Until Penrose's discovery, crystals based on fivefold symmetry were believed to be impossible to construct. My bibliography gives only a small sampling of books and papers about quasicrystals that have been published in recent years.

In 1993 John Horton Conway made a significant breakthrough when he discovered a convex solid called a biprism that tiles space only aperiodically. (Aperiodic has replaced nonperiodic as a term for a tile that tiles only in a nonperiodic way.) A few years earlier Peter Schmitt, at the University of Vienna, found a nonconvex solid that fills space aperiodically, though in a trivial fashion. Conway's subtler solid is described and pictured in Keith Devlin's *The Language of Mathematics* (W. H. Freeman, 1998, pp. 219–20; paperback, 2000). Doris Schattschneider kindly supplied me with the pattern shown in Figure 7.16 for constructing the Conway solid.

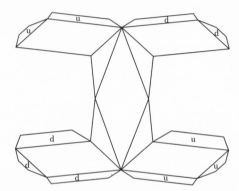

To assemble, score on all interior lines, then cut around the outline of the pattern. Tabs labeled *u* are to be folded up, those with *d* are to be folded down. Two prisms are then assembled, one on each side of the common rhombus face.

Figure 7.16. Conway's biprism. The central rhombus (which is inside the model when assembled) has sides of length 2 and short diagonal length $\sqrt{2}$. The small angle of the rhombus is arcos(3/4) ≈ 41.4 degrees. Two triangular prisms are built on this common rhombus face. The diagonal of the prism parallelogram face has length $\sqrt{(33/4)} = 2.87$. When assembled, the vertices of the rhombus that is a common face of the two prisms are the poles of 2 twofold rotation raxes.

As Devlin explains, Conway's biprism fills space in layers. Every layer is periodic, but each adjacent layer must be rotated by an irrational angle that forces aperiodicity. Unfortunately neither of the two aperiodic solids leads to the construction of a flat tile that covers the plane aperiodically. Finding such a tile or proving it nonexistent is the top unsolved problem in tiling theory.

In 1997 Penrose sued England's Kimberly-Clark Company for putting his tiling pattern on their quilted toilet paper without his permission.

See *Time,* May 5, 1997, page 26, for the story and a picture of the paper's pattern. I don't know the outcome of the lawsuit.

For a long time Penrose tiles were unavailable for purchase. Happily they are now on the market in a variety of forms that can be obtained from Kadon Enterprises, 1227 Lorrene Drive, Pasadena, MD 21122.

Bibliography

Nonperiodic tiling

E.P.M. Beenker, "Algebraic Theory of Nonperiodic Tilings of the Plane by Two Simple Building Blocks: A Square and a Rhombus," Eindhoven University of Technology, the Netherlands, Department of Mathematics and Computer Science, T. H. Report 82-WSK-04, September 1982.

R. Berger "The Undecidability of the Domino Problem," *Memoirs of the American Mathematical Society,* No. 66, 1966.

N. G. de Bruijn, "Algebraic Theory of Penrose's Nonperiodic Tilings of the Plane," Koninklijke Nederlandse Akademie van Wetenschappen Proceedings Ser. A, Vol. 84, 1981, pp. 39–66.

M. Goldberg, "Central Tessellations," *Scripta Mathematica,* Vol. 21, 1955, pp. 253–60.

B. Grünbaum and G. C. Shephard, "Some Problems on Plane Tilings," *The Mathematical Gardner,* D. A Klarner (ed.), Prindle, Weber & Schmidt, 1981.

B. Grünbaum and G. C. Shephard, *Tilings and Patterns,* W. H. Freeman, 1986.

R. K. Guy, "The Penrose Pieces," *Bulletin of the London Mathematical Society,* Vol. 8, 1976, pp. 9–10.

I. Hargittai, "Quasicrystal Structure in Bad Ragaz," *The Mathematical Intelligencer,* 44, 1992, pp. 58–59.

M. Naylor, "Nonperiodic Tiling," *The Mathematics Teacher,* 92, 1999, pp. 34–39.

K. Miyazaki, *On Some Periodical and Non-Periodical Honeycombs,* A Kobe University monograph, 1977.

R. Penrose, "The Role of Aesthetics in Pure and Applied Mathematical Research," *Bulletin of the Institute of Mathematics and Its Applications,* Vol. 10, 1974, pp. 266–71.

R. Penrose, "Pentaplexity: A Class of Nonperiodic Tilings of the Plane," *Eureka,* 39, 1978, pp. 16–22. Reprinted in *The Mathematical Intelligencer,* Vol. 2, 1979, pp. 32–37 and in *Geometrical Combinatorics,* F. C. Holroyd and R. J. Wilson, (eds.), Pitman, 1984.

R. Penrose, "Remarks on Tiling," *The Mathematics of Long-Range Aperiodic Order,* R.V. Moody (ed.), Kluwer, 1997, pp. 467–97.

I. Peterson, "From Here to Infinity," *Islands of Truth,* W. H. Freeman, 1990, pp. 86–95.

I. Peterson, "Clusters and Decagons," *Science News,* 150, 1996, pp. 232–33.

C. Radin, *Miles of Tiles,* American Mathematical Society, 1999.

R. M. Robinson, "Undecidability and Nonperiodicity for Tilings of the Plane," *Inventiones Mathematicae,* Vol. 12, 1971, pp. 177–209.

T. Sibley and S. Wagon, "Rhombic Penrose Tiling Can Be 3-Colored," *American Mathematical Monthly,* vol. 107, March 2000, pp. 251–53.

S. Wagon, "Penrose Tiles," *Mathematica in Action,* W. H. Freeman, 1993, pp. 108–17.

H. Wang, "Games, Logic and Computers," *Scientific American,* November 1965, pp. 98–106.

Quasicrystals

M. W. Browne, "Puzzling Crystals Plunge Scientists into Uncertainty," *The New York Times,* July 30, 1985, p. C1 ff.

M. W. Browne, "Potential Seen in New Quasicrystals," *The New York Times,* September 5, 1989.

M. W. Browne, " 'Impossible' Form of Matter Takes Spotlight in Study of Solids," *The New York Times,* September 5, 1989.

M. W. Browne, "New Data Help Explain Crystals That Defy Nature," *The New York Times,* November 24, 1998.

L. Danzer, "Three-Dimensional Analogs of the Planar Penrose Tiling and Quasicrystals," *Discrete Mathematics,* 76, 1989, pp. 1–7.

N. G. de Bruijn, "A Riffle Shuffle Card Trick and Its Relation to Quasicrystal Theory," *Nieuw Archief voor Wiskunde,* Vol. 5, November 1987.

A. Goldman et al., "Quasicrystalline Materials," *American Scientist,* 8, 1996, pp. 230–41.

I. Hargittai (ed.), *Fivefold Symmetry, World Scientific, 1991.*

J. Horgan, "Quasicrystals: Rules of the Game," *Science,* 247, 1990, pp. 1020–22.

C. Jano, *Quasicrystals: A Primer,* Oxford Science, 1994.

M. V. Jaric (ed.), *Introduction to Quasicrystals,* Vol. 1, *Aperiodicity and Order,* Academic Press, 1988 and Vol. 2, *Introduction to the Mathematics of Quasicrystals,* Academic Press, 1989; M. V. Jaric and D. Gratias (eds.), Vol. 3, *Extended Icosahedral Structures,* Academic Press, 1990.

M. La Brecque, "Opening the Door to Forbidden Symmetries," *Mosaic,* Vol. 18, 1987–1988, pp. 2–23.

D. Levine, "Quasicrystals: A New Class of Ordered Structure," Ph.D. thesis, University of Pennsylvania, 1986.

R. Mosseri and J. F. Sadoc, "Pauling's Model Not Universally Accepted," *Nature,* Vol. 319, 1986, p. 104. Reply to Pauling's 1985 paper.

D. R. Nelson, "Quasicrystals," *Scientific American,* August 1986, pp. 42–51.

D. R. Nelson and B. I. Halperin, "Pentagonal and Icosahedral Order in Rapidly Cooled Metals," *Science,* Vol. 229, 1985, pp. 235–38.

J. Patera, *Quasicrystals and Discrete Geometry,* American Mathematical Society, 1998.

L. Pauling, "Apparent Icosahedral Symmetry is Due to Directed Multiple Twinning of Cubic Crystals," *Nature,* Vol. 317, 1985, p. 512.

L. Pauling, "Icosahedral Symmetry," *Science,* Vol. 239, 1988, pp. 963–64. This letter from Linus Pauling continues his attack on quasicrystals, with a reply by Paul Steinhardt.

I. Peterson, "The Fivefold Way for Crystals," *Science News,* Vol. 127, 1985, pp. 188–89. Reprinted in Peterson's collection *The Mathematical Tourist,* W. H. Freeman, 1988, pp. 200–212.

I. Peterson, "Tiling to Infinity," *Science News,* Vol. 134, July 16, 1988, p. 42.

I. Peterson, "Shadows and Symmetries," *Science News,* 140, 1991, pp. 408–19.

M. Senechal, *Quasicrystals and Geometry,* Cambridge University Press, 1995.

M. Senechal and J. Taylor, "Quasicrystal: The View From Les Houches," *Mathematical Intelligencer,* 12, 1990, pp. 54–63.

P. J. Steinhardt, "Quasicrystals," *American Scientist,* November/December 1986, pp. 586–597. Its bibliography lists more than 40 technical papers.

P. Steinhardt, "Crazy Crystals," *New Scientist,* January 25, 1997, pp. 32–35.

P. Steinhardt and S. Ostlund (eds.), *The Physics of Quasicrystals,* World Scientific, 1987.

P. W. Stepheus and A. I. Goldman, "The Structure of Quasicrystals," *Scientific American,* April 1991, pp. 44ff.

D. P. Vincenzo and P. J. Steinhardt (eds.), *Quasicrystals: The State of the Art,* World Scientific, 1991.

H. C. Von Baeyer, "Impossible Crystals," *Discover,* February 1990, pp. 69–78.

D. Voss, "Smash the System," *New Scientist,* 27, 1999, pp. 42–46.

Chapter 8

The Wonders of a Planiverse

*Planiversal scientists are not a very
common breed.*

—Alexander Keewatin Dewdney

A*s* far as anyone knows the only existing universe is the one
we live in, with its three dimensions of space and one of time. It is not
hard to imagine, as many science-fiction writers have, that intelligent
organisms could live in a four-dimensional space, but two dimensions
offer such limited degrees of freedom that it has long been assumed intelligent 2-space life forms could not exist. Two notable attempts have
nonetheless been made to describe such organisms.

In 1884 Edwin Abbott Abbott, a London clergyman, published his
satirical novel *Flatland.* Unfortunately the book leaves the reader almost entirely in the dark about Flatland's physical laws and the technology developed by its inhabitants, but the situation was greatly
improved in 1907 when Charles Howard Hinton published *An Episode
of Flatland.* Although written in a flat style and with cardboard characters, Hinton's story provided the first glimpses of the possible science
and technology of the two-dimensional world. His eccentric book is,
alas, long out of print, but you can read about it in the chapter "Flatlands" in my book *The Unexpected Hanging and Other Mathematical
Diversions* (Simon & Schuster, 1969).

In "Flatlands" I wrote: "It is amusing to speculate on two-
dimensional physics and the kinds of simple mechanical devices that
would be feasible in a flat world." This remark caught the attention of
Alexander Keewatin Dewdney, a computer scientist at the University
of Western Ontario. Some of his early speculations on the subject were
set down in 1978 in a university report and in 1979 in "Exploring the
Planiverse," an article in *Journal of Recreational Mathematics* (Vol. 12,
No. 1, pp. 16–20; September). Later in 1979 Dewdney also privately
published "Two-dimensional Science and Technology," a 97-page tour
de force. It is hard to believe, but Dewdney actually lays the ground-

work for what he calls a planiverse: a possible two-dimensional world. Complete with its own laws of chemistry, physics, astronomy, and biology, the planiverse is closely analogous to our own universe (which he calls the steriverse) and is apparently free of contradictions. I should add that this remarkable achievement is an amusing hobby for a mathematician whose serious contributions have appeared in some 30 papers in technical journals.

Dewdney's planiverse resembles Hinton's in having an earth that he calls (as Hinton did) Astria. Astria is a dislike planet that rotates in planar space. The Astrians, walking upright on the rim of the planet, can distinguish east and west and up and down. Naturally there is no north or south. The "axis" of Astria is a point at the center of the circular planet. You can think of such a flat planet as being truly two-dimensional or you can give it a very slight thickness and imagine it as moving between two frictionless planes.

As in our world, gravity in a planiverse is a force between objects that varies directly with the product of their masses, but it varies inversely with the *linear* distance between them, not with the square of that distance. On the assumption that forces such as light and gravity in a planiverse move in straight lines, it is easy to see that the intensity of such forces must vary inversely with linear distance. The familiar textbook figure demonstrating that in our world the intensity of light varies inversely with the square of distance is shown at the top of Figure 8.1. The obvious planar analogue is shown at the bottom of the illustration.

To keep his whimsical project from "degenerating into idle speculation" Dewdney adopts two basic principles. The "principle of similarity" states that the planiverse must be as much like the steriverse as possible: a motion not influenced by outside forces follows a straight line, the flat analogue of a sphere is a circle, and so on. The "principle of modification" states that in those cases where one is forced to choose between conflicting hypotheses, each one equally similar to a steriversal theory, the more fundamental one must be chosen and the other must be modified. To determine which hypothesis is more fundamental Dewdney relies on the hierarchy in which physics is more fundamental than chemistry, chemistry is more fundamental than biology, and so on.

To illustrate the interplay between levels of theory Dewdney considers the evolution of the planiversal hoist in Figure 8.2. The engineer

Figure 8.1.

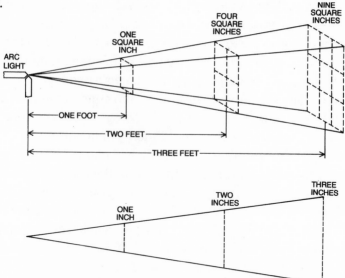

who designed it first gave it arms thinner than those in the illustration, but when a metallurgist pointed out that planar materials fracture more easily than their 3-space counterparts, the engineer made the arms thicker. Later a theoretical chemist, invoking the principles of similarity and modification at a deeper level, calculated that the planiversal molecular forces are much stronger than had been suspected, and so the engineer went back to thinner arms.

The principle of similarity leads Dewdney to posit that the planiverse

Figure 8.2.

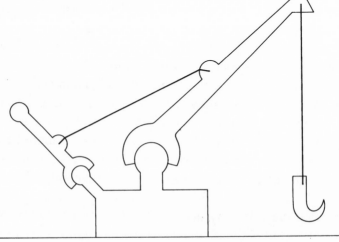

PLANE GEOMETRY

is a three-dimensional continuum of space–time containing matter composed of molecules, atoms, and fundamental particles. Energy is propagated by waves, and it is quantized. Light exists in all its wavelengths and is refracted by planar lenses, making possible planiversal eyes, planiversal telescopes, and planiversal microscopes. The planiverse shares with the steriverse such basic precepts as causality; the first and second laws of thermodynamics; and laws concerning inertia, work, friction, magnetism, and elasticity.

Dewdney assumes that his planiverse began with a big bang and is currently expanding. An elementary calculation based on the inverse-linear gravity law shows that regardless of the amount of mass in the planiverse the expansion must eventually halt, so that a contracting phase will begin. The Astrian night sky will of course be a semicircle along which are scattered twinkling points of light. If the stars have proper motions, they will continually be occulting one another. If Astria has a sister planet, it will over a period of time occult every star in the sky.

We can assume that Astria revolves around a sun and rotates, thereby creating day and night. In a planiverse, Dewdney discovered, the only stable orbit that continually retraces the same path is a perfect circle. Other stable orbits roughly elliptical in shape are possible, but the axis of the ellipse rotates in such a way that the orbit never exactly closes. Whether planiversal gravity would allow a moon to have a stable orbit around Astria remains to be determined. The difficulty is due to the sun's gravity, and resolving the question calls for work on the planar analogue of what our astronomers know as the three-body problem.

Dewdney analyzes in detail the nature of Astrian weather, using analogies to our seasons, winds, clouds, and rain. An Astrian river would be indistinguishable from a lake except that it might have faster currents. One peculiar feature of Astrian geology is that water cannot flow around a rock as it does on the earth. As a result rainwater steadily accumulates behind any rock on a slope, tending to push the rock downhill: the gentler the slope is, the more water accumulates and the stronger the push is. Dewdney concludes that given periodic rainfall the Astrian surface would be unusually flat and uniform. Another consequence of the inability of water to move sideways on Astria is that it would become trapped in pockets within the soil, tending to create large areas of treacherous quicksand in the hollows of the planet. One hopes, Dewdney writes, that rainfall is infrequent on Astria. Wind too would have much severer effects on Astria than on the earth because like rain it cannot "go around" objects.

Dewdney devotes many pages to constructing a plausible chemistry for his planiverse, modeling it as much as possible on three-dimensional matter and the laws of quantum mechanics. Figure 8.3 shows Dewdney's periodic table for the first 16 planiversal elements. Because the first two are so much like their counterparts in our world, they are called hydrogen and helium. The next 10 have composite names to suggest the steriversal elements they most resemble; for example, lithrogen combines the properties of lithium and nitrogen. The next four are named after Hinton, Abbott, and the young lovers in Hinton's novel, Harold Wall and Laura Cartwright.

ATOMIC NUMBER	NAME	SYMBOL	SHELL STRUCTURE									VALENCE
			1s	2s	2p	3s	3p	3d	4s	4p ...	
1	HYDROGEN	H	1								1
2	HELIUM	He	2								2
3	LITROGEN	Lt	2	1							1
4	BEROXYGEN	Bx	2	2							2
5	FLUORON	Fl	2	2	1						3
6	NEOCARBON	Nc	2	2	2						4
7	SODALINUM	Sa	2	2	2	1					1
8	MAGNILICON	Mc	2	2	2	2					2
9	ALUPHORUS	Ap	2	2	2	2	1				3
10	SULFICON	Sp	2	2	2	2	2				4
11	CHLOPHORUS	Cp	2	2	2	2	2	1			5
12	ARGOFUR	Af	2	2	2	2	2	2			6
13	HINTONIUM	Hn	2	2	2	2	2	2	1		1
14	ABBOGEN	Ab	2	2	2	2	2	2	2		2
15	HAROLDIUM	Wa	2	2	2	2	2	2	2	1	3
16	LAURANIUM	La	2	2	2	2	2	2	2	2	4

Figure 8.3.

In the flat world atoms combine naturally to form molecules, but of course only bonding that can be diagrammed by a planar graph is allowed. (This result follows by analogy from the fact that intersecting bonds do not exist in steriversal chemistry.) As in our world, two asymmetric molecules can be mirror images of each other, so that neither one can be "turned over" to become identical with the other. There are striking parallels between planiversal chemistry and the behavior of steriversal monolayers on crystal surfaces (see J. G. Dash, "Two-dimensional Matter," *Scientific American,* May 1973). In our world molecules can form 230 distinct crystallographic groups, but in the planiverse they can form only 17. I am obliged to pass over Dewdney's speculations about the diffusion of molecules, electrical and magnetic laws, analogues of Maxwell's equations, and other subjects too technical to summarize here.

Dewdney assumes that animals on Astria are composed of cells that cluster to form bones, muscles, and connective tissues similar to those found in steriversal biology. He has little difficulty showing how these bones and muscles can be structured to move appendages in such a way that the animals can crawl, walk, fly, and swim. Indeed, some of these movements are easier in a planiverse than in our world. For example, a steriversal animal with two legs has considerable difficulty balancing while walking, whereas in the planiverse if an animal has both legs on the ground, there is no way it can fall over. Moreover, a flying planiversal animal cannot have wings and does not need them to fly; if the body of the animal is aerodynamically shaped, it can act as a wing (since air can go around it only in the plane). The flying animal could be propelled by a flapping tail.

Calculations also show that Astrian animals probably have much lower metabolic rates than terrestrial animals because relatively little heat is lost through the perimeter of their body. Furthermore, animal bones can be thinner on Astria than they are on the earth, because they have less weight to support. Of course, no Astrian animal can have an open tube extending from its mouth to its anus, because if it did, it would be cut in two.

In the appendix to his book *The Structure and Evolution of the Universe* (Harper, 1959) G. J. Whitrow argues that intelligence could not evolve in 2-space because of the severe restrictions two dimensions impose on nerve connections. "In three or more dimensions," he writes, "any number of [nerve] cells can be connected with [one another] in pairs without intersection of the joins, but in two dimensions the maximum number of cells for which this is possible is only four." Dewdney easily demolishes this argument, pointing out that if nerve cells are allowed to fire nerve impulses through "crossover points," they can form flat networks as complex as any in the steriverse. Planiversal minds would operate more slowly than steriversal ones, however, because in the two-dimensional networks the pulses would encounter more interruptions. (There are comparable results in the theory of two-dimensional automatons.)

Dewdney sketches in detail the anatomy of an Astrian female fish with a sac of unfertilized eggs between its two tail muscles. The fish has an external skeleton, and nourishment is provided by the internal circulation of food vesicles. If a cell is isolated, food enters it through a membrane that can have only one opening at a time. If the cell is in contact with other cells, as in a tissue, it can have more than one opening at a time because the surrounding cells are able to keep it intact. We can of course see every internal organ of the fish or of any other planiversal life form, just as a four-dimensional animal could see all our internal organs.

Dewdney follows Hinton in depicting his Astrian people schematically, as triangles with two arms and two legs. Hinton's Astrians, however, always face in the same direction: males to the east and females to the west. In both sexes the arms are on the front side, and there is a single eye at the top of the triangle, as shown in Figure 8.4. Dewdney's Astrians are bilaterally symmetrical, with an arm, a leg, and an eye on each side, as shown in the illustration's center. Hence these Astrians, like terrestrial birds or horses, can see in opposite directions. Naturally the only way for one Astrian to pass another is to crawl or leap over him. My conception of an Astrian bug-eyed monster is shown at the right in the illustration. This creature's appendages serve as either arms or legs, depending on which way it is facing, and its two eyes provide binocular vision. With only one eye an Astrian would have a largely one-dimensional visual world, giving him a rather narrow perception of reality. On the other hand, parts of objects in the planiverse might be distinguished by their color, and an illusion of depth might be created by the focusing of the lens of the eye.

Figure 8.4.

On Astria building a house or mowing a lawn requires less work than it does on the earth because the amount of material involved is considerably smaller. As Dewdney points out, however, there are still formidable problems to be dealt with in a two-dimensional world: "Assuming that the surface of the planet is absolutely essential to support life-giving plants and animals, it is clear that very little of the Astrian surface can be disturbed without inviting the biological destruction of the planet. For example, here on earth we may build a modest highway through the middle of several acres of rich farmland and destroy no more than a small percentage of it. A corresponding highway on Astria will destroy *all* the 'acreage' it passes over. . . . Similarly, extensive cities would quickly use up the Astrian countryside. It would seem that the only alternative for the Astrian technological society is to go underground." A typical subterranean house with a living room, two bedrooms, and a storage room is shown in Figure 8.5. Collapsible chairs and tables are stored in recesses in the floors to make the rooms easier to walk through.

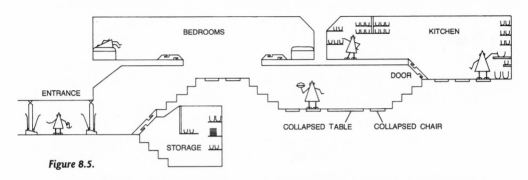

Figure 8.5.

The many simple three-dimensional mechanical elements that have obvious analogues on Astria include rods, levers, inclined planes, springs, hinges, ropes, and cables (see Figure 8.6, *top*). Wheels can be rolled along the ground, but there is no way to turn them on a fixed axle. Screws are impossible. Ropes cannot be knotted; but by the same token, they never tangle. Tubes and pipes must have partitions, to keep their sides in place, and the partitions have to be opened (but never all of them at once) to allow anything to pass through. It is remarkable that in spite of these severe constraints many flat mechanical devices can be built that will work. A faucet designed by Dewdney is shown in Figure 8.6, *bottom*. To operate it the handle is lifted. This action pulls the valve away from the wall of the spout, allowing the water to flow out. When the handle is released, the spring pushes the valve back.

The device shown in Figure 8.7 serves to open and close a door (or a wall). Pulling down the lever at the right forces the wedge at the bottom to the left, thereby allowing the door to swing upward (carrying the wedge and the levers with it) on a hinge at the top. The door is opened

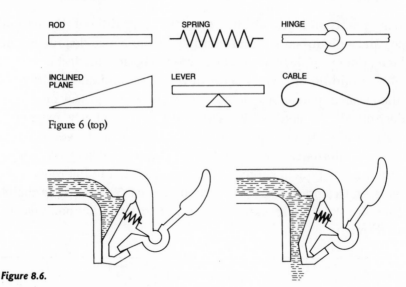

ROD   SPRING   HINGE

INCLINED PLANE   LEVER   CABLE

Figure 6 (top)

Figure 8.6.

from the left by pushing up on the other lever. The door can be lowered from either side and the wedge can be moved back to stabilize the wall by moving a lever in the appropriate direction. This device and the faucet are both mechanisms with permanent planiversal hinges: circular knobs that rotate inside hollows but cannot be removed from them.

Figure 8.8 depicts a planiversal steam engine whose operation parallels that of a steriversal engine. Steam under pressure is admitted into the cylinder of the engine through a sliding valve that forms one of its walls *(top)*. The steam pressure causes a piston to move to the right until steam can escape into a reservoir chamber above it. The subsequent loss of pressure allows the compound leaf spring at the right of the cylinder to drive the piston back to the left *(bottom)*. The sliding valve is closed as the steam escapes into the reservoir, but as the piston moves back it reopens, pulled to the right by a spring-loaded arm.

Figure 8.9 depicts Dewdney's ingenious mechanism for unlocking a door with a key. This planiversal lock consists of three slotted tumblers *(a)* that line up when a key is inserted *(b)* so that their lower halves move as a unit when the key is pushed *(c)*. The pushing of the key is transmitted through a lever arm to the master latch, which pushes down on a slave latch until the door is free to swing to the right *(d)*. The bar on the lever arm and the lip on the slave latch make the lock difficult to pick. Simple and compound leaf springs serve to return all the

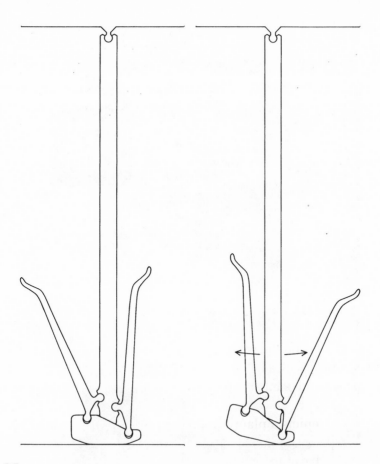

Figure 8.7.

parts of the lock except the lever arm to their original positions when the door is opened and the key is removed. When the door closes, it strikes the bar on the lever arm, thereby returning that piece to its original position as well. This flat lock could actually be employed in the steriverse; one simply inserts a key without twisting it.

"It is amusing to think," writes Dewdney, "that the rather exotic design pressures created by the planiversal environment could cause us to think about mechanisms in such a different way that entirely novel solutions to old problems arise. The resulting designs, if steriversally practical, are invariably space-saving."

Thousands of challenging planiversal problems remain unsolved. Is there a way, Dewdney wonders, to design a two-dimensional windup motor with flat springs or rubber bands that would store energy? What

Figure 8.8.

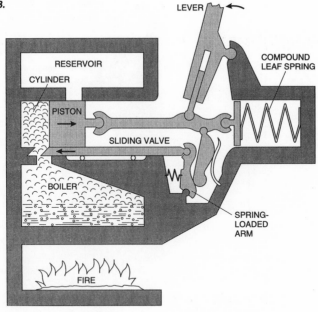

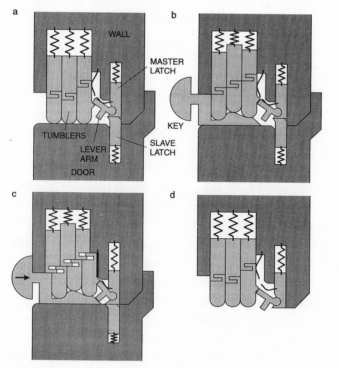

Figure 8.9.

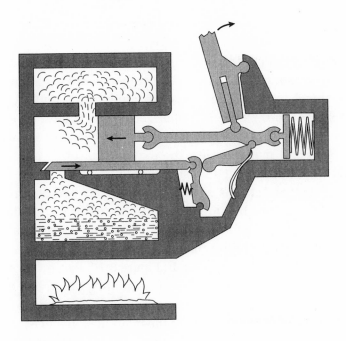

is the most efficient design for a planiversal clock, telephone, book, typewriter, car, elevator, or computer? Will some machines need a substitute for the wheel and axle? Will some need electric power?

There is a curious pleasure in trying to invent machines for what Dewdney calls "a universe both similar to and yet strangely different from ours." As he puts it, "from a small number of assumptions many phenomena seem to unfurl, giving one the sense of a kind of separate existence of this two-dimensional world. One finds oneself speaking, willy-nilly, of *the* planiverse as opposed to *a* planiverse. . . . [For] those who engage in it positively, there is a kind of strange enjoyment, like [that of] an explorer who enters a land where his own perceptions play a major role in the landscape that greets his eyes."

Some philosophical aspects of this exploration are not trivial. In constructing a planiverse one sees immediately that it cannot be built without a host of axioms that Leibniz called the "compossible" elements of any possible world, elements that allow a logically consistent structure. Yet as Dewdney points out, science in our universe is based mainly on observations and experiments, and it is not easy to find any underlying axioms. In constructing a planiverse we have nothing to observe. We can only perform *gedanken* experiments (thought experiments) about what might be observed. "The experimentalist's loss," observes Dewdney, "is the theoretician's gain."

A marvelous exhibit could be put on of working models of planiversal machines, cut out of cardboard or sheet metal, and displayed on a surface that slopes to simulate planiversal gravity. One can also imagine beautiful cardboard exhibits of planiversal landscapes, cities, and houses. Dewdney has opened up a new game that demands knowledge of both science and mathematics: the exploration of a vast fantasy world about which at present almost nothing is known.

It occurs to me that Astrians would be able to play two-dimensional board games but that such games would be as awkward for them as three-dimensional board games are for us. I imagine them, then, playing a variety of linear games on the analogue of our 8 × 8 chessboard. Several games of this type are shown in Figure 8.10. Part *(a)* shows the start of a checkers game. Pieces move forward only, one cell at a time, and jumps are compulsory. The linear game is equivalent to a game of regular checkers with play confined to the main diagonal of a standard board. It is easy to see how the second player wins in rational play and how in misère, or "giveaway," checkers the first player wins just as easily. Linear checkers games become progressively harder to analyze as longer boards are introduced. For example, which player wins standard linear checkers on the 11-cell board when each player starts with checkers on the first four cells at his end of the board?

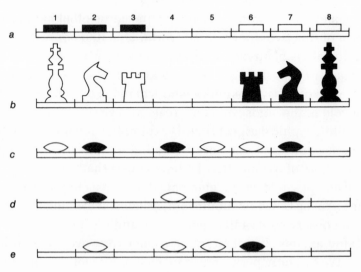

Figure 8.10.

Part *(b)* in the illustration shows an amusing Astrian analogue of chess. On a linear board a bishop is meaningless and a queen is the same as a rook, so the pieces are limited to kings, knights, and rooks. The only rule modification needed is that a knight moves two cells in either direction and can jump an intervening piece of either color. If the game is played rationally, will either White or Black win or will the game end in a draw? The question is surprisingly tricky to answer.

Linear go, played on the same board, is by no means trivial. The version I shall describe, called pinch, was invented in 1970 by James Marston Henle, a mathematician at Smith College.

In the game of pinch players take turns placing black and white stones on the cells of the linear board, and whenever the stones of one player surround the stones of the other, the surrounded stones are removed. For example, both sets of white stones shown in part *(c)* of Figure 8.10 are surrounded. Pinch is played according to the following two rules.

Rule 1: No stone can be placed on a cell where it is surrounded unless that move serves to surround a set of enemy stones. Hence in the situation shown in part *(d)* of the illustration, White cannot play on cells 1, 3, or 8, but he can play on cell 6 because this move serves to surround cell 5.

Rule 2: A stone cannot be placed on a cell from which a stone was removed on the last play if the purpose of the move is to surround something. A player must wait at least one turn before making such a move. For example, in part *(e)* of the illustration assume that Black plays on cell 3 and removes the white stones on cells 4 and 5. White cannot play on cell 4 (to surround cell 3) for his next move, but he may do so for any later move. He can play on cell 5, however, because even though a stone was just removed from that cell, the move does not serve to surround anything. This rule is designed to decrease the number of stalemates, as is the similar rule in go.

Two-cell pinch is a trivial win for the second player. The three- and four-cell games are easy wins for the first player if he takes the center in the three-cell game and one of the two central cells in the four-cell one. The five-cell game is won by the second player and the six- and seven-cell games are won by the first player. The eight-cell game jumps to such a high level of complexity that it becomes very exciting to play. Fortunes often change rapidly, and in most situations the winning player has only one winning move.

The Wonders of a Planiverse

Addendum

My column on the planiverse generated enormous interest. Dewdney received some thousand letters offering suggestions about flatland science and technology. In 1979 he privately printed *Two-Dimensional Science and Technology,* a monograph discussing these new results. Two years later he edited another monograph, *A Symposium of Two-Dimensional Science and Technology.* It contained papers by noted scientists, mathematicians, and laymen, grouped under the categories of physics, chemistry, astronomy, biology, and technology. *Newsweek* covered these monographs in a two-page article, "Life in Two Dimensions" (January 18, 1980), and a similar article, "Scientific Dreamers' Worldwide Cult," ran in Canada's *Maclean's* magazine (January 11, 1982). *Omni* (March 1983), in an article on "Flatland Redux," included a photograph of Dewdney shaking hands with an Astrian.

In 1984 Dewdney pulled it all together in a marvelous work, half nonfiction and half fantasy, titled *The Planiverse* and published by Poseidon Press, an imprint of Simon & Schuster. That same year he took over the mathematics column in *Scientific American,* shifting its emphasis to computer recreations. Several collections of his columns have been published by W. H. Freeman: *The Armchair Universe* (1987), *The Turing Omnibus* (1989), and *The Magic Machine* (1990).

An active branch of physics is now devoted to planar phenomena. It involves research on the properties of surfaces covered by a film one molecule thick, and a variety of two-dimensional electrostatic and electronic effects. Exploring possible flatlands also relates to a philosophical fad called "possible worlds." Extreme proponents of this movement actually argue that if a universe is logically possible—that is, free of logical contradictions—it is just as "real" as the universe in which we flourish.

In *Childhood's End* Arthur Clarke describes a giant planet where intense gravity has forced life to evolve almost flat forms with a vertical thickness of one centimeter.

The following letter from J. Richard Gott III, an astrophysicist at Princeton University, was published in *Scientific American* (October 1980):

I was interested in Martin Gardner's article on the physics of Flatland, because for some years I have given the students in my general relativity class the problem of deriving the theory of general relativity for Flatland. The results are surprising. One does *not* obtain the Flatland analogue of Newtonian theory (masses with gravitational fields falling off like $1/r$) as the weak-field limit. General relativity in Flatland predicts no gravitational waves and no action at a distance. A planet in Flatland would produce no gravitational effects beyond its own radius. In our four-dimensional space–time the energy momentum tensor has 10 independent components, whereas the Riemann curvature tensor has 20 independent components. Thus it is possible to find solutions to the vacuum field equations $G_{\mu\nu} = 0$ (where all components of the energy momentum tensor are zero) that have a nonzero curvature. Black-hole solutions and the gravitational-field solution external to a planet are examples. This allows gravitational waves and action at a distance. Flatland has a three-dimensional space–time where the energy momentum tensor has six independent components and the Riemann curvature tensor also has only six independent components. In the vacuum where all components of the energy momentum tensor are zero all the components of the Riemann curvature tensor must also be zero. No action at a distance or gravity waves are allowed.

Electromagnetism in Flatland, on the other hand, behaves just as one would expect. The electromagnetic field tensor in four-dimensional space–time has six independent components that can be expressed as vector E and B fields with three components each. The electromagnetic field tensor in a three-dimensional space–time (Flatland) has three independent components: a vector E field with two components and a scalar B field. Electromagnetic radiation exists, and charges have electric fields that fall off like $1/r$.

Two more letters, published in the same issue, follow. John S. Harris, of Brigham Young University's English Department, wrote:

As I examined Alexander Keewatin Dewdney's planiversal devices in Martin Gardner's article on science and technology in a two-dimensional universe, I was struck with the similarity of the mechanisms to the lockwork of the Mauser military pistol of 1895. This remarkable automatic pistol (which had many later variants) had no pivot pins or screws in its functional parts. Its entire operation was through sliding cam surfaces and two-dimensional sockets (called hinges by Dewdney). Indeed, the lockwork of a great many firearms, particularly those of the 19th century, follows essentially planiversal principles. For examples see the cutaway drawings in *Book of Pistols and Revolvers* by W.H.B. Smith.

Gardner suggests an exhibit of machines cut from cardboard, and that

The Wonders of a Planiverse

is exactly how the firearms genius John Browning worked. He would sketch the parts of a gun on paper or cardboard, cut out the individual parts with scissors (he often carried a small pair in his vest pocket), and then would say to his brother Ed, "Make me a part like this." Ed would ask, "How thick, John?" John would show a dimension with his thumb and forefinger, and Ed would measure the distance with calipers and make the part. The result is that virtually every part of the 100 or so Browning designs is essentially a two-dimensional shape with an added thickness.

This planiversality of Browning designs is the reason for the obsolescence of most of them. Dewdney says in his enthusiasm for the planiverse that "such devices are invariably space-saving." They are also expensive to manufacture. The Browning designs had to be manufactured by profiling machines: cam-following vertical milling machines. In cost of manufacture such designs cannot compete with designs that can be produced by automatic screw-cutting lathes, by broaching machines, by stamping, or by investment casting. Thus although the Browning designs have a marvelous aesthetic appeal, and although they function with delightful smoothness, they have nearly all gone out of production. They simply got too expensive to make.

Stefan Drobot, a mathematician at Ohio State University, had this to say:

In Martin Gardner's article he and the authors he quotes seem to have overlooked the following aspect of a "planiverse": any communication by means of a wave process, acoustic or electromagnetic, would in such a universe be impossible. This is a consequence of the Huygens principle, which expresses a mathematical property of the (fundamental) solutions of the wave equation. More specifically, a sharp impulse-type signal (represented by a "delta function") originating from some point is propagated in a space of three spatial dimensions in a manner essentially different from that in which it is propagated in a space of two spatial dimensions. In three-dimensional space the signal is propagated as a sharp-edged spherical wave without any trail. This property makes it possible to communicate by a wave process because two signals following each other in a short time can be distinguished.

In a space with two spatial dimensions, on the other hand, the fundamental solution of the wave equation represents a wave that, although it too has a sharp edge, has a trail of theoretically infinite length. An observer at a fixed distance from the source of the signal would perceive the oncoming front (sound, light, etc.) and then would keep perceiving it, although the intensity would decrease in time. This fact would make communication by any wave process impossible because it would not allow two signals following each other to be distinguished. More practically such communication would take much more time. This letter could not be read in the planiverse, although it is (almost) two-dimensional.

My linear checkers and chess prompted many interesting letters. Abe Schwartz assured me that on the 11-cell checker field Black also wins if the game is give-away. I. Richard Lapidus suggested modifying linear chess by interchanging knight and rook (the game is a draw), by adding more cells, by adding pawns that capture by moving forward one space, or by combinations of the three modifications. If the board is long enough, he suggested duplicating the pieces—two knights, two rooks—and adding several pawns, allowing a pawn a two-cell start option as in standard chess. Peter Stampolis proposed sustituting for the knight two pieces called "kops" because they combine features of knight and bishop moves. One kop moves only on white cells, the other moves on black.

Of course many other board games lend themselves to linear forms, for example, Reversi (also called Othello), or John Conway's Phutball, described in the two-volume *Winning Ways* written by Elwyn Berlekamp, Richard Guy, and John Conway.

Burr puzzles are wooden take-apart puzzles often referred to as Chinese puzzles. Dewdney's *Scientific American* column (October 1985) describes a clever planar version of a burr puzzle designed by Jeffrey Carter. It comes apart only after a proper sequence of pushes and pulls.

A graduate student in physics, whose name I failed to record, had a letter in *Science News,* December 8, 1984, protesting the notion that research in planar physics was important because it led to better understanding of three-dimensional physics. On the contrary, he wrote, planar research is of great interest for its own sake. He cited the quantum Hall effect and the whole field of microelectronics which is based on two-dimensional research and has many technological applications.

Answers

In 11-cell linear checkers (beginning with Black on cells 1, 2, 3, and 4 and White on cells 8, 9, 10, and 11) the first two moves are forced: Black to 5 and White to 7. To avoid losing, Black then goes to 4, and White must respond by moving to 8. Black is then forced to 3 and

White to 9. At this point Black loses with a move to 2 but wins with a move to 6. In the latter case White jumps to 5, and then Black jumps to 6 for an easy end-game victory.

On the eight-cell linear chessboard White can win in at most six moves. Of White's four opening moves, R×R is an instant stalemate and the shortest possible game. R-5 is a quick loss for White if Black plays R×R. Here White must respond with N-4, and then Black mates on his second move with R×N. This game is one of the two "fool's mates," or shortest possible wins. The R-4 opening allows Black to mate on his second or third move if he responds with N-5.

White's only winning opening is N-4. Here Black has three possible replies:

1. R×N. In this case White wins in two moves with R×R.
2. R-5. White wins with K-2. If Black plays R-6, White mates with N×R. If Black takes the knight, White takes the rook, Black moves N-5, and White mates by taking Black's knight.
3. N-5. This move delays Black's defeat the longest. In order to win White must check with N×R, forcing Black's king to 7. White moves his rook to 4. If Black plays K×N, White's king goes to 2, Black's K-7 is forced, and White's R×N wins. If Black plays N-3 (check), White moves the king to 2. Black can move only the knight. If he plays N-1, White mates with N-8. If Black plays N-5, White's N-8 forces Black's K×N, and then White mates with R×N.

The first player also has the win in eight-cell pinch (linear go) by opening on the second cell from an end, a move that also wins the six- and seven-cell games. Assume that the first player plays on cell 2. His unique winning responses to his opponent's plays on 3, 4, 5, 6, 7, and 8 are respectively 5, 7, 7, 7, 5, and 6. I leave the rest of the game to the reader. It is not known whether there are other winning opening moves. James Henle, the inventor of pinch, assures me that the second player wins the nine-cell game. He has not tried to analyze boards with more than nine cells.

Bibliography

D. Burger, *Sphereland: A Fantasy about Curved Spaces and an Expanding Universe,* T. Y. Crowell, 1965.

A. K. Dewdney, *The Planiverse,* Poseidon, 1984; Copernicus, 2000.

A. K. Dewdney, *200 Percent of Nothing: An Eye Opening Tour Through the Twists and Turns of Math Abuse and Innumeracy,* Wiley, 1994.

A. K. Dewdney, *Introductory Computer Science: Bits of Theory, Bytes of Practice,* W. H. Freeman, 1996.

A. K. Dewdney, "The Planiverse Project: Then and Now," *Mathematical Intelligencer,* Vol. 22, No. 1, 2000, pp. 46–48.

C. H. Hinton, *An Episode of Flatland,* Swan Sonnenschein & Co., 1907.

R. Jackiw, "Lower Dimensional Gravity," *Nuclear Physics,* Vol. B252, 1985, pp. 343–56.

I. Peterson, "Shadows From a Higher Dimension," *Science News,* Vol. 126, November 3, 1984, pp. 284–85.

A. Square and E. A. Abbott, *Flatland: A Romance of Many Dimensions,* Dover Publications, Inc., 1952. Several other paperback editions are currently in print.

I. Stewart, *Flatterland,* Perseus Publishing, 2001.

P. J. Stewart, "Allegory Through the Computer Class: Sufism in Dewdney's Planiverse," *Sufi,* Issue 9, Spring 1991, pp. 26–30.

III

Solid Geometry
and Higher
Dimensions

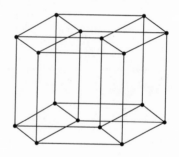

> *Rosy's instant acceptance of our model at*
> *first amazed me. I had feared that her sharp,*
> *stubborn mind, caught in her self-made an-*
> *tihelical trap, might dig up irrelevant results*
> *that would foster uncertainty about the cor-*
> *rectness of the double helix. Nonetheless,*
> *like almost everyone else, she saw the ap-*
> *peal of the base pairs and accepted the fact*
> *that the structure was too pretty not to be*
> *true.*   —James D. Watson, *The Double Helix*

A straight sword will fit snugly into a straight scabbard. The same is true of a sword that curves in the arc of a circle: it can be plunged smoothly into a scabbard of the same curvature. Mathematicians sometimes describe this property of straight lines and circles by calling them "self-congruent" curves; any segment of such a curve can be slid along the curve, from one end to the other, and it will always "fit."

Is it possible to design a sword and its scabbard that are *not* either straight or curved in a circular arc? Most people, after giving this careful consideration, will answer no, but they are wrong. There is a third curve that is self-congruent: the circular helix. This is a curve that coils around a circular cylinder in such a way that it crosses the "elements" of the cylinder at a constant angle. Figure 9.1 makes this clear. The elements are the vertical lines that parallel the cylinder's axis; A is the constant angle with which the helix crosses every element. Because of the constant curvature of the helix a helical sword would screw its way easily in and out of a helical scabbard.

Actually the straight line and the circle can be regarded as limiting cases of the circular helix. Compress the curve until the coils are very close together and you get a tightly wound helix resembling a Slinky toy; if angle A increases to 90 degrees, the helix collapses into a circle. On the other hand, if you stretch the helix until angle A becomes zero, the helix is transformed into a straight line. If parallel rays of light shine perpendicularly on a wall, a circular helix held before the wall with its axis parallel to the rays will cast on the wall a shadow that is

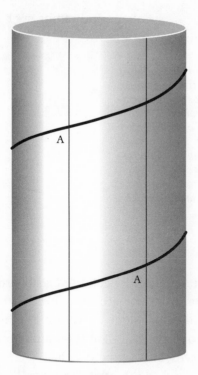

Figure 9.1. Circular helix on cylinder

a single circle. If the helix is held at right angles to the rays, the shadow is a sine curve. Other kinds of projections produce the cycloid and other familiar curves.

Every helix, circular or otherwise, is an asymmetric space curve that differs from its mirror image. We shall use the term "right-handed" for the helix that coils clockwise as it "goes away," in the manner of an ordinary wood screw or a corkscrew. Hold such a corkscrew up to a mirror and you will see that its reflection, in the words of Lewis Carroll's Alice, "goes the other way." The reflection is a left-handed corkscrew. Such a corkscrew actually can be bought as a practical joke. So unaccustomed are we to left-handed screw threads that a victim may struggle for several minutes with such a corkscrew before he realizes that he has to turn it counterclockwise to make it work.

Aside from screws, bolts, and nuts, which are (except for special purposes) standardized as right-handed helices, most manmade helical structures come in both right and left forms: candy canes, circular staircases, rope and cable made of twisted strands, and so on. The same variations in handedness are found in conical helices (curves that spiral around cones), including bedsprings and spiral ramps such as the inverted conical ramp in Frank Lloyd Wright's Guggenheim Museum in New York City.

Not so in nature! Helical structures abound in living forms, from the simplest virus to parts of the human body, and in almost every case the genetic code carries information that tells each helix precisely "which way to go." The genetic code itself is carried by a double-stranded helical molecule of DNA, its two right-handed helices twining around each other like the two snakes on the staff of Hermes. Moreover, since Linus Pauling's pioneer work on the helical structure of protein molecules, there has been increasing evidence that every giant protein molecule found in nature has a "backbone" that coils in a right-handed helix. In both nucleic acid and protein, the molecule's backbone is a chain made up of units each one of which is an asymmetric structure of the same handedness. Each unit, so to speak, gives an additional twist to the chain, in the same direction, like the steps of a helical staircase.

Larger helical structures in animals that have bilateral symmetry usually come in mirror-image pairs, one on each side of the body. The horns of rams, goats, antelopes, and other mammals are spectacular examples (see Figure 9.2). The cochlea of the human ear is a conical helix that is left-handed in the left ear and right-handed in the right. A curious exception is the tooth of the narwhal, a small whale that flourishes in arctic waters. This whimsical creature is born with two teeth in its upper jaw. Both teeth remain permanently buried in the jaw of the female narwhal, and so does the right tooth of the male. But the male's left tooth grows straight forward, like a javelin, to the ridiculous length of eight or nine feet—more than half the animal's length from snout to tail! Around this giant tooth are helical grooves that spiral forward in a counterclockwise direction (see Figure 9.3). On the rare occasions when both teeth grow into tusks, one would expect the right tooth to spiral clockwise. But no, it too is always left-handed. Zoologists disagree on how this could come about. Sir D'Arcy Thompson, in his book *On Growth and Form,* defends his own theory that the whale swims with a slight screw motion to the right. The inertia of its huge tusk would produce a torque at the base of the tooth that might cause it to rotate counterclockwise as it grows (see J. T. Bonner, "The Horn of the Unicorn," *Scientific American,* March 1951).

Whenever a single helix is prominent in the structure of any living

Figure 9.2. Helical horns of the Pamir sheep have opposite handedness.

Figure 9.3. Helical grooves of the narwhal tooth are always left-handed.

plant or animal, the species usually confines itself to a helix of a specific handedness. This is true of countless forms of helical bacteria as well as of the spermatozoa of all higher animals. The human umbilical cord is a triple helix of one vein and two arteries that invariably coil to the left. The most striking instances are provided by the conical helices of the shells of snails and other mollusks. Not all spiral shells have a handedness. The chambered nautilus, for instance, coils on one plane; like a spiral nebula, it can be sliced into identical left and right halves. But there are thousands of beautiful molluscan shells that are either left- or right-handed (see Figure 9.4). Some species are always left-handed and some are always right-handed. Some go one way in one locality and the other way in another. Occasional "sports" that twist the wrong way are prized by shell collectors.

A puzzling type of helical fossil known as the devil's corkscrew *(Daemonelix)* is found in Nebraska and Wyoming. These huge spirals, six feet or more in length, are sometimes right-handed and sometimes left-handed. Geologists argued for decades over whether they are fossils of extinct plants or helical burrows made by ancestors of the beaver. The beaver theory finally prevailed after remains of small prehistoric beavers were found inside some of the corkscrews.

In the plant world helices are common in the structure of stalks, stems, tendrils, seeds, flowers, cones, leaves—even in the spiral arrangement of leaves and branches around a stalk. The number of

Solid Geometry and Higher Dimensions

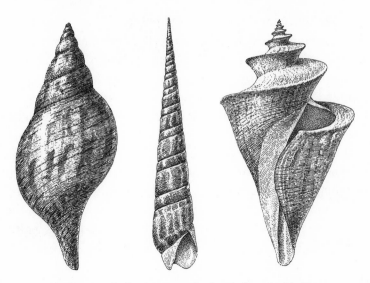

Figure 9.4. Three molluscan shells that are right-handed conical helices

turns made along a helical path, as you move from one leaf to the leaf directly above it, tends to be a number in the familiar Fibonacci series: 1, 2, 3, 5, 8, 13, . . . (Each number is the sum of the preceding two numbers.) A large literature in the field known as "phyllotaxy" (leaf arrangement) deals with the surprising appearance of the Fibonacci numbers in botanical phenomena of this sort.

The helical stalks of climbing plants are usually right-handed, but thousands of species of twining plants go the other way. The honeysuckle, for instance, is always left-handed; the bindweed (a family that includes the morning glory) is always right-handed. When the two plants tangle with each other, the result is a passionate, violent embrace that has long fascinated English poets. "The blue bindweed," wrote Ben Jonson in 1617, "doth itself enfold with honeysuckle." And Shakespeare, in *A Midsummer Night's Dream,* has Queen Titania speak of her intention to embrace Bottom the Weaver (who has been transformed into a donkey) by saying: "Sleep thou, and I will wind thee in my arms./ . . . So doth the woodbine the sweet honeysuckle/Gently entwist." In Shakespeare's day "woodbine" was a common term for bindweed. Because it later came to be applied exclusively to honeysuckle many commentators reduced the passage to absurdity by supposing that Titania was speaking of honeysuckle twined with honeysuckle. Awareness of the opposite handedness of bindweed and honeysuckle heightens, of course, the meaning of Titania's metaphor.

The Helix

More recently, a charming song called "Misalliance," celebrating the love of the honeysuckle for the bindweed, has been written by the British poet and entertainer Michael Flanders and set to music by his friend Donald Swann. With Flanders' kind permission the entire song is reproduced on the opposite page. (Readers who would like to learn the tune can hear it sung by Flanders and Swann on the Angel recording of *At the Drop of a Hat,* their hilarious two-man revue that made such a hit in London and New York.) Note that Flanders' honeysuckle is right-handed and his bindweed, left-handed. It is a matter of convention whether a given helix is called left- or right-handed. If you look at the point of a right-handed wood screw, you will see the helix moving toward you counterclockwise, so that it can just as legitimately be called left-handed. Flanders simply adopts the convention opposite to the one taken here.

The entwining of two circular helices of opposite handedness is also involved in a remarkable optical-illusion toy that was sold in this country in the 1930s. It is easily made by twisting together a portion of two wire coils of opposite handedness (see Figure 9.6). The wires must be soldered to each other at several points to make a rigid structure. The illusion is produced by pinching the wire between thumb and forefinger of each hand at the left and right edges of the central overlap. When the hands are moved apart, the fingers and thumbs slide along the wire, causing it to rotate and create a barber's-pole illusion of opposite handedness on each side. This is continuously repeated. The wire seems to be coming miraculously out of the inexhaustible meshed portion. Since the neutrino and antineutrino are now known to travel with screw motions of opposite handedness, I like to think of this toy as demonstrating the endless production of neutrinos and their mirror-image particles.

On the science toy market at the time I write (2000) is a device called "Spinfinity." It consists of two aluminum wire helices that are intertwined. When rotated you see the two helices moving in opposite directions.

The helical character of the neutrino's path results from the fusion of its forward motion (at the speed of light) with its "spin." Helical paths of a similar sort are traced by many inanimate objects and living things: a point on the propeller of a moving ship or plane, a squirrel running up or down a tree, Mexican free-tailed bats gyrating counterclockwise when they emerge from caves at Carlsbad, NM. Conically helical paths

MISALLIANCE

The fragrant Honeysuckle spirals clockwise to the sun
And many other creepers do the same.
But some climb counterclockwise, the Bindweed does, for one,
Or *Convolvulus*, to give her proper name.

Rooted on either side a door, one of each species grew,
And raced toward the window ledge above.
Each corkscrewed to the lintel in the only way it knew,
Where they stopped, touched tendrils, smiled and fell in love.

Said the right-handed Honeysuckle
To the left-handed Bindweed:
"Oh, let us get married,
If our parents don't mind. We'd
Be loving and inseparable.
Inextricably entwined, we'd
Live happily ever after,"
Said the Honeysuckle to the Bindweed.

To the Honeysuckle's parents it came as a shock.
"The Bindweeds," they cried, "are inferior stock.
They're uncultivated, of breeding bereft.
We twine to the right and they twine to the left!"

Said the countercockwise Bindweed
To the clockwise Honeysuckle:
"We'd better start saving—
Many a mickle maks a muckle—
Then run away for a honeymoon
And hope that our luck'll
Take a turn for the better,"
Said the Bindweed to the Honeysuckle.

A bee who was passing remarked to them then:
"I've said it before, and I'll say it again:
Consider your offshoots, if offshoots there be.
They'll never receive any blessing from me."

Poor little sucker, how will it learn
When it is climbing, which way to turn?
Right—left—what a disgrace!
Or it may go straight up and fall flat on its face!

Said the right-hand-thread Honeysuckle
To the left-hand-thread Bindweed:
"It seems that against us all fate has combined.
Oh my darling, oh my darling,
Oh my darling Columbine
Thou art lost and gone forever,
We shall never intertwine."

Together they found them the very next day
They had pulled up their roots and just shriveled away,
Deprived of that freedom for which we must fight,
To veer to the left or to veer to the right!

MICHAEL FLANDERS

Figure 9.5.

Figure 9.6. Helical toy that suggests the production of neutrinos

are taken by whirlpools, water going down a drain, tornadoes, and thousands of other natural phenomena.

Writers have found helical motions useful on the metaphorical level. The progress of science is often likened to an inverted conical spiral: the circles growing larger and larger as science probes further into the unknown, always building upward on the circles of the past. The same spiral, a dark, bottomless whirlpool into which an individual or humanity is sliding, has also been used as a symbol of pessimism and despair. This is the metaphor that closes Norman Mailer's book *Advertisements for Myself.* "Am I already on the way out?" he asks. Time for Mailer is a conical helix of water flushing down a cosmic drain, spinning him off "into the spiral of star-lit empty waters."

And now for a simple helix puzzle. A rotating barber's pole consists of a cylinder on which red, white, and blue helices are painted. The cylinder is four feet high. The red stripe cuts the cylinder's elements (vertical lines) with a constant angle of 60 degrees. How long is the red stripe?

The problem may seem to lack sufficient information for determining the stripe's length; actually it is absurdly easy when approached properly.

Addendum

In my *Second Scientific American Book of Mathematical Puzzles and Diversions* (1961) I introduced the following brainteaser involving two helices of the same handedness:

THE TWIDDLED BOLTS

Two identical bolts are placed together so that their helical grooves intermesh (Figure 9.7). If you move the bolts around each other as you would twiddle your thumbs, holding each bolt firmly by the head so that it does not rotate and twiddling them in the direction shown, will the heads (a) move inward, (b) move outward, or (c) remain the same distance from each other? The problem should be solved without resorting to actual test.

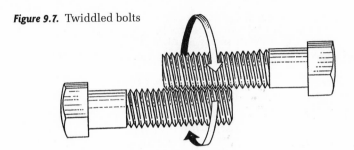

Figure 9.7. Twiddled bolts

Two similar helices are also involved in a curious patent by someone named Socrates Scholfield of Providence, RI. I cannot now recall how I came across it. A picture of the device, from the patent's first page, is shown in Figure 9.8. It is to be used in classrooms for demonstrating the nature of God. One helix represents good; the other, evil. As you can see, they are holessly intertwined. The use of the device is given in detail in five pages of the patent.

I know nothing about Mr. Scholfield except that in 1907 he published a 59-page booklet titled *The Doctrine of Mechanicalism*. I tried to check it in the New York Public Library many years ago, but the library has lost its copy. I once showed Scholfield's patent to the philosopher Charles Hartshorne, one of my teachers at the University of Chicago. I expected him to find the device amusing. To my surprise, Hartshorne solemnly read the patent's pages and pronounced them admirable.

The Slinky toy furnishes an interesting problem in physics. If you stand on a chair, holding one end of Slinky so that the helix hangs straight down, then drop the toy, what happens? Believe it or not, the lower end of Slinky doesn't move until the helix has come together, then it falls with the expected rate. Throughout the experiment the toy's center of gravity descends with the usual acceleration.

Answers

If a right triangle is wrapped around any type of cylinder, the base of the triangle going around the base of the cylinder, the triangle's hypotenuse will trace a helix on the cylinder. Think of the red stripe of the barber's pole as the hypotenuse of a right triangle, then "unwrap" the triangle from the cylinder. The triangle will have angles of 30 and 60 degrees. The hypotenuse of such a triangle must be twice the alti-

1,087,186.

Patented Feb. 17, 1914.

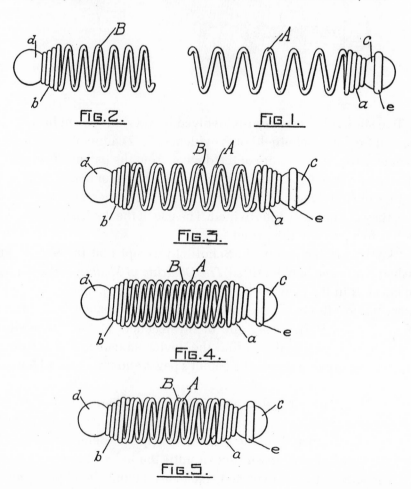

Figure 9.8.

tude. (This is easily seen if you place two such triangles together to form an equilateral triangle.) In this case the altitude is four feet, so that the hypotenuse (red stripe) is eight feet.

The interesting part of this problem is that the length of the stripe is independent not only of the diameter of the cylinder but also of the shape of its cross section. The cross section can be an irregular closed curve of any shape whatever; the answer to the problem remains the same.

The twiddled bolts move neither inward nor outward. They behave like someone walking up a down escalator, always staying in the same place.

Bibliography

T. A. Cook, *The Curves of Life*, New York: Henry Holt, 1914; Dover, 1979.

H.S.M. Coxeter, *Introduction to Geometry*, New York: John Wiley and Sons, 1961; Second edition, 1989.

F. E. Graves, "Nuts and Bolts," *Scientific American*, June 1984, pp. 136–44.

J. B. Pettigrew, *Design in Nature*, Vol. II, London: Longmans, Green, and Co., 1908.

J. D. Watson, *The Double Helix*, New York: Atheneum, 1968; Norton, 1980.

Chapter 10

Packing Spheres

Spheres of identical size can be piled and packed together in many different ways, some of which have fascinating recreational features. These features can be understood without models, but if the reader can obtain a supply of 30 or more spheres, he will find them an excellent aid to understanding. Ping-pong balls are perhaps the best for this purpose. They can be coated with rubber cement, allowed to dry, then stuck together to make rigid models.

First let us make a brief two-dimensional foray. If we arrange spheres in square formation (see Figure 10.1, *right*), the number of balls involved will of course be a square number. If we form a triangle (see Figure 10.1, *left*), the number of balls is a triangular number. These are the simplest examples of what the ancients called "figurate numbers." They were intensively studied by early mathematicians (a famous treatise on them was written by Blaise Pascal), and although little attention is paid them today, they still provide intuitive insights into many aspects of elementary number theory.

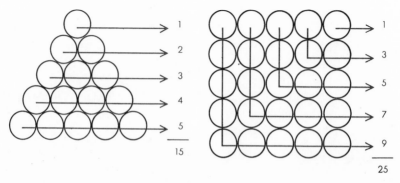

Figure 10.1. The basis of triangular numbers *(left)* and of square numbers *(right)*

For example, it takes only a glance at Figure 10.1, *left,* to see that the sum of any number of consecutive positive integers, beginning with 1, is a triangular number. A glance at Figure 10.1, *right,* shows that square numbers are formed by the addition of consecutive *odd* integers, beginning with 1. Figure 10.2 makes immediately evident an interesting theorem known to the ancient Pythagoreans: Every square number is the sum of two consecutive triangular numbers. The algebraic proof is simple. A triangular number with n units to a side is the sum of $1 + 2 + 3 + \ldots n$, and can be expressed by the formula $\frac{1}{2}n(n + 1)$. The preceding triangular number has the formula $\frac{1}{2}n(n - 1)$. If we add the two formulas and simplify, the result is n^2. Are there numbers that are simultaneously square and triangular? Yes, there are infinitely many of them. The smallest (not counting 1, which belongs to any figurate series) is 36; then the series continues: 1225, 41616, 1413721, 48024900, ... It is not so easy to devise a formula for the nth term of this series.

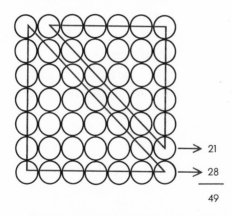

$$\begin{array}{r} 21 \\ 28 \\ \hline 49 \end{array}$$

Figure 10.2. Square and triangular numbers are related.

Three-dimensional analogies of the plane-figurate numbers are obtained by piling spheres in pyramids. Three-sided pyramids, the base and sides of which are equilateral triangles, are models of what are called the tetrahedral numbers. They form the series 1, 4, 10, 20, 35, 56, 84, . . . and can be represented by the formula $\frac{1}{6}n(n + 1)(n + 2)$, where n is the number of balls along an edge. Four-sided pyramids, with square bases and equilateral triangles for sides (i.e., half of a regular octahedron), represent the (square) pyramidal numbers 1, 5, 14, 30, 55, 91, 140, ... They have the formula $\frac{1}{6}n(n + 1)(2n + 1)$. Just as a square can be divided by a straight line into two consecutive triangles, so can a square pyramid be divided by a plane into two consecutive tetrahedral pyramids. (If you build a model of a pyramidal number, the bottom layer has to be kept from rolling apart. This can be done by placing rulers or other strips of wood along the sides.)

Many old puzzles exploit the properties of these two types of pyramidal number. For example, in making a courthouse monument out of cannon balls, what is the smallest number of balls that can first be arranged on the ground as a square, then piled in a square pyramid? The surprising thing about the answer (4,900) is that it is the *only* answer. (The proof of this is difficult and was not achieved until 1918.) Another example: A grocer is displaying oranges in two tetrahedral pyramids. By putting together the oranges in both pyramids he is able to make one large tetrahedral pyramid. What is the smallest number of oranges he can have? If the two small pyramids are the same size, the unique answer is 20. If they are different sizes, what is the answer?

Imagine now that we have a very large box, say a crate for a piano, which we wish to fill with as many golf balls as we can. What packing procedure should we use? First we form a layer packed as shown by the unshaded circles with light gray circumferences in Figure 10.3. The second layer is formed by placing balls in alternate hollows as indicated by the shaded circles with black rims. In making the third layer we have a choice of two different procedures:

Figure 10.3. In hexagonal close-packing, balls go in hollows labeled *A;* in cubic, in hollows labeled *B*.

SOLID GEOMETRY AND HIGHER DIMENSIONS

1. We place each ball on a hollow A that is directly above a ball in the first layer. If we continue in this way, placing the balls of each layer directly over those in the next layer but one, we produce a structure called hexagonal close-packing.
2. We place each ball in a hollow B, directly above a hollow in the first layer. If we follow this procedure for each layer (each ball will be directly above a ball in the third layer beneath it), the result is known as cubic close-packing. Both the square and the tetrahedral pyramids have a packing structure of this type, though on a square pyramid the layers run parallel to the sides rather than to the base.

In forming the layers of a close-packing we can switch back and forth whenever we please from hexagonal to cubic packing to produce various hybrid forms of close-packing. In all these forms—cubic, hexagonal, and hybrid—each ball touches 12 other balls that surround it, and the density of the packing (the ratio of the volume of the spheres to the total space) is $\pi/\sqrt{18} = .74048 +$, or almost 75 percent.

Is this the largest density obtainable? No denser packing is known, but in an article published in 1958 (on the relation of close-packing to froth) H.S.M. Coxeter of the University of Toronto made the startling suggestion that perhaps the densest packing has not yet been found. It is true that no more than 12 balls can be placed so that all of them touch a central sphere, but a thirteenth ball can *almost* be added. The large leeway here in the spacing of the 12 balls, in contrast to the complete absence of leeway in the close-packing of circles on a plane, suggests that there might be some form of irregular packing that would be denser than .74. No one has yet proved that no denser packing is possible, or even that 12 point-contacts for each sphere are necessary for densest packing. As a result of Coxeter's conjecture, George D. Scott of the University of Toronto made some experiments in random packing by pouring large numbers of steel balls into spherical flasks, then weighing them to obtain the density. He found that stable random-packings had a density that varied from about .59 to .63. So if there is a packing denser than .74, it will have to be carefully constructed on a pattern that no one has yet thought of.

Assuming that close-packing is the closest packing, readers may like to test their packing prowess on this exceedingly tricky little problem. The interior of a rectangular box is 10 inches on each side and 5 inches deep. What is the largest number of steel spheres 1 inch in diameter that can be packed in this space?

If close-packed circles on a plane expand uniformly until they fill the interstices between them, the result is the familiar hexagonal tiling of bathroom floors. (This explains why the pattern is so common in nature: the honeycomb of bees, a froth of bubbles between two flat surfaces almost in contact, pigments in the retina, the surface of certain diatoms and so on.) What happens when closely packed spheres expand uniformly in a closed vessel or are subjected to uniform pressure from without? Each sphere becomes a polyhedron, its faces corresponding to planes that were tangent to its points of contact with other spheres. Cubic close-packing transforms each sphere into a rhombic dodecahedron (see Figure 10.4, *top*), the 12 sides of which are congruent rhombi. Hexagonal close-packing turns each ball into a trapezo-rhombic dodecahedron (see Figure 10.4, *bottom*), six faces of which are rhombic and six, trapezoidal. If this figure is sliced in half along the gray plane and one half is rotated 60 degrees, it becomes a rhombic dodecahedron.

In 1727 the English physiologist Stephen Hales wrote in his book *Vegetable Staticks* that he had poured some fresh peas into a pot, compressed them, and had obtained "pretty regular dodecahedrons." The experiment became known as the "peas of Buffon" (because the Comte de Buffon later wrote about a similar experiment), and most biologists accepted it without question until Edwin B. Matzke, a botanist at Columbia University, repeated the experiment. Because of the irregular sizes and shapes of peas, their nonuniform consistency and the random packing that results when peas are poured into a container, the shapes of the peas after compression are too random to be identifiable. In experiments reported in 1939 Matzke compressed lead shot and found that if the spheres had been cubic close-packed, rhombic dodecahedrons were formed; but if they had been randomly packed, irregular 14-faced bodies predominated. These results have important bearing, Matzke has pointed out, on the study of such structures as foam and living cells in undifferentiated tissues.

The problem of closest packing suggests the opposite question: What is the *loosest* packing; that is, what rigid structure will have the lowest possible density? For the structure to be rigid, each sphere must touch at least four others, and the contact points must not be all in one hemisphere or all on one equator of the sphere. In his *Geometry and the*

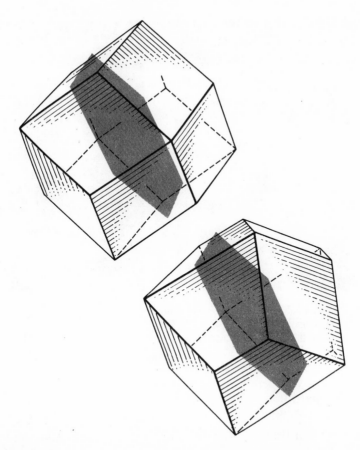

Figure 10.4. Packed spheres expand into dodecahedrons.

Imagination, first published in Germany in 1932, David Hilbert describes what was then believed to be the loosest packing: a structure with a density of .123. In the following year, however, two Dutch mathematicians, Heinrich Heesch and Fritz Laves, published the details of a much looser packing with a density of only .0555 (see Figure 10.5). Whether there are still looser packings is another intriguing question that, like the question of the closest packing, remains undecided.

Addendum

The unique answer of 4,900 for the number of balls that will form both a square and a square-based pyramid was proved by G. N. Watson in *Messenger of Mathematics,* new series, Vol. 48, 1918, pages 1–22. This had been conjectured as early as 1875 by the French math-

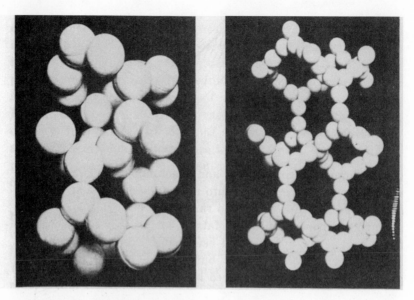

Figure 10.5. The Heesch and Laves loose-packing. Large spheres are first packed as shown on *left,* then each sphere is replaced by three smaller spheres to obtain the packing shown on *right.* It has a density of .055+.

ematician Edouard Lucas. Henry Ernest Dudeney makes the same guess in his answer to problem 138, *Amusements in Mathematics* (1917).

There is a large literature on numbers that are both triangular and square. Highlights are cited in an editorial note to problem E1473, *American Mathematical Monthly,* February 1962, page 169, and the following formula for the nth square triangular number is given:

$$\frac{(17 + 12\sqrt{2})^n + (17 - 12\sqrt{2})^n - 2}{32}.$$

The question of the densest possible regular or lattice packing of spheres has been known since 1930 for all spaces up through eight dimensions. (See *Proceedings of Symposia in Pure Mathematics,* Vol. 7, American Mathematical Society, 1963, pp. 53–71.) In 3-space, the question is answered by the regular close-packings described earlier, which have a density of .74+. But, as Constance Reid notes in her *Introduction to Higher Mathematics* (1959), when 9-space is considered, the problem takes one of those sudden, mysterious turns that so often occur in the geometries of higher Euclidean spaces. So far as I know, no one yet knows the densest lattice packing of hyperspheres in 9-space.

Figure 10.6 (*top*) is a neat "look–see" proof that the space between four circles packed with their centers on a square lattice is equal to the area of the square minus the area of a circle. Figure 10.6 *(bottom)* is a similar proof that the space between circles packed on a triangular lattice is equal to the area of the triangle minus half the circle's area.

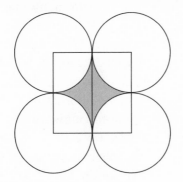 

Figure 10.6. Two look–see proofs

Kepler conjectured that the close-packing of spheres described in this chapter is the densest of all possible packings. Although this packing is known to be the densest for packings based on a lattice, the question of whether an irregular packing might be denser remains in limbo. In 1990 Wu-Yi Hsiang, at the University of California—Berkeley, published a 150-page "proof" of Kepler's conjecture. It was widely hailed as valid, but soon experts on sphere packing were finding large holes in the proof. Hsiang tried to repair them, but as of today the consensus is that his proof is false and beyond repair.

In 1998 Thomas Hales, at the University of Michigan and one of Hsiang's harshest critics, published a 250-page proof of his own. Because it relies heavily, like the famous proof of the four-color map theorem, on lengthy computer calculations, the verdict is still out over whether Kepler's conjecture has finally been verified.

Incidently, Stanislaw Ulam told me in 1972 that he suspected that spheres, in their densest packing, allow more empty space than the densest packing of any identical convex solids.

Answers

The smallest number of oranges that will form two tetrahedral pyramids of different sizes, and also one larger tetrahedral pyramid, is

680. This is a tetrahedral number that can be split into two smaller tetrahedral numbers: 120 and 560. The edges of the three pyramids are 8, 14, and 15.

A box 10 inches square and 5 inches deep can be close-packed with 1 inch-diameter steel balls in a surprising variety of ways, each accommodating a different number of balls. The maximum number, 594, is obtained as follows: Turn the box on its side and form the first layer by making a row of five, then a row of four, then of five, and so on. It is possible to make 11 rows (6 rows of five each, 5 rows of four each), accommodating 50 balls and leaving a space of more than .3 inch to spare. The second layer also will take 11 rows, alternating four and five balls to a row, but this time the layer begins and ends with four-ball rows, so that the number of balls in the layer is only 49. (The last row of four balls will project .28+ inch beyond the edge of the first layer, but because this is less than .3 inch, there is space for it.) Twelve layers (with a total height of 9.98+ inches) can be placed in the box, alternating layers of 50 balls with layers of 49, to make a grand total of 594 balls.

Bibliography

Kepler's Conjecture

B. Cipra, "Gaps in a Sphere Packing Proof," *Science,* Vol. 259, 1993, p. 895.

T. C. Hales, "The Status of the Kepler Conjecture," *Mathematical Intelligencer,* Vol. 16, No. 3, 1994, pp. 47–58.

T. C. Hales, "Cannonballs and Honeycombs," *Notices of the American Mathematical Society,* Vol. 47, April 2000, pp. 440–49.

J. Horgan, "Mathematicians Collide Over a Claim About Packing Spheres," *Scientific American,* December 1994, pp. 32–34.

D. Mackenzie, "The Proof is in the Packing," *American Scientist,* Vol. 86, November/December, 1998, pp. 524–25.

S. Singh, "Mathematics 'Proves' What the Grocer Always Knew," *The New York Times,* August 25, 1998, p. F3.

I. Stewart, "Has the Sphere Packing Problem Been Solved?" *New Scientist,* Vol. 134, May 2, 1992, p. 16.

H. Wu, "On the Sphere Packing Problem and the Proof of Kepler's Conjecture," *International Journal of Mathematics,* Vol. 4, No. 5, 1998, pp. 739–831.

H. Wu, "A Rejoinder to Hales's Article," *Mathematical Intelligencer,* Vol. 17, No. 1, 1995, pp. 35–42.

Chapter 11

Spheres and Hyperspheres

A circle is the locus of all points on the plane at a given distance from a fixed point on the plane. Let's extend this to Euclidean spaces of all dimensions and call the general n-sphere the locus of all points in n-space at a given distance from a fixed point in n-space. In a space of one dimension (a line) the 1-sphere consists of two points at a given distance on each side of a center point. The 2-sphere is the circle, the 3-sphere is what is commonly called a sphere. Beyond that are the hyperspheres of 4, 5, 6, . . . dimensions.

Imagine a rod of unit length with one end attached to a fixed point. If the rod is allowed to rotate only on a plane, its free end will trace a unit circle. If the rod is allowed to rotate in 3-space, the free end traces a unit sphere. Assume now that space has a fourth coordinate, at right angles to the other three, and that the rod is allowed to rotate in 4-space. The free end then generates a unit 4-sphere. Hyperspheres are impossible to visualize; nevertheless, their properties can be studied by a simple extension of analytic geometry to more than three coordinates. A circle's Cartesian formula is $a^2 + b^2 = r^2$, where r is the radius. The sphere's formula is $a^2 + b^2 + c^2 = r^2$. The 4-sphere's formula is $a^2 + b^2 + c^2 + d^2 = r^2$, and so on up the ladder of Euclidean hyperspaces.

The "surface" of an n-sphere has a dimensionality of $n - 1$. A circle's "surface" is a line of one dimension, a sphere's surface is two-dimensional, and a 4-sphere's surface is three-dimensional. Is it possible that 3-space is actually the hypersurface of a vast 4-sphere? Could such forces as gravity and electromagnetism be transmitted by the vibrations of such a hypersurface? Many late-19th-century mathematicians and physicists, both eccentric and orthodox, took such suggestions seriously. Einstein himself proposed the surface of a 4-sphere as a model of the cosmos, unbounded and yet finite. Just as

Flatlanders on a sphere could travel the straightest possible line in any direction and eventually return to their starting point, so (Einstein suggested) if a spaceship left the earth and traveled far enough in any one direction, it would eventually return to the earth. If a Flatlander started to paint the surface of the sphere on which he lived, extending the paint outward in ever widening circles, he would reach a halfway point at which the circles would begin to diminish, with himself on the *inside,* and eventually he would paint himself into a spot. Similarly, in Einstein's cosmos, if terrestrial astronauts began to map the universe in ever-expanding spheres, they would eventually map themselves into a small globular space on the opposite side of the hypersphere.

Many other properties of hyperspheres are just what one would expect by analogy with lower-order spheres. A circle rotates around a central point, a sphere rotates around a central line, a 4-sphere rotates around a central *plane.* In general the axis of a rotating n-sphere is a space of $n - 2$. (The 4-sphere is capable, however, of a peculiar double rotation that has no analogue in 2- or 3-space: it can spin simultaneously around two fixed planes that are perpendicular to each other.) The projection of a circle on a line is a line segment, but every point on the segment, with the exception of its end points, corresponds to two points on the circle. Project a sphere on a plane and you get a disk, with every point inside the circumference corresponding to two points on the sphere's surface. Project a 4-sphere on our 3-space and you get a solid ball with every internal point corresponding to two points on the 4-sphere's hypersurface. This too generalizes up the ladder of spaces.

The same is true of cross sections. Cut a circle with a line and the cross section is a 1-sphere, or a pair of points. Slice a sphere with a plane and the cross section is a circle. Slice a 4-sphere with a 3-space hyperplane and the cross section is a 3-sphere. (You can't divide a 4-sphere into two pieces with a 2-plane. A hyperapple, sliced down the middle by a 2-plane, remains in one piece.) Imagine a 4-sphere moving slowly through our space. We see it first as a point and then as a tiny sphere that slowly grows in size to its maximum cross section, then slowly diminishes and disappears.

A sphere of any dimension, made of sufficiently flexible material, can be turned inside out through the next-highest space. Just as we can twist a thin rubber ring until the outside rim becomes the inside, so a hypercreature could seize one of our tennis balls and turn it inside out through his space. He could do this all at once or he could start at one spot on the ball, turn a tiny portion first, then gradually enlarge it until the entire ball had its inside outside.

One of the most elegant of the formulas that generalize easily to spheres of all dimensions is the formula for the radii of the maximum number of mutually touching n-spheres. On the plane, no more than four circles can be placed so that each circle touches all the others, with every pair touching at a different point. There are two possible situations (aside from degenerate cases in which one circle has an infinite radius and so becomes a straight line): either three circles surround a smaller one (Figure 11.1, *left*) or three circles are inside a larger one (Figure 11.1, *right*). Frederick Soddy, the British chemist who received a Nobel prize in 1921 for his discovery of isotopes, put it this way in the first stanza of *The Kiss Precise,* a poem that appeared in *Nature* (Vol. 137, June 20, 1936, p. 1021):

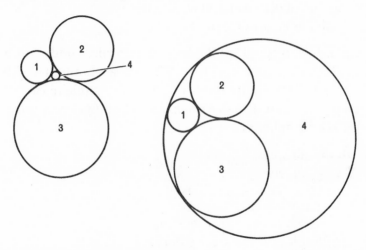

Figure 11.1. Find the radius of the fourth circle.

> For pairs of lips to kiss maybe
> Involves no trigonometry.
> 'Tis not so when four circles kiss
> Each one the other three.
> To bring this off the four must be
> As three in one or one in three.
> If one in three, beyond a doubt
> Each gets three kisses from without.
> If three in one, then is that one
> Thrice kissed internally.

Soddy's next stanza gives the simple formula. His term "bend" is what is usually called the circle's curvature, the reciprocal of the radius. (Thus a circle of radius 4 has a curvature or "bend" of 1/4.) If a circle is touched on the inside, as it is in the case of the large circle enclosing the other three, it is said to have a concave bend, the value of which is preceded by a minus sign. As Soddy phrased all this:

> *Four circles to the kissing come.*
> *The smaller are the benter.*
> *The bend is just the inverse of*
> *The distance from the center.*
> *Though their intrigue left Euclid dumb*
> *There's now no need for rule of thumb.*
> *Since zero bend's a dead straight line*
> *And concave bends have minus sign,*
> The sum of the squares of all four bends
> Is half the square of their sum.

Letting *a, b, c, d* stand for the four reciprocals, Soddy's formula is $2(a^2 + b^2 + c^2 + d^2) = (a + b + c + d)^2$. The reader should have little difficulty computing the radii of the fourth kissing circle in each illustration. In the poem's third and last stanza this formula is extended to five mutually kissing spheres:

> *To spy out spherical affairs*
> *An oscular surveyor*
> *Might find the task laborious,*
> *The sphere is much the gayer,*
> *And now besides the pair of pairs*
> *A fifth sphere in the kissing shares.*
> *Yet, signs and zero as before,*
> *For each to kiss the other four*
> The square of the sum of all five bends
> Is thrice the sum of their squares.

The editors of *Nature* reported in the issue for January 9, 1937 (Vol. 139, p. 62), that they had received several fourth stanzas generalizing Soddy's formula to *n*-space, but they published only the following, by Thorold Gosset, an English barrister and amateur mathematician:

> *And let us not confine our cares*
> *To simple circles, planes and spheres,*
> *But rise to hyper flats and bends*

Where kissing multiple appears.
In n-ic space the kissing pairs
Are hyperspheres, and Truth declares—
As n + 2 such osculate
Each with an n + 1-fold mate.
The square of the sum of all the bends
Is *n* times the sum of their squares.

In simple prose, for *n*-space the maximum number of mutually touching spheres is $n + 2$, and *n* times the sum of the squares of all bends is equal to the square of the sum of all bends. It later developed that the formula for four kissing circles had been known to René Descartes, but Soddy rediscovered it and seems to have been the first to extend it to spheres.

Note that the general formula even applies to the three mutually touching two-point "spheres" of 1-space: two touching line segments "inside" a third segment that is simply the sum of the other two. The formula is a great boon to recreational mathematicians. Puzzles about mutually kissing circles or spheres yield readily to it. Here is a pretty problem. Three mutually kissing spherical grapefruits, each with a radius of three inches, rest on a flat counter. A spherical orange is also on the counter under the three grapefruits and touching each of them. What is the radius of the orange?

Problems about the packing of unit spheres do not generalize easily as one goes up the dimensional ladder; indeed, they become increasingly difficult. Consider, for instance, the problem of determining the largest number of unit spheres that can touch a unit sphere. For circles the number is six (see Figure 11.2). For spheres it is 12, but this was not proved until 1874. The difficulty lies in the fact that when 12 spheres are arranged around a thirteenth, with their centers at the corners of an imaginary icosahedron (see Figure 11.3), there is space between every pair. The waste space is slightly more than needed to accommodate a thirteenth sphere if only the 12 could be shifted around and properly packed. If the reader will coat 14 ping-pong balls with rubber cement, he will find it easy to stick 12 around one of them, and it will not be at all clear whether or not the thirteenth can be added without undue distortions. An equivalent question (can the reader see why?) is: Can 13 paper circles, each covering a 60-degree arc of a great circle on a sphere, be pasted on that sphere without overlapping?

H.S.M. Coxeter, writing on "The Problem of Packing a Number of

Figure 11.2. Six unit circles touch a seventh.

Figure 11.3. Twelve unit spheres touch a thirteenth.

SOLID GEOMETRY AND HIGHER DIMENSIONS

Equal Nonoverlapping Circles on a Sphere" (in *Transactions of the New York Academy of Sciences,* Vol. 24, January 1962, pp. 320–31), tells the story of what may be the first recorded discussion of the problem of the 13 spheres. David Gregory, an Oxford astronomer and friend of Isaac Newton, recorded in his notebook in 1694 that he and Newton had argued about just this question. They had been discussing how stars of various magnitudes are distributed in the sky and this had led to the question of whether or not one unit sphere could touch 13 others. Gregory believed they could. Newton disagreed. As Coxeter writes, "180 years were to elapse before R. Hoppe proved that Newton was right." Simpler proofs have since been published, the latest in 1956 by John Leech, a British mathematician.

How many unit hyperspheres in 4-space can touch a unit hypersphere? It is not yet known if the answer is 24 or 25. Nor is it known for any higher space. For spaces 4 through 8 the densest possible packings are known only if the centers of the spheres form a regular lattice. These packings give 24, 40, 72, 126, and 240 for the "kissing number" of spheres that touch another.

Why the difficulty with 9-space? A consideration of some paradoxes involving hypercubes and hyperspheres may cast a bit of dim light on the curious turns that take place in 9-space. Into a unit square one can pack, from corner to diagonally opposite corner, a line with a length of $\sqrt{2}$. Into a unit cube one can similarly pack a line of $\sqrt{3}$. The distance between opposite corners of an n-cube is $\sqrt{n}$, and since square roots increase without limit, it follows that a rod of *any* size will pack into a unit n-cube if n is large enough. A fishing pole 10 feet long will fit diagonally in the one-foot 100-cube. This also applies to objects of higher dimension. A cube will accommodate a square larger than its square face. A 4-cube will take a 3-cube larger than its cubical hyperface. A 5-cube will take larger squares and cubes than any cube of lower dimension with an edge of the same length. An elephant or the entire Empire State Building will pack easily into an n-cube with edges the same length as those of a sugar cube if n is sufficiently large.

The situation with respect to an n-sphere is quite different. No matter how large n becomes, an n-sphere can never contain a rod longer than twice its radius. And something very queer happens to its n-volume as n increases. The area of the unit circle is, of course, π. The volume of the unit sphere is 4.1+. The unit 4-sphere's hypervolume is 4.9+. In 5-space the volume is still larger, 5.2+, then in 6-space it decreases to 5.1+, and thereafter steadily declines. Indeed, as n approaches infinity the hypervolume of a unit n-sphere approaches zero! This leads to many unearthly results. David Singmaster, writing "On Round Pegs in Square Holes and Square Pegs in Round Holes" (*Mathematics Magazine,* Vol. 37, November 1964, pp. 335–37), decided that a round peg fits better in a square hole than vice versa because the ratio of the area of a circle to a circumscribing square ($\pi/4$) is larger than the ratio of a square inscribed in a circle ($2/\pi$). Similarly, one can show that a ball fits better in a cube than a cube fits in a ball, although the difference between ratios is a bit smaller. Singmaster found that the difference continues to decrease through 8-space and then reverses: in 9-space the ratio of n-ball to n-cube is *smaller* than the ratio of n-cube to n-ball. In other words, an n-ball fits better in an n-cube than an n-cube fits in an n-ball if and only if n is 8 or less.

The same 9-space turn occurs in an unpublished paradox discovered by Leo Moser. Four unit circles will pack into a square of side 4 (see Figure 11.4). In the center we can fit a smaller circle of radius $\sqrt{2}$ – 1. Similarly, eight unit spheres will pack into the corners of a cube of side 4 (see Figure 11.5). The largest sphere that will fit into the center has a radius of $\sqrt{3}$ – 1. This generalizes in the obvious way: In a 4-cube of side 4 we can pack 16 unit 4-spheres and a central 4-sphere of radius $\sqrt{4}$ – 1, which equals 1, so that the central sphere now is the same size as the others. In general, in the corners of an n-cube of side 4 we can pack 2^n unit n-spheres and presumably another sphere of radius $\sqrt{n}$ – 1 will fit at the center. But see what happens when we come to 9-space: the central hypersphere has a radius of $\sqrt{9}$ – 1 = 2, which is equal to half the hypercube's edge. The central sphere cannot be larger than this in any higher n-cube because it now fills the hypercube. No longer is the central hypersphere *inside* the spheres that surround the center of every hyperface, yet there is space at $2^9 = 512$ corners to take 512 unit 9-spheres!

A related unpublished paradox, also discovered by Moser, concerns n-dimensional chessboards. All the black squares of a chessboard are enclosed with circumscribed circles (see Figure 11.6). Assume that each cell is of side 2 and area 4. Each circle has a radius of $\sqrt{2}$ and an area of 2π. The area in each white cell that is left white (is not enclosed by a circle) is $8 - 2\pi = 1.71+$. In the analogous situation for a cubical chessboard, the black cubical cells of edge 2 are surrounded by spheres. The volume of each black cell is 8 and the volume of each sphere,

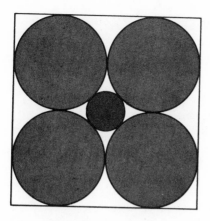

Figure 11.4. Four circles around one of radius $\sqrt{2} - 1$

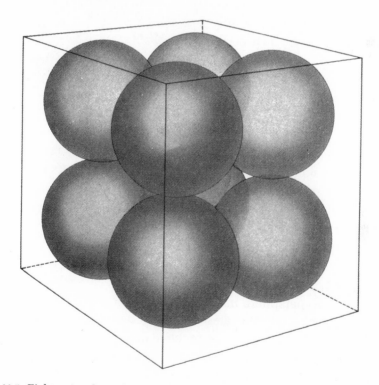

Figure 11.5. Eight unit spheres leave room for one with a radius of $\sqrt{3} - 1$.

which has a radius of $\sqrt{3}$, is $4\pi\sqrt{3}$, but the volume of the unenclosed portion of each white cube is not so easy to calculate because the six surrounding spheres intersect one another.

Consider now the four-dimensional lattice of hypercubes of edge 2

Spheres and Hyperspheres

Figure 11.6. Leo Moser's hyperchessboard problem

with cells alternately colored as before so that each cell is surrounded by eight hypercubes of opposite color. Around each black hypercell is circumscribed a hypersphere. What is the hypervolume of the unenclosed portion within each white cell? The surprising answer can be determined quickly without knowing the formula for the volume of a hypersphere.

Addendum

As far as I know, the latest results on the number of unit spheres in n-dimensional space that can touch a single unit sphere have been proved only when $n = 3$, 8, and 24. As we have seen, when $n = 3$, the number is 12; when $n = 8$, the number is 240; and when $n = 24$, the number is 196,560. In 4-space the lower and upper bounds are 24 and 25. In 5-space the bounds are 40–46; in 6-space, 72–82; and in 7-space, 126–140. A new proof of the 3-space case was announced in January 2000 by Sean T. McLaughlin of the University of Michigan.

The ongoing search for the densest possible sphere packings in n-dimensional space is now the topic of a vast literature, mainly because such packing lattices are closely related to the construction of error-correcting codes—efficient ways of transmitting information with a minimum number of errors. The definitive treatise on the topic, by John Horton Conway and N.J.A. Sloane (1988), runs 663 pages and has a bibliography of 1,500 entries!

The densest packing of unit spheres, either lattice-based or irregular, is known only for the circle, regarded as a two-dimensional "sphere." Densest lattice packings are known only for n-dimensional spheres when $n = 2$ through 8. Beyond $n = 8$ there are only conjectures about such packings. In some cases irregular packings have been discovered that are denser than lattice packings. The most famous of higher space packings is a very dense lattice in 24 dimensions, known as the Leech lattice after the British Mathematician John Leech who discovered it. Each sphere in this remarkable structure touches 196,560 others!

Answers

The first problem was to determine the sizes of two circles, each of which touches three mutually tangent circles with radii of one, two, and three units. Using the formula given in the chapter,

$$2\left(1 + \frac{1}{4} + \frac{1}{9} + \frac{1}{x^2}\right) = \left(1 + \frac{1}{2} + \frac{1}{3} + \frac{1}{x}\right)^2,$$

where x is the radius of the fourth circle, one obtains a value of 6/23 for the radius of the smaller circle, 6 for the larger one.

The second problem concerned three grapefruits with three-inch radii and an orange, all resting on a counter and mutually touching. What size is the orange? The plane on which they rest is considered a fifth sphere of infinite radius that touches the other four. Since it has zero curvature it drops out of the formula relating the reciprocals of the radii of five mutually touching spheres. Letting x be the radius of the orange, we write the equation,

$$3\left(\frac{1}{3^2} + \frac{1}{3^2} + \frac{1}{3^2} + \frac{1}{x^2}\right) = \left(\frac{1}{3} + \frac{1}{3} + \frac{1}{3} + \frac{1}{x}\right)^2,$$

which gives x a value of one inch.

The problem can, of course, be solved in other ways. When it appeared as problem 43 in the *Pi Mu Epsilon Journal*, November 1952,

Leon Bankoff solved it this way, with R the radius of each large sphere and r the radius of the small sphere:

"The small sphere, radius r, touches the table at a point equidistant from the contacts of each of the large spheres with the table. Hence it lies on the circumcenter of an equilateral triangle, the side of which is $2R$. Then $(R + r)$ is the hypotenuse of a right triangle, the altitude of which is $(R - r)$ and the base of which is $2R\sqrt{3}/3$. So

$$(R + r)^2 = (R - r)^2 + 4R^2/3, \text{ or } r = R/3."$$

The answer to Leo Moser's paradox of the hypercubic chessboard in four-dimensional space is that *no* portion of a white cell remains unenclosed by the hyperspheres surrounding each black cell. The radius of each hypersphere is $\sqrt{4}$, or 2. Since the hypercubic cells have edges of length 2, we see at once that each of the eight hyperspheres around a white cell will extend all the way to the center of that cell. The eight hyperspheres intersect one another, leaving no portion of the white cell unenclosed.

Bibliography

The Kiss Precise

W. S. Brown, "The Kiss Precise," *American Mathematical Monthly*, Vol. 76, June 1969, pp. 661–63.

D. Pedoe, "On a Theorem in Geometry," *American Mathematical Monthly*, Vol. 76, June/July 1967, pp. 627–40.

Maximum Spheres that Touch a Sphere

H.S.M. Coxeter, "An Upper Bound for the Number of Equal Nonoverlapping Spheres That Can Touch Another of the Same Size," *Proceedings of Symposia in Pure Mathematics*, Vol. 7, 1963, pp. 53–71.

J. Leech, "The Problem of the Thirteen Spheres," *The Mathematical Gazette*, Vol. 240, February 1956, pp. 22–23.

A. M. Odlyzko and N.J.A. Sloane, "New Bounds on the Number of Unit Spheres That Can Touch a Unit Sphere in n Dimensions," *Journal of Combinatorial Theory*, Ser. A, Vol. 26, March 1979, pp. 210–14.

J. C. Prepp, "Kepler's Spheres and Rubik's Cube," *Mathematics Magazine*, Vol. 61, October 1988, pp. 231–39.

I. Stewart, "The Kissing Number," *Scientific American*, February 1992, pp. 112–15.

A. J. Wasserman, "The Thirteen Spheres Problem," *Eureka*, Spring 1978, pp. 46–49.

N-space Packing

J. H. Conway and N.J.A. Sloane, *Leech Lattice, Sphere Packings, and Related Topics,* Springer-Verlag, 1984.

J. H. Conway and N.J.A. Sloane, *Sphere Packings, Lattices, and Groups,* Springer-Verlag, 1988.

I. Peterson, "Curves For a Tighter Fit," *Science News,* Vol. 137, May 19, 1990, pp. 316–17.

N.J.A. Sloane, "The Packing of Spheres," *Scientific American,* January 1984, pp. 116–25.

Chapter 12

The Church of
the Fourth Dimension

*"Could I but rotate my arm out of the limits
set to it," one of the Utopians had said to
him, "I could thrust it into a thousand
dimensions."* —H. G. Wells, *Men Like Gods*

Alexander Pope once described London as a "dear, droll, dis-
tracting town." Who would disagree? Even with respect to recreational
mathematics, I have yet to make an imaginary visit to London without
coming on something quite extra-ordinary. Once, for instance, I was
reading the London *Times* in my hotel room a few blocks from Pic-
cadilly Circus when a small advertisement caught my eye:

> Weary of the world of three dimensions? Come worship with us Sunday
> at the Church of the Fourth Dimension. Services promptly at 11 A.M., in
> Plato's grotto. Reverend Arthur Slade, Minister.

An address was given. I tore out the advertisement, and on the following Sunday morning rode the Underground to a station within walking
distance of the church. There was a damp chill in the air and a light
mist was drifting in from the sea. I turned the last corner, completely
unprepared for the strange edifice that loomed ahead of me. Four enormous cubes were stacked in one column, with four cantilevered cubes
jutting in four directions from the exposed faces of the third cube from
the ground. I recognized the structure at once as an unfolded hypercube. Just as the six square faces of a cube can be cut along seven lines
and unfolded to make a two-dimensional Latin cross (a popular floor
plan for medieval churches), so the eight cubical hyperfaces of a four-
dimensional cube can be cut along seventeen squares and "unfolded"
to form a three-dimensional Latin cross.

A smiling young woman standing inside the portal directed me to a
stairway. It spiraled down into a basement auditorium that I can only
describe as a motion-picture theater combined with a limestone cavern.
The front wall was a solid expanse of white. Formations of translucent
pink stalactites glowed brightly on the ceiling, flooding the grotto with

a rosy light. Huge stalagmites surrounded the room at the sides and back. Electronic organ music, like the score of a science-fiction film, surged into the room from all directions. I touched one of the stalagmites. It vibrated beneath my fingers like the cold key of a stone xylophone.

The strange music continued for 10 minutes or more after I had taken a seat, then slowly softened as the overhead light began to dim. At the same time I became aware of a source of bluish light at the rear of the grotto. It grew more intense, casting sharp shadows of the heads of the congregation on the lower part of the white wall ahead. I turned around and saw an almost blinding point of light that appeared to come from an enormous distance.

The music faded into silence as the grotto became completely dark except for the brilliantly illuminated front wall. The shadow of the minister rose before us. After announcing the text as Ephesians, Chapter 3, verses 17 and 18, he began to read in low, resonant tones that seemed to come directly from the shadow's head: ". . . that ye, being rooted and grounded in love, may be able to comprehend with all saints what is the breadth, and length, and depth, and height. . . ."

It was too dark for note-taking, but the following paragraphs summarize accurately, I think, the burden of Slade's remarkable sermon.

Our cosmos—the world we see, hear, feel—is the three-dimensional "surface" of a vast, four-dimensional sea. The ability to visualize, to comprehend intuitively, this "wholly other" world of higher space is given in each century only to a few chosen seers. For the rest of us, we must approach hyperspace indirectly, by way of analogy. Imagine a Flatland, a shadow world of two dimensions like the shadows on the wall of Plato's famous cave (*Republic,* Chapter 7). But shadows do not have material substance, so it is best to think of Flatland as possessing an infinitesimal thickness equal to the diameter of one of its fundamental particles. Imagine these particles floating on the smooth surface of a liquid. They dance in obedience to two-dimensional laws. The inhabitants of Flatland, who are made up of these particles, cannot conceive of a third direction perpendicular to the two they know.

We, however, who live in 3-space can see every particle of Flatland. We see inside its houses, inside the bodies of every Flatlander. We can touch every particle of their world without passing our finger through their space. If we lift a Flatlander out of a locked room, it seems to him a miracle.

In an analogous way, Slade continued, our world of 3-space floats on the quiet surface of a gigantic hyperocean; perhaps, as Einstein once suggested, on an immense hypersphere. The four-dimensional thickness of our world is approximately the diameter of a fundamental particle. The laws of our world are the "surface tensions" of the hypersea. The surface of this sea is uniform, otherwise our laws would not be uniform. A slight curvature of the sea's surface accounts for the slight, constant curvature of our space–time. Time exists also in hyperspace. If time is regarded as our fourth coordinate, then the hyperworld is a world of five dimensions. Electromagnetic waves are vibrations on the surface of the hypersea. Only in this way, Slade emphasized, can science escape the paradox of an empty space capable of transmitting energy.

What lies outside the sea's surface? The wholly other world of God! No longer is theology embarrassed by the contradiction between God's immanence and transcendence. Hyperspace touches every point of 3-space. God is closer to us than our breathing. He can see every portion of our world, touch every particle without moving a finger through our space. Yet the Kingdom of God is completely "outside" 3-space, in a direction in which we cannot even point.

The cosmos was created billions of years ago when God poured (Slade paused to say that he spoke metaphorically) on the surface of the hypersea an enormous quantity of hyperparticles with asymmetric three-dimensional cross sections. Some of these particles fell into 3-space in right-handed form to become neutrons, the others in left-handed form to become antineutrons. Pairs of opposite parity annihilated each other in a great primeval explosion, but a slight preponderance of hyperparticles happened to fall as neutrons and this excess remained. Most of these neutrons split into protons and electrons to form hydrogen. So began the evolution of our "one-sided" material world. The explosion caused a spreading of particles. To maintain this expanding universe in a reasonably steady state, God renews its matter at intervals by dipping his fingers into his supply of hyperparticles and flicking them toward the sea. Those which fall as antineutrons are annihilated, those which fall as neutrons remain. Whenever an antiparticle is created in the laboratory, we witness an actual "turning over" of an asymmetric particle in the same way that one can reverse in 3-space an asymmetric two-dimensional pattern of cardboard. Thus the production of antiparticles provides an empirical proof of the reality of 4-space.

Slade brought his sermon to a close by reading from the recently discovered Gnostic Gospel of Thomas: "If those who lead you say to you: Behold the kingdom is in heaven, then the birds will precede you. If they say to you that it is in the sea, then the fish will precede you. But the kingdom is within you and it is outside of you."

Again the unearthly organ music. The blue light vanished, plunging the cavern into total blackness. Slowly the pink stalactites overhead began to glow, and I blinked my eyes, dazzled to find myself back in 3-space.

Slade, a tall man with iron-gray hair and a small dark mustache, was standing at the grotto's entrance to greet the members of his congregation. As we shook hands I introduced myself and mentioned my *Scientific American* column. "Of course!" he exclaimed. "I have some of your books. Are you in a hurry? If you wait a bit, we'll have a chance to chat."

After the last handshake Slade led me to a second spiral stairway of opposite handedness from the one on which I had descended earlier. It carried us to the pastor's study in the top cube of the church. Elaborate models, 3-space projections of various types of hyperstructures, were on display around the room. On one wall hung a large reproduction of Salvador Dali's painting "Corpus Hypercubus." In the picture, above a flat surface of checkered squares, floats a three-dimensional cross of eight cubes; an unfolded hypercube identical in structure with the church in which I was standing.

"Tell me, Slade," I said, after we were seated, "is this doctrine of yours new or are you continuing a long tradition?"

"It's by no means new," he replied, "though I *can* claim to have established the first church in which hyperfaith serves as the cornerstone. Plato, of course, had no conception of a geometrical fourth dimension, though his cave analogy clearly implies it. In fact, every form of Platonic dualism that divides existence into the natural and supernatural is clearly a nonmathematical way of speaking about higher space. Henry More, the 17th-century Cambridge Platonist, was the first to regard the spiritual world as having four spatial dimensions. Then along came Immanuel Kant, with his recognition of our space and time as subjective lenses, so to speak, through which we view only a thin slice of transcendent reality. After that it is easy to see how the concept of higher space provided a much needed link between modern science and traditional religions."

"You say 'religions,' " I put in. "Does that mean your church is not Christian?"

"Only in the sense that we find essential truth in all the great world faiths. I should add that in recent decades the Continental Protestant theologians have finally discovered 4-space. When Karl Barth talks about the 'vertical' or 'perpendicular' dimension, he clearly means it in a four-dimensional sense. And of course in the theology of Karl Heim there is a full, explicit recognition of the role of higher space."

"Yes," I said. "I recently read an interesting book called *Physicist and Christian,* by William G. Pollard (executive director of the Oak Ridge Institute of Nuclear Studies and an Episcopal clergyman). He draws heavily on Heim's concept of hyperspace."

Slade scribbled the book's title on a note pad. "I must look it up. I wonder if Pollard realizes that a number of late-19th-century Protestants wrote books about the fourth dimension. A. T. Schofield's *Another World,* for example (it appeared in 1888) and Arthur Willink's *The World of the Unseen* (subtitled "An Essay on the Relation of Higher Space to Things Eternal"; published in 1893). Of course modern occultists and spiritualists have had a field day with the notion. Peter D. Ouspensky, for instance, has a lot to say about it in his books, although most of his opinions derive from the speculations of Charles Howard Hinton, an American mathematician. Whately Carington, the English parapsychologist, wrote an unusual book in 1920—he published it under the byline of W. Whately Smith—on *A Theory of the Mechanism of Survival.*"

"Survival after death?"

Slade nodded. "I can't go along with Carington's belief in such things as table tipping being accomplished by an invisible four-dimensional lever, or clairvoyance as perception from a point in higher space, but I regard his basic hypothesis as sound. Our bodies are simply three-dimensional cross sections of our higher four-dimensional selves. Obviously a man is subject to all the laws of this world, but at the same time his experiences are permanently recorded—stored as information, so to speak—in the 4-space portion of his higher self. When his 3-space body ceases to function, the permanent record remains until it can be attached to a new body for a new cycle of life in some other 3-space continuum."

"I like that," I said. "It explains the complete dependence of mind on body in this world, at the same time permitting an unbroken continuity between this life and the next. Isn't this close to what William James struggled to say in his little book on immortality?"

"Precisely. James, unfortunately, was no mathematician, so he had to express his meaning in nongeometrical metaphors."

"What about the so-called demonstrations of the fourth dimension by certain mediums," I asked. "Wasn't there a professor of astrophysics in Leipzig who wrote a book about them?"

I thought I detected an embarrassed note in Slade's laugh. "Yes, that was poor Johann Karl Friedrich Zöllner. His book *Transcendental Physics* was translated into English in 1881, but even the English copies are now quite rare. Zöllner did some good work in spectrum analysis, but he was supremely ignorant of conjuring methods. As a consequence he was badly taken in, I'm afraid, by Henry Slade, the American medium."

"Slade?" I said with surprise.

"Yes, I'm ashamed to say we're related. He was my great-uncle. When he died, he left a dozen fat notebooks in which he had recorded his methods. Those notebooks were acquired by the English side of my family and handed down to me."

"This excites me greatly," I said. "Can you demonstrate any of the tricks?"

The request seemed to please him. Conjuring, he explained, was one of his hobbies, and he thought that the mathematical angles of several of Henry's tricks would be of interest to my readers.

From a drawer in his desk Slade took a strip of leather, cut as shown at the left in Figure 12.1, to make three parallel strips. He handed me a ballpoint pen with the request that I mark the leather in some way to prevent later substitution. I initialed a corner as shown. We sat on opposite sides of a small table. Slade held the leather under the table for a few moments, then brought it into view again. It was braided exactly as shown at the right in the illustration! Such braiding would be easy to accomplish if one could move the strips through hyperspace. In 3-space it seemed impossible.

Slade's second trick was even more astonishing. He had me examine

Figure 12.1. Slade's leather strip
—braided in hyperspace?

a rubber band of the wide, flat type shown at the left in Figure 12.2. This was placed in a matchbox, and the box was securely sealed at both ends with cellophane tape. Slade started to place it under the table, then remembered he had forgotten to have me mark the box for later identification. I drew a heavy X on the upper surface.

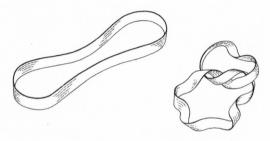

Figure 12.2. Slade's rubber band —knotted in hyperspace?

"If you like," he said, "you yourself may hold the box under the table."

I did as directed. Slade reached down, taking the box by its other end. There was a sound of movement and I could feel that the box seemed to be vibrating slightly.

Slade released his grip. "Please open the box."

First I inspected the box carefully. The tape was still in place. My mark was on the cover. I slit the tape with my thumbnail and pushed open the drawer. The elastic band—*mirabile dictu*—was tied in a simple knot as shown at the right in Figure 12.2.

"Even if you managed somehow to open the box and switch bands," I said, "how the devil could you get a rubber band like this?"

Slade chuckled. "My great-uncle was a clever rascal."

I was unable to persuade Slade to tell me how either trick was done. The reader is invited to think about them before he reads this chapter's answer section.

We talked of many other things. When I finally left the Church of the Fourth Dimension, a heavy fog was swirling through the wet streets of London. I was back in Plato's cave. The shadowy forms of moving cars, their headlights forming flat elliptical blobs of light, made me think of some familiar lines from the Rubáiyát of a great Persian mathematician:

> We are no other than a moving row
> Of magic shadow-shapes that come and go
> Round with the sun-illumined lantern held
> In midnight by the Master of the Show.

Addendum

Although I spoke in the first paragraph of this chapter of an "imaginary visit" to London, when the chapter first appeared in *Scientific American* several readers wrote to ask for the address of Slade's church. The Reverend Slade is purely fictional, but Henry Slade the medium was one of the most colorful and successful mountebanks in the history of American spiritualism. See my article on Slade in *The Encyclopedia of the Paranormal,* edited by Gordon Stein, Prometheus Books, 1996, and my remarks about Slade's fourth-dimensional trickery in *The New Ambidextrous Universe* (W. H. Freeman, 1990).

At the time I wrote about the Church of the Fourth Dimension no eminent physicist had ever contended that there might actually be spaces "out there," higher than our familiar 3-space. (The use of a fourth dimension in relativity theory was no more than a way of handling time in the theory's equations.) Now, however, particle physicists are in a euphoric state over a theory of superstrings in which fundamental particles are not modeled as geometrical points, but as extremely tiny closed loops, of great tensile strength, that vibrate in higher spaces. These higher spaces are "compacted"—curled up into tight little structures too small to be visible or even to be detected by today's atom smashers.

Some physicists regard these higher spaces as mere artifices of the mathematics, but others believe they are just as real as the three spaces we know and love. (On superstrings, see my *New Ambidextrous Uni-*

verse.) It is the first time physicists have seriously entertained the notion of higher spaces that are physically real. This may have stimulated the publication of recent books on the fourth and higher dimensions. I particularly recommend Thomas F. Banchoff's *Beyond the Third Dimension*, if for no other reason than for its wondrous computer graphics.

In recent years the theory of superstrings has been extended to membrane or brane theory in which the strings are attached to vibrating surfaces that live in higher space dimensions. Some theorists are speculating that our universe is an enormous brane floating in hyperspace along with countless other "island universes," each with its own set of laws. A possibility looms that these other universes could be detected by their gravity seeping into our universe and generating the elusive invisible dark matter which physicists suspect furnishes 90 percent of the matter in our universe. For these wild speculations see George Johnson's mind-boggling article, "Physicists Finally Find a Way to Test Superstring Theory," in *The New York Times*, April 4, 2000.

Membrane theory is also called M-theory, the M standing for membrane, mystery, magic, marvel, and matrix. See "Magic, Mystery, and Matrix," a lecture by Edward Witten, the world's top expert on superstrings, published in *Notices of the AMS* (American Mathematical Society), Vol. 45, November 1998, and his 1998 videotape *M-Theory*, available from the AMS.

Answers

Slade's method of braiding the leather strip is familiar to Boy Scouts in England and to all those who make a hobby of leathercraft. Many readers wrote to tell me of books in which this type of braiding is described: George Russell Shaw, *Knots, Useful and Ornamental* (p. 86); Constantine A. Belash, *Braiding and Knotting* (p. 94); Clifford Pyle, *Leather Craft as a Hobby* (p. 82); Clifford W. Ashley, *The Ashley Book of Knots* (p. 486); and others. For a full mathematical analysis, see J.A.H. Shepperd, "Braids Which Can Be Plaited with Their Threads Tied Together at Each End," *Proceedings of the Royal Society*, A, Vol. 265, 1962, pages 229–44.

There are several ways to go about making the braid. Figure 12.3 was drawn by reader George T. Rab of Dayton, OH. By repeating this procedure one can extend the braid to any multiple of six crossings. Another procedure is simply to form the six-cross plat in the upper half of the strip by braiding in the usual manner. This creates a mirror image of the plat in the lower half. The lower plat is easily removed by one hand while the upper plat is held firmly by the other hand. Both procedures can be adapted to leather strips with more than three strands. If stiff leather is used, it can be made pliable by soaking it in warm water.

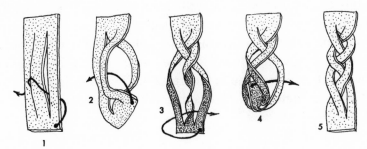

Figure 12.3. Slade's first trick

Slade's trick of producing a knot in a flat rubber band calls first for the preparation of a knotted band. Obtain a rubber ring of circular cross section and carefully carve a portion of it flat as shown in Figure 12.4. Make three half twists in the flat section *(middle drawing),* then continue carving the rest of the ring to make a flat band with three half twists *(last drawing).* Mel Stover of Winnipeg, Canada, suggests that this can best be done by stretching the ring around a wooden block, freezing the ring, then flattening it with a home grinding tool. When the final band is cut in half all the way around, it forms a band twice as large and tied in a single knot.

Figure 12.4. Slade's second trick

A duplicate band of the same size, but unknotted, must also be obtained. The knotted band is placed in a matchbox and the ends of the box are sealed with tape. It is now necessary to substitute this matchbox for the one containing the unknotted band. I suspect that Slade did

The Church of the Fourth Dimension   159

this when he started to put the box under the table, then "remembered" that I had not yet initialed it. The prepared box could have been stuck to the underside of the table with magician's wax. It would require only a moment to press the unprepared box against another dab of wax, then take the prepared one. In this way the switch occurred *before* I marked the box. The vibrations I felt when Slade and I held the box under the table were probably produced by one of Slade's fingers pressing firmly against the box and sliding across it.

Fitch Cheney, mathematician and magician, wrote to tell about a second and simpler way to create a knotted elastic band. Obtain a hollow rubber torus—they are often sold as teething rings for babies—and cut as shown by the dotted line in Figure 12.5. The result is a wide endless band tied in a single knot. The band can be trimmed, of course, to narrower width.

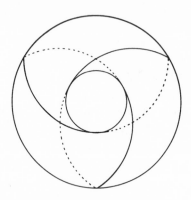

Figure 12.5. A second way to produce a knotted rubber band

It was Stover, by the way, who first suggested to me the problem of tying a knot in an elastic band. He had been shown such a knotted band by magician Winston Freer. Freer said he knew *three* ways of doing it.

Bibliography

T. F. Banchoff, *Beyond the Third Dimension*, New York: W. H. Freeman, 1990.

M. Gardner, "Group Theory and Braids," *New Mathematical Diversions from Scientific American*, Chapter 2, New York: Simon and Schuster, 1966. Reissued by the University of Chicago Press, 1983.

K. Heim, *Christian Faith and Natural Science*, New York: Harper, 1953; paperback, 1957.

C. H. Hinton, *The Fourth Dimension*, London: Swan Sonnenschein, 1904; Allen & Unwin, 1951.

H. P. Manning, *Geometry of Four Dimensions*, New York: Macmillan, 1914; Dover Publications (paperback), 1956.

H. P. Manning, *The Fourth Dimension Simply Explained*, New York: Munn, 1910; New York: Dover Publications (paperback), 1960.

E. H. Neville, *The Fourth Dimension*, Cambridge, England: Cambridge University Press, 1921.

C. Pickover, *Surfing Through Hyperspace*, England: Oxford University Press, 1999.

R. Rucker (ed.), *Speculations on the Fourth Dimension*, New York: Dover, 1980.

R. Rucker, *The Fourth Dimension: Toward a Geometry of Higher Reality*, Boston: Houghton-Mifflin, 1984.

J.A.H. Shepperd, "Braids Which Can Be Plaited with Their Threads Tied Together at Each End," *Proceedings of the Royal Society, A*, Vol. 265, 1962, pp. 229–44.

D.M.Y. Sommerville, *An Introduction to the Geometry of N Dimensions*, London: Methuen, 1929; New York: Dover Publications (paperback), 1958.

Chapter 13

Hypercubes

> *The children were vanishing.*
> *They went in fragments, like thick smoke in a*
> *wind, or like movement in a distorting mirror.*
> *Hand in hand they went, in a direction Para-*
> *dine could not understand. . . .*
>
> —LEWIS PADGETT, from "Mimsy Were the Borogoves"

*T*he direction that Paradine, a professor of philosophy, could not understand is a direction perpendicular to each of the three coordinates of space. It extends into 4-space in the same way a chess piece extends upward into 3-space with its axis at right angles to the *x* and *y* coordinates of the chessboard. In Padgett's great science fiction story, Paradine's children find a wire model of a tesseract (a hypercube of four dimensions) with colored beads that slide along the wires in curious ways. It is a toy abacus that had been dropped into our world by a 4-space scientist tinkering with a time machine. The abacus teaches the children how to think four-dimensionally. With the aid of some cryptic advice in Lewis Carroll's *Jabberwocky* they finally walk out of 3-space altogether.

Is it possible for the human brain to visualize four-dimensional structures? The 19th-century German physicist Hermann von Helmholtz argued that it is, provided the brain is given proper input data. Unfortunately our experience is confined to 3-space and there is not the slightest scientific evidence that 4-space actually exists. (Euclidean 4-space must not be confused with the non-Euclidean four-dimensional space–time of relativity theory, in which time is handled as a fourth coordinate.) Nevertheless, it is conceivable that with the right kind of mathematical training a person might develop the ability to visualize a tesseract. "A man who devoted his life to it," wrote Henri Poincaré, "could perhaps succeed in picturing to himself a fourth dimension."

Charles Howard Hinton, an eccentric American mathematician who once taught at Princeton University and who wrote a popular book called *The Fourth Dimension*, devised a system of using colored blocks for making 3-space models of sections of a tesseract. Hinton believed that by playing many years with this "toy" (it may have suggested the toy in Padgett's story), he had acquired a dim intuitive grasp of 4-space. "I do not like to speak positively," he wrote, "for I might occasion a loss of time on the part of others, if, as may very well be, I am mistaken. But for my own part, I think there are indications of such an intuition. . . ."

Hinton's colored blocks are too complicated to explain here (the fullest account of them is in his 1910 book, *A New Era of Thought*). Perhaps, however, by examining some of the simpler properties of the tesseract we can take a few wobbly first steps toward the power of visualization Hinton believed he had begun to achieve.

Let us begin with a point and move it a distance of one unit in a straight line, as shown in Figure 13.1(*a*). All the points on this unit line can be identified by numbering them from 0 at one end to 1 at the other. Now move the unit line a distance of one unit in a direction perpendicular to the line (*b*). This generates a unit square. Label one corner 0, then number the points from 0 to 1 along each of the two lines that meet at the zero corner. With these *x* and *y* coordinates we can now label every point on the square with an ordered *pair* of numbers. It is just as easy to visualize the next step. Shift the square a unit distance in a direction at right angles to both the *x* and the *y* axes (*c*). The result is a unit cube. With *x*, *y*, *z* coordinates along three edges that meet at a corner, we can label every point in the cube with an ordered *triplet* of numbers.

Although our visual powers boggle at the next step, there is no logical reason why we cannot assume that the cube is shifted a unit distance in a direction perpendicular to all three of its axes *(d)*. The space generated by such a shift is a 4-space unit hypercube—a tesseract—with four mutually perpendicular edges meeting at every corner. By choosing a set of such edges as *w*, *x*, *y*, *z* axes, one might label every point in the hypercube with an ordered *quadruplet* of numbers. Analytic geometers can work with these ordered quadruplets in the same way they work with ordered pairs and triplets to solve problems in plane and solid geometry. In this fashion Euclidean geometry can be extended to higher spaces with dimensions represented by any positive integer. Each space is Euclidean but each is topologically distinct: a square cannot be continuously deformed to a straight line, a cube deformed to a square, a hypercube to a cube, and so on.

Accurate studies of figures in 4-space can be made only on the basis of an axiomatic system for 4-space, or by working analytically with the *w*, *x*, *y*, *z* equations of the four-coordinate system. But the tesseract is

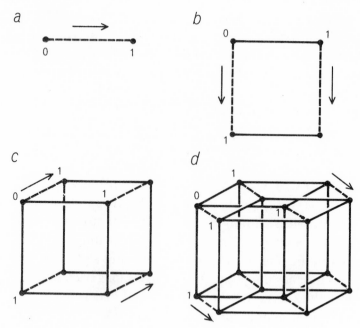

Figure 13.1. Steps toward generating a hypercube

such a simple 4-space structure that we can guess many of its properties by intuitive, analogical reasoning. A unit line has two end points. When it is moved to generate a square, its ends have starting and stopping positions and therefore the number of corners on the square is twice the number of points on the line, or four. The two moving points generate two lines, but the unit line has a start and a stop position and so we must add two more lines to obtain four as the number of lines bounding the square.

In similar fashion, when the square is moved to generate a cube, its four corners have start and stop positions and therefore we multiply four by two to arrive at eight corners on the cube. In moving, each of the four points generates a line, but to those four lines we must add the square's four lines at its start and the four lines at its stop, making 4 + 4 + 4 = 12 edges on the cube. The four lines of the moving square generate four new faces, to which the start and stop faces are added, making 4 + 1 + 1 = 6 faces on the cube's surface.

Now suppose the cube is pushed a unit distance in the direction of a fourth axis at right angles to the other three, a direction in which we cannot point because we are trapped in 3-space. Again each corner of the cube has start and stop positions, so that the resulting tesseract has

$2 \times 8 = 16$ corners. Each point generates a line, but to these eight lines we must add the start and stop positions of the cube's 12 edges to make $8 + 12 + 12 = 32$ unit lines on the hypercube. Each of the cube's 12 edges generates a square, but to those 12 squares we must add the cube's six squares before the push and the six after the push, making $12 + 6 + 6 = 24$ squares on the tesseract's hypersurface.

It is a mistake to suppose the tesseract is bounded by its 24 squares. They form only a skeleton of the hypercube, just as the edges of a cube form its skeleton. A cube is bounded by square faces and a hypercube is bounded by cubical faces. When the cube is pushed, each of its squares moves a unit distance in an unimaginable direction at right angles to its face, thereby generating another cube. To the six cubes generated by the six moving squares we must add the cube before it is pushed and the same cube after it is pushed, making eight in all. These eight cubes form the hypercube's hypersurface.

The chart in Figure 13.2 gives the number of elements in "cubes" of spaces one through four. There is a simple, surprising trick by which this chart can be extended downward to higher n-cubes. Think of the nth line as an expansion of the binomial $(2x + 1)^n$. For example, the line segment of 1-space has two points and one line. Write this as $2x + 1$ and multiply it by itself:

$$
\begin{array}{r}
2x + 1 \\
2x + 1 \\
\hline
4x^2 + 2x \\
2x + 1 \\
\hline
4x^2 + 4x + 1
\end{array}
$$

n-SPACE	POINTS	LINES	SQUARES	CUBES	TESSERACTS
0	1	0	0	0	0
1	2	1	0	0	0
2	4	4	1	0	0
3	8	12	6	1	0
4	16	32	24	8	1

Figure 13.2. Elements of structures analogous to the cube in various dimensions

Note that the coefficients of the answer correspond to the chart's third line. Indeed, each line of the chart, written as a polynomial and multiplied by $2x + 1$, gives the next line. What are the elements of a 5-space cube? Write the tesseract's line as a fourth-power polynomial and multiply it by $2x + 1$:

$$
\begin{array}{r}
16x^4 + 32x^3 + 24x^2 + 8x + 1 \\
2x + 1 \\
\hline
32x^5 + 64x^4 + 48x^3 + 16x^2 + 2x \\
16x^4 + 32x^3 + 24x^2 + 8x + 1 \\
\hline
32x^5 + 80x^4 + 80x^3 + 40x^2 + 10x + 1
\end{array}
$$

The coefficients give the fifth line of the chart. The 5-space cube has 32 points, 80 lines, 80 squares, 40 cubes, 10 tesseracts, and one 5-space cube. Note that each number on the chart equals twice the number above it plus the number diagonally above and left.

If you hold a wire skeleton of a cube so that light casts its shadow on a plane, you can turn it to produce different shadow patterns. If the light comes from a point close to the cube and the cube is held a certain way, you obtain the projection shown in Figure 13.3. The network of this flat pattern has all the topological properties of the cube's skeleton. For example, a fly cannot walk along all the edges of a cube in a continuous path without going over an edge twice, nor can it do this on the projected flat network.

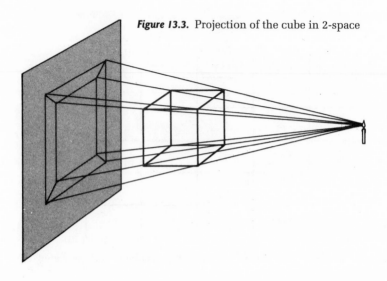

Figure 13.3. Projection of the cube in 2-space

SOLID GEOMETRY AND HIGHER DIMENSIONS

Figure 13.4 is the analogous projection in 3-space of the edges of a tesseract; more accurately, it is a plane projection of a three-dimensional model that is in turn a projection of the hypercube. All the elements of the tesseract given by the chart are easily identified in the model, although six of the eight cubes suffer perspective distortions just as four of the cube's square faces are distorted in its projection on the plane. The eight cubes are the large cube, the small interior cube, and the six hexahedrons surrounding the small cube. (Readers should also try to find the eight cubes in Figure 13.1(*d*)—a projection of the tesseract, from a different angle, into another 3-space model.) Here again the topological properties of both models are the same as those of the edges of the tesseract. In this case a fly *can* walk along all the edges without traversing any edge twice. (In general the fly can do this only on hypercubes in even spaces, because only in even spaces do an even number of edges meet at each vertex.)

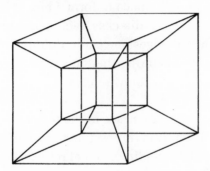

Figure 13.4. Projection of the tesseract in 3-space

Many properties of unit hypercubes can be expressed in simple formulas that apply to hypercubes of all dimensions. For example, the diagonal of a unit square has a length of $\sqrt{2}$. The longest diagonals on the unit cube have a length of $\sqrt{3}$. In general a diagonal from corner to opposite corner on a unit cube in n-space is $\sqrt{n}$.

A square of side x has an area of x^2 and a perimeter of $4x$. What size square has an area equal to its perimeter? The equation $x^2 = 4x$ gives x a value of 4. The unique answer is therefore a square of side 4. What size cube has a volume equal to its surface area? After the reader has answered this easy question he should have no difficulty answering two more: (1) What size hypercube has a hypervolume (measured by unit hypercubes) equal to the volume (measured by unit cubes) of its hypersurface? (2) What is the formula for the edge of an n-cube whose n-volume is equal to the $(n-1)$-volume of its "surface"?

Puzzle books often ask questions about cubes that are easily asked about the tesseract but not so easily answered. Consider the longest line that will fit inside a unit square. It is obviously the diagonal, with a length of $\sqrt{2}$. What is the largest square that will fit inside a unit cube? If the reader succeeds in answering this rather tricky question, and if he learns his way around in 4-space, he might try the more difficult problem of finding the largest cube that can be fitted into a unit tesseract.

An interesting combinatorial problem involving the tesseract is best approached, as usual, by first considering the analogous problems for the square and cube. Cut open one corner of a square (see top drawing in Figure 13.5) and its four lines can be unfolded as shown to form a one-dimensional figure. Each line rotates around a point until all are in the same 1-space. To unfold a cube, think of it as formed of squares joined at their edges; cut seven edges and the squares can be unfolded (bottom drawing) until they all lie in 2-space to form a hexomino (six unit squares joined at their edges). In this case each square rotates around an edge. By cutting different edges one can unfold the cube to make different hexomino shapes. Assuming that an asymmetric hexomino and its mirror image are the same, how many different hexominoes can be formed by unfolding a cube?

The eight cubes that form the exterior surface of the tesseract can be cut and unfolded in similar fashion. It is impossible to visualize how a 4-space person might "see" (with three-dimensional retinas?) the hollow tesseract. Nevertheless, the eight cubes that bound it are true surfaces in the sense that the hyperperson can touch any point inside any cube with the point of a hyperpin without the pin's passing through any other point in the cube, just as we, with a pin, can touch any point inside any square face of a cube without the pin's going through any other point on that face. Points are "inside" a cube only to *us*. To a hyperperson every point in each cubical "face" of a tesseract is directly exposed to his vision as he turns the tesseract in his hyperfingers.

Even harder to imagine is the fact that a cube in 4-space will rotate around any of its *faces*. The eight cubes that bound the tesseract are joined at their faces. Indeed, each of the 24 squares in the tesseract is a joining spot for two cubes, as can easily be verified by studying the 3-space models. If 17 of these 24 squares are cut, separating the pair of cubes attached at that spot, and if these cuts are made at the right places, the eight cubes will be free to rotate around the seven uncut

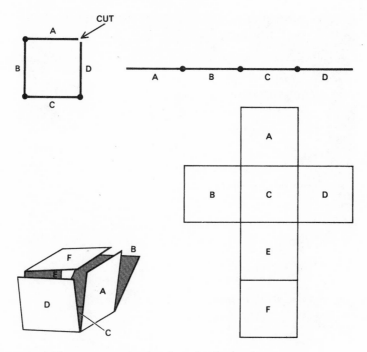

Figure 13.5. Unfolding a square *(top)* and a cube *(bottom)*

squares where they remain attached until all eight are in the same 3-space. They will then form an order-8 polycube (eight cubes joined at their faces).

Salvador Dali's painting "Corpus Hypercubus" (Figure 13.6 owned by the Metropolitan Museum of Art) shows a hypercube unfolded to form a cross-shaped polycube analogous to the cross-shaped hexomino. Observe how Dali has emphasized the contrast between 2-space and 3-space by suspending his polycube above a checkerboard and by having a distant light cast shadows of Christ's arms. By making the cross an un-folded tesseract Dali symbolizes the orthodox Christian belief that the death of Christ was a metahistorical event, taking place in a region transcendent to our time and 3-space and seen, so to speak, only in a crude, "unfolded" way by our limited vision. The use of Euclidean 4-space as a symbol of the "wholly other" world has long been a favorite theme of occultists such as P. D. Ouspensky as well as of several lead-ing Protestant theologians, notably the German theologian Karl Heim.

On a more mundane level the unfolded hypercube provides the gim-mick for Robert A. Heinlein's wild story "—And He Built a Crooked

Hypercubes

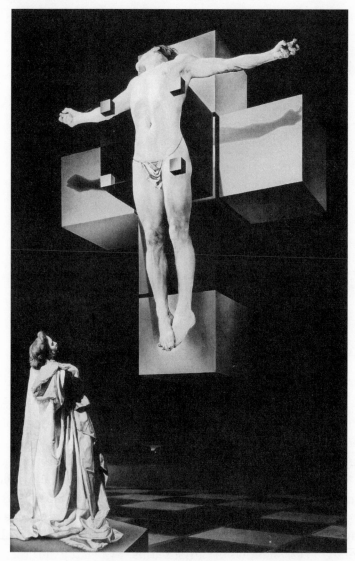

Figure 13.6. Salvador Dali's *Corpus Hypercubus,* 1954

House," which can be found in Clifton Fadiman's anthology *Fantasia Mathematica.* A California architect builds a house in the form of an unfolded hypercube, an upside-down version of Dali's polycube. When an earthquake jars the house, it folds itself up into a hollow tesseract. It appears as a single cube because it rests in our space on its cubical face just as a folded cardboard cube, standing on a plane, would appear to Flatlanders as a square. There are some remarkable adventures inside the tesseract and some unearthly views through its windows before the house, jarred by another earthquake, falls out of our space altogether.

The notion that part of our universe might fall out of 3-space is not so crazy as it sounds. The eminent American physicist J. A. Wheeler has a perfectly respectable "dropout" theory to explain the enormous energies that emanate from quasi-stellar radio sources, or quasars. When a giant star undergoes gravitational collapse, perhaps a central mass is formed of such incredible density that it puckers space–time into a blister. If the curvature is great enough, the blister could pinch together at its neck and the mass could fall out of space–time, releasing energy as it vanishes.

But back to hypercubes and one final question. How many different order-8 polycubes can be produced by unfolding a hollow hypercube into 3-space?

Addendum

Hiram Barton, a consulting engineer of Etchingham, Sussex, England, had the following grim comments to make about Hinton's colored cubes:

Dear Mr. Gardner:

A shudder ran down my spine when I read your reference to Hinton's cubes. I nearly got hooked on them myself in the nineteen-twenties. Please believe me when I say that they are completely mind-destroying. The only person I ever met who had worked with them seriously was Francis Sedlak, a Czech neo-Hegelian philosopher (he wrote a book called *The Creation of Heaven and Earth*) who lived in an Oneida-like community near Stroud, in Gloucestershire.

As you must know, the technique consists essentially in the sequential visualizing of the adjoint internal faces of the poly-colored unit cubes making up the large cube. It is not difficult to acquire considerable facility in this, but the process is one of autohypnosis and, after a while, the sequences begin to parade themselves through one's mind of their own accord. This is pleasurable, in a way, and it was not until I went to see Sedlak in 1929 that I realized the dangers of setting up an autonomous process in one's own brain. For the record, the way out is to establish consciously a countersystem differing from the first in that the core cube shows different colored faces, but withdrawal is slow and I wouldn't recommend anyone to play around with the cubes at all.

The problem of pushing a larger cube through a hole in a smaller cube is known as Prince Rupert's problem. The question seems to have first been asked by Prince Rupert (1619–1682), a nephew of England's King Charles. If you hold a cube so one corner points directly toward you, you will see a regular hexagon. The largest square that will go into a cube has a face that can be inscribed within this hexagon. Note that two of the interior square's sides can be drawn on the outside of the cube. The other two edges are within the cube.

Apparently I was the first to pose the analogous problem of determining the largest cube that would go inside a hypercube. Over the years I received many letters from readers who claimed to have answered this question. Most of the letters were too technical for me to understand, and wildy different results were claimed.

Richard Guy and Richard Nowakowski, in their feature on unsolved problems (*American Mathematical Monthly*, Vol. 104, December 1997, pp. 967–69) report on a lengthy proof by Kay R. Pechenick DeVicci of Moorestown, NJ, that the desired cube's edge is 1.007434775. . . . It is the square root of 1.014924 . . ., the smaller root of $4x^4 - 28x^3 - 7x^2 + 16x + 16$. Her paper, which includes a generalization to the largest m-cube in an n-cube, has not been published. Guy and his two collaborators on *Unsolved Problems in Geometry* (1991), Section B4, refer to DeVicci's work as having solved the problem for the 3-cube in a 4-cube, but they refer to the more general problem of n dimensions as unsolved.

When I began editing columns for this anthology I searched my files for all the letters I had received on this problem. One writer believed the answer was simply the unit cube of side 1, which turned out to be quite close to the truth.

In going more carefully through my correspondence on this problem I was astonished to find that no less than four readers had obtained the same answer as Da Vicci—a cube with an edge of 1.00743. . . . I here list them alphabetically along with the year of their letter: Hermann Baer, Post Gilboa, Israel (1974); Eugen I. Bosch, Washington, D.C. (1966); G. de Josselin de Jong, Delft, The Netherlands (1971); and Kay R. Pechenick, Lafayette Hill, PA (1983).

The problem of finding all the ways a hypercube can be "unfolded" to make distinct polycubes (unit cubes joined at their faces) was solved by Peter Turney in his 1984 paper (see bibliography). Using graph theory, Turney found 261 unfoldings. His method extends easily to hypercubes of any number of dimensions.

In the *Journal of Recreational Mathematics* (Vol. 15, No. 2. 1982–83, p. 146) Harry Nelson showed how the 11 polyominoes that fold into a cube would fit inside a 9×9 square and also inside a 7×11 rectangle. It is not known if they will fit inside a smaller rectangle.

Answers

A tesseract of side x has a hypervolume of x^4. The volume of its hypersurface is $8x^3$. If the two magnitudes are equal, the equation gives x a value of 8. In general an n-space "cube" with an n-volume equal to the $(n-1)$-volume of its "surface" is an n-cube of side $2n$.

The largest square that can be fitted inside a unit cube is the square shown in Figure 13.7. Each corner of the square is a distance of ¼ from a corner of the cube. The square has an area of exactly 9/8 and a side that is three-fourths of the square root of 2. Readers familiar with the old problem of pushing the largest possible cube through a square hole in a smaller cube will recognize this square as the cross section of the limiting size of the square hole. In other words, a cube of side not quite three-fourths of the square root of 2 can be pushed through a square hole in a unit cube.

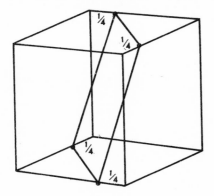

Figure 13.7. Packing a square in a cube

Figure 13.8 shows the 11 different hexominoes that fold into a cube. They form a frustrating set, because they will not fit together to make any of the rectangles that contain 66 unit squares, but perhaps there are some interesting patterns they *will* form.

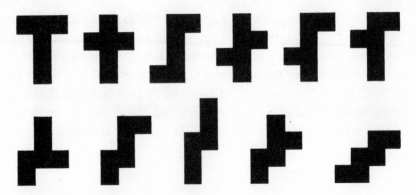

Figure 13.8. The 11 hexominoes that fold into cubes

Bibliography

H.S.M. Coxeter, *Regular Polytopes,* Dover, 3e, 1973.

M. Gardner, "And He Built Another Crooked House," *Puzzles From Other Worlds,* Vintage, 1984.

K. Heim, *Christian Faith and Natural Science,* Harper Torchbook, 1957.

C. H. Hinton, *A New Era of Thought,* Swan Sonnenschein, 1888.

C. H. Hinton, *The Fourth Dimension,* Allen & Unwin, 1904.

R. A. Jacobson, "Paths of Minimal Length Within Hypercubes," *The American Mathematical Monthly,* Vol. 73, October 1966, pp. 868–72.

H. P. Manning, *Geometry of Four Dimensions,* Dover, 1956.

H. P. Manning, *The Fourth Dimension Simply Explained,* Dover, 1960.

E. H. Neville, *The Fourth Dimension,* Cambridge University Press, 1921.

R. Rucker (ed), *Speculations on the Fourth Dimension: Selected Writings of C. H. Hinton,* Dover, 1980.

R. Rucker, *Infinity and the Mind,* Birkäuser, 1982.

R. Rucker, *The Fourth Dimension,* Houghton Mifflin, 1984.

D.M.Y. Sommerville, *An Introduction to the Geometry of N Dimensions,* Dover, 1958.

W. S. Tevis, Jr., "The Ifth of Oofth," *Galaxy Science Fiction,* April, 1957, pp. 59–69. A wild and funny tale about a five-dimensional cube that distorts space and time.

P. Turney, "Unfolding the Tesseract," *Journal of Recreational Mathematics,* Vol. 17, No. 1, 1984–85, pp. 1–16.

Chapter 14

Non-Euclidean Geometry

"Lines that are parallel
  meet at Infinity!"
Euclid repeatedly,
heatedly,
    urged.
Until he died,
and so reached that vicinity:
in it he
found that the damned things
    diverged.

—Piet Hein, *Grooks VI*

*E*uclid's *Elements* is dull, long-winded, and does not make explicit the fact that two circles can intersect, that a circle has an outside and an inside, that triangles can be turned over, and other assumptions essential to his system. By modern standards Bertrand Russell could call Euclid's fourth proposition a "tissue of nonsense" and declare it a scandal that the *Elements* was still used as a textbook.

On the other hand, Euclid's geometry was the first major effort to organize the subject as an axiomatic system, and it seems hardly fair to find fault with him for not anticipating all the repairs made when David Hilbert and others formalized the system. There is no more striking evidence of Euclid's genius than his realization that his notorious fifth postulate was not a theorem but an axiom that had to be accepted without proof.

Euclid's way of stating the postulate was rather cumbersome, and it was recognized early that it could be given the following simpler form: Through a point on a plane, not on a given straight line, only one line is parallel to the given line. Because this is not quite as intuitively obvious as Euclid's other axioms mathematicians tried for 2,000 years to remove the postulate by making it a theorem that could be established on the basis of Euclid's other axioms. Hundreds of proofs were attempted. Some eminent mathematicians thought they had succeeded, but it always turned out that somewhere in their proof an assumption had been made that either was equivalent to the parallel postulate or required the postulate.

For example, it is easy to prove the parallel postulate if you assume that the sum of the angles of every triangle equals two right angles. Unfortunately you cannot prove this assumption without using the parallel postulate. An early false proof, attributed to Thales of Miletus, rests on the existence of a rectangle, that is, a quadrilateral with four right angles. You cannot prove, however, that rectangles exist without using the parallel postulate! In the 17th century John Wallis, a renowned English mathematician, believed he had proved the postulate. Alas, he failed to realize that his assumption that two triangles can be similar but not congruent cannot be proved without the parallel postulate. Long lists can be made of other assumptions, all so intuitively obvious that they hardly seem worth asserting, and all equivalent to the parallel postulate in the sense that they do not hold unless the postulate holds.

In the early 19th century trying to prove the postulate became something of a mania. In Hungary, Farkas Bolyai spent much of his life at the task, and in his youth he discussed it often with his German friend Karl Friedrich Gauss. Farkas' son János became so obsessed by the problem that his father was moved to write in a letter: "For God's sake, I beseech you, give it up. Fear it no less than sensual passions because it too may take all your time and deprive you of your health, peace of mind and happiness in life."

János did not give it up, and soon he became persuaded not only that the postulate was independent of the other axioms but also that a consistent geometry could be created by assuming that through the point an infinity of lines were parallel to the given line. "Out of nothing I have created a new universe," he proudly wrote to his father in 1823.

Farkas at once urged his son to let him publish these sensational claims in an appendix to a book he was then completing. "If you have really succeeded, it is right that no time be lost in making it public, for two reasons: first, because ideas pass easily from one to another who can anticipate its publication; and secondly, there is some truth in this, that many things have an epoch in which they are found at the same time in several places, just as the violets appear on every side in spring. Also every scientific struggle is just a serious war, in which I cannot say when peace will arrive. Thus we ought to conquer when we are able, since the advantage is always to the first comer."

János' brief masterpiece did appear in his father's book, but as it happened the publication of the book was delayed until 1832. The Russian mathematician Nikolai Ivanovitch Lobachevski had beat him to it by disclosing details of the same strange geometry (later called by Felix Klein hyperbolic geometry) in a paper of 1829. What is worse, when Farkas sent the appendix to his old friend Gauss, the Prince of Mathematicians replied that if he praised the work, he would only be praising himself, inasmuch as he had worked it all out many years earlier but had published nothing. In other letters he gave his reason. He did not want to arouse an "outcry" among the "Boeotians," by which he meant his conservative colleagues. (In ancient Athens the Boeotians were considered unusually stupid.)

Crushed by Gauss's response, János even suspected that his father might have leaked his marvelous discovery to Gauss. When he later learned of Lobachevski's earlier paper, he lost interest in the topic and published nothing more. "The nature of real truth of course cannot but be one and the same in Marcos-Vasarhely as in Kamchatka and on the moon," he wrote, resigned to having published too late to win the honor for which he had so passionately hoped.

In some ways the story of the Italian Jesuit Giralamo Saccheri is even sadder than that of Bolyai. As early as 1733, in a Latin book called *Euclid Cleared of All Blemish,* Saccheri actually constructed both types of non-Euclidean geometry (we shall come to the second type below) without knowing it! Or so it seems. At any rate Saccheri refused to believe either geometry was consistent, but he came so close to accepting them that some historians think he pretended to disbelieve them just to get his book published. "To have claimed that a non-Euclidean system was as 'true' as Euclid's," writes Eric Temple Bell (in a chapter on Saccheri in *The Magic of Numbers*), "would have been a foolhardy invitation to repression and discipline. The Copernicus of Geometry therefore resorted to subterfuge. Taking a long chance, Saccheri denounced his own work, hoping by this pious betrayal to slip his heresy past the censors."

I cannot resist adding two anecdotes about the Bolyais. János was a cavalry officer (mathematics had always been strictly a recreation) known for his swordsmanship, his skill on the violin, and his hot temper. He is said to have once challenged 13 officers to duels, provided that after each victory he would be allowed to play to the loser a piece on his violin. The elder Bolyai is reported to have been buried at his own request under an apple tree, with no monument, to commemorate history's three most famous apples: the apple of Eve, the golden apple

Paris gave Venus as a beauty-contest prize, and the falling apple that in-spired Isaac Newton.

Before the 19th century had ended it became clear that the parallel postulate not only was independent of the others but also that it could be altered in two opposite ways. If it was replaced (as Gauss, Bolyai, and Lobachevski had proposed) by assuming an infinite number of "ul-traparallel" lines through the point, the result would be a new geome-try just as elegant and as "true" as Euclid's. All Euclid's other postulates remain valid; a "straight" line is still a geodesic, or shortest line. In this hyperbolic space all triangles have an angle sum less than 180 degrees, and the sum decreases as triangles get larger. All similar polygons are congruent. The circumference of any circle is greater than pi times the diameter. The measure of curvature of the hyperbolic plane is negative (in contrast to the zero curvature of the Euclidean plane) and every-where the same. Like Euclidean geometry, hyperbolic geometry gener-alizes to 3-space and all higher dimensions.

The second type of non-Euclidean geometry, which Klein names "el-liptic," was later developed simultaneously by the German mathe-matician Georg Friedrich Bernhard Riemann and the Swiss mathematician Ludwig Schläfli. It replaces the parallel postulate with the assumption that through the point *no* line can be drawn parallel to the given line. In this geometry the angle sum of a triangle is always more than 180 degrees, and the circumference of a circle is always less than pi times the diameter. Every geodesic is finite and closed. The lines in every pair of geodesics cross.

To prove consistency for the two new geometries various Euclidean models of each geometry were found showing that if Euclidean geom-etry is consistent, so are the other two. Moreover, Euclidean geometry has been "arithmetized," proving that if arithmetic is consistent, so too is Euclid's geometry. We now know, thanks to Kurt Gödel, that the con-sistency of arithmetic is not provable in arithmetic, and although there are consistency proofs for arithmetic (such as the famous proof by Ger-hard Gentzen in 1936), no such proof has yet been found that can be considered entirely constructive by an intuitionist (see A. Calder, "Con-structive Mathematics," *Scientific American,* October 1979). God ex-ists, someone once said, because mathematics is consistent, and the Devil exists because we are not able to prove it.

The various metaproofs of arithmetic's consistency, as Paul C. Rosen-bloom has put it, may not have eliminated the Devil, but they have reduced the size of hell almost to zero. In any case no mathematician today expects arithmetic (therefore also Euclidean and non-Euclidean geometries) ever to produce a contradiction. Curiously, Lewis Carroll was one of the last mathematicians to doubt non-Euclidean geometry. "It is a strange paradox," the geometer H.S.M. Coxeter has written, "that he, whose Alice in Wonderland could alter her size by eating a little cake, was unable to accept the possibility that the area of a triangle could remain finite when its sides tend to infinity."

What Coxeter had in mind can be grasped by studying M. C. Escher's *Circle Limit III,* reproduced in Figure 14.1. This 1959 woodcut (one of Escher's rare works with several colors in one picture) is a tessellation based on a Euclidean model of the hyperbolic plane that was constructed by Henri Poincaré. In Poincaré's ingenious model every point on the Euclidean plane corresponds to a point inside (but not on) the circle's circumference. Beyond the circle there is, as Escher put it, "absolute nothingness."

Figure 14.1.

Imagine that Flatlanders live on this model. As they move outward from the center their size seems to us to get progressively smaller, although they are unaware of any change because all their measuring instruments similarly get smaller. At the boundary their size would become zero, but they can never reach the boundary. If they proceed toward it with uniform velocity, their speed (to us) steadily decreases, although to them it seems constant. Thus their universe, which we see as being finite, is to them infinite. Hyperbolic light follows geodesics, but because its velocity is proportional to its distance from the boundary it takes paths that we see as circular arcs meeting the boundary at right angles.

In this hyperbolic world a triangle has a maximum finite area, as is shown in Figure 14.2, although its three "straight" sides go to infinity in hyperbolic length and its three angles are zero. You must not think of Escher's mosaic as being laid out on a sphere. It is a circle enclosing an infinity of fish—Coxeter calls it a "miraculous draught"—that get progressively smaller as they near the circumference. In the hyperbolic plane, of which the picture is only a model, the fish are all identical in size and shape. It is important to remember that the creatures of a hyperbolic world would not change in shape as they moved about, light would not change in speed, and the universe would be infinite in all directions.

Figure 14.2.

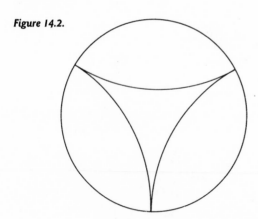

The curved white lines in Escher's woodcut do not, as many people have supposed, model hyperbolic geodesics. The lines are called equidistant curves or hypercycles. Each line has a constant perpendicular distance (measured hyperbolically) from the hyperbolic straight line that joins the arc's ends. Note that along each white curve fish of the same color swim head to tail. If you consider all the points where four fins meet, these points are the vertexes of a regular tiling of the hyperbolic plane by equilateral triangles with angles of 45 degrees. The centers of the triangles are the points where three left fins meet and three mouths touch three tails. The 45-degree angles make it possible for eight triangles to surround each vertex, where in a Euclidean tiling by equilateral triangles only six triangles can surround each vertex.

Escher and Coxeter had corresponded from the time they met in 1954, and Escher's interest in tilings of the hyperbolic plane had been aroused by the illustrations in a 1957 paper on crystal symmetry that Coxeter had written and sent to him. In a lovely article titled "The Non-Euclidean Symmetry of Escher's Picture 'Circle Limit III' " (in the journal *Leonardo,* Vol. 12, 1979, pp. 19–25) Coxeter shows that each white arc meets the boundary at an angle of almost 80 degrees. (The precise value is $2^{7/4} + 2^{5/4}$ arc secants.) Coxeter considers *Circle Limit III* the most mathematically sophisticated of all Escher's pictures. It even anticipated a discovery Coxeter did not make until five years after the woodcut was finished!

Elliptic geometry is roughly modeled by the surface of a sphere. Here Euclidean straight lines become great circles. Clearly no two can be parallel, and it is easy to see that triangles formed by arcs of great circles must have angles that add up to more than two right angles. The hyperbolic plane is similarly modeled by the saddle-shaped surface of a pseudosphere, generated by rotating a tractrix about its asymptote.

It is a misuse of the word "crank" to apply it to mathematicians who erred in thinking, before the independence of the parallel postulate was established, that they had proved the postulate. The same cannot be said of those amateurs of later decades who could not understand the proofs of the postulate's independence or who were too egotistical to try. Augustus De Morgan, in his classic compendium of eccentric mathematics, *A Budget of Paradoxes,* introduces us to Britain's most indefatigable 19th-century parallel-postulate prover, General Perronet Thompson. Thompson kept issuing revisions of his many proofs (one was based on the equiangular spiral), and although De Morgan did his best to dissuade him from his futile efforts, he was unsuccessful. Thompson also wanted to replace the tempered scale of the piano with an octave of 40 notes.

The funniest of the American parallel-postulate provers was the Very Reverend Jeremiah Joseph Callahan, then president of Duquesne University in Pittsburgh. In 1931, when Father Callahan announced he had trisected the angle, *Time* gave the story sober treatment and ran his photograph. The following year Callahan published his major work, *Euclid or Einstein: A Proof of the Parallel Theory and a Critique of Metageometry* (Devon-Adair, 1932), a 310-page treatise in which he ascended to heights of argumentem ad hominem. Einstein is "fuddled," he "has not a logical mind," he is in a "mental fog," he is a "careless thinker." "His thought staggers, and reels, and stumbles, and falls, like a blind man rushing into unknown territory." "Sometimes one feels like laughing," Callahan wrote, "and sometimes one feels a little irritated. . . . But there is no use expecting Einstein to reason."

What Callahan found so irritating was Einstein's adoption of a generalized non-Euclidean geometry, formulated by Riemann, in which the curvature of physical space varies from point to point depending on the influence of matter. One of the great revolutions brought about by relativity theory was the discovery that an enormous overall simplification of physics is obtained by assuming physical space to have this kind of non-Euclidean structure.

It is now commonplace (how astonished, and I think delighted, Kant would have been by the notion!) to recognize that all geometric systems are equally "true" in the abstract but that the structure of physical space must be determined empirically. Gauss himself thought of triangulating three mountain peaks to see if their angles added up to two right angles. It is said he actually made such a test, with inconclusive results. Although experiments can prove physical space is non-Euclidean, it is a curious fact that there is no way to prove it is Euclidean! Zero curvature is a limiting case, midway between elliptic and hyperbolic curvatures. Since all measurement is subject to error, the deviation from zero could always be too slight for detection.

Poincaré held the opinion that if optical experiments seemed to show physical space was non-Euclidean, it would be best to preserve the simpler Euclidean geometry of space and assume that light rays do not follow geodesics. Many mathematicians and physicists, including Russell, agreed with Poincaré until relativity theory changed their mind. Alfred North Whitehead was among the few whose mind was never changed. He even wrote a book on relativity, now forgotten, in which he argued for preserving a Euclidean universe (or at least one of constant curvature) and modifying the physical laws as necessary. (For a discussion of Whitehead's controversy with Einstein, see Robert M.

Palter's *Whitehead's Philosophy of Science,* University of Chicago Press, 1960.)

Physicists are no longer disturbed by the notion that physical space has a generalized non-Euclidean structure. Callahan was not merely disturbed; he was also convinced that all non-Euclidean geometries are self-contradictory. Einstein, poor fellow, did not know how easy it is to prove the parallel postulate. If you are curious about how Callahan did it, and about his elementary error, see D. R. Ward's paper in *The Mathematical Gazette* (Vol. 17, May 1933, pp. 101–4).

Like their cousins who trisect the angle, square the circle, and find simple proofs of Fermat's last theorem, the parallel-postulate provers are a determined breed. A more recent example is William L. Fischer of Munich, who in 1959 published a 100-page *Critique of Non-Euclidean Geometry.* Ian Stewart exposed its errors in the British journal *Manifold* (No. 12, Summer 1972, pp. 14–21). Stewart quotes from a letter in which Fischer accuses establishment mathematicians of suppressing his great work and orthodox journals of refusing to review it: "The university library at Cambridge refused even to put my booklet on file. . . . I had to write to the vice-chancellor to overcome this boycott."

There are, of course, no sharp criteria for distinguishing crank mathematics from good mathematics, but then neither are there sharp criteria for distinguishing day from night, life from non-life, and where the ocean ends and the shore begins. Without words for parts of continuums we could not think or talk at all. If you, dear reader, have a way to prove the parallel postulate, don't tell me about it!

Addendum

Imagine a small circle around the north pole of the earth. If it keeps expanding, it reaches a maximum size at the equator, after which it starts to contract until it finally becomes a point at the south pole. In similar fashion, an expanding sphere in four-dimensional elliptical space reaches a maximum size, then contracts to a point.

In addition to the three geometries described in this chapter, there is what Bolyai called "absolute geometry" in which theorems are true in all three. It is astonishing that the first 28 theorems of Euclid's *Elements* are in this category, along with other novel theorems that Bolyai showed to be independent of the parallel postulate.

I was surprised to see in a 1984 issue of *Speculations in Science and Technology* (Vol. 7, pp. 207–16), a defense of Father Callahan's proof of the parallel postulate! The authors are Richard Hazelett, vice president of the Hazelett Strip-Casting Corporation, Colchester, VT, and Dean E. Turner, who teaches at the University of North Colorado, in Greeley. Hazelett is a mechanical engineer with master's degrees from the University of Texas and Boston University. Taylor, an ordained minister in the Disciples of Christ Church, has a doctorate from the University of Texas.

It is easy to understand why both men do not accept Einstein's general theory of relativity. Indeed, they have edited a book of papers attacking Einstein. Titled *The Einstein Myth and the Ives Papers,* it was published in 1979 by Devin-Adair.

In an earlier column on geometrical fallacies, reprinted in *Wheels, Life, and Other Mathematical Amusements* (1983), I discussed in detail Father Callihan's false "proof" of the parallel postulate. For other false proofs, see Underwood Dudley's book cited in the bibliography.

In the *Encyclopaedia Britannica's* tenth edition Bertrand Russell contributed an article on "Geometry, Non-Euclidean." It was revised for the eleventh edition (1910) with Whitehead's name added as coauthor. They argued that at present, on grounds of simplicity, if an experiment ever contradicted Euclidean geometry it would be preferable to question the experiment rather than give up Euclidean geometry. Of course they wrote this before relativity theory proved otherwise. Poincaré's objection to non-Euclidean geometry was unqualified. Russell changed his mind in the light of relativity theory, but if Whitehead ever changed his mind I've been unable to find evidence for it.

Bibliography

R. Bonola, *Non-Euclidean Geometry,* Open Court, 1912; Dover, 1955.

E. Breitenberger, *"Gauss's Geodesy and the Axiom of Parallels,"* Archive for History of Exact Sciences, Vol. 31, 1984, pp. 273–89.

H.S.M. Coxeter, *Regular Compound Tessellations of the Hyperbolic Plane,* Proceedings of the Royal Society, A, Vol. 278, 1964, pp. 147–67.

H.S.M. Coxeter, *Non-Euclidean Geometry,* 5e, University of Toronto Press, 1965.

H.S.M. Coxeter, *The Non-Euclidean Symmetry of Escher's Picture 'Circle Limit III.'* Leonardo, Vol. 12, 1979, pp. 19–25.

U. Dudley, *Euclid's Fifth Postulate, Mathematical Cranks,* Mathematical Association of America, 1992, pp. 137–58.

S. Gindikin, *The Wonderland of Poincaria, Quantum,* November/December 1992, pp. 21–28.

S. H. Gould, *The Origin of Euclid's Axioms, Mathematical Gazette,* Vol. 46, December, 1962, pp. 269–90.

M. J. Greenberg, *Euclidean and Non-Euclidean Geometries: Development and History, 3e,* W. H. Freeman, 1994.

S. Kulczycki, *Non-Euclidean Geometry,* Macmillan, 1961.

W. W. Maiers, *Introduction to Non-Euclidean Geometry, Mathematics Teacher,* November 1964, pp. 457–61.

R. M. Palter, *Whitehead's Philosophy of Science,* University of Chicago Press, 1960. Reviewed by Adolf Grünbaum in *The Philosophy Review,* Vol. 71, April 1962.

D. Pedoe, *Non-Euclidean Geometry, New Scientist,* No. 219, January 26, 1981, pp. 206–7.

J. F. Rigby, *Some Geometrical Aspects of a Maximal Three-Coloured Triangle-Free Graph, Journal of Combinatorial Theory, Series B,* Vol. 34, June 1983, pp. 313–22.

D.M.Y. Sommerville, *The Elements of Non-Euclidean Geometry,* Dover, 1958.

J. W. Withers, *Euclid's Parallel Postulate: Its Nature, Validity, and Place in Geometrical Systems,* Open Court, 1905.

H. E. Wolfe, *Non-Euclidean Geometry,* Henry Holt, 1945.

IV
Symmetry

MAURITS
Escher

Chapter 15   *Rotations and Reflections*

A geometric figure is said to be symmetrical if it remains unchanged after a "symmetry operation" has been performed on it. The larger the number of such operations, the richer the symmetry. For example, the capital letter *A* is unchanged when reflected in a mirror placed vertically beside it. It is said to have vertical symmetry. The capital *B* lacks this symmetry but has horizontal symmetry: it is unchanged in a mirror held horizontally above or below it. *S* is neither horizontally nor vertically symmetrical but remains the same if rotated 180 degrees (twofold symmetry). All three of these symmetries are possessed by *H, I, O,* and *X. X* is richer in symmetry than *H* or *I* because, if its arms cross at right angles, it is also unchanged by quarter-turns (fourfold symmetry). *O,* in circular form, is the richest letter of all. It is unchanged by any type of rotation or reflection.

Because the earth is a sphere toward the center of which all objects are drawn by gravity, living forms have found it efficient to evolve shapes that possess strong *vertical* symmetry combined with an obvious lack of horizontal or rotational symmetry. In making objects for his use man has followed a similar pattern. Look around and you will be struck by the number of things you see that are essentially unchanged in a vertical mirror: chairs, tables, lamps, dishes, automobiles, airplanes, office buildings—the list is endless. It is this prevalence of vertical symmetry that makes it so difficult to tell when a photograph has been reversed, unless the scene is familiar or contains such obvious clues as reversed printing or cars driving on the wrong side of the road. On the other hand, an upside-down photograph of almost anything is instantly recognizable as inverted.

The same is true of works of graphic art. They lose little, if anything, by reflection, but unless they are completely nonrepresentational no careless museum director is likely to hang one upside down. Of course, abstract paintings are often inverted by accident. *The New York Times Magazine* (October 5, 1958) inadvertently both reversed and inverted a picture of an abstraction by Piet Mondrian, but only readers who knew the painting could possibly have noticed it. In 1961, at the New York Museum of Modern Art, Matisse's painting, *Le Bateau,* hung upside down for 47 days before anyone noticed the error.

So accustomed are we to vertical symmetry, so unaccustomed to seeing things upside down, that it is extremely difficult to imagine what most scenes, pictures, or objects would look like inverted. Landscape artists have been known to check the colors of a scene by the undignified technique of bending over and viewing the landscape through their legs. Its upside-down contours are so unfamiliar that colors can be seen uncontaminated, so to speak, by association with familiar shapes. Thoreau liked to view scenes this way and refers to such a view of a pond in Chapter 9 of *Walden.* Many philosophers and writers have found symbolic meaning in this vision of a topsy-turvy landscape; it was one of the favorite themes of G. K. Chesterton. His best mystery stories (in my opinion) concern the poet-artist Gabriel Gale (in *The Poet and the Lunatics*), who periodically stands on his hands so that he can "see the landscape as it really is: with the stars like flowers, and the clouds like hills, and all men hanging on the mercy of God."

The mind's inability to imagine things upside down is essential to the surprise produced by those ingenious pictures that turn into something entirely different when rotated 180 degrees. Nineteenth-century political cartoonists were fond of this device. When a reader inverted a drawing of a famous public figure, he would see a pig or jackass or something equally insulting. The device is less popular today, although *Life* for September 18, 1950, reproduced a remarkable Italian poster on which the face of Garibaldi became the face of Stalin when viewed upside down. Children's magazines sometimes reproduce such upside-down pictures, and now and then they are used as advertising gimmicks. The back cover of *Life* for November 23, 1953, depicted an Indian brave inspecting a stalk of corn. Thousands of readers probably failed to notice that when this picture was inverted it became the face of a man, his mouth watering at the sight of an open can of corn.

I know of only four books that are collections of upside-down drawings. Peter Newell, a popular illustrator of children's books who died in 1924, published two books of color plates of scenes that undergo amusing transformations when inverted: *Topsys & Turvys* (1893) and *Topsys and Turvys Number 2* (1894). In 1946 a London publisher issued a collection of 15 astonishing upside-down faces drawn by Rex Whistler, an English muralist who died in 1944. The book has the richly symmetrical title of *¡OHO!* (Its title page is reproduced in Figure 15.1.)

Figure 15.1. Invertible faces on the title page of Whistler's invertible book

The technique of upside-down drawing was carried to unbelievable heights in 1903 and 1904 by a cartoonist named Gustave Verbeek. Each week he drew a six-panel color comic for the Sunday "funny paper" of the *New York Herald.* One took the panels in order, reading the captions beneath each picture; then one turned the page upside down and continued the story, reading a new set of captions and taking the same six panels in reverse order! (See Figure 15.2.) Verbeek managed to achieve continuity by means of two chief characters called Little Lady Lovekins and Old Man Muffaroo. Each became the other when inverted. How Verbeek managed to work all this out week after week without going mad surpasses all understanding. A collection of 25 of his comics was published by G. W. Dillingham in 1905 under the title of *The Upside-Downs of Little Lady Lovekins and Old Man Muffaroo.* The book is extremely rare.

The 90-degree rotation is less frequently used in art play, perhaps because it is easier for the mind to anticipate results. If done artfully, however, it can be effective. An example is a landscape by the 17th-

Figure 15.2. A typical upside-down cartoon by Gustave Verbeek

century Swiss painter Matthäus Merian that becomes a man's profile when the picture is given a quarter-turn counterclockwise. The rabbit–duck in Figure 15.3 is the best-known example of a quarter-turn picture. Psychologists have long used it for various sorts of testing. Harvard philosopher Morton White once reproduced a rabbit–duck drawing in a magazine article to symbolize the fact that two historians can survey the same set of historical facts but see them in two essentially different ways.

Our lifelong conditioning in the way we see things is responsible for a variety of startling upside-down optical illusions. All astronomers know the necessity of viewing photographs of the moon's surface so that sunlight appears to illuminate the craters from above rather than below. We are so unaccustomed to seeing things illuminated from below that when such a photograph of the moon is inverted, the craters instantly appear to be circular mesas rising above the surface. One of the most amusing illusions of this same general type is shown in Figure 15.4. The missing slice of pie is found by turning the picture upside down. Here again the explanation surely lies in the fact that we almost always see plates and pies from above and almost never from below.

Upside-down faces could not be designed, of course, if it were not for

Figure 15.3. A quarter-turn clockwise makes the duck into a rabbit.

Figure 15.4. Where is the missing slice?

the fact that our eyes are not too far from midway between the top of the head and the chin. School children often amuse themselves by turning a history book upside down and penciling a nose and mouth on the forehead of some famous person.

When this is done on an actual face, using eyebrow pencil and lipstick, the effect becomes even more grotesque. It was a popular party pastime of the late 19th century. The following account is from an old book entitled *What Shall We Do Tonight?*

> The severed head always causes a sensation and should not be suddenly exposed to the nervous. . . . A large table, covered with a cloth sufficiently long to reach to the floor all around and completely hide all beneath, is placed in the center of the room. . . . A boy with soft silky hair, rather long, being selected to represent the *head,* must lie upon his back under the table entirely concealed, excepting that portion of his face above the bridge of his nose. The rest is under the tablecloth.
>
> His hair must now be carefully combed down, to represent whiskers, and a face must be painted . . . upon the cheeks and forehead; the false eyebrows, nose and mouth, with mustache, must be strongly marked with black water color, or India ink, and the real eyebrows covered with a little powder or flour. The face should also be powdered to a deathlike pallor. . . .
>
> The horror of this illusion may be intensified by having a subdued light in the room in which the exhibition has been arranged. This conceals in a great degree any slight defects in the "making-up" of the head. . . .

Needless to add, the horror is heightened when the "head" suddenly opens its eyes, blinks, stares from side to side, wrinkles its cheeks (forehead).

The physicist Robert W. Wood (author of *How to Tell the Birds from the Flowers*) invented a funny variation of the severed head. The face is viewed upside down as before, but now it is the forehead, eyes, and nose that are covered, leaving only the mouth and chin exposed. Eyes and nose are drawn on the chin to produce a weird little pinheaded creature with a huge, flexible mouth. The stunt was a favorite of Paul Winchell, the television ventriloquist. He wore a small dummy's body on his head to make a figure that he called Ozwald, while television camera techniques inverted the screen to bring Ozwald right side up. In 1961 an Ozwald kit was marketed for children, complete with the dummy's body and a special mirror with which to view one's own face upside down.

It is possible to print or even write in longhand certain words in such a way that they possess twofold symmetry. The Zoological Society of San Diego, for instance, publishes a magazine called *ZOONOOZ*, the name of which is the same upside down. The longest sentence of this type that I have come across is said to be a sign by a swimming pool designed to read the same when viewed by athletes practicing handstands: NOW NO SWIMS ON MON. (See Figure 15.5.)

sketch reproduced by courtesy of the artist, John McClellan

Figure 15.5. An invertible sign

It is easy to form numbers that are the same upside down. As many have noticed, 1961 is such a number. It was the first year with twofold symmetry since 1881, the last until 6009, and the twenty-third since the year 1. Altogether there are 38 such years between A.D. 1 and A.D. 10000 (according to a calculation made by John Pomeroy), with the longest interval between 1961 and 6009. J. F. Bowers, writing in the *Mathemati-*

cal Gazette for December 1961, explains his clever method of calculating that by A.D. 1000000 exactly 198 invertible years will have passed. The January 1961 issue of *Mad* featured an upside-down cover with the year's numerals in the center and a line predicting that the year would be a mad one.

Some numbers, for example 7734 (when the 4 is written so that it is open at the top), become words when inverted; others can be written to become words when reflected. With these quaint possibilities in mind, the reader may enjoy tackling the following easy problems:

1. Oliver Lee, age 44, who lives at 312 Main Street, asked the city to give his car a license plate bearing the number 337-31770. Why?

2. Prove the sum in Figure 15.6 to be correct.

$$
\begin{array}{r}
3414 \\
340 \\
74813 \\
\hline
43374813
\end{array}
$$

Figure 15.6. Is the sum correct?

3. Circle six digits in the group below that will add up to exactly 21.

1	1	1
3	3	3
5	5	5
9	9	9

4. A basket contains more than half a dozen eggs. Each egg is either white or brown. Let x be the number of white eggs, and y be the number of brown. The sum of x and y, turned upside down, is the product of x and y. How many eggs are in the basket?

Addendum

George Carlson, art editor of *John Martin's Book,* a monthly magazine for children that flourished in the 1920s, contributed some dozen upside-down pictures to the magazine. Many other examples can be found in several books on optical illusions and related pictures by England's Keith Kay. In 1980 Lothrop published *The Turn About, Think About, Look About Book,* by Beau Gardner. It features pictures to be viewed both upside down and after quarter-turns.

In 1963 George Naimark privately published, under the imprint of Rajah Press, a collection of Verbeek's comic strips. It was followed by an edition in Japan that was fully colored.

Salvador Dali painted a number of pictures that changed when inverted or rotated 90 degrees. His frontispiece for *The Maze,* by Maurice Yves Sandoz (1945), is the face of a man. Upside down it becomes a frog on the rim of a bowl. In 1969 he designed an ashtray to be given away on Air-India flights. Three elephant heads surrounding the tray become swans when the tray is inverted. See a full-page ad for the gift in *The New York Times,* June 27, 1969, p. 38. A 1937 painting by Dali shows

Figure 15.7. Rotate the top picture 90 degrees to the left, and the bottom landscape 90 degrees to the right.

three swans on a lake. Their reflections in the water are three elephants. The landscape is reproduced in color in Richard Gregory's book on mirrors, *Mirrors in Mind* (1997).

Landscapes that turn into faces when rotated 90 degrees were popular among German painters of the Renaissance. Two examples are reproduced in Figure 15.7.

Answers

1. The number 337-31770 upside down spells "Ollie Lee."

2. Hold the sum to a mirror.

3. Turn the picture upside down, circle three 6's and three 1's to make a total of 21.

4. The basket has nine white eggs and nine brown eggs. When the sum, 18, is inverted, it becomes 81, the product. Had it not been specified that the basket contained more than six eggs, three white and three brown would have been another answer.

Chapter 16

The Amazing
Creations of Scott Kim

Scott Kim's *Inversions,* published in 1981 by Byte Publications, is one of the most astonishing and delightful books ever printed. Over the years Kim has developed the magical ability to take just about any word or short phrase and letter it in such a way that it exhibits some kind of striking geometrical symmetry. Consider Kim's lettering of my name in Figure 16.1. Turn it upside down and presto! It remains exactly the same!

Figure 16.1.

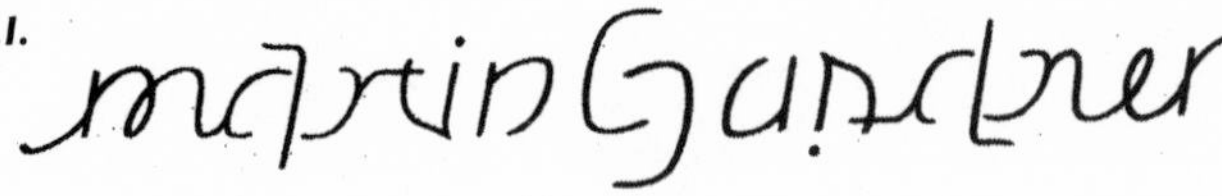

Students of curious wordplay have long recognized that short words can be formed to display various types of geometrical symmetry. On the Rue Mozart in Paris a clothing shop called "New Man" has a large sign lettered "NeW MaN" with the *e* and the *a* identical except for their orientation. As a result the entire sign has upside-down symmetry. The names VISTA (the magazine of the United Nations Association), ZOONOOZ (the magazine of the San Diego zoo) and NISSIN (a Japanese manufacturer of camera flash equipment) are all cleverly designed so that they have upside-down symmetry.

BOO HOO, DIOXIDE, EXCEEDED, and DICK COHEN DIED 10 DEC 1883 all have mirror symmetry about a horizontal axis. If you hold them upside down in front of a mirror, they appear unchanged. One day in a supermarket my sister was puzzled by the name on a box of crackers, "spep oop," until she realized that a box of "doo dads" was on the shelf upside down. Wallace Lee, a magician in North Carolina, liked to amuse friends by asking if they had ever eaten any "ittaybeds," a word he printed on a piece of paper like this:

ı┼┼ƏYbeds

After everyone said no, he would add:

"Of course, they taste much better upside down."

Many short words in conventional typefaces turn into other words when they are inverted. MOM turns into WOW and "up" becomes the abbreviation "dn." SWIMS remains the same. Other words have mirror symmetry about a vertical axis, such as "bid" (and "pig" if the *g* is drawn as a mirror image of the *p*). Here is an amusing way to write "minimum" so that it is the same when it is rotated 180 degrees:

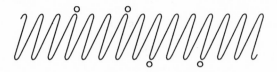

It is Kim who has carried this curious art of symmetrical calligraphy to heights not previously known to be possible. By ingeniously distorting letters, yet never so violently that one cannot recognize a word or phrase, Kim has produced incredibly fantastic patterns. His book is a collection of such wonders, interspersed with provocative observations on the nature of symmetry, its philosophical aspects, and its embodiment in art and music as well as in wordplay.

Kim is no stranger to my *Scientific American* columns. He is of Korean descent, born in the U.S., who in 1981 was doing graduate work in computer science at Stanford University. He was in his teens when he began to create highly original problems in recreational mathematics. Some that have been published in *Scientific American* include his "lost-king tours" (April 1977), the problem of placing chess knights on the corners of a hypercube (February 1978), his solution to "boxing a box" (February 1979), and his beautifully symmetrical "*m*-pire map" given in Chapter 6 of my *Last Recreations.* In addition to a remarkable ability to think geometrically (not only in two and three dimensions but also in 4-space and higher spaces) Kim is a classical pianist who for years could not decide between pursuing studies in mathematics or in music. In the early 1980s he was intensely interested in the use of computers for designing typefaces, a field pioneered by his friend and mentor at Stanford, the computer scientist Donald E. Knuth.

For several years Kim's talent for lettering words to give them unexpected symmetries was confined to amusing friends and designing family Christmas cards. He would meet a stranger at a party, learn his or her name, then vanish for a little while and return with the name neatly drawn so that it would be the same upside down. His 1977 Christmas card, with upside-down symmetry, is shown in Figure 16.2. (Lester and Pearl are his father and mother; Grant and Gail are his brother and sister.) The following year he found a way to make "Merry Christmas, 1978," mirror-symmetrical about a horizontal axis, and in 1979 he made the mirror axis vertical. (See Figures 16.3 and 16.4.)

Figure 16.2.

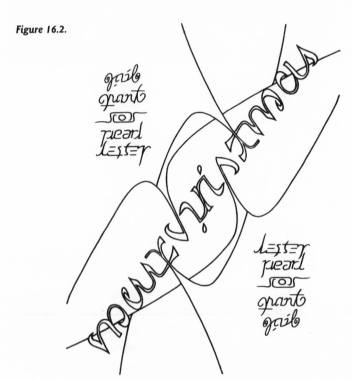

For a wedding anniversary of his parents Kim designed a cake with chocolate and vanilla frosting in the pattern shown in Figure 16.5. ("Lester" is in black, "Pearl" is upside down in white.) This is Kim's "figure and ground" technique. You will find another example of it in *Gödel, Escher, Bach: An Eternal Golden Braid*, the Pulitzer-prize-winning book by Kim's good friend Douglas R. Hofstadter. Speaking of

Figure 16.3.

Figure 16.4

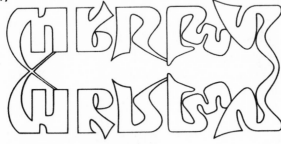

Figure 16.5

Figure 16.6.

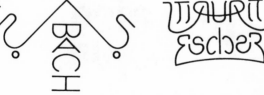

Kurt Gödel, J. S. Bach, and M. C. Escher, Figure 16.6 shows how Kim has given each name a lovely mirror symmetry. In Figure 16.7 Kim has lettered the entire alphabet in such a way that the total pattern has left–right symmetry.

Kim's magic calligraphy came to the attention of Scot Morris, an editor at *Omni.* Morris devoted a page of his popular column on games to Kim's work in *Omni's* September 1979 issue, and he announced a reader's contest for similar patterns. Kim was hired to judge the thousands of entries that came in. You will find the beautiful prizewinners in *Omni's* April 1980 issue and close runners-up in Morris' columns for May and November of the same year.

All the patterns in Kim's book are his own. A small selection of a few

Figure 16.7.

Figure 16.8.

FALSE   UPSIDE
               DOWN

Communication   MAN
                       WOMAN

infinity infinity

infinity infinity

infinity infinity

more is given in Figure 16.8 to convey some notion of the amazing variety of visual tricks Kim has up his sleeve.

I turn now to two unusual mathematical problems originated by Kim, both of which are still only partly solved. In 1975, when Kim was in high school, he thought of the following generalization of the old problem of placing eight queens on a chessboard so that no queen attacks another. Let us ask, said Kim, for the maximum number of queens that can be put on the board so that each queen attacks exactly n other queens. As in chess, we assume that a queen cannot attack through another queen.

When n is 0, we have the classic problem. Kim was able to prove that when n is 1, 10 queens is the maximum number. (A proof is in *Journal of Recreational Mathematics,* Vol. 13, No. 1, 1980–81, p. 61.) A pleasing solution is shown in Figure 16.9 *top.* The middle illustration shows a maximal solution of 14 queens when n is 2, a pattern Kim described in a letter as being "so horribly asymmetric that it has no right to exist." There are only conjectures for the maximum when n is 3 or 4. Kim's best result of 16 queens for $n = 3$ has the ridiculously simple solution shown in Figure 16.9, *bottom,* but there is no known proof that 16 is maximum. For $n = 4$ Kim's best result is 20 queens. Can you place 20 queens on a chessboard so each queen attacks exactly four other queens?

The problem can of course be generalized to finite boards of any size, but Kim has a simple proof based on graph theory that on no finite board, however large, can n have a value greater than 4. For $n = 1$ Kim has shown that the maximum number of queens cannot exceed the largest integer less than or equal to $4k/3$, where k is the number of squares along an edge of the board. For $n = 2$ he has a more difficult proof that the maximum number of queens cannot exceed $2k - 2$, and that this maximum is obtainable on all even-order boards.

Kim's problem concerning polycube snakes has not previously been published, and he and I would welcome any light that readers can throw on it. First we must define a snake. It is a single connected chain of identical unit cubes joined at their faces in such a way that each cube (except for a cube at the end of a chain) is attached face to face to exactly two other cubes. The snake may twist in any possible direction, provided no internal cube abuts the face of any cube other than its two immediate neighbors. The snake may, however, twist so that any number of its cubes touch along edges or at corners. A polycube snake may be finite in length, having two end cubes that are each fastened to only

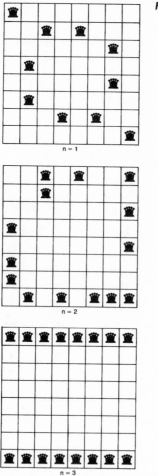

Figure 16.9.

one cube, or it may be finite and closed so that it has no ends. A snake may also have just one end and be infinite in length, or it may be infinite and endless in both directions.

We now ask a deceptively simple question. What is the smallest number of snakes needed to fill all space? We can put it another way. Imagine space to be completely packed with an infinite number of unit cubes. What is the smallest number of snakes into which it can be dissected by cutting along the planes that define the cubes?

If we consider the two-dimensional analogue of the problem (snakes made of unit squares), it is easy to see that the answer is two. We simply intertwine two spirals of infinite one-ended flat snakes, one gray, one white, as in Figure 16.10.

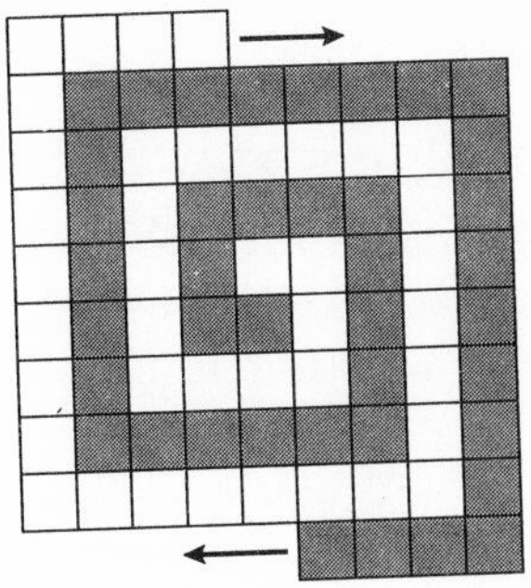

Figure 16.10.

The question of how to fill three-dimensional space with polycube snakes is not so easily answered. Kim has found a way of twisting four infinitely long one-ended snakes (it is convenient to think of them as being each a different color) into a structure of interlocked helical shapes that fill all space. The method is too complicated to explain in a limited space; you will have to take my word that it can be done.

Can it be done with three snakes? Not only is this an unanswered question but also Kim has been unable to prove that it cannot be done with two! "A solution with only two snakes," he wrote in a letter, "would constitute a sort of infinite three-dimensional yin-yang symbol: the negative space left by one snake would be the other snake. It is the beauty of such an entwining, and the possibility of building a model large enough to crawl through, that keeps me searching for a solution."

The problem can of course be generalized to snakes made of unit cubes in any number of dimensions. Kim has conjectured that in a space of n dimensions the minimum number of snakes that completely fill it is $2(n - 1)$, but the guess is still a shaky one.

I once had the pleasure of explaining the polycube-snake problem to John Horton Conway, the Cambridge mathematician now at Princeton University. When I concluded by saying Kim had not yet shown that two snakes could not tile three-dimensional space, Conway instantly said, "But it's obvious that—" He checked himself in mid-sentence, stared into three-space for a minute or two, then exclaimed, "It's *not* obvious!"

I have no idea what passed through Conway's mind. I can only say that if the impossibility of filling three-space with two snakes is not obvious to Conway or to Kim, it probably is not obvious to anyone else.

Addendum

Dozens of readers sent examples of printed words and even sentences that are unreversed in a mirror or which change to other words. Several readers noticed that A TOYOTA, not only is a palindrome, but when written vertically is unaltered by reflection. Jim Scott sent the photograph reproduced here as Figure 16.11. The word IBGUG on the box of toothpaste is IPANA upside down in a mirror. Kim wrote that a friend was puzzled over the words "sped deos" until he discovered he was reading "soap pads" upside down.

Figure 16.11.

I discovered that the following garbled sentence:

MOM

top

OTTO

A

got

reads properly when you see it in a mirror.

David Morice published this two-stanza "poem" in *Wordways* (November 1987, p. 235).

DICK HID
CODEBOOK +
DOBIE KICKED
HOBO—OH HECK—I DECIDED
I EXCEEDED ID—I BOXED
HICK—ODD DODO—EH KID
DEBBIE CHIDED—HOCK CHECKBOOK
ED—BOB BEDDED CHOICE CHICK
HO HO—HE ECHOED—OH OH
DOBIE ICED HOODED IBEX
I COOKED OXHIDE COD
EDIE HEEDED COOKBOOK +
ED
DECKED
BOB

To read the second stanza, hold the poem upside down in front of a mirror.

Donald Knuth, Ronald Graham, and Oren Patashnik, in their marvelous book *Concrete Mathematics* (the word is a blend of *Con*tinuous and Dis*crete* mathematics), published by Addison-Wesley in 1989, introduce their readers to the "umop-apisdn" function. Rotate the word 180 degrees to see what it means in English.

One conjecture about the origin of the expression "Mind your *ps* and *qs*" is that printers often confused the two letters when they were in lower case. A more plausible theory is that British tavern owners had to mind their pints and quarts.

In his autobiography *Arrow in the Blue,* Arthur Koestler recalls meeting many science cranks when he was a science editor in Berlin. One was a man who had invented a new alphabet. Each letter had fourfold rotational symmetry. This, he proclaimed, made it possible for four people, seated on the four sides of a table, to simultaneously read a book or newspaper at the table's center.

Have you heard about the dyslexic atheist who didn't believe in dog? Or D.A.M.N., an organization of National Mothers Against Dyslexia?

I could easily write another chapter about the amazing Scott Kim. He received his Ph.D. in Computers and Graphic Design, at Stanford University, working under Donald Knuth. At a curious gathering of mathematicians, puzzle buffs, and magicians in Atlanta in 1995, Kim demonstrated how your fingers can model the skeleton of a tetrahedron and a cube, and how they can form a trefoil knot of either handedness. He also played an endless octave on a piano, each chord rising up the scale yet never going out of hearing range, and proved he could whistle one tune and hum another at the same time. During the Atlanta gathering, he and his friends Karl Schaffer and Erik Stern of the Dr. Schaffer & Mr. Stern Dance Ensemble presented a dance performance titled "Dances for the Mind's Eye." Choreographed by the three performers, the performance was based throughout on mathematical symmetries.

Among books illustrated by Kim are my *Aha! Gotcha* (W. H. Freeman, 1982) and Ilan Vardi's *Illustrated Computational Recreations in Mathematica* (Addison-Wesley, 1991). Together with Ms. Robin Samelson, Kim produced *Letterform and Illusion,* a computer disk with an accompanying 48-page book of programs designed for use with Claris's MacPaint. In 1994 Random House published Kim's *Puzzle Workout,* a collection of 42 brilliant puzzles reprinted from his puzzle column in *New Media Magazine.* It is the only book of puzzles known to me in which every single puzzle is totally original with the author.

Scott Kim's queens problem brought many letters from readers who sent variant solutions for $n = 2$, 3, and 4 on the standard chessboard, as well as proofs for maximum results and unusual ways to vary the problem. The most surprising letters came from Jeffrey Spencer, Kjell Rosquist, and William Rex Marshall. Spencer and Rosquist, writing in 1981, each independently bettered by one Kim's 20-queen solution for $n = 4$ on the chessboard. Figure 16.12 shows how each placed 21 queens. It is not unique. Writing in 1989 from Dunedin, New Zealand, Marshall sent 36 other solutions!

Marshall also went two better than Kim's chessboard pattern for $n = 3$. He sent nine ways that 18 queens can each attack three others on the chessboard. The solution shown in Figure 16.13 is of special interest because only three queens are not on the perimeter. Marshall found a simple pigeonhole proof that for the order-8 board, $n = 4$, 21 queens is indeed maximum. His similar proof shows 18 maximum for $n = 3$. More generally, he showed that for $n = 4$, with k the order of the board, the maximum is $3k - 3$ for k greater than 5. When $n = 3$, Marshall proved that the maximum number of queens is the largest even number less than or

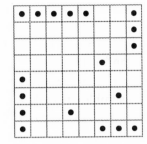

Figure 16.12.

Figure 16.13.

equal to $(12k - 4)/5$. For $n = 2$, he found that Kim's formula of $2k - 2$ applies to all boards larger than order 2, not just to boards of even order.

Perhaps it is worth noting that when $n = 4$, no queen can occupy a corner cell because there is no way it can attack more than three other queens. Dean Hoffman sent a simple proof that n cannot exceed 4. Consider the topmost queen in the leftmost occupied row. At the most it can attack four other queens.

In 1991 Peter Hayes sent a letter from Melbourne, Australia, in which he independently obtained the same results, including all proofs, as those obtained by William Marshall. They were published in a paper titled "A Problem of Chess Queens," in the *Journal of Recreational Mathematics,* Vol. 24, No. 4, 1992, pages 264–71.

In 1996 I received a second letter from William Marshall. He sent me the results of his computer program which provided complete solutions of Kim's chess problem for k (the order of the board) = 1 through

K	$N = 1$	$N = 2$	$N = 3$	$N = 4$
3	0	4	2	0
4	5	2	4	0
5	0	1	31	0
6	2	1	304 (307)	1
7	138 (149)	5	2	3
8	47 (49)	2	9	40
9	1	15	755	655
10	12,490 (12,897)	3	39,302	16,573

Figure 16.14.

The Amazing Creations of Scott Kim

Figure 16.15.

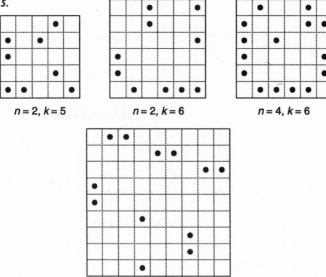

$n = 2, k = 5$ $n = 2, k = 6$ $n = 4, k = 6$

$n = 1, k = 9$

Figure 16.16.   **Figure 16.17**

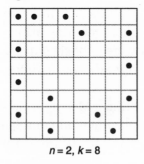   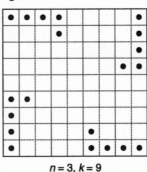

$n = 2, k = 8$ $n = 3, k = 9$

9, and n (number of attacked queens) = 1 through 4. A chart extending these results to $k = 10$ is shown in Figure 16.14.

Note that in four cases there are unique solutions. These are shown in Figure 16.15. Figure 16.16 shows a second solution for $n = 2$, $k = 8$, and Figure 16.17 is an elegant solution for $n = 3$, $k = 9$ found among the 755 patterns produced by Marshall's program.

Dr. Koh Chor Jin, a physicist at the National University of Singapore, sent a clever proof that given a finite volume of space it is possible to cover it with two of Kim's cube-connected snakes. However, as Kim pointed out, Jin's construction does not approach all of space as a limit as the volume of space increases. Each time you wish to enlarge his construction, it has to be modified. Kim is convinced that tiling all of space with two snakes is impossible, but for three snakes the question remains open.

In the early 1990s Kim began contributing a puzzle page to *Discover*. It is now a regular feature. Ian Stewart devoted his September 1999 *Scientific American* column on recreational mathematics to the mathematical dances choreographed by Kim and his associates Karl Schaffer and Erik Stern.

Answer

Readers were asked to place 20 queens on a chessboard so that each queen attacks exactly four others. A solution is given in Figure 16.18.

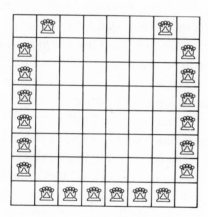

Figure 16.18.

Chapter 17

The Art of
M. C. Escher

*What I give form to in daylight is only one
percent of what I've seen in darkness.*

—M. C. ESCHER

*T*here is an obvious but superficial sense in which certain
kinds of art can be called mathematical art. Op art, for instance, is
"mathematical," but in a way that is certainly not new. Hard-edged,
rhythmic, decorative patterns are as ancient as art itself, and even the
modern movement toward abstraction in painting began with the geometric forms of the cubists. When the French Dadaist painter Hans Arp
tossed colored paper squares in the air and glued them where they fell,
he linked the rectangles of cubism to the globs of paint slung by the
later "action" painters. In a broad sense even abstract expressionist art
is mathematical, since randomness is a mathematical concept.

This, however, expands the term "mathematical art" until it becomes
meaningless. There is another and more useful sense of the term that
refers not to techniques and patterns but to a picture's subject matter.
A representational artist who knows something about mathematics can
build a composition around a mathematical theme in the same way
that Renaissance painters did with religious themes or Russian painters
do with political themes. No living artist has been more successful
with this type of "mathematical art" than Maurits C. Escher of the
Netherlands.

"I often feel closer to mathematicians than to my fellow artists," Escher has written, and he has been quoted as saying, "All my works are
games. Serious games." His lithographs, woodcuts, wood engravings,
and mezzotints can be found hanging on the walls of mathematicians
and scientists in all parts of the world. There is an eerie, surrealist aspect to some of his work, but his pictures are less the dreamlike fantasies of a Salvador Dali or a René Magritte than they are subtle
philosophical and mathematical observations intended to evoke what
the poet Howard Nemerov, writing about Escher, called the "mystery,

absurdity, and sometimes terror" of the world. Many of his pictures concern mathematical structures that have been discussed in books on recreational mathematics, but before we examine some of them, a word about Escher himself.

He was born in Leeuwarden in Holland in 1898, and as a young man he studied at the School of Architecture and Ornamental Design in Haarlem. For 10 years he lived in Rome. After leaving Italy in 1934 he spent two years in Switzerland and five in Brussels, then settled in the Dutch town of Baarn where he and his wife lived until his death in 1972. Although he had a successful exhibit in 1954 at the Whyte Gallery in Washington, he was much better known in Europe than here. A large collection of his work is now owned by the National Gallery of Art in Washington, D.C.

Among crystallographers Escher is best known for his scores of ingenious tessellations of the plane. Designs in the Alhambra reveal how expert the Spanish Moors were in carving the plane into periodic repetitions of congruent shapes, but the Mohammedan religion forbade them to use the shapes of living things. By slicing the plane into jigsaw patterns of birds, fish, reptiles, mammals, and human figures, Escher has been able to incorporate many of his tessellations into a variety of startling pictures.

In *Reptiles,* the lithograph shown in Figure 17.1, a little monster crawls out of the hexagonal tiling to begin a brief cycle of 3-space life that reaches its summit on the dodecahedron; then the reptile crawls back again into the lifeless plane. In *Day and Night,* the woodcut in Figure 17.2, the scenes at the left and the right are not only mirror images but also almost "negatives" of each other. As the eye moves up the center, rectangular fields flow into interlocking shapes of birds, the black birds flying into daylight, the white birds flying into night. In the circular wood-cut *Heaven and Hell* (Figure 17.3) angels and devils fit together, the similar shapes becoming smaller farther from the center and finally fading into an infinity of figures, too tiny to be seen, on the rim. Good, Escher may be telling us, is a necessary background for evil, and vice versa. This remarkable tessellation is based on a well-known Euclidean model, devised by Henri Poincaré, of the non-Euclidean hyperbolic plane; the interested reader will find it explained in H.S.M. Coxeter's *Introduction to Geometry* (Wiley, 1961), pages 282–90.

If the reader thinks that patterns of this kind are easy to invent, let him try it! "While drawing I sometimes feel as if I were a spiritualist

Figure 17.1. *Reptiles,* lithograph, 1943

Mickelson Gallery, Washington

Figure 17.2. *Day and Night,* woodcut, 1938

Figure 17.3. *Heaven and Hell,* woodcut, 1960

medium," Escher has said, "controlled by the creatures I am conjuring up. It is as if they themselves decide on the shape in which they choose to appear. . . . The border line between two adjacent shapes having a double function, the act of tracing such a line is a complicated business. On either side of it, simultaneously, a recognizability takes shape. But the human eye and mind cannot be busy with two things at the same moment and so there must be a quick and continuous jumping from one side to the other. But this difficulty is perhaps the very moving-spring of my perseverance."

It would take a book to discuss all the ways in which Escher's fantastic tessellations illustrate aspects of symmetry, group theory, and crystallographic laws. Indeed, such a book has been written by Caroline H. MacGillavry of the University of Amsterdam: *Symmetry Aspects of M. C. Escher's Periodic Drawings.* This book, published in Utrecht for the International Union of Crystallography, reproduces 41 of Escher's tessellations, many in full color.

Figures 17.4 and 17.5 illustrate another category of Escher's work, a play with the laws of perspective to produce what have been called "impossible figures." In the lithograph *Belvedere,* observe the sketch of the cube on a sheet lying on the checked floor. The small circles mark two spots where one edge crosses another. In the skeletal model held by the seated boy, however, the crossings occur in a way that is not realizable in 3-space. The belvedere itself is made up of impossible structures. The youth near the top of the ladder is outside the belvedere but the base of the ladder is inside. Perhaps the man in the dungeon has lost his mind trying to make sense of the contradictory structures in his world.

The lithograph *Ascending and Descending* derives from a perplexing impossible figure that first appeared in an article, "Impossible Objects: A Special Type of Visual Illusion," by L. S. Penrose, a British geneticist, and his son, the mathematician Roger Penrose (*British Journal of Psychology,* February 1958). The monks of an unknown sect are engaged in a daily ritual of perpetually marching around the impossible stairway on the roof of their monastery, the outside monks climbing, the inside monks descending. "Both directions," comments Escher, "though not without meaning, are equally useless. Two refractory individuals refuse to take part in this 'spiritual exercise.' They think they know better than their comrades, but sooner or later they will admit the error of their nonconformity."

Many Escher pictures reflect an emotional response of wonder to the forms of regular and semiregular solids. "In the midst of our often chaotic society," Escher has written, "they symbolize in an unrivaled manner man's longing for harmony and order, but at the same time their perfection awes us with a sense of our own helplessness. Regular polyhedrons have an absolutely nonhuman character. They are not inventions of the human mind, for they existed as crystals in the earth's crust long before mankind appeared on the scene. And in regard to the spherical shape—is the universe not made up of spheres?"

The lithograph *Order and Chaos* (Figure 17.6) features the "small stellated dodecahedron," one of the four "Kepler–Poinsot polyhedrons" that, together with the five Platonic solids, make up the nine possible "regular polyhedrons." It was first discovered by Johannes Kepler, who called it "urchin" and drew a picture of it in his *Harmonices mundi*

Figure 17.4. *Belvedere,* lithograph, 1958

The Art of M. C. Escher 217

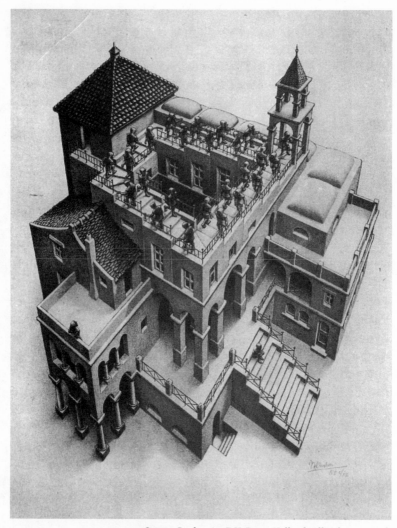

Figure 17.5. *Ascending and Descending,* lithograph, 1960

(*Harmony of the World*), a fantastic numerological work in which basic ratios found in music and the forms of regular polygons and polyhedrons are applied to astrology and cosmology. Like the Platonic solids, Kepler's urchin has faces that are equal regular polygons, and it has equal angles at its vertices, but its faces are not convex and they intersect one another. Imagine each of the 12 faces of the dodecahedron (as in the picture *Reptiles*) extended until it becomes a pentagram, or five-

Figure 17.6. *Order and Chaos,* lithograph, 1950

pointed star. These 12 intersecting pentagrams form the small stellated dodecahedron. For centuries mathematicians refused to call the pentagram a "polygon" because its five edges intersect, and for similar reasons they refused to call a solid such as this a "polyhedron" because its faces intersect. It is amusing to learn that as late as the middle of the 19th century the Swiss mathematician Ludwig Schläfli, although he recognized some face-intersecting solids as being polyhedrons, refused to call this one a "genuine" polyhedron because its 12 faces, 12 vertices, and 30 edges did not conform to Leonhard Euler's famous polyhedral formula, $F + V = E + 2$. (It *does* conform if it is reinterpreted as a solid with 60 triangular faces, 32 vertices, and 90 edges, but in this interpretation it cannot be called "regular" because its faces are isosceles triangles.) In *Order and Chaos* the beautiful symmetry of this solid, its

The Art of M. C. Escher   *219*

points projecting through the surface of an enclosing bubble, is thrown into contrast with an assortment of what Escher has described as "useless, cast-off, and crumpled objects."

The small stellated dodecahedron is sometimes used as a shape for light fixtures. Has any manufacturer of Christmas tree ornaments, I wonder, ever sold it as a three-dimensional star to top a Christmas tree? A cardboard model is not difficult to make. H. M. Cundy and A. P. Rollett, in *Mathematical Models* (Oxford University Press, revised edition, 1961), advise one not to try to fold it from a net but to make a dodecahedron and then cement a five-sided pyramid to each face. Incidentally, every line segment on the skeleton of this solid is (as Kepler observed) in golden ratio to every line segment of next-larger length. The solid's polyhedral dual is the "great dodecahedron," formed by the intersection of 12 regular pentagons. For details about the Kepler–Poinsot star polyhedrons the reader is referred to the book by Cundy and Rollett and to Coxeter's *Regular Polytopes* (Dover, 1973).

The lithograph *Hand with Reflecting Globe* (Figure 17.7) exploits a reflecting property of a spherical mirror to dramatize what philosopher Ralph Barton Perry liked to call the "egocentric predicament."

Figure 17.7. *Hand with Reflecting Globe,* 1935

All any person can possibly know about the world is derived from what enters his skull through various sense organs; there is a sense in which one never experiences anything except what lies within the circle of his own sensations and ideas. Out of this "phenomenology" he constructs what he believes to be the external world, including those other people who appear to have minds in egocentric predicaments like his own. Strictly speaking, however, there is no way he can prove that anything exists except himself and his shifting sensations and thoughts. Escher is seen staring at his own reflection in the sphere. The glass mirrors his surroundings, compressing them inside one perfect circle. No matter how he moves or twists his head, the point midway between his eyes remains exactly at the center of the circle. "He cannot get away from that central point," says Escher. "The ego remains immovably the focus of his world."

Escher's fascination with the playthings of topology is expressed in a number of his pictures. At the top of the woodcut *Knots* (Figure 17.8) we see the two mirror-image forms of the trefoil knot. The knot at top left is made with two long flat strips that intersect at right angles. This double strip was given a twist before being joined to itself. Is it a single one-sided band that runs twice around the knot, intersecting itself, or does it consist of two distinct but intersecting Möbius bands? The large knot below the smaller two has the structure of a four-sided tube that has been given a quarter-twist so that an ant walking inside, on one of the central paths, would make four complete circuits through the knot before it returned to its starting point.

The wood engraving *Three Spheres* (Figure 17.9), a copy of which is owned by New York's Museum of Modern Art, appears at first to be a sphere undergoing progressive topological squashing. Look more carefully, however, and you will see that it is something quite different. Can the reader guess what Escher, with great verisimilitude, is depicting here?

Addendum

When Escher died in 1972, at the age of 73, he was just beginning to become world-famous; not only among mathematicians and scientists (who were the first to appreciate him), but also with the public at large, especially with the young counterculture. The Escher cult is still growing. You see his pictures everywhere: on the covers of mathematical textbooks, on albums of rock music, on psychedelic posters

The Art of M. C. Escher

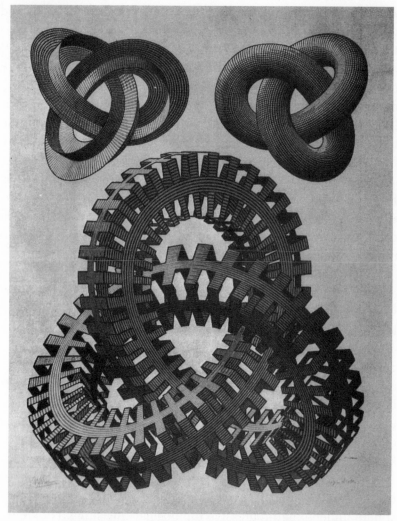

Figure 17.8. Knots, woodcut, 1965

that glow under black light, even on T-shirts. When I first reproduced an Escher picture in my column for April 1961 (and *Scientific American* ran one of his bird tesselations on the cover), I purchased from Escher only one print, a woodcut. For a mere $40 to $60 each I could have bought scores of pictures that now would each be worth thousands. But who could then have anticipated the astonishing growth of Escher's fame?

Figure 17.9. *Three Spheres,* wood engraving, 1945

After Escher died his son George gave me a Persian leather painting, owned by his father, which is framed and hanging in my library. It is a circle of horses on which each horse's head is the head of three horse bodies.

Answer

Three Spheres is a picture of three flat disks, each painted to simulate a sphere. The bottom disk is flat on a table. The middle disk is bent at right angles along a diameter. The top disk stands vertically on the horizontal half of the middle one. Clues are provided by a fold line in the middle disk and by identical shading on the three pseudospheres.

Bibliography

This highly selective list of references is limited to books and articles in English. For an extensive bibliography, including foreign references and motion pictures, see *Visions of Symmetry*, the definitive work on Escher by Doris Schattschneider.

H.S.M. Coxeter, "Angels and Devils," *The Mathematical Gardner*, D. Klarner (ed.), Wadsworth, 1981.

H.S.M. Coxeter et al. (eds.), *M. C. Escher: Art and Science*, Elsevier, 1986.

H.S.M. Coxeter, "The Trigonometry of Escher's Woodcut 'Circle Limit III,' " *The Mathematical Intelligencer*, Vol. 18, No. 4, 1996, pp. 42–46.

B. Ernst, *The Magic Mirror of M. C. Escher*, Random House, 1976; Ballantine, 1976.

M. C. Escher, *Escher on Escher*, Abrams, 1989. (Translated by Karin Ford.)

L. Glasser, "Teaching Symmetry," *The Journal of Chemical Education*, Vol. 44, September 1967, pp. 502–11.

J. L. Locher (ed.), *The World of M. C. Escher*, Abrams, 1971.

J. L. Locher (ed.), *M. C. Escher: His Life and Complete Graphic Work*, Abrams, 1982. Reviewed by H.S.M. Coxeter in *The Mathematical Intelligencer*, Vol. 7, No. 1, 1985, pp. 59–69.

C. H. MacGillavry, *Symmetry Aspects of M. C. Escher's Periodic Drawings*, A. Oosthoek's Uitgeversmaatschappij NV, 1965. Reprinted as *Fantasy and Symmetry— The Periodic Drawings of M. C. Escher*, Abrams, 1976

H. Nemerov, "The Miraculous Transformations of Maurits Cornelis Escher," *Artists's Proof*, Vol. 3, Fall/Winter 1963–64, pp. 32–39.

E. R. Ranucci, "Master of Tesselations: M. C. Escher, 1898–1972," *Mathematics Teacher*, Vol. 67, April 1974, pp. 299–306.

J. C. Rush, "On the Appeal of M. C. Escher's Pictures," *Leonardo*, Vol. 12, 1979, pp. 48–50.

D. Schattschneider, "The Pólya Escher Connection," *Mathematics Magazine*, Vol. 60, December 1987, pp. 293–98.

D. Schattschneider, *Visions of Symmetry*, W. H. Freeman, 1990.

D. Schattschneider, "Escher's Metaphors," *Scientific American*, November 1994, pp. 66–71.

D. Schattschneider and W. Walker, *M. C. Escher: Kaleidocycles*, Ballantine, 1977; revised edition, Pomegranate Artbooks, 1987.

M. Severin, "The Dimensional Experiments of M. C. Escher," *Studio*, February 1951, pp. 50–53.

J. L. Teeters, "How to Draw Tesselations of the Escher Type," *Mathematics Teacher*, Vol. 67, April 1974, pp. 307–10.

M. L. Teuber, "Sources of Ambiguity in the Prints of Maurits C. Escher," *Scientific American*, Vol. 231, July 1974, pp. 90–104. See also the correspondence on this article in Vol. 232, January 1975, pp. 8–9.

K. Wilkie, "Escher: The Journey to Infinity," *Holland Herald*, Vol. 9, No. 1, 1974, pp. 20–43.

V
Topology

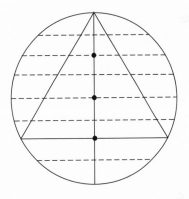

Chapter 18

Klein Bottles and Other Surfaces

Three jolly sailors from Blaydon-on-Tyne
They went to sea in a bottle by Klein.
Since the sea was entirely inside the hull
The scenery seen was exceedingly dull.

> —Frederick Winsor,
> *The Space Child's Mother Goose*

*T*o a topologist a square sheet of paper is a model of a two-sided surface with a single edge. Crumple it into a ball and it is still two-sided and one-edged. Imagine that the sheet is made of rubber. You can stretch it into a triangle or circle, into any shape you please, but you cannot change its two-sidedness and one-edgedness. They are topological properties of the surface, properties that remain the same regardless of how you bend, twist, stretch, or compress the sheet.

Two other important topological invariants of a surface are its chromatic number and Betti number. The chromatic number is the maximum number of regions that can be drawn on the surface in such a way that each region has a border in common with every other region. If each region is given a different color, each color will border on every other color. The chromatic number of the square sheet is 4. In other words, it is impossible to place more than four differently colored regions on the square so that any pair has a boundary in common. The term "chromatic number" also designates the minimum number of colors sufficient to color any finite map on a given surface. It is now known that 4 is the chromatic number, in this map-coloring sense, for the square, tube, and sphere, and for all other surfaces considered in this chapter, the chromatic number is the same under both definitions.

The Betti number, named after Enrico Betti, a 19th-century Italian physicist, is the maximum number of cuts that can be made without dividing the surface into two separate pieces. If the surface has edges, each cut must be a "crosscut": one that goes from a point on an edge to another point on an edge. If the surface is closed (has no edges), each cut must be a "loop cut": a cut in the form of a simple closed curve. Clearly the Betti number of the square sheet is 0. A crosscut is certain to produce two disconnected pieces.

If we make a tube by joining one edge of the square to its opposite edge, we create a model of a surface topologically distinct from the square. The surface is still two-sided but now there are two separate edges, each a simple closed curve. The chromatic number remains 4 but the Betti number has changed to 1. A crosscut from one edge to the other, although it eliminates the tube, allows the paper to remain in one piece.

A third type of surface, topologically the same as the surface of a sphere or cube, is made by folding the square in half along a diagonal and then joining the edges. The surface continues to be two-sided but all edges have been eliminated. It is a closed surface. The chromatic number continues to be 4. The Betti number is back to 0: any loop cut obviously creates two pieces.

Things get more interesting when we join one edge of the square to its opposite edge but give the surface a half-twist before doing so. You might suppose that this cannot be done with a square piece of paper, but it is easily managed by folding the square twice along its diagonals, as shown in Figure 18.1. Tape together the pair of edges indicated by the arrow in

Figure 18.1. Möbius surface constructed with a square

the last drawing. The resulting surface is the familiar Möbius strip, first analyzed by A. F. Möbius, the 19th-century German astronomer who was one of the pioneers of topology. The model will not open out, so it is hard to see that it is a Möbius strip, but careful inspection will convince you that it is. The surface is one-sided and one-edged, with a Betti number of 1. Surprisingly, the chromatic number has jumped to 6. Six regions, of six different colors, can be placed on the surface so that each region has a border in common with each of the other five.

When both pairs of the square's opposite edges are joined, without twisting, the surface is called a torus. It is topologically equivalent to the surface of a doughnut or a cube with a hole bored through it. Figure 18.2 shows how a flat, square-shaped model of a torus is easily made by folding the square twice, taping the edges as shown by the solid gray line in the second drawing and the arrows in the last. The torus is two-sided, closed (no-edged), and has a chromatic number of 7 and a Betti number of 2. One way to make the two cuts is first to make

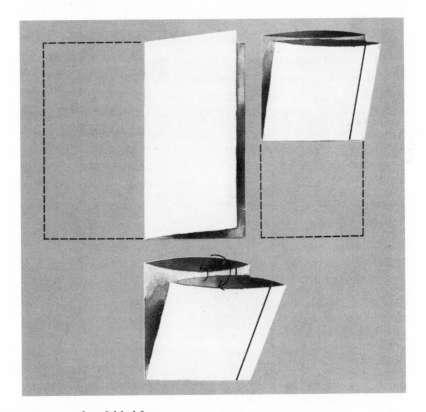

Figure 18.2. Torus surface folded from a square

Klein Bottles and Other Surfaces

a loop cut where you joined the last pair of edges (this reduces the torus to a tube) and then a crosscut where you joined the first pair. Both cuts, strictly speaking, are loop cuts when they are marked on the torus surface. It is only because you make one cut before the other that the second cut becomes a crosscut.

It is hard to anticipate what will happen when the torus model is cut in various ways. If the entire model is bisected by being cut in half either horizontally or vertically, along a center line parallel to a pair of edges, the torus surface receives two loop cuts. In both cases the resulting halves are tubes. If the model is bisected by being cut in half along either diagonal, each half proves to be a square. Can the reader find a way to give the model two loop cuts that will produce two separate bands interlocked like two rings of a chain?

Many different surfaces are closed like the surface of a sphere and a torus, yet one-sided like a Möbius strip. The easiest one to visualize is a surface known as the Klein bottle, discovered in 1882 by Felix Klein (1849–1925), a great German mathematician. An ordinary bottle has an outside and inside in the sense that if a fly were to walk from one side to the other, it would have to cross the edge that forms the mouth of the bottle. The Klein bottle has no edges, no inside or outside. Its volume, therefore, is zero. What seems to be its inside is continuous with its outside, like the two apparent "sides" of a Möbius surface.

Unfortunately it is not possible to construct a Klein bottle in three-dimensional space without self-intersection of the surface. Figure 18.3

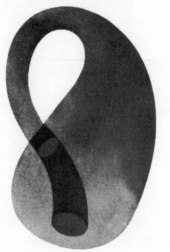

Figure 18.3. Klein bottle: a closed surface with no inside or outside

shows how the bottle is traditionally depicted. Imagine the lower end of a tube stretched out, bent up, and plunged through the tube's side, then joined to the tube's upper mouth. In an actual model made, say, of glass there would be a hole where the tube intersects the side. You must disregard this defect and think of the hole as being covered by a continuation of the bottle's surface. There is no hole, only an intersection of surfaces. This self-intersection is necessary because the model is in 3-space. If we conceive of the surface as being embedded in 4-space, the self-intersection can be eliminated entirely. The Klein bottle is one-sided, no-edged, and has a Betti number of 2 and a chromatic number of 6.

Daniel Pedoe, a mathematician at the University of Minnesota, is the author of *The Gentle Art of Mathematics*. It is a delightful book, but on page 84 Professor Pedoe slips into a careless bit of dogmatism. He describes the Klein bottle as a surface that is a challenge to the glass blower, but one "which cannot be made with paper." Now, it is true that at the time he wrote this apparently no one had tried to make a paper Klein bottle, but that was before Stephen Barr, a science-fiction writer and an amateur mathematician of Woodstock, NY, turned his attention to the problem. Barr quickly discovered dozens of ways to make paper Klein bottles. Here I will describe a variation of my own that is made from a paper tube. The tube can be a sealed envelope with its left and right edges cut open.

The steps are given in Figure 18.4. First, make a tube by folding the square in half and joining the right edges with a strip of tape as shown (Step 1). Cut a slot about a quarter of the distance from the top of the tube (Step 2), cutting only through the thickness of paper nearest you. This corresponds to the "hole" in the glass model. Fold the model in half along the broken line A. Push the lower end of the tube up through the slot (Step 3) and join the edges all the way around the top of the model (Step 4) as indicated by the arrows. It is not difficult to see that this flat, square model is topologically identical with the glass bottle shown in Figure 18.3. In one way it is superior: there is no actual hole. True, you have a slot where the surface self-intersects, but it is easy to imagine that the edges of the slot are joined so that the surface is everywhere edgeless and continuous.

Moreover, it is easy to cut this paper model and demonstrate many of the bottle's astonishing properties. Its Betti number of 2 is demonstrated by cutting the two loops formed by the two pairs of taped edges. If you cut the bottle in half vertically, you get two Möbius bands, one

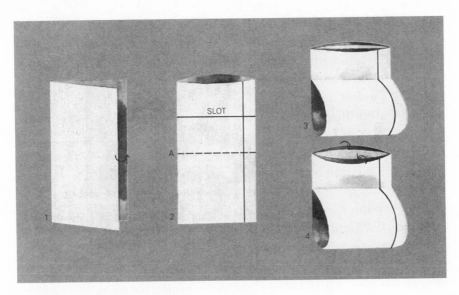

Figure 18.4. Folding a Klein bottle from a square

a mirror image of the other. This is best demonstrated by making a tall, thin model (see Figure 18.5) from a tall, thin rectangle instead of a square. When you slice it in half along the broken line (actually this is one long loop cut all the way around the surface), you will find that each half opens out into a Möbius strip. Both strips are partially self-intersecting, but you can slide each strip out of its half-slot and close the slot, which is not supposed to be there anyway.

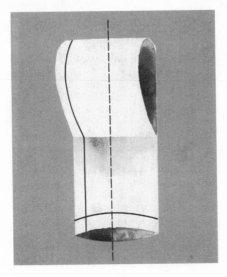

Figure 18.5. Bisected bottle makes two Möbius strips

If the bottle can be cut into a pair of Möbius strips, of course the reverse procedure is possible, as described in the following anonymous limerick:

> *A mathematician named Klein*
> *Thought the Möbius band was divine.*
> * Said he: "If you glue*
> * The edges of two,*
> *You'll get a weird bottle like mine."*

Surprisingly, it is possible to make a single loop cut on a Klein bottle and produce not two Möbius strips but only one. A great merit of Barr's paper models is that problems like this can be tackled empirically. Can the reader discover how the cut is made?

The Klein bottle is not the only simple surface that is one-sided and no-edged. A surface called the projective plane (because of its topological equivalence to a plane studied in projective geometry) is similar to the Klein bottle in both respects as well as in having a chromatic number of 6. As in the case of the Klein bottle, a model cannot be made in 3-space without self-intersection. A simple Barr method for folding such a model from a square is shown in Figure 18.6. First cut the square

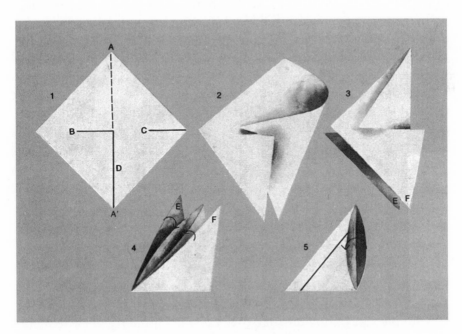

Figure 18.6. Folding a cross-cap and projective plane from a square

Klein Bottles and Other Surfaces

along the solid black lines shown in Step 1. Fold the square along the diagonal A-A', inserting slot C into slot B (Steps 2 and 3). You must think of the line where the slots interlock as an abstract line of self-intersection. Fold up the two bottom triangular flaps E and F, one on each side (Step 4), and tape the edges as indicated.

The model is now what topologists call a cross-cap, a self-intersecting Möbius strip with an edge that can be stretched into a circle without further self-intersection. This edge is provided by the edges of cut D, originally made along the square's diagonal. Note that unlike the usual model of a Möbius strip, this one is symmetrical: neither right- nor left-handed. When the edge of the cross-cap is closed by taping it (Step 5), the model becomes a projective plane. You might expect it to have a Betti number of 2, like the Klein bottle, but it does not. It has a Betti number of 1. No matter how you loop-cut it, the cut produces either two pieces or a piece topologically equivalent to a square sheet that cannot be cut again without making two pieces. If you remove a disk from anywhere on the surface of the projective plane, the model reverts to a cross-cap.

Figure 18.7 summarizes all that has been said. The square diagrams in the first column show how the edges join in each model. Sides of the same color join each to each, with the direction of their arrows coinciding. Corners labeled with the same letter are corners that come together. Broken lines are sides that remain edges in the finished model. Next to the chromatic number of each model is shown one way the surface can be mapped to accommodate the maximum number of colors. It is instructive to color each sheet as shown, coloring the regions on both sides of the paper (as though the paper were cloth through which the colors soaked), because you must think of the sheet as having zero thickness. An inspection of the final model will show that each region does indeed border on every other one.

Addendum

Although I was not the first to model Klein bottles with paper (credit for this goes to Stephen Barr—see bibliography), my contribution was to show how easily a model can be made from an envelope and cut in half to make two Möbius strips of opposite handedness.

A simple way to demonstrate the one-sided property of a Klein bottle is to punch a hole in a model and insert a piece of string. No matter

SURFACE	CHROMATIC NUMBER		SIDES	EDGES	BETTI NUMBER
SQUARE (OR DISK)	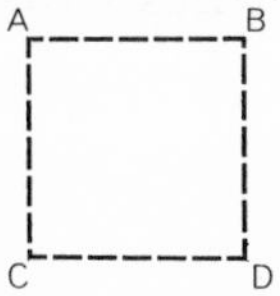	4	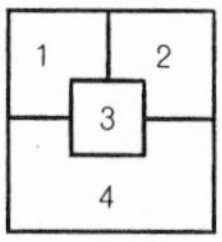 2	1	0
TUBE	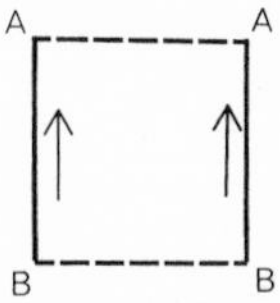	4	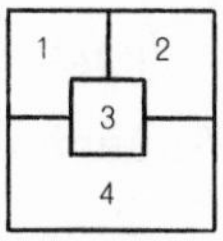 2	2	1
SPHERE	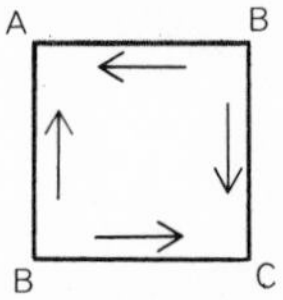	4	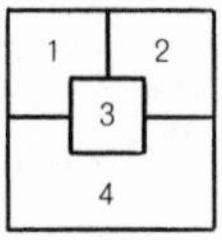 2	0	0
MÖBIUS STRIP	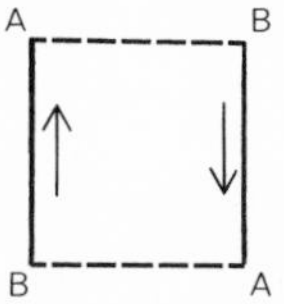	6	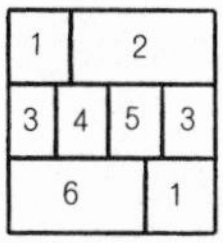 1	1	1
TORUS	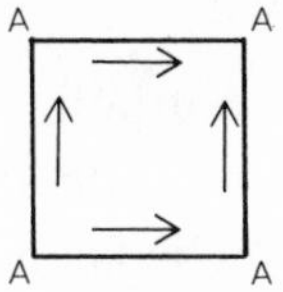	7	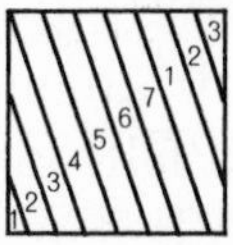 2	0	2
KLEIN BOTTLE	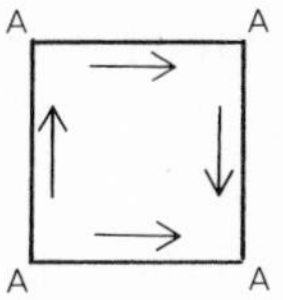	6	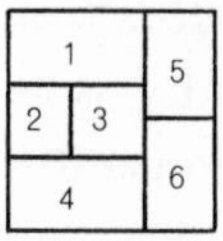 1	0	2
PROJECTIVE PLANE	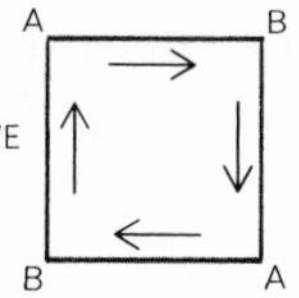	6	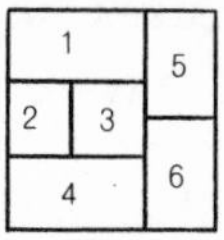 1	0	1

Figure 18.7. Topological invariants of seven basic surfaces

Klein Bottles and Other Surfaces

where you make the hole, you can always tie the ends of the string together.

The Klein bottle continues to intrigue limerick writers. Here are three I have encountered:

> *Topologists try hard to floor us*
> *With a nonorientable torus.*
> *The bottle of Klein*
> *They say is divine*
> *But it is so exceedingly porous.*

—Anonymous

> *A geometrician named Klein*
> *Thought the Möbius band asinine.*
> *"Though its outside is in,*
> *Still it's ugly as sin;*
> *It ain't round like that bottle o'mine!"*

—M. M. H. Coffee and J. J. Zeltmacher, Jr.

> *An anti-strong-drinker named Klein*
> *Invented a bottle for wine.*
> *"There's no stopper," he cried,*
> *"And it has no inside,*
> *So the grapes have to stay on the vine!"*

—James Albert Lindon

I confess that I have made use of a Klein bottle in two works of fiction. Professor Slapenarski falls into a Klein bottle and disappears at the end of my mathematically flawed story "The Island of Five Colors" (you'll find it in Clifton Fadiman's anthology *Fantasia Mathematica*), and again in my novel *Visitors from Oz* (1999) where it is used as a device for transporting Dorothy, Scarecrow, and Tin Woodman from Oz (now in a parallel world) to Central Park in Manhattan.

If you would like to own a glass Klein bottle, a firm called Acme, 6270 Colby Street, Oakland, CA 94618, has five handsome glass Klein bottles for sale, of various sizes, shapes, and prices. The firm can also be reached on the Internet at www.kleinbottle.com.

Answers

The torus-cutting problem is solved by first ruling three parallel lines on the unfolded square (see Figure 18.8). When the square is

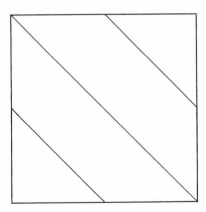

Figure 18.8. Solution to the torus-cutting problem

folded into a torus, as explained, the lines make two closed loops. Cutting these loops produces two interlocked bands, each two-sided with two half-twists.

How does one find a loop cut on the Klein bottle that will change the surface to a single Möbius strip? On both left and right sides of the narrow rectangular model described you will note that the paper is creased along a fold that forms a figure-eight loop. Cutting only the left loop transforms the model into a Möbius band; cutting only the right loop produces an identical band of opposite handedness.

What happens if both loops are cut? The result is a two-sided, two-edged band with four half-twists. Because of the slot the band is cut apart at one point, so that you must imagine the slot is not there. This self-intersecting band is mirror-symmetrical, neither right- nor left-handed. You can free the band of self-intersection by sliding it carefully out of the slot and taping the slot together. The handedness of the resulting band (that is, the direction of the helices formed by its edges) depends on whether you slide it out to the right or the left. This and the previous cutting problems are based on paper models that were invented by Stephen Barr and are described in his *Experiments in Topology* (Crowell, 1964).

Bibliography

M. A. Armstrong, *Basic Topology,* Springer-Verlag, 1983.

B. H. Arnold, *Intuitive Concepts in Elementary Topology,* Prentice-Hall, 1962.

S. Barr, *Experiments in Topology,* Crowell, 1964.

W. G. Chinn and N. E. Steenrod, *First Concepts of Topology.* Mathematical Association of America, 1966.

F. H. Crowe, *Introduction to Topology,* Saunders, 1989.

G. K. Francis, *A Topological Picture Book,* Springer-Verlag, 1987.

D. W. Hall and G. L. Spencer, *Elementary Topology,* Wiley, 1955.

W. Lietzmann, *Visual Topology,* Chatto and Windus, 1965.

M. J. Mansfield, *Introduction to Topology,* Van Nostrand, 1963; Krieger, 1972.

W. S. Massey, *Algebraic Topology: An Introduction,* Harcourt, Brace, 1967.

B. Mendelsohn, *Introduction to Topology,* Blackie, 1963.

E. M. Patterson, *Topology,* Oliver and Boyd, 1956.

V. V. Prasolov, *Intuitive Topology,* Providence, RI: American Mathematical Society, 1995.

I. Stewart, "The Topological Dressmaker," *Scientific American,* July 1993, pp. 110–12.

I. Stewart, "Glass Klein Bottles," *Scientific American,* March 1998, pp. 100–1. See also the note at the end of Stewart's December 1998 column.

I. Stewart, "Tangling With Topology," *Scientific American,* April 1999, pp. 123–24.

A. W. Tucker and H. S. Bailey, "Topology," *Scientific American,* January 1950, pp. 18–24.

R. L. Wilder, "Topology: Its Nature and Significance," *Mathematics Teacher,* October 1962, pp. 462–75.

Chapter 19

Knots

*"A knot!" said Alice, always ready to make
herself useful, and looking anxiously about
her. "Oh, do let me help to undo it!"*

—*Alice in Wonderland*, Chapter 3

*T*o a topologist knots are closed curves embedded in three-dimensional space. It is useful to model them with rope or cord and to diagram them as projections on a plane. If it is possible to manipulate a closed curve—of course, it must not be allowed to pass through itself—so that it can be projected on a plane as a curve with no crossing points then the knot is called trivial. In ordinary discourse one would say the curve is not knotted. "Links" are two or more closed curves that cannot be separated without passing one through another.

The study of knots and links is now a flourishing branch of topology that interlocks with algebra, geometry, group theory, matrix theory, number theory, and other branches of mathematics. Some idea of its depth and richness can be had from reading Lee Neuwirth's excellent article "The Theory of Knots" in *Scientific American* (June 1979). Here we shall be concerned only with some recreational aspects of knot theory: puzzles and curiosities that to be understood require no more than the most elementary knowledge of the topic.

Let's begin with a question that is trivial but that can catch even mathematicians off guard. Tie an overhand knot in a piece of rope as is shown in Figure 19.1. If you think of the ends of the rope as being joined, you have tied what knot theorists call a trefoil knot. It is the simplest of all knots in the sense that it can be diagrammed with a minimum of three crossings. (No knot can have fewer crossings except the trivial knot that has none.) Imagine that end *A* of the rope is passed through the loop from behind and the ends are pulled. Obviously the knot will dissolve. Now suppose the end is passed *twice* through the loop as is indicated by the broken line. Will the knot dissolve when the ends of the rope are pulled?

Most people guess that it will form another knot. Actually the knot

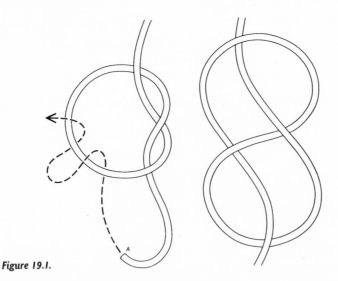

Figure 19.1.

dissolves as before. The end must go *three* times through the loop to produce another knot. If you try it, you will see that the new trefoil created in this way is not the same as the original. It is a mirror image. The trefoil is the simplest knot that cannot be changed to its mirror image by manipulating the rope.

The next simplest knot, the only one with a minimum of four crossings, is the figure eight at the right in Figure 19.1. In this form it is easily changed to its mirror image. Just turn it over. A knot that can be manipulated to make its mirror image is called amphicheiral because like a rubber glove it can be made to display either handedness. After the figure eight the next highest amphicheiral knot has six crossings, and it is the only 6-knot of that type. Amphicheiral knots become progressively scarcer as crossing numbers increase.

A second important way to divide knots into two classes is to distinguish between alternating and nonalternating knots. An alternating knot is one that can be diagrammed so that if you follow its curve in either direction, you alternately go over and under at the crossings. Alternating knots have many remarkable properties not possessed by nonalternating knots.

Still another important division is into prime and composite knots. A prime knot is one that cannot be manipulated to make two or more separated knots. For example, the square knot and the granny knot are not prime because each can be changed to two side-by-side trefoils. The square knot is the "product" of two trefoils of opposite handed-

TOPOLOGY

ness. The granny is the product of two trefoils of the same handedness, and therefore (unlike the square knot) it is not amphicheiral. Both knots are alternating. As an easy exercise, see if you can sketch a square knot with six (the minimum) alternating crossings.

All prime knots of seven or fewer crossings are alternating. Among the 8-knots only the three in Figure 19.2 are nonalternating. No matter how long you manipulate a rope model of one of these knots, you will never get it to lie flat in the form of an alternating diagram. The knot at top right is a bowline. The bottom knot is a torus knot as explained below.

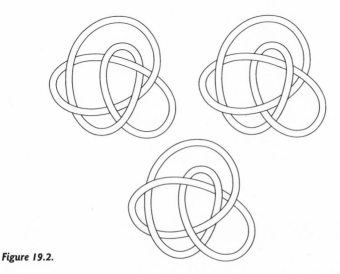

Figure 19.2.

A fourth basic binary division of knots is into the invertible and noninvertible. Imagine an arrow painted on a knotted rope to give a direction to the curve. If it is possible to manipulate the rope so that the structure remains the same but the arrow points the other way, the knot is invertible. Until the mid-1960s one of the most vexing unsolved problems in knot theory was whether noninvertible knots exist. All knots of seven or fewer crossings, and all but one 8-knot and four 9-knots had earlier been found invertible by manipulating rope models. It was in 1963 that Hale F. Trotter, now at Princeton University, announced in the title of a surprising paper "Non-invertible Knots Exist" (*Topology,* Vol. 2, No. 4, December 1963, pp. 275–80).

Trotter described an infinite family of pretzel knots that will not invert. A pretzel knot is one that can be drawn, without any crossings, on the surface of a pretzel (a two-hole torus). It can be drawn as shown in

Figure 19.3 as a two-strand braid that goes around two "holes," or it can be modeled by the edge of a sheet of paper with three twisted strips. If the braid surrounds just one hole, it is called a torus knot because it can be drawn without crossings on the surface of a doughnut.

Figure 19.3.

Trotter found an elegant proof that all pretzel knots are noninvertible if the crossing numbers for the three twisted strips are distinct odd integers with absolute values greater than 1. Positive integers indicate braids that twist one way and negative integers indicate an opposite twist. Later Trotter's student Richard L. Parris showed in his unpublished Ph.D. thesis that the absolute values can be ignored provided the signed values are distinct and that these conditions are necessary as well as sufficient for noninvertible pretzels. Thus the simplest noninvertible pretzel is the one shown. Its crossing numbers of 3, –3, and 5 make it an 11-knot.

It is now known that the simplest noninvertible knot is the amphicheiral 8-knot in Figure 19.4. It was first proved noninvertible by Akio Kawauchi in *Proceedings of the Japan Academy* (Vol. 55, Series A, No. 10, December 1979, pp. 399–402). According to Richard Hartley, in "Identifying Non-invertible Knots" (*Topology*, Vol. 22, No. 2, 1983, pp. 137–45), this is the only noninvertible knot of eight crossings, and there are only two such knots of nine crossings and 33 of 10. All 36 of these knots had earlier been declared noninvertible by John Horton Conway, but only on the empirical grounds that he had not been able to invert them. The noninvertible knots among the more than 550 knots with 11 crossings had not yet been identified.

In 1967 Conway published the first classification of all prime knots with 11 or fewer crossings. (A few minor errors were corrected in a later printing.) You will find clear diagrams for all prime knots through 10 crossings, and all links through nine crossings, in Dale Rolfsen's valu-

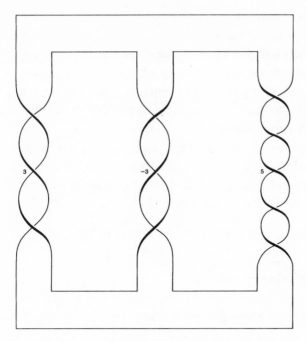

Figure 19.4.

able 1990 book *Knots and Links.* There are no knots with 1 or 2 crossings, one with 3, one with 4, two with 5, three with 6, seven with 7, 21 with 8 crossings, 49 with 9, 165 with 10, and 552 with 11, for a total of 801 prime knots with 11 or fewer crossings. At the time I write, the classification has been extended through 16 crossings.

There are many strange ways to label the crossings of a knot, then derive an algebraic expression that is an invariant for all possible diagrams of that knot. One of the earliest of such techniques produces what is called a knot's Alexander polynomial, named after the American mathematician James W. Alexander who discovered it in 1928. Conway later found a beautiful new way to compute a "Conway polynomial" that is equivalent to the Alexander one.

For the unknotted knot with no crossings the Alexander polynomial is 1. The expression for the trefoil knot of three crossings is $x^2 - x + 1$, regardless of its handedness. The figure-eight knot of four crossings has the polynomial $x^2 - 3x + 1$. The square knot, a product of two trefoils, has an Alexander polynomial of $(x^2 - x + 1)^2$, the square of the trefoil's expression. Unfortunately, a granny knot has the same polynomial. If two knot diagrams give different polynomials, they are sure to be different knots, but the converse is not true. Two knots may have the same polynomial yet not be the same. Finding a way to give any knot an expression that applies to all diagrams of that knot, and only that knot, is the major unsolved problem in knot theory.

Although there are tests for deciding whether any given knot is trivial, the methods are complex and tedious. For this reason many problems that are easy to state are not easy to resolve except by working empirically with rope models. For instance, is it possible to twist an elastic band around a cube so that each face of the cube has an under–over crossing as shown in Figure 19.5. To put it another way, can you tie a cord around a cube in this manner so that if you slip the cord off the cube, the cord will be unknotted?

Figure 19.5.

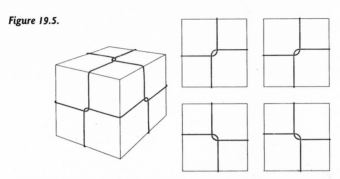

Note that on each face the crossing must take one of the four forms depicted in the illustration. This makes $4^6 = 4,096$ ways to wrap the cord. The wrapping can be diagrammed as a 12-knot, with six pairs of crossings, each pair of which can have one of four patterns. The problem was first posed by Horace W. Hinkle in *Journal of Recreational Mathematics* in 1978. In a later issue (Vol. 12, No. 1, 1979–80, pp. 60–62) Karl Scherer showed how symmetry considerations reduce the number of essentially different wrappings to 128. Scherer tested each wrapping empirically and found that in every case the cord is knotted. This has yet to be confirmed by others, and no one has so far found a simpler way to attack the problem. The impossibility of getting the desired wrapping with an unknotted cord seems odd, because it is easy to twist a rubber band around a cube to put the under-over crossings on just two or four faces (all other faces being straight crossings), and seemingly impossible to do it on just one face, three faces, or five faces. One would therefore expect six to be possible, but apparently it is not. It may also be impossible to get the pattern even if two, three, or four rubber bands are used.

Figure 19.6 depicts a delightful knot-and-link puzzle that was sent to me by its inventor, Majunath M. Hegde, then a mathematics student in India. The rope's ends are tied to a piece of furniture, say a chair. Note that the two trefoil knots form a granny. The task is to manipulate the rope and ring so that the ring is moved to the upper knot as is indicated by the broken line. All else must remain identical.

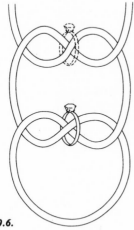

Figure 19.6.

It is easy to do if you have the right insight. Of course, the rope must not be untied from the chair, nor are you allowed to open a knot and pass the chair through it. It will help if you think of the ends of the rope as being permanently fastened to a wall.

The trick of dissolving or creating knots by passing a person through a loop was actually used by fake mediums in the days when it was fashionable to relate psychic phenomena to the fourth dimension. Knots in closed curves are possible only in 3-space. In 4-space all knots dissolve. If you could toss an unknotted loop of rope to a creature in 4-space, it could tie any knot in the loop and toss it back to you with the knot permanently formed. There was a popular theory among physicists who believed in spiritualism that mediums had the power to move objects in and out of higher spaces. Some mediums, such as the American mountebank Henry Slade, exploited this theory by pretending to put knots into closed loops of cord. Johann Karl F. Zöllner, an Austrian physicist, devoted an entire book to Slade and hyperspace. Its English translation, *Transcendental Physics* (Arno Press, 1976), is worth reading as striking testimony to the ease with which an intelligent physicist can be gulled by a clever conjurer.

In another instance, psychic investigators William Cox and John Richards exhibited a stop-action film that purported to show two leather rings becoming linked and unlinked inside a fish tank. "Later examination showed no evidence that the rings were severed in any way," wrote *National Enquirer* when it reported this "miracle" on October 27, 1981. I was then reminded of an old conjuring stage joke. The performer announces that he has magically transported a rabbit from one opaque box to another. Then before opening either box he says that he will magically transport the rabbit back again.

It is easy, by the way, to fabricate two linked "rubber bands." Just draw them linked on the surface of a baby's hollow rubber teething ring and carefully cut them out. Two linked wood rings, each of a different wood, can be carved if you insert one ring into a notch cut into a tree, then wait many years until the tree grows around and through it. Because the trefoil is a torus knot, it too is easily cut from a teething ring.

The trick I am about to describe was too crude for Slade, but less clever mediums occasionally resorted to it. You will find it explained, along with other knot-tying swindles, in Chapter 2 of Hereward Carrington's *The Physical Phenomena of Spiritualism, Fraudulent and Genuine* (H. B. Turner & Co., Boston, 1907). One end of a very long piece of rope is tied to the wrist of one guest and the other end is tied to the wrist of another guest. After the seance, when the lights are turned on, several knots are in the rope. How do they get there?

The two guests stand side by side when the lights go out. In the dark the medium (or an accomplice) makes a few large coils of rope, then passes them carefully over the head and body of one of the guests. The coils lie flat on the floor until later, when the medium casually asks that guest to step a few feet to one side. This frees the coils from the person, allowing the medium to pull them into a sequence of tight knots at the center of the rope. Stepping to one side seems so irrelevant to the phenomenon that no one remembers it. Ask the guest himself a few weeks later whether he changed his position, and he will vigorously and honestly deny it.

Roger Penrose, the British mathematician and physicist, once showed me an unusual trick involving the mysterious appearance of a knot. Penrose invented it when he was in grade school. It is based on what in crocheting, sewing, and embroidery is called a chain stitch. Begin the chain by trying a trefoil knot at one end of a long piece of heavy cord or thin rope and hold it with your left hand as in step 1 in Figure 19.7. With your right thumb and finger take the cord at *A* and pull down a loop as in step 2. Reach through the loop, take the cord at *B,* and pull down another loop (step 3). Again reach forward through the lowest loop, take the cord at *D,* and pull down another loop (step 4). Continue in this way until you have formed as long a chain as possible.

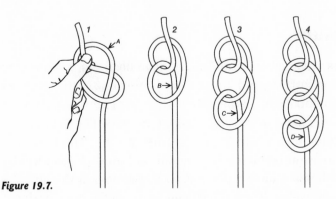

Figure 19.7.

With your right hand holding the lower end of the chain, pull the chain taut. Ask someone to select any link he likes and then pinch the link between his thumb and forefinger. Pull on both ends of the cord. All links dissolve, as expected, but when he separates his finger and thumb, there is a tight knot at precisely the spot he pinched!

Joel Langer, a mathematician at Case Western Reserve University, made a remarkable discovery. He found a way of constructing what he calls "jump knots" out of stainless-steel wire. The wire is knotted and then its ends are bonded. When it is manipulated properly, it can be pressed flat to form a braided ring. Release pressure on the ring; tension in the wire causes it to spring suddenly into a symmetrical three-dimensional shape. It is now a frustrating puzzle to collapse the wire back to its ring form.

In 1981 Langer and his associate Sharon O'Neil formed a company they called Why Knots. It made and sold three handsome jump knots: the Figure Eight, the Chinese Button Knot, and the Mathematician's Loop. When you slide one of these wire knots out of its square envelope, it pops into an elegant hanging ornament. The figure eight is the easiest to put back into its envelope. The Chinese button knot (so called because it is a form widely used in China for buttons on nightclothes) is more difficult. The mathematician's loop is the most difficult.

These shapes make it easier to understand how the 18th-century physicists could have developed a theory, respectable in its day, that molecules are different kinds of knots into which vortex rings of ether (today read "space–time") get themselves tied. Indeed, it was just such speculation that led the Scottish physicist Peter Guthrie Tait to study topology and conduct the world's first systematic investigation of knot theory.

Addendum

Enormous advances in knot theory have been made since this chapter was written in 1983, and knot theory is now one of the most exciting and active branches of mathematics. Dozens of new polynomials for classifying knots have been discovered. One is called the Homfly after the last initials of its six independent discoverers. The most significant new expression is the Jones polynomial found in 1984 by the New Zealand mathematician Vaughan F. R. Jones, now at the University of California, Berkeley. It has since been improved and generalized by Louis Kauffman and others. Although these new polynomials are surprisingly simple and powerful, no one has yet come up with an algebraic technique for distinguishing all knots. Knots with different polynomials are different, but it is still possible that two distinct knots will have the same expression.

The Alexander polynomial does not decide between mirror-image knots, and as we have seen, it does not distinguish the square knot from the granny. The Jones polynomial provides both distinctions. So far, it is not clear just why the Jones and the other new polynomials work. "They're magic" is how Joan Birman, a knot expert at Barnard College, put it.

The most amazing development in recent knot theory was the discovery that the best way to understand the Jones polynomial was in terms of statistical mechanics and quantum theory! Sir Michael Atiyah now retired from the University of Edinburgh, was the first to see these connections, then Edward Witten, at the Institute for Advanced Study in Princeton, did the pioneer work in developing the connections. Knot theory now has surprising applications to superstrings, a theory that explains basic particles by treating them as tiny loops, and to quantum field theory. There is intense interaction between today's physicists and topologists. Discoveries in physics are leading to new discoveries in topology, and vice versa. No one can predict where it will all lead.

Another unexpected application of knot theory is in broadening our understanding of the structure and properties of large molecules such as polymers, and especially the behavior of DNA molecules. DNA strands can become horribly knotted and linked, unable to replicate until they are untied or unlinked by enzymes called topoisom-erases. To straighten out a DNA strand, enzymes have to slice them so they can pass through themselves or another strand, then splice the ends to-gether again. The number of times this must occur to undo a knot or linkage of DNA determines the speed with which the DNA unknots or unlinks.

There is a delightful three-color test for deciding if a knot diagram represents a knot. Draw the diagram, then see if you can color its "arcs" (portions of the line between two crossings) with three colors so that ei-ther all three colors meet at each crossing or there is only one color at each crossing, and provided at least one crossing shows all three col-ors. If you can do this, the line is knotted. If you can't, the line may or may not be knotted. The three-coloring can also be used to prove that two knots are different.

In 1908 the German mathematician Heinrich Tietze conjectured that two knots are identical if and only if their complements—the topolog-ical structure of the space in which they are embedded—are identical. His conjecture was proved in 1988 by two American mathematicians, Cameron M. Gordon and John E. Luecke. A knot's complement is a structure in 3-space, in contrast to the knot which is one-dimensional. Its topological structure is more complicated than the knot's, but of course it contains complete information about the knot. The theorem fails for links. Two links that are not the same can have identical com-plements.

Associated with each knot's complement is a group. Like the poly-nomials, which can be extracted from the group, two knots can have the same group yet not be the same knots. An anonymous poet summed up the situation this way in the British periodical *Manifold* (Summer 1972):

> *A knot and*
> *another*

> *knot may*
> *not be the*
> *same knot, though*
> *the knot group of*
> *the knot and the*
> *other knot's*
> *knot group*
> *differ not; BUT*
> *if the knot group*
> *of a knot*
> *is the knot group*
> *of the not*
> *knotted*
> *knot,*
> *the knot is*
> *not*
> *knotted.*

The American philosopher Charles Peirce, in a section on knots in his *New Elements of Mathematics* (Volume 2, Chapter 4), shows how the Borromean rings (three rings linked in such a way that although they can't be separated, no two rings are linked) can be cut from a three-hole torus. Peirce also shows how to cut the figure-eight knot and the bowline knot from a two-hole torus.

Richard Parris called attention to the fact that not all of the 4,096 ways to wrap string around the cube, in the problem I posed, are knots. Most of them are links of two, three, or four separate loops.

Conway has proved that if you draw a complete graph for seven points located anywhere in space—that is, draw a line connecting each pair of points—the lines will form at least one knot.

For the most recent results in knot theory there is now a *Journal of Knot Theory and Its Ramifications.*

Answers

Figure 19.8 shows how a square knot can be changed to an alternating knot of six crossings. Simply flip dotted arc *a* over to make arc *b*.

Figure 19.9 shows one way to solve the ring-and-granny-knot puzzle. First make the lower knot small, then slide it (carrying the ring with it)

Figure 19.8.

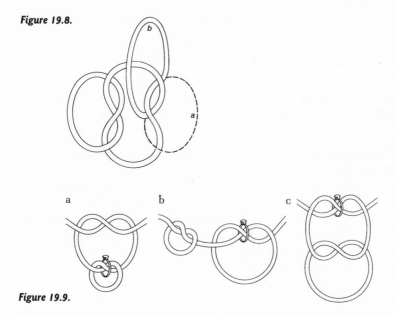

Figure 19.9.

up and through the higher knot *(a)*. Open it. Two trefoil knots are now side by side *(b)*. Make the ringless knot small, then slide it through and down the other knot. Open it up and you have finished *(c)*.

Bibliography

Books

C. C. Adams, *The Knot Book*. W. H. Freeman, 1994.

M. Atiyah, The Geometry and Physics of Knots, Cambridge University Press, 1990.

R. H. Crowell and R. H. Fox, *Introduction to Knot Theory,* Blaisdell, 1963; Springer-Verlag, 1977.

D. W. Farmer and T. B. Stanford, *Knots and Surfaces,* American Mathematical Society, 1998.

L. Kauffman, *On Knots,* Princeton University Press, 1987.

L. Kauffman, *Knots and Physics,* World Scientific, 1991.

T. Kohno, *New Developments in the Theory of Knots,* World Scientific, 1990.

C. Livingston, Knot Theory, Washington, DC: Mathematical Association of America, 1993.

K. Murasugi, *Knot Theory and Its Applications,* Birkhauser, 1999.

V. V. Pracolov, *Knots, Links, Braids, and 3-Manifolds,* Providence, RI: American Mathematical Society, 1997.

D. Rolfsen, *Knots and Links,* Publish or Perish, 1976, Second edition, 1990.

J. C. Turner and P. van de Griend, *The History and Science of Knots,* World Scientific, 1996.

D.J.A. Walsh, *Complexity Knots, Colourings and Counting,* Cambridge University Press, 1993.

Papers

Out of hundreds of papers on knot theory published since 1980, I have selected only a few that have appeared since 1990.

C. C. Adams, "Tilings of Space by Knotted Tiles," *Mathematical Intelligencer,* Vol. 17, No. 2, 1995, pp. 41–51.

P. Andersson "The Color Invariant for Knots and Links," *American Mathematical Monthly,* Vol. 102, May 1995, pp. 442–48.

M. Atiyah, "Geometry and Physics," *The Mathematical Gazette,* March 1996, pp. 78–82.

J. S. Birman "Recent Developments in Braid and Link Theory," *The Mathematical Intelligencer,* Vol. 13 1991, pp. 57–60.

B. Cipra, "Knotty Problems—and Real-World Solutions," *Science,* Vol. 255, January 24, 1992, pp. 403–4.

B. Cipra, "From Knot to Unknot," *What's Happening in the Mathematical Sciences,* Vol 2, 1994, pp. 8–13.

O. T. Dasbach and S. Hougardy, "Does the Jones Polynomial Detect Unknottedness?" *Experimental Mathematics,* Vol. 6, 1997, pp. 51–56.

S. Harris and G. Quenell, "Knot Labelings and Knots Without Labelings." *Mathematical Intelligencer,* Vol. 21, No. 2, 1999, pp. 51–56.

B. Hayes, "Square Knots," *American Scientist,* November/December, 1997, pp. 506–10.

J. Hoste, M. Thistlewaite, and J. Weeks, "The First 1,701,936 Knots," *Mathematical Intelligencer,* Vol. 20, No. 4, 1998, pp. 33–48.

V. F. R. Jones, "Knot Theory and Statistical Mechanics," *Scientific American,* November 1990, pp. 98–103.

R. Matthews, "Knots Landing," *New Scientist,* February 1, 1997, pp. 42–43.

W. Menasco and L. Rudolph, "How Hard is it to Untie a Knot?" *American Scientist,* Vol. 83, January/February 1995, pp. 38–50.

I. Peterson, "Knotty Views," *Science News,* Vol. 141, March 21, 1992, pp. 186–7.

A. Sosinsky, "Braids and Knots," *Quantum,* January/February 1995, pp. 11–15.

A. Sosinsky, "Knots, Links, and Their Polynomials," *Quantum,* July/August 1995, pp. 9–14.

I. Stewart, "Knots, Links and Videotape," *Scientific American,* January 1994, pp. 152–54.

D. W. Sumners, "Untangling DNA," *The Mathematical Intelligencer,* Vol. 12, 1990, pp. 71–80.

D. W. Sumners, "Lifting the Curtain: Using Topology to Probe the Hidden Action of Enzymes," *Notices of the American Mathematical Society,* Vol. 42, May 1995, pp. 528–31.

O. Viro, "Tied Into Knot Theory," *Quantum,* May/June 1998, pp. 18–20.

Knot Theory and Quantum Mechanics

M. Atiyah, *The Geometry and Physics of Knots,* Cambridge University Press, 1990.

M. L. Ge and C. N. Yang (eds), *Braid Groups, Knot Theory and Statistical Mechanics,* World Scientific, 1989.

V. F. R. Jones, "Knot Theory and Statical Mechanics," *Scientific American,* November 1990, pp. 98–103.

E. Witten, "Quantum Field Theory and the Jones Polynomial," *Communications in Mathematical Physics,* Vol. 121, No. 3, 1989, pp. 351–99.

Chapter 20

Doughnuts:
Linked and Knotted

As you ramble on through life, brother,
  Whatever be your goal,
Keep your eye upon the doughnut
  And not upon the hole!

—ANON.

A torus is a doughnut-shaped surface generated by rotating a circle around an axis that lies on the plane of the circle but does not intersect the circle. Small circles, called meridians, can be drawn around the torus with radii equal to that of the generating circle. Circles of varying radii that go around the hole or center of the torus on parallel planes are called parallels (see Figure 20.1). Both meridians and parallels on a torus are infinite in number. There are two other less obvious infinite sets of "oblique" circles. Can you find them? Members of one set do not intersect one another, whereas any member of one set twice intersects any member of the other.

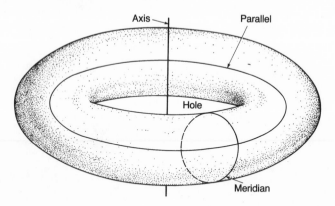

Figure 20.1. The torus

To a topologist, concerned only with properties that do not alter when a figure is elastically deformed, a torus is topologically equivalent to the surface of such objects as a ring, a bagel, a life preserver, a button with one hole, a coffee cup, a soda straw, a rubber band, a sphere

254

with one handle, a cube with one hole through it, and so on. Think of these surfaces as a thin membrane that can be stretched or compressed as much as one wishes. Each can be deformed until it becomes a perfect toroidal surface. In what follows, "torus" will mean any surface topologically equivalent to a torus.

A common misunderstanding about topology is the belief that a rubber model of a surface can always be deformed in three-dimensional space to make any topologically equivalent model. This often is not the case. A Möbius strip, for example, has a handedness in 3-space that cannot be altered by twisting and stretching. Handedness is an extrinsic property it acquires only when embedded in 3-space. Intrinsically it has no handedness. A 4-space creature could pick up a left-handed strip, turn it over in 4-space and drop it back in our space as a right-handed model.

A similar dichotomy applies to knots in closed curves. Tie a single overhand (or trefoil) knot in a piece of rope and join the ends. The surface of the rope is equivalent to a knotted torus. It has a handedness, and no amount of fiddling with the rope can change the parity. Intrinsically the rope is not even knotted. A 4-space creature could take from us an unknotted closed piece of rope and, without cutting it, return it to us as knotted in either left or right form. All the properties of knots are extrinsic properties of toruses (or, if you prefer, one-dimensional curves that may be thought of as toruses whose meridians have shrunk to points) that are embedded in 3-space.

It is not always easy to decide intuitively if a given surface in 3-space can be elastically deformed to a different but topologically equivalent surface. A striking instance, discussed nearly 50 years ago (see Albert W. Tucker and Herbert S. Bailey, Jr., "Topology," *Scientific American,* January 1950), concerns a rubber torus with a hole in its surface. Can it be turned inside out to make a torus of identical shape? The answer is yes. It is hard to do with a rubber model (such as an inner tube), but a model made of wool reverses readily. Stephen Barr, in his *Second Miscellany of Puzzles* (Macmillan, 1969), recommends making it from a square piece of cloth. Fold the cloth in half and sew together opposite edges to make a tube. Now sew the ends of the tube together to make a torus that is square shaped when flattened. For ease in reversing, the surface hole is a slot cut in the outer layer of cloth (shown by the broken line in Figure 20.2).

After the cloth torus is turned inside out, it is exactly the same shape

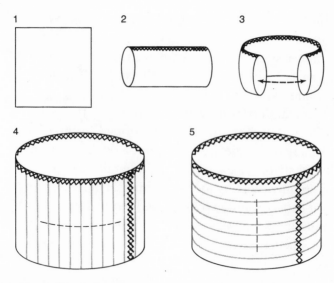

Figure 20.2. Reversible cloth torus

as before, except that what were formerly meridians have become parallels, and vice versa. To make the switch visible, sew or ink on the model a meridian of one color and a parallel of another so that both colors are visible from either side of the cloth. In 1958 Mrs. Eunice Hakala sent me a model she had made by cutting off the ribbed top of a sock and joining the tube's ends. The ribbing provides a neat set of parallels that turn into meridians after the torus is reversed.

Let us complicate matters by considering a torus tied in a trefoil knot. If we ignore handedness, there are only two such toruses: one with an external knot and one with an internal knot (see Figure 20.3(a) and (b)). A way to visualize the internally knotted torus is to imagine that the externally knotted torus on the left is sliced open along a meridian outside the knot. One end is turned back, as though reversing a sock; then the tube is expanded and drawn over the entire knot, and its ends are joined once more. Or imagine a solid wood cube with a hole bored through it that, instead of going straight, ties a knot before it emerges on the opposite side. The surface of such a cube is topologically equivalent to an internally knotted torus.

You might suppose that a torus could be simultaneously knotted externally and internally, but it can't be done. One kind of torus seems to have both an outside and an inside knot (see Figure 20.3(c)). Actually both knots are humbugs. Untying the outer knot simultaneously unties

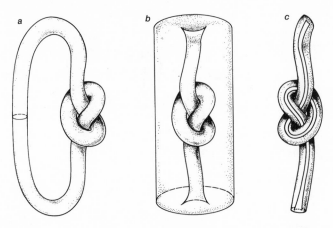

Figure 20.3. Torus with outside knot *(a)*, inside knot *(b)*, and pseudoknots *(c)*

the inner one, proving that the model is topologically the same as an unknotted torus—its hole elongated like the hole of a garden hose.

Although an outside-knotted torus is intrinsically identical with an inside-knotted one, it is not possible to deform one to the other when it is embedded in 3-space. If there is a hole in the side of an outside-knotted torus, can the torus be reversed in 3-space to put the knot inside? In the answer section I shall show how R. H. Bing (1915–1986) answered this question with a simple sketch.

A similar but harder problem was solved by Bing in his paper "Mapping a 3-Sphere onto a Homotopy 3-Sphere," in *Topology Seminar, Wisconsin, 1965,* edited by Bing and R. J. Bean (Princeton University Press, 1966). Imagine a cube with two straight holes (see Figure 20.4*(a)*). Its surface is topologically the same as a two-hole doughnut. We can also have a cube with two holes, one straight, one knotted (Figure 20.4*(b)*). It is not possible in 3-space to deform the second cube so that the knot dissolves and the model looks like the first one. A third cube has one straight hole and one knotted hole with the knot around the straight hole (Figure 20.4*(c)*). Can this cube be elastically deformed until it becomes the first model? It is hard to believe, but the answer is yes. Bing's proof is so elegant and simple that the diagrams for it are almost self-explanatory (see Figure 20.5). In elastic deformation a hole can be moved any distance over a surface without altering the surface's topology. As the hole moves, the surface merely stretches in back and shrinks in front. In Bing's proof the knotted tube is drawn as a single line to make the proof easier to follow. The hole, at the base of this tube

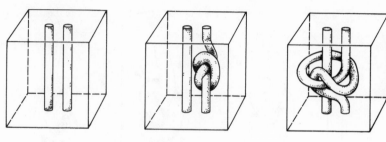

Figure 20.4. Three varieties of a two-hole torus

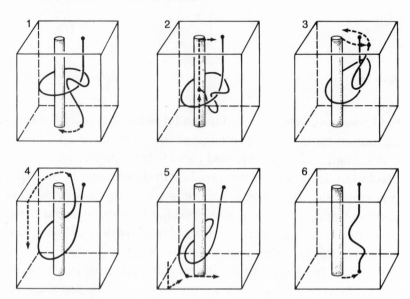

Figure 20.5. R. H. Bing's proof

is moved over the cube's surface, as indicated by the arrows, dragging the tube along with it. It goes left to the base of the other tube, climbs that tube's side, moves to the right across the top of the cube, circles its top hole counterclockwise, continues left around the other hole, over the cube's front edge, down the front face, around the lower edge to the cube's bottom face, and then across that face to the position it formerly occupied. It is easy to see that the tube attached to this hole has been untied. Naturally the procedure is reversible. If you had a sufficiently pliable doughnut surface with two holes, you could manipulate it until one hole became a knot tied around the other.

Topologists worried for decades about whether two separate knots side by side on a closed rope could cancel each other; that is, could the rope be manipulated until both knots dissolved? No pair of canceling knots had been found, but proving the impossibility of such a pair was another matter. It was not even possible to show that two trefoil knots of opposite handedness could not cancel. Proofs of the general case were not found until the early 1950s. One way of proving it is explained by Ralph H. Fox in "A Quick Trip through Knot Theory," in *Topology of 3-Manifolds and Related Topics,* edited by M. K. Fort, Jr. (Prentice-Hall, 1963). It is a reductio ad absurdum proof that unfortunately involves the sophisticated concept of an infinity of knots on a closed curve and certain assumptions about infinite sets that must be carefully specified to make the proof rigorous.

When John Horton Conway, a former University of Cambridge mathematician now at Princeton University, was in high school, he hit on a simpler proof that completely avoids infinite sets of knots. Later he learned that essentially the same proof had been formulated earlier, but I have not been able to determine by whom. Here is Conway's version as he explained it years ago in a letter. It is a marvelous example of how a knotted torus can play an unexpected role in proving a fundamental theorem of modern knot theory.

Conway's proof, like the one for the infinite knots, is a reductio ad absurdum. We begin by imagining that a closed string passes through the opposite walls of a room (see Figure 20.6). Since we shall be concerned only with what happens inside the room, we can forget about the string outside and regard it as being attached to the side walls. On the string are knots *A* and *B*. Each is assumed to be genuine in the sense that it cannot be removed by manipulating the string if it is the only knot on the string. It also is assumed that the two knots will cancel each other when both are on the same closed curve. The proof applies to pairs of knots of any kind whatever, but here we show the knots as simple trefoils of opposite parity. If the knots can cancel, it means that the string can be manipulated until it stretches straight from wall to wall. Think of the string as being elastic to provide all the needed slack for such an operation. In the center figure we introduce an elastic torus around the string. Note that the tube "swallows" knot *A* but "circumnavigates" knot *B* (Conway's terminology). Any parallel drawn on this tube, on the section between the walls, obviously must be knotted in the same way as knot *B*. Indeed, it can be shown that any line on the tube's surface, stretching from wall to wall and never crossing itself at any spot on the tube's surface, will be knotted like knot *B*.

"Now," writes Conway, "comes the crunch." Perform on the string

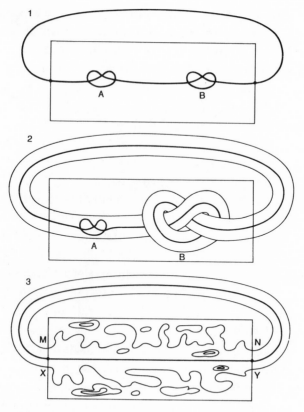

Figure 20.6. John Horton Conway's proof

the operation that we assumed would dissolve both knots. This can be done without breaking the tube. Because the string is never allowed to pass through itself during the deformation, we can always push the tube's wall aside if it gets in the way. The third drawing in Figure 20.6 shows the final result. The string is unknotted. The tube may have reached a horribly complicated shape impossible to draw. Consider a vertical plane passing through the straight string and cutting the twisted tube. We can suppose that the tube's cross section will look something like what is shown with the possibility of various "islands," but there will necessarily be two lines, *XY* and *MN*, from wall to wall that do not cross themselves at any point on the vertical plane. Each line will be unknotted. Moreover, each line also is a curve that does not cross itself on the tube's surface. As we have seen, all such lines were (before the deformation) knotted like knot *B*. The deformation has therefore removed a knot equivalent to knot *B* from each of these two lines. Therefore knot *B,* alone on a line, can be removed by manipulating that line. But knot *B,* by definition, is a genuine knot that cannot be so removed. We have contradicted an assumption. If two knots on a string can cancel, neither knot (since the same proof can be applied to knot *A*) can be genuine. Both must really have been pseudoknots.

Although a one-hole torus can be embedded in 3-space in only three ways (outside knot, inside knot, no knot), a two-hole torus has so many bizarre forms that the number is, I believe, not yet known. In some cases it can be reduced to a simpler form by deformation. For example, a tube-through-hole is equivalent to an ordinary two-hole doughnut (see Figure 20.7*(a)*), but what about the other two figures (Figures 20.7*(b)* and *(c)*)? They are among several dozen monstrosities sketched by Piet Hein in a moment of meditation on two-hole toruses. In Figure 20.7*(b)* an inside knot goes through an outside one, and in Figure 20.7*(c)* an outside knot goes through a hole. Is it possible, by deformation, to dissolve the inside knot of Figure 20.7*(b)* and the outside knot of Figure 20.7*(c)?*

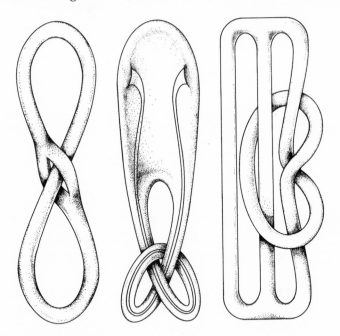

Figure 20.7. Two-hole toruses

With more complicated pairs of two-holers embedded in 3-space, proofs that one can be deformed to the other are not so easy. As one of Piet Hein's "grooks" puts it:

There are doughnuts and doughnuts
with knots and with no knots
and many a doughnut
so nuts that we know not.

Here are three more toroidally knotty questions.

1. How many closed curves can be drawn on a torus, each a trefoil knot of the same handedness, so that no two curves cross each other at any point?

2. If two closed curves are drawn on a torus so that each forms a trefoil knot but the knots are of opposite parity, what is the minimum number of points at which the two curves will intersect each other?

3. Show how to cut a solid two-hole doughnut with one slice of a knife so that the result is a solid outside-knotted torus. The "slice" is not, of course, planar. More technically, show how to remove from a two-hole doughnut a section topologically equivalent to a disk so that what remains is a solid knotted torus. (This amusing result was discovered by John Stallings in 1957 and communicated to me by James Stasheff.)

Addendum

In studying the properties of topological surfaces, one must always keep in mind the distinction between intrinsic properties, independent of the space in which the surface is embedded, and properties that arise from the embedding. The "complement" of a surface consists of all the points in the embedding space that are not in the surface. For example, a torus with no knot, one with an outside knot, and one with an inside knot all have identical intrinsic properties. No two have topologically identical complements; hence, no two are equivalent in their extrinsic topological properties.

John Stillwell, a mathematician at Monash University, Australia, sent several fascinating letters in which he showed how an unknotted torus with any number of holes—such toruses are equivalent to the surfaces of spheres with handles—could be turned inside out through a surface hole. He was not sure whether a knotted torus, even with only one hole, could be turned inside out through a hole in its surface.

Many beautiful, counterintuitive problems involving links and knots in toruses have been published. See Rolfsen's book, cited in the bibliography, especially the startling problem on page 95, where he shows that the surface on the left of Figure 20.8 is topologically equivalent to the surface shown on the right.

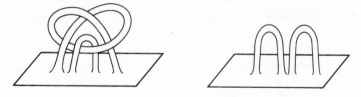

Figure 20.8. The surface on the left can be continuously deformed to the surface on the right.

In my opening paragraph I mentioned the four sets of circles that are contained in a torus. For readers who have difficulty finding the two systems of "oblique" circles, they are explained on pages 132–33 in H.S.M. Coxeter's classic *Introduction to Geometry.* (Wiley, 1961). In a footnote Coxeter says that drawings of all four systems of circles are given in Hermann Schmidt, *Die Inversion und ihre Anwendungen* (Oldenbourg, Munich, 1950, p. 52).

Here are four other toroidal puzzles that I have taken from other books of mine. The first is from *Penrose Tiles to Trapdoor Ciphers,* the second is from *Last Recreations,* and the third and fourth are from *Science Fiction Puzzle Tales.*

1. Stillwell thought of the following problem. Two toruses, *A* and *B,* are linked as is shown in Figure 20.9. There is a "mouth" (a hole) in *B.* We can stretch, compress, and deform either torus as radically as we please, but of course no tearing is allowed. Can *B* swallow *A?* At the finish *B* must have its original shape, although it will be larger, and *A* must be entirely inside it.

Figure 20.9. Can torus *B* swallow torus *A?*

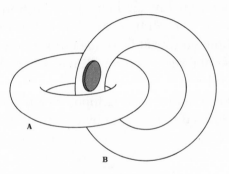

2. Two topologists were discussing at lunch the two linked surfaces shown at the left in Figure 20.10, which one of them had drawn on a paper napkin. You must not think of these objects as solids, like ropes or solid rubber rings. They are the surfaces of toruses, one surface of genus 1 (one hole), the other of genus 2 (two holes).

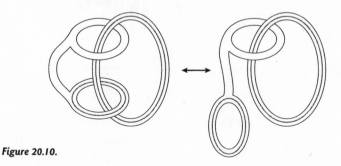

Figure 20.10.

Thinking in the mode of "rubber-sheet geometry," assume that the surfaces in the illustration can be stretched or shrunk in any desired way provided there is no tearing or sticking together of separate parts. Can the two-hole torus be deformed so that one hole becomes unlinked as is shown at the right in the illustration?

Topologist X offers the following impossibility proof. Paint a ring on each torus as is shown by the black lines. At the left the rings are linked. At the right they are unlinked.

"You will agree," says X, "that it is impossible by continuous deformation to unlink two linked rings embedded in three-dimensional space. It therefore follows that the transformation is impossible."

"But it doesn't follow at all," says Y.

Who is right? I am indebted to Herbert Taylor for discovering and sending this mystifying problem.

3. Figure 20.11 depicts a familiar linkage known as the Borromean rings. No two rings are linked, yet all three are linked. It is impossible to separate them without cutting at least one torus. Show how it is possible for any number of toruses to form a linked structure even though no two of them are linked.

4. Are the two structures shown in Figure 20.12 topologically the same? That is, can one be transformed to the other by a continuous deformation?

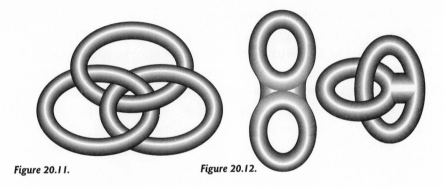

Figure 20.11.   **Figure 20.12.**

Answers

R. H. Bing shows how an internally knotted torus can be reversed through a hole to produce an externally knotted torus (see Figure 20.13). A small hole, *h,* is enlarged to cover almost the entire side of the cylinder, leaving only the shaded strip on the right. The top and bottom disks of the cylinder are flipped over, and the hole is shrunk to its original size.

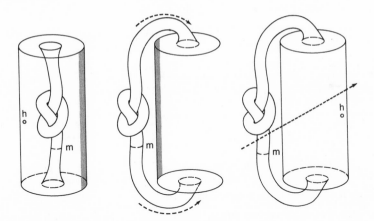

Figure 20.13. Solution to torus-reversed problem

As in reversing the unknotted torus through a hole, the deformation interchanges meridians and parallels. You might not at first think so because the circle, *m,* appears the same in all three pictures. The fact is, however, that initially it is a parallel circling the torus's elongated hole, whereas after the reversal it is a meridian. Moreover, after the reversal the torus's original hole is no longer through the knotted tube, which

Doughnuts: Linked and Knotted   265

is now closed at both ends. As indicated by the arrow, the hole is now surrounded by the knotted tube.

Piet Hein's two-hole torus, with an internal knot passing through an external one, is easily shown to be the same as a two-holer with only an external knot. Simply slide one end of the inside knot around the outside knot (in the manner explained earlier) and back to its starting point. This unties the internal knot. Piet Hein's two-holer, with the external knot going through a hole, can be unknotted by the deformation shown in Figure 20.14.

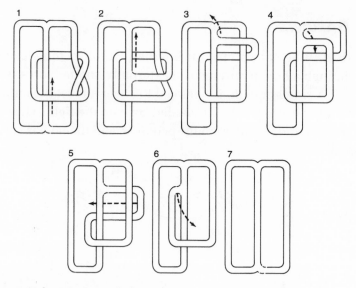

Figure 20.14. Unknotting a two-hole torus

Answers to the first three toroidal questions are as follows:

1. An infinity of noncrossing closed curves, each knotted with the same handedness, can be drawn on a torus (see Figure 20.15, *top*). If a torus surface is cut along any of these curves, the result is a two-sided, knotted band.
2. Two closed curves on a torus, knotted with opposite handedness, will intersect each other at least 12 times.
3. A rotating slice through a solid two-hole doughnut is used to produce a solid that is topologically equivalent to a solid, knotted torus (see Figure 20.15, *bottom*). Think of a short blade as moving downward and rotating one and a half turns as it descends. If the blade does not turn at all, two solid toruses result. A half-turn produces one solid,

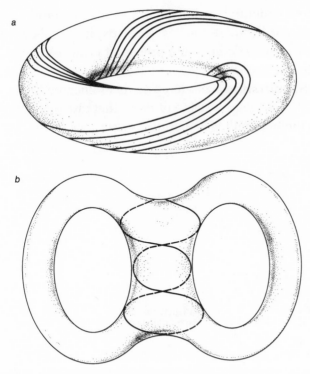

Figure 20.15. Knotted, nonintersecting curves on a torus *(a)* and rotating slice through a two-hole torus *(b)*

unknotted torus. One turn produces two solid, unknotted, linked toruses. Readers may enjoy investigating the general case of *n* half-turns.

Here are solutions to the four puzzles given in the addendum:

1. One torus can be inside another in two topologically distinct ways: the inside torus may surround the hole of the outside torus or it may not.

 If two toruses are linked and one has a "mouth," it cannot swallow the other so that the eaten torus is inside in the second sense. This result can be proved by drawing a closed curve on each torus in such a way that the two curves are linked in a simple manner. No amount of deformation can unlink the two curves. If one torus could swallow the other in the manner described, however, it could disgorge the eaten torus through its mouth and the two toruses would be unlinked. This result would also unlink the two closed curves. Because unlinking is impossible, cannibalism of this kind is also impossible.

The torus with the mouth can, however, swallow the other one so that the eaten torus is inside in the first sense explained above. Figure 20.16 shows how it is done. In the process it is necessary for the cannibal torus to turn inside out.

A good way to understand what happens is to imagine that torus A is shrunk until it becomes a stripe of paint that circles B. Turn A inside out through its mouth. The painted stripe goes inside, but in

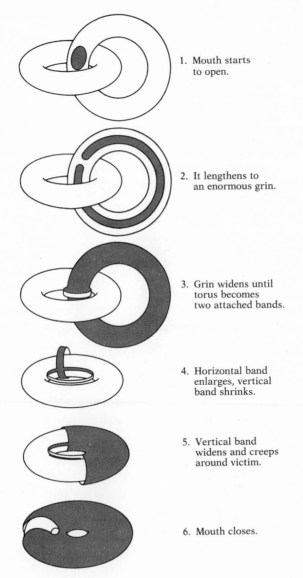

1. Mouth starts to open.

2. It lengthens to an enormous grin.

3. Grin widens until torus becomes two attached bands.

4. Horizontal band enlarges, vertical band shrinks.

5. Vertical band widens and creeps around victim.

6. Mouth closes.

Figure 20.16. How one torus eats another

doing so it ends up circling *B*'s hole. Expand the stripe back to a torus and you have the final picture of the sequence.

2. Figure 20.17 shows how a continuous deformation of the two-hole torus will unlink one of its holes from the single-hole torus. The argument for the impossibility of this task fails because if a ring is painted around one hole (as is shown by the black line), the ring becomes distorted in such a way that after the hole is unlinked the painted ring remains linked through the one-hole torus.

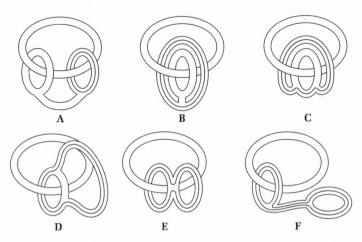

Figure 20.17.

For a mind-boggling selection of similar problems involving linked toruses, see Herbert Taylor's article "Bicycle Tubes Inside Out," in *The Mathematical Gardner* (Wadsworth, 1981), edited by David Klarner.

3. Figure 20.18 shows a circular chain that obviously can be enlarged to include any number of links. No two toruses are linked. If one link is cut and removed, all the others are free of one another.

4. The two forms are topologically identical. To prove this, imagine the linked form deformed to a sphere with two "handles" as shown in Figure 20.19.

Now imagine the base of one handle being moved over the surface, by shrinking the surface in front and stretching it in back, along the path shown by the dotted line. This links the two handles. The structure is now easily altered to correspond with the linked form of the toroids.

Doughnuts: Linked and Knotted

Figure 20.18.

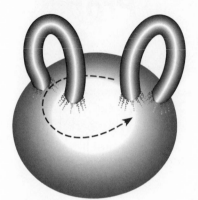

Figure 20.19.

Bibliography

B. H. Arnold, *Intuitive Concepts in Elementary Topology,* Prentice Hall, 1962.

M. Grayson, B. Kitchens, and G. Zettler, "Visualizing Toral Automorphisms," *The Mathematical Intelligencer,* Vol. 15, No. 1, 1993, pp. 6365.

J. Tanton, "A Dozen Questions About a Donut," *Math Horizons,* November 1998, pp. 26–31.

VI
Probability

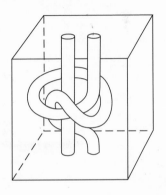

Chapter 21

Probability and Ambiguity

Charles Sanders Peirce once observed that in no other branch of mathematics is it so easy for experts to blunder as in probability theory. History bears this out. Leibniz thought it just as easy to throw 12 with a pair of dice as to throw 11. Jean le Rond d'Alembert, the great 18th-century French mathematician, could not see that the results of tossing a coin three times are the same as tossing three coins at once, and he believed (as many amateur gamblers persist in believing) that after a long run of heads, a tail is more likely.

Today, probability theory provides clear, unequivocal answers to simple questions of this sort, but only when the experimental procedure involved is precisely defined. A failure to do this is a common source of confusion in many recreational problems dealing with chance. A classic example is the problem of the broken stick. If a stick is broken at random into three pieces, what is the probability that the pieces can be put together in a triangle? This cannot be answered without additional information about the exact method of breaking to be used.

One method is to select, independently and at random, two points from the points that range uniformly along the stick, then break the stick at these two points. If this is the procedure to be followed, the answer is 1/4, and there is a neat way of demonstrating it with a geometrical diagram. We draw an equilateral triangle, then connect the midpoints of the sides to form a smaller shaded equilateral triangle in the center (see Figure 21.1). If we take any point in the large triangle and draw perpendiculars to the three sides, the sum of these three lines will be constant and equal to the altitude of the large triangle. When this point, like point *A,* is *inside* the shaded triangle, no one of the three perpendiculars will be longer than the sum of the other two.

Therefore the three line-segments will form a triangle. On the other hand, if the point, like point *B,* is *outside* the shaded triangle, one perpendicular is sure to be longer than the sum of the other two, and consequently no triangle can be formed with the three line-segments.

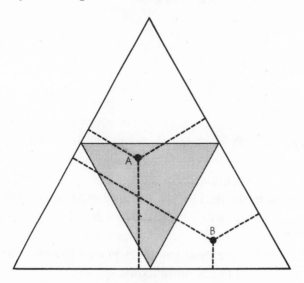

Figure 21.1. If a stick is broken in three pieces, the probability is 1/4 that they will form a triangle.

We now have a neat geometrical analogy to the problem of the broken stick. The sum of the three perpendiculars corresponds to the length of the stick. Each point on the large triangle represents a unique way of breaking the stick, the three perpendiculars corresponding to the three broken pieces. The probability of breaking the stick favorably is the same as the probability of selecting a point at random and finding that its three perpendiculars will form a triangle. As we have seen, this happens only when the point is inside the shaded triangle. Since this area is one-fourth the total area, the probability is 1/4.

Suppose, however, that we interpret in a different way the statement "break a stick at random into three pieces." We break the stick at random, we select randomly one of the two pieces, and we break that piece at random. What are the chances that the three pieces will form a triangle?

The same diagram will provide the answer. If after the first break we choose the smaller piece, no triangle is possible. What happens when we pick the larger piece? Let the vertical perpendicular in the diagram represent the smaller piece. In order for this line to be smaller than the sum of the other two perpendiculars, the point where the lines meet cannot be inside the small triangle at the top of the diagram. It must range uniformly over the lower three triangles. The shaded triangle continues to represent favorable points, but now it is only one-third the area under consideration. The chances, therefore, are 1/3 that when we break the larger piece, the three pieces will form a triangle. Since our chance of picking the larger piece is 1/2, the answer to the original question is the product of 1/2 and 1/3, or 1/6.

Geometrical diagrams of this sort must be used with caution because they too can be fraught with ambiguity. For example, consider this problem discussed by Joseph Bertrand, a famous French mathematician. What is the probability that a chord drawn at random inside a circle will be longer than the side of an equilateral triangle inscribed in the circle?

We can answer as follows. The chord must start at some point on the circumference. We call this point A, then draw a tangent to the circle at A, as shown in the top illustration of Figure 21.2. The other end of the chord will range uniformly over the circumference, generating an infinite series of equally probable chords, samples of which are shown on the illustration as broken lines. It is clear that only those chords that cut across the triangle are longer than the side of the triangle. Since the angle of the triangle at A is 60 degrees, and since all possible chords lie within a 180-degree range, the chances of drawing a chord larger than the side of the triangle must be 60/180, or 1/3.

Now let us approach the same problem a bit differently. The chord we draw must be perpendicular to one of the circle's diameters. We draw the diameter, then add the triangle as shown in the illustration at bottom left of Figure 21.2. All chords perpendicular to this diameter will pass through a point that ranges uniformly along the diameter. Samples of these chords are again shown as broken lines. It is not hard to prove that the distance from the center of the circle to A is half the radius. Let B mark the midpoint on the other side of the diameter. It is now easy to see that only those chords crossing the diameter between A and B will be longer than the side of the triangle. Since AB is half the diameter, we obtain an answer to our problem of 1/2.

Here is a third approach. The midpoint of the chord will range uniformly over the entire space within the circle. A study of the illustration at bottom right of Figure 21.2 will convince you that only chords

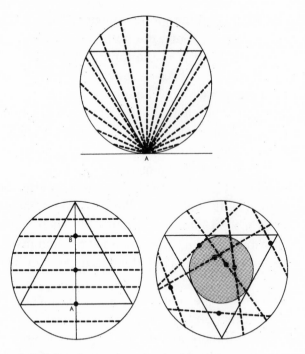

Figure 21.2. Probability that random chord is longer than side of inscribed equilateral triangle is proved to be 1/3 *(top)*, 1/2 *(left)*, and 1/4 *(right)*.

whose midpoints lie within the smaller shaded circle are longer than the side of the triangle. The area of the small circle is exactly one-fourth the area of the large circle, so the answer to our problem now appears to be 1/4.

Which of the three answers is right? Each is correct in reference to a certain mechanical procedure for drawing a random chord. Examples of the three procedures are as follows:

1. Two spinners are mounted at the center of a circle. They rotate independently. We spin them, mark the two points at which they stop, and connect the points with a straight line. The probability that this line will be longer than the side of the inscribed triangle is 1/3.

2. A large circle is chalked on the sidewalk. We roll a broom handle toward it, from a distance of 50 feet, until the handle stops somewhere on the circle. The probability that it will mark a chord longer than the side of the triangle is 1/2.

3. We paint a circle with molasses and wait until a fly lights on it, then

we draw the chord on which the fly is the midpoint. The probability that this chord is longer than the side of the triangle is 1/4.

Each of these procedures is a legitimate method of obtaining a "random chord." The problem as originally stated, therefore, is ambiguous. It has no answer until the meaning of "draw a chord at random" is made precise by a description of the procedure to be followed. Apparently nothing resembling any of the three procedures is actually adopted by most people when they are asked to draw a random chord. In an interesting unpublished paper entitled "The Human Organism as a Random Mechanism" Oliver L. Lacey, professor of psychology at the University of Alabama, reports on a test which showed the probability to be much better than 1/2 that a subject would draw a chord longer than the side of the inscribed triangle.

Another example of ambiguity arises from a failure to specify the randomizing procedure. Readers were told that Mr. Smith had two children, at least one of whom was a boy, and were asked to calculate the probability that both were boys. Many readers correctly pointed out that the answer depends on the procedure by which the information "at least one is a boy" is obtained. If from all families with two children, at least one of whom is a boy, a family is chosen at random, then the answer is 1/3. But there is another procedure that leads to exactly the same statement of the problem. From families with two children, one family is selected at random. If both children are boys, the informant says "at least one is a boy. If both are girls, he says "at least one is a girl." And if both sexes are represented, he picks a child at random and says "at least one is a . . . ," naming the child picked. When *this* procedure is followed, the probability that both children are of the same sex is clearly 1/2. (This is easy to see because the informant makes a statement in each of the four cases—BB, BG, GB, GG—and in half of these cases both children are of the same sex.)

The following wonderfully confusing little problem involving three prisoners and a warden is even more difficult to state unambiguously. Three men—A, B, and C—were in separate cells under sentence of death when the governor decided to pardon one of them. He wrote their names on three slips of paper, shook the slips in a hat, drew out one of them, and telephoned the warden, requesting that the name of the lucky man be kept secret for several days. Rumor of this reached prisoner A. When the warden made his morning rounds, A tried to persuade the warden to tell him who had been pardoned. The warden refused.

"Then tell me," said A, "the name of one of the others who will be executed. If B is to be pardoned, give me C's name. If C is to be pardoned, give me B's name. And if I'm to be pardoned, flip a coin to decide whether to name B or C."

"But if you see me flip the coin," replied the wary warden, "you'll know that you're the one pardoned. And if you see that I don't flip a coin, you'll know it's either you or the person I don't name."

"Then don't tell me now," said A. "Tell me tomorrow morning."

The warden, who knew nothing about probability theory, thought it over that night and decided that if he followed the procedure suggested by A, it would give A no help whatever in estimating his survival chances. So next morning he told A that B was going to be executed.

After the warden left, A smiled to himself at the warden's stupidity. There were now only two equally probable elements in what mathematicians like to call the "sample space" of the problem. Either C would be pardoned or himself, so by all the laws of conditional probability, his chances of survival had gone up from 1/3 to 1/2.

The warden did not know that A could communicate with C, in an adjacent cell, by tapping in code on a water pipe. This A proceeded to do, explaining to C exactly what he had said to the warden and what the warden had said to him. C was equally overjoyed with the news because he figured, by the same reasoning used by A, that his own survival chances had also risen to 1/2.

Did the two men reason correctly? If not, how should each have calculated his chances of being pardoned?

Addendum

In giving the second version of the broken stick problem I could hardly have picked a better illustration of the ease with which experts can blunder on probability computations, and the dangers of relying on a geometrical diagram. My solution was taken from William A. Whitworth's *DCC Exercises in Choice and Chance,* Problem 677; the same answer will be found in many other older textbooks on probability. It is entirely wrong!

In the first version of the problem, in which the two breaking points are simultaneously chosen, the representative point on the diagram ranges uniformly over the large triangle, permitting a comparison of areas to obtain a correct answer. In the second version, in which the stick is broken, then the larger piece is broken, Whitworth assumed that the point on the diagram ranged uniformly over the three lower triangles. It doesn't. There are more points within the central triangle than in the other two.

Let the length of the stick be 1 and x be the length of the smallest piece after the first break. To obtain pieces that will form a triangle, the larger segment must be broken within a length equal to $1 - x$. Therefore the probability of obtaining a triangle is $x/1 - x$. We now have to average all values of x, from 0 to $1/2$, to obtain a value for this expression. It proves to be $-1 + 2 \log 2$, or .386. Since the probability is $1/2$ that the larger piece will be picked for breaking, we multiply .386 by $1/2$ to obtain .193, the answer to the problem. This is a trifle larger than $1/6$, the answer obtained by following Whitworth's reasoning.

A large number of readers sent very clear analyses of the problem. In the above summary, I followed a solution sent by Mitchell P. Marcus of Binghamton, NY. Similar solutions were received from Edward Adams, Howard Grossman, Robert C. James, Gerald R. Lynch, G. Bach and R. Sharp, David Knaff, Norman Geschwind, and Raymond M. Redheffer. Professor Redheffer, at the University of California, is coauthor (with Ivan S. Sokolnikoff) of *Mathematics of Physics and Modern Engineering* (McGraw-Hill, 1958), in which will be found (p. 636) a full discussion of the problem. See also *Ingenious Mathematical Problems and Methods* by L. A. Graham (Dover, 1959, Problem 32) for other methods of solving the problem's first version.

Frederick R. Kling, John Ross, and Norman Cliff, all with the Educational Testing Service, Princeton, NJ, also sent a correct solution of the problem's second version. At the close of their letter they asked which of the following three hypotheses was most probable:

1. Mr. Gardner honestly blundered.

2. Mr. Gardner deliberately blundered in order to test his readers.

3. Mr. Gardner is guilty of what is known in the mathematical world as keeping up with the d'Alemberts.

The answer: number three.

Answers

The answer to the problem of the three prisoners is that A's chances of being pardoned are 1/3, and that C's chances are 2/3.

Regardless of who is pardoned, the warden can give A the name of a man, other than A, who will die. The warden's statement therefore has no influence on A's survival chances; they continue to be 1/3.

The situation is analogous to the following card game. Two black cards (representing death) and a red card (the pardon) are shuffled and dealt to three men: A, B, C (the prisoners). If a fourth person (the warden) peeks at all three cards, then turns over a black card belonging to either B or C, what is the probability that A's card is red? There is a temptation to suppose it is 1/2 because only two cards remain face-down, one of which is red. But since a black card can always be shown for B or C, turning it over provides no information of value in betting on the color of A's card.

This is easy to understand if we exaggerate the situation by letting death be represented by the ace of spades in a full deck. The deck is spread, and A draws a card. His chance of avoiding death is 51/52. Suppose now that someone peeks at the cards, then turns face up 50 cards that do not include the ace of spades. Only two face-down cards are left, one of which must be the ace of spades, but this obviously does not lower A's chances to 1/2. It doesn't because it is always possible, if one looks at the faces of the 51 cards, to find 50 that do not include the ace of spades. Finding them and turning them face up, therefore, has no effect on A's chances. Of course if 50 cards are turned over at random, and none prove to be the ace of spades, then the chance that A drew the death card *does* rise to 1/2.

What about prisoner C? Since either A or C must die, their respective probabilities for survival must add up to 1. A's chances to live are 1/3; therefore C's chances must be 2/3. This can be confirmed by considering the four possible elements in our sample space, and their respective initial probabilities:

1. C is pardoned, warden names B (probability 1/3).
2. B is pardoned, warden names C (probability 1/3).
3. A is pardoned, warden names B (probability 1/6).
4. A is pardoned, warden names C (probability 1/6).

Only cases 1 and 3 apply when it becomes known that B will die. The chances that it is case 1 are 1/3, or twice the chances (1/6) that it is case 3, so C's survival chances are two to one, or 2/3. In the card-game model this means that there is a probability of 2/3 that C's card is red.

This problem of the three prisoners brought a flood of mail, pro and con; happily, all objections proved groundless. Sheila Bishop of East Haven, CT, sent the following well-thought-out analysis:

> Sirs:
>
> I was first led to the conclusion that A's reasoning was incorrect by the following paradoxical situation. Suppose the original conversation between A and the warden had taken place in the same way, but now suppose that just as the warden was approaching A's cell to tell him that B would be executed, the warden fell down a manhole or was in some other way prevented from delivering the message.
>
> A could then reason as follows: "Suppose he was about to tell me that B would be executed. Then my chance of survival would be 1/2. If, on the other hand, he was going to tell me that C would be executed, then my chances would still be 1/2. Now I know as a certain fact that he would have told me one of those two things; therefore, either way, my survival chances are bound to be 1/2." Following this line of thought shows that A could have figured his chances to be 1/2 without ever asking the warden anything!
>
> After a couple of hours I finally arrived at this conclusion: Consider a large number of trios of prisoners all in this same situation, and in each group let A be the one who talks to the warden. If there are $3n$ trios altogether, then in n of them A will be pardoned, in n B will be pardoned, and in n C will be pardoned. There will be $3n/2$ cases in which the warden will say, "B will be executed." In n of these cases C will go free and in $n/2$ cases A will go free; C's chances are twice as good as A's. Hence A's and C's chances of survival and 1/3 and 2/3 respectively. . . .

Lester R. Ford, Jr., and David N. Walker, both with the Arizona office of General Analysis Corporation, felt that the warden has been unjustly maligned:

> Sirs:
>
> We are writing to you on behalf of the warden, who is a political appointee and therefore unwilling to enter into controversial matters in his own behalf.
>
> You characterize him in a slurring manner as "The warden, who knew

nothing about probability theory, . . ." and I feel that a grave injustice is being done. Not only are you incorrect (and possibly libelous), but I can personally assure you that his hobby for many years has been mathematics, and in particular, probability theory. His decision to answer A's question, while based on a humanitarian attempt to brighten the last hours of a condemned man (for, as we all now know, it was C who received the pardon), was a decision completely compatible with his instructions from the governor.

The only point on which he is open to criticism (and on this he has already been reprimanded by the governor) is that he was unable to prevent A from communicating with C, thereby permitting C to more accurately estimate *his* chances of survival. Here too, no great damage was done, since C failed to make proper use of the information.

If you do not publish both a retraction and an apology, we shall feel impelled to terminate our subscription.

Addendum

The problem of the two boys, as I said, must be very carefully stated to avoid ambiguity that prevents a precise answer. In my *Aha, Gotcha* I avoided ambiguity by imagining a lady who owned two parrots—one white, one black. A visitor asks the owner, "Is one bird a male?" The owner answers yes. The probability both parrots are male is 1/3. Had the visitor asked, "Is the dark bird a male?", a yes answer would have raised the probability that both birds are male to 1/2.

Richard E. Bedient, a mathematician at Hamilton College, described the prisoner's paradox in a poem that appeared in *The American Mathematical Monthly*, Vol. 101, March 1994, page 249:

THE PRISONER'S PARADOX REVISITED

Awaiting the dawn sat three prisoners wary,
A trio of brigands named Tom, Dick and Mary.
Sunrise would signal the death knoll of two,
Just one would survive, the question was who.

Young Mary sat thinking and finally spoke.
To the jailer she said, "You may think this a joke"
But it seems that my odds of surviving 'til tea,
Are clearly enough just one out of three.

But one of my cohorts must certainly go,
Without question, that's something I already know.

Telling the name of one who is lost,
Can't possibly help me. What could it cost?"

The shriveled old jailer himself was no dummy,
He thought, "But why not?" and pointed to Tommy.
"Now it's just Dick and I" Mary chortled with glee,
"One in two are my chances, and not one in three!"

Imagine the jailer's chagrin, that old elf,
She'd tricked him, or had she? Decide for yourself.

When I introduced the three prisoners paradox in my October 1959 column, I received a raft of letters from mathematicians who believed my solution was invalid. The number of such letters, however, was small compared to the thousands of letters Marilyn vos Savant received when she gave a version of the problem in her popular *Parade* column for September 9, 1990.

Ms. Savant's version of the paradox was based on a then-popular television show called *Let's Make a Deal,* hosted by Monty Hall. Imagine three doors, Marilyn wrote, to three rooms. Behind one door is a prize car. Behind each of the other two doors is a goat. A guest on the show is given a chance to win the prize by selecting the door with the car. If she chooses at random, clearly the probability she will select the prize door is 1/3. Now suppose, that after the guest's selection is voiced, Monty Hall, who knows what is behind each door, opens one door to disclose a goat. Two closed doors remain. One might reason that because the car is now behind one of just two doors, the probability the guest had chosen the correct door has risen to 1/2. Not so! As Marilyn correctly stated, it remains 1/3. Because Monty can always open a door with a goat, his opening such a door conveys no new information that alters the 1/3 probability.

Now comes an even more counterintuitive result. If the guest switches her choice from her initial selection to the other closed door, her chances of winning rise to 2/3. This should be obvious if one grants that the probability remains 1/3 for the first selection. The car must be behind one of the two doors, therefore the probabilities for each door must add to 1, or certainty. If one door has a probability of 1/3 being correct, the other door must have a 2/3 probability.

Marilyn was flooded with letters from irate readers, many accusing her of being ignorant of elementary probability theory and many from professional mathematicians. So awesome was the mail, and so con-

Probability and Ambiguity   283

troversial, that *The New York Times,* on July 21, 1991, ran a front page, lengthy feature about the flap. The story, written by John Tierney, was titled "Behind Monty Hall's Doors: Puzzle, Debate and Answer?" (See also letters about the feature in *The Times,* August 11, 1991.)

The red-faced mathematicians, who were later forced to confess they were wrong, were in good company. Paul Erdös, one of the world's greatest mathematicians, was among those unable to believe that switching doors doubled the probability of success. Two recent biographies of the late Erdös reveal that he could not accept Marilyn's analysis until his friend Ron Graham, of Bell Labs, patiently explained it to him.

The Monty Hall problem, as it came to be known, generated many articles in mathematical journals. I list some of them in this chapter's bibliography.

Bibliography

The Three Prisoners Paradox

D. H. Brown, "The Problem of the Three Prisoners," *Mathematics Teacher,* February 1966, pp. 181–82.

R. Falk, "A Closer Look at the Probabilities of the Notorious Three Prisoners." *Cognition,* Vol. 43, 1992, pp. 197–223.

S. Ichikawa and H. Takeichi, "Erroneous Beliefs in Estimating Posterior Probability," *Bahaviormetrika,* No. 27, 1990, pp. 59–73.

S. Shimojo and S. Ichikano, "Intuitive Reasoning About Probability: Theoretical and Experiential Analysis of Three Prisoners," *Cognition,* Vol. 32, 1989, pp. 1–24.

N. Starr, "A Paradox in Probability Theory," *Mathematics Teacher,* February 1973, pp. 166–68.

The Monty Hall Problem

E. Barbeau, "Fallicies, Flaws, and Flimflams," *The College Mathematics Journal,* Vol. 26, May 1995, pp. 132–84.

A. H. Bohl, M. J. Liberatore, and R. L. Nydick, "A Tale of Two Goats . . . and a Car," *Journal of Recreational Mathematics,* Vol. 27, 1995, pp. 1–9.

J. P. Georges and T. V. Craine, "Generalizing Monty's Dilemma," *Quantum,* March/April 1995, pp. 17–21.

L. Gillman, "The Car and the Goats." *American Mathematical Monthly,* Vol. 99, 1992, pp. 3–7.

A. Lo Bello, "Ask Marilyn: The Mathematical Controversy in *Parade* Magazine," Mathematical Gazette, Vol, 75, October 1991, pp. 271–72.

J. Paradis, P. Viader, and L. Bibiloni, "A Mathematical Excursion: From the Three-door Problem to a Cantortype Set." *American Mathematical Monthly,* Vol. 106, March 1999, pp. 241–51.

J. M. Shaughnessy and T. Dick, "Monte's Dilemma: Should You Stick or Switch?" *Mathematics Teacher,* April 1991, pp. 252–56.

M. vos Savant, "Ask Marilyn," *Parade,* Septembr 9, 1990; December 2, 1990; February 17, 1991; July 7, 1991; September 8, 1991; October 13 1991; January 5, 1992; January 26, 1992.

Chapter 22

Nontransitive Dice and Other Paradoxes

*P*robability theory abounds in paradoxes that wrench common sense and trap the unwary. In this chapter we consider a startling new paradox involving the relation called transitivity and a group of paradoxes stemming from the careless application of what is called the principle of indifference.

Transitivity is a binary relation such that if it holds between *A* and *B* and between *B* and *C*, it must also hold between *A* and *C*. A common example is the relation "heavier than." If *A* is heavier than *B* and *B* is heavier than *C*, then *A* is heavier than *C*. The three sets of four dice shown "unfolded" in Figure 22.1 were designed by Bradley Efron, a statistician at Stanford University, to dramatize some discoveries about a general class of probability paradoxes that violate transitivity. With any of these sets of dice you can operate a betting game so contrary to intuition that experienced gamblers will find it almost impossible to comprehend even after they have completely analyzed it.

The four dice at the top of the illustration are numbered in the simplest way that provides the winner with the maximum advantage. Allow someone to pick any die from this set. You then select a die from the remaining three. Both dice are tossed and the person who gets the highest number wins. Surely, it seems, if your opponent is allowed the first choice of a die before each contest, the game must either be fair or favor your opponent. If at least two dice have equal and maximum probabilities of winning, the game is fair because if he picks one such die, you can pick the other; if one die is better than the other three, your opponent can always choose that die and win more than half of the contests. This reasoning is completely wrong. The incredible truth is that regardless of which die he picks you can always pick a die that has a 2/3 probability of winning, or two-to-one odds in your favor!

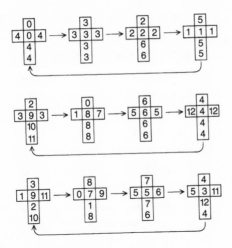

Figure 22.1. Nontransitive dice

The paradox (insofar as it violates common sense) arises from the mistaken assumption that the relation "more likely to win" must be transitive between pairs of dice. This is not the case with any of the three sets of dice. In each set the relation "more likely to win" is indicated by an arrow that points to the losing die. Die *A* beats *B*, *B* beats *C*, *C* beats *D*—and *D* beats *A!* In the first set the probability of winning with the indicated die of each pair is 2/3. This is easily verified by listing the 36 possible throws of each pair, then checking the 24 cases in which one die bears the highest number.

The other two sets of four dice, also designed by Efron, have the same nontransitive property but fewer numbers are repeated in order to make an analysis of the dice more difficult. In the second set the probability of winning with the indicated die is also 2/3. Because ties are possible with the third set it must be agreed that ties will be broken by rolling again. With this procedure the winning probability for each of the four pairings in the third set is 11/17, or .647.

It has been proved, Efron writes, that 2/3 is the greatest possible advantage that can be achieved with four dice. For three sets of numbers the maximum advantage is .618, but this cannot be obtained with dice because the sets must have more than six numbers. If more than four sets are used (numbers to be randomly selected within each set), the possible advantage approaches a limit of 3/4 as the number of sets increases.

A fundamental principle in calculating probabilities such as dice throws is one that goes back to the beginnings of classical probability theory in the 18th century. It was formerly called "the principle of insufficient reason" but is now known as "the principle of indifference," a crisper phrase coined by John Maynard Keynes in *A Treatise on Probability*. (Keynes is best known as an economist, but his book on probability has become a classic. It had a major influence on the inductive logic of Rudolf Carnap.) The principle is usually stated as follows: If you have no grounds whatever for believing that any one of n mutually exclusive events is more likely to occur than any other, a probability of $1/n$ is assigned to each.

For example, you examine a die carefully and find nothing that favors one side over another, such as concealed loads, noncubical shape, beveling of certain edges, stickiness of certain sides, and so on. You assume that there are six equally probable ways the cube can fall; therefore you assign a probability of 1/6 to each. If you toss a penny, or play the Mexican game of betting on which of two sugar cubes a fly will alight on first, your ignorance of any possible bias prompts you to assign a probability of 1/2 to each of the two outcomes. In none of these samples do you feel obligated to make statistical, empirical tests. The probabilities are assigned a priori. They are based on symmetrical features in the structures and forces involved. The die is a regular solid, the probability of the penny's balancing on its edge is virtually zero, there is no reason for a fly to prefer one sugar cube to another, and so on. Ultimately, of course, your analysis rests on empirical grounds, since only experience tells you, say, that a weighted die face would affect the odds, whereas a face colored red (with the others blue) would not.

Some form of the principle of indifference is indispensable in probability theory, but it must be carefully qualified and applied with extreme caution to avoid pitfalls. In many cases the traps spring from a difficulty in deciding which are the equally probable cases. Suppose, for instance, you shuffle a packet of four cards—two red, two black—and deal them face down in a row. Two cards are picked at random, say by placing a penny on each. What is the probability that those two cards are the same color?

One person reasons: "There are three equally probable cases. Either both cards are black, both are red, or they are different colors. In two cases the cards match, therefore the matching probability is 2/3."

"No," another person counters, "there are *four* equally probable cases. Either both cards are black, both are red, card x is black and y is red, or x is red and y is black. More simply, the cards either match or they do not. In each way of putting it the matching probability clearly is 1/2."

The fact is that both people are wrong. (The correct probability will be given in the Answer Section. Can the reader calculate it?) Here the errors arise from a failure to identify correctly the equally probable cases. There are, however, more confusing paradoxes—actually fallacies—in which the principle of indifference seems intuitively to be applicable, whereas it actually leads straight to a logical contradiction. Cases such as these result when there are no positive reasons for believing n events to be equally probable and the assumption of equiprobability is therefore based entirely, or almost entirely, on ignorance.

For example, someone tells you: "There is a cube in the next room whose size has been selected by a randomizing device. The cube's edge is not less than one foot or more than three feet." How would you estimate the probability that the cube's edge is between one and two feet as compared with the probability that it is between two and three feet? In your total ignorance of additional information, is it not reasonable to invoke the principle of indifference and regard each probability as 1/2?

It is not. If the cube's edge ranges between one and two feet, its volume ranges between 1^3, or one, cubic foot and 2^3, or eight, cubic feet. But in the range of edges from two to three feet, the volume ranges between 2^3 (eight) and 3^3 (27) cubic feet—a range almost three times the other range. If the principle of indifference applies to the two ranges of edges, it is violated by the equivalent ranges of volume. You were not told how the cube's "size" was randomized, and since "size" is ambiguous (it could mean either the cube's edge or its volume) you have no clues to guide your guessing. If the cube's edge was picked at random, the principle of indifference does indeed apply. It is also applicable if you are told that the cube's volume was picked at random, but of course you then have to assign a probability of 1/2 to each of the two ranges from one to 14 and from 14 to 27 cubic feet, and to the corresponding ranges for the cube's edge. If the principle applies to the edge, it cannot apply to the volume without contradiction, and vice versa. Since you do not know how the size was selected, any application of the principle is meaningless.

Carnap, in attacking an uncritical use of the principle in Harold Jeffreys' *Theory of Probability,* gives the following example of its misuse.

You know that every ball in an urn is blue, red, or yellow, but you know nothing about how many balls of each color are in the urn. What is the probability that the first ball taken from the urn will be blue? Applying the principle of indifference, you say it is 1/2. The probability that it is not blue must also be 1/2. If it is not blue, it must be red or yellow, and because you know nothing about the number of red or yellow balls, those colors are equally probable. Therefore you assign to red a probability of 1/4. On the other hand, if you begin by asking for the probability that the first ball will be red, you must give red a probability of 1/2 and blue a probability of 1/4, which contradicts your previous estimates.

It is easy to prove along similar lines that there is life on Mars. What is the probability that there is simple plant life on Mars? Since arguments on both sides are about equally cogent, we answer 1/2. What is the probability that there is simple animal life on Mars? Again, 1/2. Now we seem forced to assert that the probability of there being "either plant or animal life" on Mars is 1/2 + 1/2 = 1, or certainty, which is absurd. The philosopher Charles Sanders Peirce gave a similar argument that seems to show that the hair of inhabitants on Saturn had to be either of two different colors. It is easy to invent others.

In the history of metaphysics the most notorious misuse of the principle surely was by Blaise Pascal, who did pioneer work on probability theory, in a famous argument that became known as "Pascal's wager." A few passages from the original and somewhat lengthy argument (in Pascal's *Pensées,* Thought 233) are worth quoting:

> "God is, or he is not." To which side shall we incline? Reason can determine nothing about it. There is an infinite gulf fixed between us. A game is playing at the extremity of this infinite distance in which heads or tails may turn up. What will you wager? There is no reason for backing either one or the other, you cannot reasonably argue in favor of either. . . .
>
> Yes, but you must wager. . . . Which will you choose? . . . Let us weigh the gain and the loss in choosing "heads" that God is. . . . If you gain, you gain all. If you lose, you lose nothing. Wager, then, unhesitatingly that he is.

Lord Byron, in a letter, rephrased Pascal's argument effectively: "Indisputably, the firm believers in the Gospel have a great advantage over all others, for this simple reason—that, if true, they will have their reward hereafter; and if there be no hereafter, they can be but with the infidel in his eternal sleep, having had the assistance of an exalted hope through life, without subsequent disappointment, since (at the worst for them) out of nothing nothing can arise, not even sorrow." Similar passages can be found in many contemporary books of religious apologetics.

Pascal was not the first to insist in this fashion that faith in Christian orthodoxy was the best bet. The argument was clearly stated by the 4th-century African priest Arnobius the Elder, and non-Christian forms of it go back to Plato. This is not the place, however, to go into the curious history of defenses and criticisms of the wager. I content myself with mentioning Denis Diderot's observation that the wager applies with equal force to other major faiths such as Islam. The mathematically interesting aspect of all of this is that Pascal likens the outcome of his bet to the toss of a coin. In other words, he explicitly invokes the principle of indifference to a situation in which its application is mathematically senseless.

The most subtle modern reformulation of Pascal's wager is by William James, in his famous essay *The Will to Believe,* in which he argues that philosophical theism is a better gamble than atheism. In a still more watered-down form it is even used occasionally by humanists to defend optimism against pessimism at a time when the extinction of the human race seems as likely in the near future as its survival.

"While there is a chance of the world getting through its troubles," says the narrator of H. G. Wells's little read novel *Apropos of Dolores,* "I hold that a reasonable man has to behave as though he was sure of it. If at the end your cheerfulness is not justified, at any rate you will have been cheerful."

Addendum

The following letter, from S. D. Turner, contains some surprising information:

> Your bit about the two black and two red cards reminds me of an exercise I did years ago, which might be called *N*-Card Monte. A few cards, half red, half black, or nearly so, are shown face up by the pitchman, then shuffled and dealt face down. The sucker is induced to bet he can pick two of the same color.
>
> The odds will always be against him. But because the sucker will

make erroneous calculations (like the 2/3 and 1/2 in your 2:2 example), or for other reasons, he will bet. The pitchman can make a plausible spiel to aid this: "Now, folks, you don't need to pick two blacks, and you don't need to pick two reds. If you draw either pair you win!"

The probability of getting two of the same color, where there are R reds and B blacks, is:

$$P = \frac{R^2 + B^2 - (R + B)}{(R + B)(R + B - 1)}$$

This yields the figures in the table [see Figure 22.2], one in lowest-terms fractions, the other in decimal. Only below and to the left of the stairstep line does the sucker get an even break or better. But no pitchman would bother with odds more favorable to the sucker than the 1/3 probability for 2:2, or possibly the 2/5 for 3:3.

Surprisingly, the two top diagonal lines are identical. That is, if you are using equal reds and blacks, odds are not changed if a card is removed before the two are selected! In your example of 2:2, the proba-

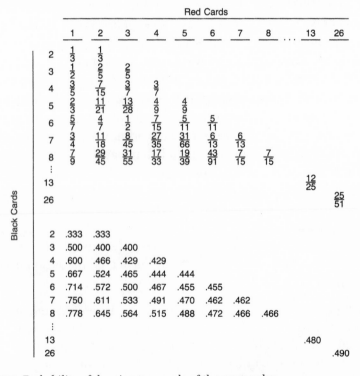

Figure 22.2. Probability of drawing two cards of the same color

bility is 1/3 and it is also 1/3 when starting with 2:1 (as is evident because the one card not selected can be any one of the three). The generality of this can be shown thus: If $B = R$ and $B = R - 1$ are substituted into the above equation, the result in each case is $R - 1/2R - 1$.

Some readers sent detailed explanations of why the arguments behind the fallacies that I described were wrong, apparently not realizing that these fallacies were intended to be howlers based on the misuse of the principle of indifference. Several readers correctly pointed out that although Pascal did invoke the principle of indifference by referring to a coin flip in his famous wager, the principle is not essential to his argument. Pascal posits an infinite gain for winning a bet in which the loss (granting his assumptions) would always be finite regardless of the odds.

Efron's nontransitive dice aroused almost as much interest among magicians as among mathematicians. It was quickly perceived that the basic idea generalized to k sets of n-sided dice, such as dice in the shapes of regular octahedrons, dodecahedrons, icosahedrons, or cylinders with n flat sides. The game also can be modeled by k sets of n-sided tops, spinners with n numbers on each dial, and packets of n playing cards.

Karl Fulves, in his magic magazine *The Pallbearers Review* (January 1971) proposed using playing cards to model Efron's dice. He suggested the following four packets: 2, 3, 4, 10, J, Q; 1, 2, 8, 9, 9, 10; 6, 6, 7, 7, 8, 8; and 4, 5, 5, 6, Q, K. Suits are irrelevant. First player selects a packet, shuffles it, and draws a card. Second player does the same with another packet. If the chosen cards have the same value, they are replaced and two more cards drawn. Ace is low, and high card wins. This is based on Efron's third set of dice where the winning probability, if the second player chooses properly, is 11/17. To avoid giving away the cyclic sequence of packets, each could be placed in a container (box, cup, tray, etc.) with the containers secretly marked. Before each play, the containers would be randomly mixed by the first player while the second player turned his back. Containers with numbered balls or counters could of course be substituted for cards.

In the same issue of *The Pallbearers Review* cited above, Columbia University physicist Shirley Quimby proposed a set of four dice with the following faces:

3, 4, 5, 20, 21, 22
1, 2, 16, 17, 18, 19
10, 11, 12, 13, 14, 15
6, 7, 8, 9, 23, 24

Note that numbers 1 through 24 are used just once each in this elegant arrangement. The dice give the second player a winning probability of 2/3. If modeled with 24 numbered cards, the first player would select one of the four packets, shuffle, then draw a card. The second player would do likewise, and high card wins.

R.C.H. Cheng, writing from Bath University, England, proposed a novel variation using a single die. On each face are numbers 1 through 6, each numeral a different color. Assume that the colors are the rainbow colors red, orange, yellow, green, blue, and purple. The chart below shows how the numerals are colored on each face.

FACE	RED	ORANGE	YELLOW	GREEN	BLUE	PURPLE
A	1	2	3	4	5	6
B	6	1	2	3	4	5
C	5	6	1	2	3	4
D	4	5	6	1	2	3
E	3	4	5	6	1	2
F	2	3	4	5	6	1

The game is played as follows: The first player selects a color, then the second player selects another color. The die is rolled and the person whose color has the highest value wins. It is easy to see from the chart that if the second player picks the adjacent color on the right—the sequence is cyclic, with red to the "right" of purple—the second player wins five out of six times. In other words, the odds are 5 to 1 in his favor!

To avoid giving away the sequence of colors, the second player should occasionally choose the second color to the right, where his winning odds are 4 to 2, or the color third to the right where the odds are even. Perhaps he should even, on rare occasions, take the fourth or fifth color to the right where odds against him are 4 to 2 and 5 to 1 respectively. Mel Stover has suggested putting the numbers and colors on a six-sided log instead of a cube.

This, too, models nicely with 36 cards, formed in six piles, each bearing a colored numeral. The chart's pattern is obvious, and easily applied to n^2 cards, each with numbers 1 through n, and using n different colors. In presenting it as a betting game you should freely display the faces of each packet to show that all six numbers and all six colors are represented. Each packet is shuffled and placed face down. The first player is "generously" allowed first choice of a color and to select any packet. The color with the highest value in that packet is the winner. In the general case, as Cheng pointed out in his 1971 letter, the second player can always choose a pile that gives him a probability of winning equal to $(n - 1)/n$.

A simpler version of this game uses 16 playing cards. The four packets are

> AS, JH, QC, KD
> KS, AH, JC, QD
> QS, KH, AC, JD
> JS, QH, KC, AD

Ace here is high, and the cyclic sequence of suits is spades, hearts, clubs, diamonds. The second player wins with 3 to 1 odds by choosing the next adjacent suit, and even odds if he goes to the next suit but one.

These betting games are all variants of nontransitive voting paradoxes, about which there is extensive literature.

Answers

The probability that two randomly selected cards, from a set of two red and two black cards, are the same color is 1/3. If you list the 24 equally probable permutations of the four cards, then pick any two positions (for example, second and fourth cards), you will find eight cases in which the two cards match in color. One way to see that this probability of 8/24 or 1/3 is correct is to consider one of the two chosen cards. Assume that it is red. Of the remaining three cards only one is red, and so the probability that the second chosen card will be red is 1/3. Of course, the same argument applies if the first card is black. Most people guess that the odds are even, when actually they are 2:1 in favor of the cards' having different colors.

Bibliography

R. Carnap, "Statistical and Inductive Probability," *The Structure of Scientific Thought,* E. H. Madden (ed.), Houghton Mifflin, 1960.

J. M. Keynes, *A Treatise on Probability,* Macmillan, 1921; Harper & Row (paperback), 1962.

On Pascal's Wager:

J. Cargile. "Pascal's Wager," *Philosophy: The Journal of the Royal Institute of Philosophy,* Vol. 41, July 1966, pp. 250–57.

I. Hacking, *The Emergence of Probability: A Philosophical Study of Early Ideas About Probability, Induction, and Statistical Inference,* Cambridge University Press, 1976.

M.B. Turner, "Deciding for God—the Bayesian Support of Pascal's Wager," *Philosophy and Phenomenological Research,* Vol. 29, September 1968, pp. 84–90.

On Nontransitive Dice

G. Finnell, "A Dice Paradox," *Epilogue,* July 1971, pp. 2–3. This is another magic magazine published by Karl Fulves.

R. P. Savage, "The Paradox of Nontransitive Dice," *American Mathematical Monthly,* Vol. 101, May 1994, pp. 429–36.

Chapter 23

More Nontransitive Paradoxes

*I have just so much logic, as to be able to see . . .
that for me to be too good for you, and for you to
be too good for me, cannot be true at once, both
ways.*

—Elizabeth Barrett, in a letter to Robert Browning.

When a relation *R* that is true with respect to *xRy* and *yRz* also holds for *xRz*, the relation is said to be transitive. For example, "less than" is transitive among all real numbers. If 2 is less than π, and the square root of 3 is less than 2, we can be certain that the square root of 3 is less than π. Equality also is transitive: if *a* = *b* and *b* = *c*, then *a* = *c*. In everyday life such relations as "earlier than," "heavier than," "taller than," "inside," and hundreds of others are transitive.

It is easy to think of relations that are not transitive. If *A* is the father of *B* and *B* is the father of *C*, it is never true that *A* is the father of *C*. If *A* loves *B* and *B* loves *C*, it does not follow that *A* loves *C*. Familiar games abound in transitive rules (if poker hand *A* beats *B* and *B* beats *C*, then *A* beats *C*), but some games have nontransitive (or intransitive) rules. Consider the children's game in which, on the count of three, one either makes a fist to symbolize "rock," extends two fingers for "scissors," or all fingers for "paper." Rock breaks scissors, scissors cut paper, and paper covers rock. In this game the winning relation is not transitive.

Occasionally in mathematics, particularly in probability theory and decision theory, one comes on a relation that one expects to be transitive but that actually is not. If the nontransitivity is so counterintuitive as to boggle the mind, we have what is called a nontransitive paradox.

The oldest and best-known paradox of this type is a voting paradox sometimes called the Arrow paradox after Kenneth J. Arrow because of its crucial role in Arrow's "impossibility theorem," for which he shared a Nobel prize in economics in 1972. In *Social Choice and Individual Values,* Arrow specified five conditions that almost everyone agrees are essential for any democracy in which social decisions are based on individual preferences expressed by voting. Arrow proved that the five conditions are logically inconsistent. It is not possible to devise a voting system that will not, in certain instances, violate at least one of the five essential conditions. In short, a perfect democratic voting system is in principle impossible.

As Paul A. Samuelson has put it: "The search of the great minds of recorded history for the perfect democracy, it turns out, is the search for a chimera, for a logical self-contradiction. . . . Now scholars all over the world—in mathematics, politics, philosophy, and economics—are trying to salvage what can be salvaged from Arrow's devastating discovery that is to mathematical politics what Kurt Gödel's 1931 impossibility-of-proving-consistency theorem is to mathematical logic."

Let us approach the voting paradox by first considering a fundamental defect of our present system for electing officials. It frequently puts in office a man who is cordially disliked by a majority of voters but who has an enthusiastic minority following. Suppose 40 percent of the voters are enthusiastic supporters of candidate A. The opposition is split between 30 percent for B and 30 percent for C. A is elected even though 60 percent of the voters dislike him.

One popular suggestion for avoiding such consequences of the split vote is to allow voters to rank all candidates in their order of preference. Unfortunately, this too can produce irrational decisions. The matrix in Figure 23.1 (left) displays the notorious voting paradox in its simplest form. The top row shows that a third of the voters prefer candidates A, B, and C in the order ABC. The middle row shows that another third

RANK ORDER

VOTERS	1	2	3
⅓	A	B	C
⅓	B	C	A
⅓	C	A	B

TEAMS	D	E	F
A	8	1	6
B	3	5	7
C	4	9	2

Figure 23.1. The voting paradox (left) and the tournament paradox based on the magic square (right)

rank them *BCA*, and the bottom row shows that the remaining third rank them *CAB*. Examine the matrix carefully and you will find that when candidates are ranked in pairs, nontransitivity rears its head. Two-thirds of the voters prefer *A* to *B*, two-thirds prefer *B* to *C*, and two-thirds prefer *C* to *A*. If *A* ran against *B*, *A* would win. If *B* ran against *C*, *B* would win. If *C* ran against *A*, *C* would win. Substitute proposals for candidates and you see how easily a party in power can rig a decision simply by its choice of which paired proposals to put up first for a vote.

The paradox was recognized by the Marquis de Condorcet and others in the late 18th century and is known in France as the Condorcet effect. Lewis Carroll, who wrote several pamphlets on voting, rediscovered it. Most of the early advocates of proportional representation were totally unaware of this Achilles' heel; indeed, the paradox was not fully recognized by political theorists until the mid-1940s, when Duncan Black, a Welsh economist, rediscovered it in connection with his monumental work on committee decision making. The experts are nowhere near agreement on which of Arrow's five conditions should be abandoned in the search for the best voting system. One surprising way out, recommended by many decision theorists, is that when a deadlock arises, a "dictator" is chosen by lot to break it. Something close to this solution actually obtains in certain democracies, England for instance, where a constitutional monarch (selected by chance in the sense that lineage guarantees no special biases) has a carefully limited power to break deadlocks under certain extreme conditions.

The voting paradox can arise in any situation in which a decision must be made between two alternatives from a set of three or more. Suppose that *A*, *B*, and *C* are three men who have simultaneously proposed marriage to a girl. The rows of the matrix for the voting paradox can be used to show how she ranks them with respect to whatever three traits she considers most important, say intelligence, physical attractiveness, and income. Taken by pairs, the poor girl finds that she prefers *A* to *B*, *B* to *C*, and *C* to *A*. It is easy to see how similar conflicts can arise with respect to one's choice of a job, where to spend a vacation, and so on.

Paul R. Halmos once suggested a delightful interpretation of the matrix. Let *A*, *B*, and *C* stand for apple pie, blueberry pie, and cherry pie. A certain restaurant offers only two of them at any given meal. The rows show how a customer ranks the pies with respect to three properties, say taste, freshness, and size of slice. It is perfectly rational, says Halmos, for the customer to prefer apple pie to blueberry, blueberry to cherry, and cherry to apple. In his *Adventures of a Mathematician* (Scribner, 1976), Stanislaw Ulam speaks of having discovered the nontransitivity of such preferences when he was eight or nine and of later realizing that it prevented one from ranking great mathematicians in a linear order of relative merit.

Experts differ on how often nontransitive orderings such as this one arise in daily life, but some recent studies in psychology and economics indicate that they are commoner than one might suppose. There are even reports of experiments with rats showing that under certain conditions the pairwise choices of individual rats are nontransitive. (See Warren S. McCulloch, "A Heterarchy of Values Determined by the Topology of Nervous Nets," *Bulletin of Mathematical Biophysics* 7, 1945, pp. 89–93.)

Similar paradoxes arise in round-robin tournaments between teams. Assume that nine tennis players are ranked in ability by the numbers 1 through 9, with the best player given the number 9 and the worst given the number 1. The matrix in Figure 23.1 *(right)* is the familiar order-3 magic square. Let rows *A*, *B*, and *C* indicate how the nine players are divided into three teams with each row comprising a team. In round-robin tournaments between teams, where each member of one team plays once against each member of the others, assume that the stronger player always wins. It turns out that team *A* defeats *B*, *B* defeats *C*, and *C* defeats *A*, in each case by five games to four. It is impossible to say which team is the strongest. The same nontransitivity holds if columns *D*, *E*, and *F* of the matrix are the teams.

Many paradoxes of this type were jointly investigated by Leo Moser and J. W. Moon. Some of the Moser–Moon paradoxes underlie striking and little-known sucker bets. For example, let each row (or each column) of an order-3 magic-square matrix be a set of playing cards, say the ace, 6, and 8 of hearts for set *A*, the 3, 5, and 7 of spades for set *B*, and the 2, 4, and 9 of clubs for *C* (see Figure 23.2). Each set is randomized and placed face down on a table. The sucker is allowed to draw a card from any set, then you draw a card from a different set. The high card wins. It is easy to prove that no matter what set the sucker draws from, you can pick a set that gives you winning odds of five to four. Set *A* beats *B*, *B* beats *C*, and *C* beats *A*. The victim may even be allowed to decide each time whether the high or the low card wins. If you play low

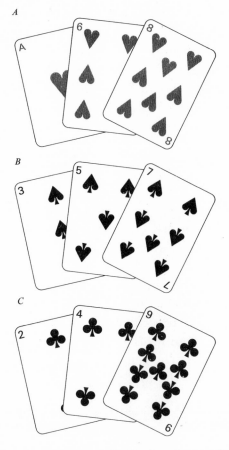

Figure 23.2. Nontransitive sucker bet based on magic square: $A \rightarrow B \rightarrow C \rightarrow A$

card wins, simply pick the winning pile with respect to a nontransitive circle that goes the other way. A good way to play the game is to use sets of cards from three decks with backs of different colors. The packet of nine cards is shuffled each time, then separated by the backs into the three sets. The swindle is, of course, isomorphic with the tennis-tournament paradox.

Nontransitivity prevails in many other simple gambling games. In some cases, such as the top designed by Andrew Lenard (see Figure 23.3), the nontransitivity is easy to understand. The lower part of the top is fixed but the upper disk rotates. Each of two players chooses a different arrow, the top is spun (in either direction), and the person whose arrow points to the section with the highest number wins. A beats B, B beats C, and C beats A, in each case with odds of two to one.

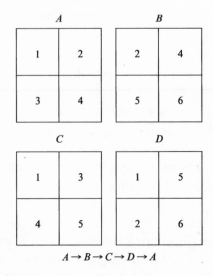

Figure 23.3. Nontransitive top

$A \longrightarrow B \longrightarrow C \longrightarrow A$

In a set of four bingo cards designed by Donald E. Knuth (see Figure 23.4), the nontransitivity is cleverly concealed. Two players each select a bingo card. Numbers from 1 through 6 are randomly drawn without replacement, as they are in standard bingo. If a called number is on a card, it is marked with a bean. The first player to complete a horizontal row wins. Here, of course, the numbers are just symbols; they can be replaced by any set of six different symbols. I leave it to the reader to work out the probabilities that show card A beats B, B beats C, C beats D, and D beats A. The game is transitive with three players, but the winning odds for the four possible triplets are surprising.

One of the most incredible of all nontransitive betting situations, dis-

A

| 1 | 2 |
| 3 | 4 |

B

| 2 | 4 |
| 5 | 6 |

C

| 1 | 3 |
| 4 | 5 |

D

| 1 | 5 |
| 2 | 6 |

$A \rightarrow B \rightarrow C \rightarrow D \rightarrow A$

Figure 23.4. Nontransitive bingo cards

covered (appropriately) by a mathematician named Walter Penney, was given as a problem in the *Journal of Recreational Mathematics* (October 1969, p. 241). It is not well known, and most mathematicians simply cannot believe it when they first hear of it. It is certainly one of the finest of all sucker bets. It can be played with a penny, or as a side bet on the reds and blacks of a roulette wheel, or in any situation in which two alternatives are randomized with equal odds. We shall assume that a penny is used. If it is flipped three times, there are eight equally probable outcomes: *HHH, HHT, HTH, HTT, THH, THT, TTH,* and *TTT.* One player selects one of these triplets, and the other player selects a different one. The penny is then flipped repeatedly until one of the chosen triplets appears as a run and wins the game. For example, if the chosen triplets are *HHT* and *THT* and the flips are *THHHT,* the last three flips show that *HHT* has won. In brief, the first triplet to appear as a run wins.

One is inclined to assume that one triplet is as likely to appear first as any other, but it takes only a moment to realize that this is not the case even with doublets. Consider the doublets *HH, HT, TH,* and *TT. HH* and *HT* are equally likely to appear first because, after the first *H* appears, it is just as likely to be followed by an *H* as by a *T.* The same reasoning shows that *TT* and *TH* are equal. Because of symmetry, *HH* = *TT* and *HT* = *TH. TH* beats *HH* with odds of three to one, however, and *HT* beats *TT* with the same probability. Consider *HT* and *TT. TT* is always preceded by *HT* except when *TT* appears on the first two flips. This happens in the long run only once in four times, and so the probability that *HT* beats *TT* is 3/4. Figure 23.5 shows the probability that *B,* the second player, will win for all pairs of doublets.

A \ B	HH	HT	TH	TT
HH		1/2	1/4	1/2
HT	1/2		1/2	3/4
TH	3/4	1/2		1/2
TT	1/2	1/4	1/2	

Figure 23.5. Probabilities of *B* winning

More Nontransitive Paradoxes

When we turn to triplets, the situation becomes much more surprising. Since it does not matter which side of a coin is designated heads, we know that $HHH = TTT$, $TTH = HHT$, $HTH = THT$, and so on. When we examine the probabilities for unequal pairs, however, we discover that the game is not transitive. No matter what triplet the first player takes, the second player can select a better one. Figure 23.6 gives the probability that B, the second player, defeats A for all possible pairings. To find B's best response to a triplet chosen by A, find A's triplet at the top, go down the column until you reach a probability (shown in gray), then move left along the row to B's triplet on the left.

B \ A	HHH	HHT	HTH	HTT	THH	THT	TTH	TTT
HHH		1/2	2/5	2/5	1/8	5/12	3/10	1/2
HHT	1/2		2/3	2/3	1/4	5/8	1/2	7/10
HTH	3/5	1/3		1/2	1/2	1/2	3/8	7/12
HTT	3/5	1/3	1/2		1/2	1/2	3/4	7/8
THH	7/8	3/4	1/2	1/2		1/2	1/3	3/5
THT	7/12	3/8	1/2	1/2	1/2		1/3	3/5
TTH	7/10	1/2	5/8	1/4	2/3	2/3		1/2
TTT	1/2	3/10	5/12	1/8	2/5	2/5	1/2	

Figure 23.6. Probabilities of B winning in a triplet game

Note that B's probability of winning is, at the worst, 2/3 (or odds of two to one) and can go as high as 7/8 (or odds of seven to one). The seven-to-one odds are easy to comprehend. Consider THH and HHH. If HHH first appears anywhere except at the start, it must be preceded by a *T,* which means that *THH* has appeared earlier. *HHH* wins, therefore, only when it appears on the first three flips. Clearly this happens only once in eight flips.

Barry Wolk of the University of Manitoba has discovered a curious rule for determining the best triplet. Let X be the first triplet chosen. Convert it to a binary number by changing each H to zero and each T to 1. Divide the number by 2, round down the quotient to the nearest integer, multiply by 5, and add 4. Express the result in binary, then convert the last three digits back to H and T.

Nontransitivity holds for all higher n-tuplets. A chart supplied by Wolk gives the winning probabilities for B in all possible pairings of quadruplets (see Figure 23.7). Like the preceding two charts and charts for all higher n-tuplets, the matrix is symmetric about the center. The upper right quadrant is the lower left quadrant upside down, and the same holds for the upper left and lower right quadrants. The probabilities for B's best responses to A are shown in gray.

In studying these figures, Wolk discovered another kind of anomaly as surprising as nontransitivity. It has to do with what are called waiting times. The waiting time for an n-tuplet is the average number of tosses, in the long run, until the specified n-tuplet appears. The longer you wait for a bus, the shorter becomes the expected waiting time. Pennies, however, have no memory, so that the waiting time for an n-tuplet is independent of all previous flips. The waiting time for H and T is 2. For doublets the waiting time is 4 for *HT* and *TH,* and 6 for *HH* and *TT.* For triplets the waiting times are 8 for *HHT, HHT, THH,* and *TTH;* 10 for *HTH* and *THT;* and 14 for *HHH* and *TTT.* None of this contradicts what we know about which triplet of a pair is likely to show first. With quadruplets, however, contradictions arise with six pairs. For example, *THTH* has a waiting time of 20 and *HTHH* has a waiting time of 18. Yet, *THTH* is more likely to turn up before *HTHH* with a probability of 9/14, or well over one-half. In other words, an event that is less frequent in the long run is likely to happen before a more frequent event. There is no logical contradiction involved here, but it does show that "average waiting time" has peculiar properties.

There are many ways to calculate the probability that one n-tuplet will precede another. You can do it by summing infinite series, by drawing tree diagrams, by recursive techniques that produce sets of linear equations, and so on. One of the strangest and most efficient techniques was devised by John Horton Conway of Princeton Univer-

B \ A	HHHH	HHHT	HHTH	HHTT	HTHH	HTHT	HTTH	HTTT	THHH	THHT	THTH	THTT	TTHH	TTHT	TTTH	TTTT
HHHH		1/2	2/5	2/5	3/10	5/12	4/11	4/11	1/16	3/8	3/8	3/8	1/4	3/8	7/22	1/2
HHHT	1/2		2/3	2/3	1/2	5/8	4/7	4/7	1/8	9/16	9/16	9/16	5/12	9/16	1/2	15/22
HHTH	3/5	1/3		1/2	3/5	5/7	1/2	1/2	5/12	5/12	9/16	9/16	5/14	1/2	7/16	5/8
HHTT	3/5	1/3	1/2		3/7	5/9	2/3	2/3	5/12	5/12	9/16	9/16	1/2	9/14	7/12	3/4
HTHH	7/10	1/2	2/5	4/7		1/2	1/2	1/2	7/12	7/12	5/14	1/2	7/16	7/16	7/16	5/8
HTHT	7/12	3/8	2/7	4/9	1/2		1/2	1/2	7/16	7/16	1/2	9/14	7/16	7/16	7/16	5/8
HTTH	7/11	3/7	1/2	1/3	1/2	1/2		1/2	1/2	1/2	9/16	5/12	7/12	7/12	7/16	5/8
HTTT	7/11	3/7	1/2	1/3	1/2	1/2	1/2		1/2	1/2	9/16	5/12	7/12	7/12	7/8	15/16
THHH	15/16	7/8	7/12	7/12	5/12	9/16	1/2	1/2		1/2	1/2	1/2	1/3	1/2	3/7	7/11
THHT	5/8	7/16	7/12	7/12	5/12	9/16	1/2	1/2	1/2		1/2	1/2	1/3	1/2	3/7	7/11
THTH	5/8	7/16	7/16	7/16	9/14	1/2	7/16	7/16	1/2	1/2		1/2	4/9	2/7	3/8	7/12
THTT	5/8	7/16	7/16	7/16	1/2	5/14	7/12	7/12	1/2	1/2	1/2		4/7	2/5	1/2	7/10
TTHH	3/4	7/12	9/14	1/2	9/16	9/16	5/12	5/12	2/3	2/3	5/9	3/7		1/2	1/3	3/5
TTHT	5/8	7/16	1/2	5/14	9/16	9/16	5/12	5/12	1/2	1/2	5/7	3/5	1/2		1/3	3/5
TTTH	15/22	1/2	9/16	5/12	9/16	9/16	9/16	1/8	4/7	4/7	5/8	1/2	2/3	2/3		1/2
TTTT	1/2	7/22	3/8	1/4	3/8	3/8	3/8	1/16	4/11	4/11	5/12	3/10	2/5	2/5	1/2	

Figure 23.7. Probabilities of B winning in a quadruplet game

sity. I have no idea why it works. It just cranks out the answer as if by magic, like so many of Conway's other algorithms.

The key to Conway's procedure is the calculation of four binary numbers that Conway calls leading numbers. Let A stand for the 7-tuplet *HHTHHHT* and B for *THHTHHH*. We want to determine the probability of B beating A. To do this, write A above A, B above B, A above B, and B above A (see Figure 23.8). Above the top tuplet of each pair a binary number is constructed as follows. Consider the first pair, AA. Look at the first letter of the top tuplet and ask yourself if the seven letters, beginning with this first one, correspond exactly to the first seven letters of the tuplet below it. Obviously they do, and so we put a 1 above the first letter. Next, look at the second letter of the top tuplet and ask if the six letters, starting with this one, correspond to the *first six* letters of the tuplet below. Clearly they do not, and so we put zero above the second letter. Do the five letters starting with the third letter of the top tuplet correspond to the first five letters of the lower tuplet? No, and so this letter also gets zero. The fourth letter gets another zero. When we check the fifth letter of the top *A*, we see that *HHT* does correspond to the first three letters of the lower *A*, and so the fifth letter gets a 1. Letters six and seven each get zero. The "*A* leading *A* number," or *AA*, is 1000100, in which each 1 corresponds to a yes answer; each zero, to a no. Translating 1000100 from binary to decimal gives us 68 as the leading number for *AA*.

Figure 23.8 shows the results of this procedure in calculating leading numbers *AA*, *BB*, *AB*, and *BA*. Whenever an *n*-tuplet is compared with itself, the first digit of the leading number must, of course, be 1. When compared with a different tuplet, the first digit may or may not be 1.

```
    1 0 0 0 1 0 0 = 68          0 0 0 0 0 0 1 = 1
A = HHTHHHT                A = HHTHHHT
A = HHTHHHT                B = THHTHHH
                                                    AA − AB : BB − BA
                                                    68 − 1 : 64 − 35
                                                    67 : 29

    1 0 0 0 0 0 0 = 64          0 1 0 0 0 1 1 = 35
B = THHTHHH                B = THHTHHH
B = THHTHHH                A = HHTHHHT
```

Figure 23.8. John Horton Conway's algorithm for calculating odds of *B*'s *n*-tuplet beating *A*'s *n*-tuplet

The odds in favor of *B* beating *A* are given by the ratio *AA—AB:BB—BA*. In this case 68—1:64—35 = 67:29. As an exercise, the reader can try calculating the odds in favor of *THH* beating *HHH*. The four leading numbers will be *AA* = 7, *BB* = 4, *AB* = 0, and *BA* = 3. Plugging these into the formula, *AA—AB:BB—BA* gives odds of 7—0:4—3, or seven to one, as expected. The algorithm works just as well on tuplets of unequal lengths, provided the smaller tuplet is not contained within the larger one. If, for example, *A* = *HH* and *B* = *HHT*, *A* obviously wins with a probability of 1.

I conclude with a problem by David L. Silverman, who was the first to introduce the Penney paradox in the problems department that he then edited for the *Journal of Recreational Mathematics* (Vol. 2, October 1969, p. 241). The reader should have little difficulty solving it by Conway's algorithm. *TTHH* has a waiting time of 16 and *HHH* has a waiting time of 14. Which of these tuplets is most likely to appear first and with what probability?

Addendum

Numerous readers discovered that Barry Wolk's rule for picking the best triplet B to beat triplet A is equivalent to putting in front of A the complement of its next-to-last symbol, then discarding the last symbol. More than half these correspondents found that the method also works for quadruplets except for the two in which H and T alternate throughout. In such cases the symbol put in front of A is the same as its next to last one.

Since October 1974, when this chapter first appeared in *Scientific American,* many papers have been published that prove Conway's algorithm and give procedures for picking the best n-tuplet for all values of n. Two important early articles are cited in the bibliography. The paper by Guibas and Odlyzko gives 26 references.

Readers David Sachs and Bryce Hurst each noted that Conway's "leading number," when an n-tuplet is compared with itself, automatically gives that tuplet's waiting time. Simply double the leading number.

Ancient Chinese philosophers (I am told) divided matter into five categories that form a nontransitive cycle: wood gives birth to fire, fire to earth, earth to metal, metal to water, and water to wood. Rudy Rucker's science-fiction story "Spacetime Donuts" (*Unearth,* Summer 1978) is based on a much more bizarre nontransitive theory. If you move down the scale of size, to several steps below electrons, you get back to the galaxies of the same universe we now occupy. Go up the scale several stages beyond our galactic clusters, and you are back to the elementary particles—not larger ones, but the very same particles that make our stars. The word "matter" loses all meaning.

The following letter was published in *Scientific American* (January 1975):

Sirs:

Martin Gardner's article on the paradoxical situations that arise from nontransitive relations may have helped me win a bet in Rome on the outcome of the Ali v. Foreman world heavyweight boxing title match in Zaïre on October 30.

Ali, though slower than in former years, and a 4-1 betting underdog, may have had a psychological and motivational advantage for that particular fight. But in addition, Gardner's mathematics might be relevant. Even though Foreman beat Frazier, who beat Ali, Ali could still beat Foreman because there may be a nontransitive relation between the three.

I ranked the three fighters against the criteria of speed, power, and technique (including psychological technique) as reported in the press, and spotted a nontransitive relation worth betting on:

	ALI	FRAZIER	FOREMAN
Speed	2	1	3
Power	3	2	1
Technique	1	3	2

Foreman's power and technique beat Frazier, but Ali's technique and speed beat Foreman. It was worth the bet. The future implications are, however, that Frazier can *still* beat Ali!

ANTHONY PIEL

Vaud, Switzerland

David Silverman (*Journal of Recreational Mathematics* 2, October 1969, p. 241) proposed a two-person game that he called "blind Penney-ante." It is based on the nontransitive triplets in a run of fair coin tosses. Each player *simultaneously* chooses a triplet without knowing his opponent's choice. The triplet that shows up first wins. What is a player's best strategy? This is not an easy problem. A full solution, based on an 8 × 8 game matrix, is given in *The College Mathematics Journal* as the answer to Problem 299 (January 1987, pp. 74–76).

Answers

Which pattern of heads and tails, *TTHH* or *HHH,* is more likely to appear first as a run when a penny is repeatedly flipped? Applying

John Horton Conway's algorithm, we find that *TTHH* is more likely to precede *HHH* with a probability of 7/12, or odds of seven to five. Some quadruplets beat some triplets with even greater odds. For example, *THHH* precedes *HHH* with a probability of 7/8, or odds of seven to one. This is easy to see. *HHH* must be preceded by a *T* unless it is the first triplet of the series. Of course, the probability of that is 1/8.

The waiting time for *TTHH* and for *THHH* is 16, compared with a waiting time of 14 for *HHH*. Both cases of the quadruplet versus the triplet, therefore, exhibit the paradox of a less likely event occurring before a more likely event with a probability exceeding 1/2.

Bibliography

Nontransitive Voting and Tournament Paradoxes

K. J. Arrow, *Social Choice and Individual Values*, Wiley, 1951; second edition, 1970.

D. Black, *The Theory of Committees and Elections*, Cambridge University Press, 1958; Kluwer Academic, 1986.

S. J. Brams, *Paradoxes in Politics*, Free Press, 1976.

R. Farquharson, *Theory of Voting*, Yale University Press, 1969.

P. C. Fishburn, "Paradoxes of Voting," *The American Political Science Review*, Vol. 68, June 1974, pp. 539–46.

D. Khahr, "A Computer Simulation of the Paradox of Voting," *American Political Science Review*, Vol. 50, June 1966, pp. 384–96.

R. D. Luce and H. Raiffa, *Games and Decision*, Wiley, 1957.

A. F. MacKay, *Arrow's Theorem: The Paradox of Social Choice*, Yale University Press, 1980.

J. W. Moon, *Topics on Tournaments*, Holt, Rinehart, and Winston, 1968.

W. H. Riker, "Voting and the Summation of Preferences: An Interpretive Bibliographical Review of Selected Developments During the Last Decade." *The American Political Science Review* 55, December 1961, pp. 900–11.

W. H. Riker, "Arrow's Theorem and Some Examples of the Paradox of Voting," *Mathematical Applications in Political Science*, Q.M. Claunch (ed.), Southern Methodist University, 1965.

Z. Usiskin, "Max–Min Probabilities in the Voting Paradox," *The Annals of Mathematical Statistics*, Vol. 35, June 1964, pp. 857–62.

Nontransitive Betting Games

M. Gardner, "Lucifer at Las Vegas," *Science Fiction Puzzle Tales*, Clarkson Potter, 1981, Problem 27.

R. Honsberger, "Sheep Fleecing with Walter Funkenbusch," *Mathematical Gems III*, Washington, DC: Mathematical Association of America, 1985.

J. C. Frauenthal and A. B. Miller, "A Coin Tossing Game," *Mathematics Magazine* 53, September 1980, pp. 239–43.

R. L. Tenney and C. C. Foster, "Nontransitive Dominance," *Mathematics Magazine* 49, May 1976, pp. 115–20.

Conway's Algorithm

S. Collings, "Coin Sequence Probabilities and Paradoxes," *Bulletin of the Institute of Mathematics and its Applications* 18, November/December 1982, pp. 227–32.

R. L. Graham, D. E. Knuth, and O. Patashnik, *Concrete Mathematics,* 2e, Addison-Wesley, 1994, pp. 403–10.

L. J. Guibas and A. M. Odlyzko, "String Overlaps, Pattern Matching, and Nontransitive Games," *Journal of Combinatorial Theory* 30, March 1981, pp. 183–208.

VII
Infinity

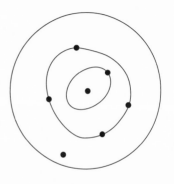

Chapter 24

Infinite Regress

Chairman of a meeting of the Society of Logicians: "Before we put the motion: 'That the motion be now put,' should we not first put the motion: 'That the motion: "That the motion be now put" be now put'?"

From an old issue of *Punch*

*T*he infinite regress, along which thought is compelled to march backward in a never ending chain of identical steps, has always aroused mixed emotions. Witness the varied reactions of critics to the central symbol of Broadway's most talked-about 1964 play, Edward Albee's *Tiny Alice*. The principal stage setting—the library of an enormous castle owned by Alice, the world's richest woman—was dominated by a scale model of the castle. Inside it lived Tiny Alice. When lights went on and off in the large castle, corresponding lights went on and off in the small one. A fire erupted simultaneously in castle and model. Within the model there was a smaller model in which a tinier Alice perhaps lived, and so on down, like a set of nested Chinese boxes. ("Hell to clean," commented the butler, whose name was Butler.) Was the castle itself, into which the play's audience peered, a model in a still larger model, and that in turn . . . ? A similar infinite nesting is the basis of E. Nesbit's short story, "The Town in the Library in the Town in the Library" (in her *Nine Unlikely Tales*); perhaps this was the source of Albee's idea.

For many of the play's spectators the endless regress of castles stirred up feelings of anxiety and despair: Existence is a mysterious, impenetrable, ultimately meaningless labyrinth; the regress is an endless corridor that leads nowhere. For theological students, who were said to flock to the play, the regress deepens an awareness of what Rudolf Otto, the German theologican, called the *mysterium tremendum:* the ultimate mystery, which one must approach with awe, fascination, humility, and a sense of "creaturehood." For the mathematician and the logician the regress has lost most of its terrors; indeed, as we shall soon see, it is a powerful, practical tool even in recreational mathematics. First, however, let us glance at some of the roles it has played in Western thought and letters.

Aristotle, taking a cue from Plato's *Parmenides,* used the regress in his famous "third man" criticism of Plato's doctrine of ideas. If all men are alike because they have something in common with Man, the ideal and eternal archetype, how (asked Aristotle) can we explain the fact that one man and Man are alike without assuming another archetype? And will not the same reasoning demand a third, fourth, and fifth archetype, and so on into the regress of more and more ideal worlds?

A similar aversion to the infinite regress underlies Aristotle's argument, elaborated by hundreds of later philosophers, that the cosmos must have a first cause. William Paley, an 18th-century English theologian, put it this way: "A chain composed of an infinite number of links can no more support itself than a chain composed of a finite number of links." A finite chain does indeed require support, mathematicians were quick to point out, but in an infinite chain *every* link hangs securely on the one above. The question of what supports the entire series no more arises than the question of what kind of number precedes the infinite regress of negative integers.

Agrippa, an ancient Greek skeptic, argued that nothing can be proved, even in mathematics, because every proof must be proved valid and its proof must in turn be proved, and so on. The argument is repeated by Lewis Carroll in his paper "What the Tortoise Said to Achilles" (*Mind,* April 1895). After finishing their famous race, which involved an infinite regress of smaller and smaller distances, the Tortoise traps his fellow athlete in a more disturbing regress. He refuses to accept a simple deduction involving a triangle until Achilles has written down an infinite series of hypothetical assumptions, each necessary to make the preceding argument valid.

F. H. Bradley, the English idealist, argued (not very convincingly) that our mind cannot grasp *any* type of logical relation. We cannot say, for example, that castle A is smaller than castle B and leave it at that, because "smaller than" is a relation to which both castles are related. Call these new relations c and d. Now we have to relate c and d to the two castles and to "smaller than." This demands four more relations, they in turn call for eight more, and so on, until the shaken reader collapses into the arms of Bradley's Absolute.

In recent philosophy the two most revolutionary uses of the regress have been made by the mathematicians Alfred Tarski and Kurt Gödel. Tarski avoids certain troublesome paradoxes in semantics by defining truth in terms of an endless regress of "metalanguages," each capable of discussing the truth and falsity of statements on the next lower level but not on its own level. As Bertrand Russell once explained it: "The man who says 'I am telling a lie of order n' is telling a lie, but a lie of order $n + 1$." In a closely related argument Gödel was able to show that there is no single, all-inclusive mathematics but only an infinite regress of richer and richer systems.

The endless hierarchy of gods implied by so many mythologies and by the child's inevitable question "Who made God?" has appealed to many thinkers. William James closed his *Varieties of Religious Experience* by suggesting that existence includes a collection of many gods, of different degrees of inclusiveness, "with no absolute unity realized in it at all. Thus would a sort of polytheism return upon us. . . ." The notion turns up in unlikely places. Benjamin Franklin, in a quaint little work called *Articles of Belief and Acts of Religion,* wrote: "For I believe that man is not the most perfect being but one, but rather that there are many degrees of beings superior to him." Our prayers, said Franklin, should be directed only to the god of our solar system, the deity closest to us. Many writers have viewed life as a board game in which we are the pieces moved by higher intelligences who in turn are the pieces in a vaster game. The prophet in Lord Dunsany's story "The South Wind" observes the gods striding through the stars, but as he worships them he sees the outstretched hand of a player "enormous over Their heads."

Graphic artists have long enjoyed the infinite regress. The striking cover of the April 1965, issue of *Scientific American* showed the magazine cover reflected in the pupil of an eye. The cover of the November 1964, *Punch* showed a magician pulling a rabbit out of a hat. The rabbit in turn is pulling a smaller rabbit out of a smaller hat, and this endless series of rabbits and hats moves up and off the edge of the page. It is not a bad picture of contemporary particle physics. It is now known that protons and neutrons are made of smaller units called quarks, and if superstring theory is correct, all particles are made of extremely tiny loops that vibrate at different frequencies. Does the universe have, as some physicists believe, infinite levels of structure?

The play within the play, the puppet show within the puppet show, the story within the story have amused countless writers. Luigi Pirandello's *Six Characters in Search of an Author* is perhaps the best-known stage example. The protagonist in Miguel de Unamuno's novel *Mist,* anticipating his death later in the plot, visits Unamuno to protest and troubles the author with the thought that he too is only the figment of a higher imagination. Philip Quarles, in Aldous Huxley's *Point Counter Point,* is writing a novel suspiciously like *Point Counter Point.* Edouard, in André Gide's *The Counterfeiters,* is writing *The Counterfeiters.* Norman Mailer's story "The Notebook" tells of an argument between the writer and his girl friend. As they argue he jots in his notebook an idea for a story that has just come to him. It is, of course, a story about a writer who is arguing with his girl friend when he gets an idea. . . .

J. E. Littlewood, in *A Mathematician's Miscellany,* recalls the following entry, which won a newspaper prize in Britain for the best piece on the topic: "What would you most like to read on opening the morning paper?"

OUR SECOND COMPETITION

The First Prize in the second of this year's competitions goes to Mr. Arthur Robinson, whose witty entry was easily the best of those we received. His choice of what he would like to read on opening his paper was headed "Our Second Competition" and was as follows: "The First Prize in the second of this year's competitions goes to Mr. Arthur Robinson, whose witty entry was easily the best of those we received. His choice of what he would like to read on opening his paper was headed 'Our Second Competition,' but owing to paper restrictions we cannot print all of it."

One way to escape the torturing implications of the endless regress is by the topological trick of joining the two ends to make a circle, not necessarily vicious, like the circle of weary soldiers who rest themselves in a bog by each sitting on the lap of the man behind. Albert Einstein did exactly this when he tried to abolish the endless regress of distance by bending three-dimensional space around to form the hypersurface of a four-dimensional sphere. One can do the same thing with time. There are Eastern religions that view history as an endless recurrence of the same events. In the purest sense one does not even think of cycles following one another, because there is no outside time by which the cycles can be counted; the *same* cycle, the *same* time go around and around. In a similar vein, there is a sketch by the Dutch artist Maurits C. Escher of two hands, each holding a pencil and sketching the other (see Figure 24.1). In *Through the Looking Glass* Alice dreams of the Red King, but the King is himself asleep and, as Twee-

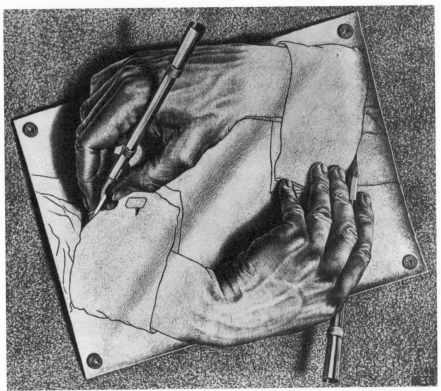

Figure 24.1. Maurits C. Escher's *Drawing Hands*

dledee points out, Alice is only a "sort of thing" in *his* dream. *Finnegans Wake* ends in the middle of a sentence that carries the reader back for its completion to the broken sentence that opens the book.

Since Fitz-James O'Brien wrote his pioneer yarn "The Diamond Lens" in 1858 almost countless writers have played with the theme of an infinite regress of worlds on smaller and smaller particles. In Henry Hasse's story "He Who Shrank" a man on a cosmic level much larger than ours is the victim of a scientific experiment that has caused him to shrink. After diminishing through hundreds of subuniverses he lingers just long enough in Cleveland to tell his story before he vanishes again, wondering how long this will go on, hoping that the levels are joined at their ends so that he can get back to his original cosmos.

Even the infinite hierarchy of gods has been bent into a closed curve by Dunsany in his wonderful tale "The Sorrow of Search." One night as the prophet Shaun is observing by starlight the four mountain gods of old—Asgool, Trodath, Skun, and Rhoog—he sees the shadowy forms of three larger gods farther up the slope. He leads his disciples up the mountain only to observe, years later, two larger gods seated at the summit, from which they point and mock at the gods below. Shaun takes his followers still higher. Then one night he perceives across the plain an enormous, solitary god looking angrily toward the mountain. Down the mountain and across the plain goes Shaun. While he is carving on rock the story of how his search has ended at last with the discovery of the ultimate god, he sees in the far distance the dim forms of four higher deities. As the reader can guess, they are Asgool, Trodath, Skun, and Rhoog.

No branch of mathematics is immune to the infinite regress. Numbers on both sides of zero gallop off to infinity. In modular arithmetics they go around and around. Every infinite series is an infinite regress. The regress underlies the technique of mathematical induction. Georg Cantor's transfinite numbers form an endless hierarchy of richer infinities. A beautiful modern example of how the regress enters into a mathematical proof is related to the difficult problem of dividing a square into other squares no two of which are alike. The question arises: Is it possible similarly to cut a cube into a finite number of smaller cubes no two of which are alike? Were it not for the deductive power of the regress, mathematicians might still be searching in vain for ways to do this. The proof of impossibility follows.

Assume that it is possible to "cube the cube." The bottom face of such a dissected cube, as it rests on a table, will necessarily be a "squared square." Consider the smallest square in this pattern. It cannot be a corner square, because a larger square on one side keeps any larger square from bordering the other side (see Figure 24.2(a)). Similarly, the smallest square cannot be elsewhere on the border, between corners, because larger squares on two sides prevent a third larger square from touching the third side (Figure 24.2(b)). The smallest square must therefore be somewhere in the pattern's interior. This in turn requires that the smallest cube touching the table be surrounded by cubes larger than itself. This is possible (Figure 24.2(c)), but it means that four walls must rise above all four sides of the small cube—preventing a larger cube from resting on top of it. Therefore on this smallest cube there must rest a set of smaller cubes, the bottoms of which will form another pattern of squares.

The same argument is now repeated. In the new pattern of squares

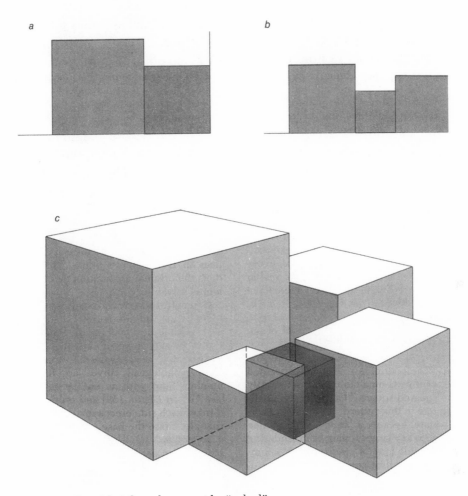

Figure 24.2. Proof that the cube cannot be "cubed"

the smallest square must be somewhere in the interior. On this smallest square must rest the smallest cube, and the little cubes on top of it will form another pattern of squares. Clearly the argument leads to an endless regress of smaller cubes, like the endless hierarchy of fleas in Dean Swift's jingle. This contradicts the original assumption that the problem is solvable.

Geometric constructions such as this one, involving an infinite regress of smaller figures, sometimes lead to startling results. Can a closed curve of infinite length enclose a finite area of, say, one square inch? Such pathological curves are infinite in number. Start with an equilateral triangle (see Figure 24.3(a)) and on the central third of each side erect a smaller equilateral triangle. Erase the base lines and you have a six-pointed star (Figure 24.3*(b)*). Repeating the construction on each of the star's 12 sides produces a 48-sided polygon (Figure 24.3*(c)*). The third step is shown in (Figure 24.3*(d)*). The limit of this infinite construction, called the snowflake curve, bounds an area 8/5 that of the original triangle. It is easy to show that successive additions of length form an infinite series that diverges; in short, the length of the snowflake's perimeter is infinite. (In 1956 W. Grey Walter, the British physiologist, published a science-fiction novel, *The Curve of the Snowflake,* in which a solid analogue of this crazy curve provides the basis for a timetravel machine.)

Figure 24.3. The snowflake curve

Here are two easy puzzles about the less-well-known square version of the snowflake, a curve that has been called the cross-stitch. On the middle third of each side of a unit square erect four smaller squares as shown at the top of Figure 24.4. The second step is shown at the bottom. (The squares will never overlap, but corners will touch.) If this procedure continues to infinity, how long is the final perimeter? How large an area does it enclose?

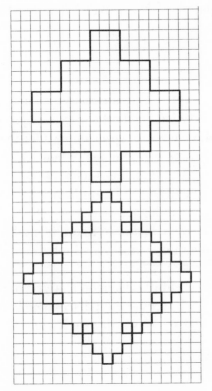

Figure 24.4. The cross-stitch curve

Addendum

Protons and neutrons are believed to be made of smaller particles called quarks. Today the leading "theory of everything" conjectures that all the well-known particles are made of still smaller units called superstrings. These inconceivably tiny loops are vibrating in a space–time of 10 dimensions. And what are superstrings made of? Apparently they are mathematical structures not made of anything. As a friend once said, the universe seems to be made of nothing, yet somehow it manages to exist.

Some top mathematical physicists, Stanislaw Ulam and David Bohm for example, defended and defend the notion that matter has infinite layers of structure in both directions. One of H. G. Wells's fantasies portrayed our universe as a molecule in a ring worn by a gigantic hand.

I described the snowflake curve before it became known as the oldest and simplest of Benoit Mandelbrot's famous fractals. For more on fractals, see this volume's Chapter 27 and Chapter 8 of my *Penrose Tiles to Trapdoor Ciphers*.

Answers

The cross-stitch curve has, like its analogue the snowflake, an infinite length. It bounds an area twice that of the original square. The drawing at the left in Figure 24.5 shows its appearance after the third construction. After many more steps it resembles (when viewed at a distance) the drawing at the right. Although the stitches seem to run diagonally, actually every line segment in the figure is vertical or horizontal. Similar constructions of pathological curves can be based on any regular polygon, but beyond the square the figure is muddied by overlapping, so that certain conventions must be adopted in defining what is meant by the enclosed area.

Samuel P. King, Jr., of Honolulu, supplied a good analysis of curves of this type, including a variant of the cross-stitch discovered by his father. Instead of erecting four squares outwardly each time, they are

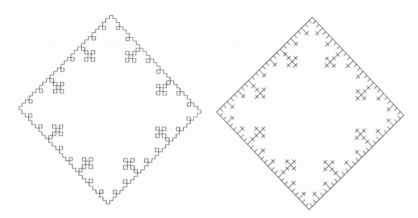

Figure 24.5. Solution to cross-stitch curve problem

erected inwardly from the sides of each square. The limit curve has an infinite length, but encloses zero area.

Bibliography

J. L. Borges, "Avatars of the Tortoise," *Labyrinths: Selected Stories and Other Writings,* New York: New Directions, 1964.

J. L. Borges, "Partial Magic in the Quixote," *Labyrinths: Selected Stories and Other Writings,* New York: New Directions, 1964.

B. W. King, "Snowflake Curves," *The Mathematics Teacher,* Vol. 57, No. 4, April, 1964, pp. 219–22.

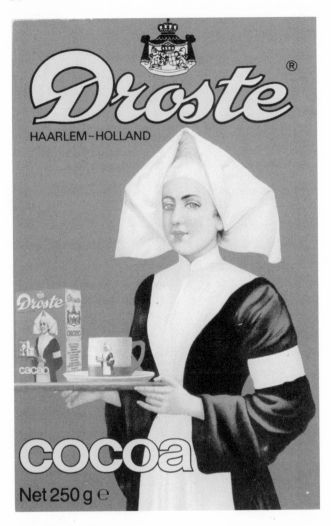

A Dutch advertisement

H. P. Owen, "Infinity in Theology and Metaphysics," *The Encyclopedia of Philosophy.* Vol. 4, New York: Crowell Collier, 1967, pp. 190–93.

J. Passmore, "The Infinite Regress," *Philosophical Reasoning,* New York: Charles Scribner's Sons, 1961.

J. E. Schneider, "A Generalization of the Von Koch Curve," *Mathematics Magazine,* Vol. 38, No. 3, May, 1965, pp. 144–47.

J. Thomson, "Infinity in Mathematics and Logic," *The Encyclopedia of Philosophy,* Vol. 4, New York: Crowell Collier, 1967, pp. 183–90.

Chapter 25

Aleph-Null and Aleph-One

A graduate student at Trinity
Computed the square of infinity.
But it gave him the fidgets
To put down the digits,
So he dropped math and took up divinity

—ANONYMOUS

*I*n 1963 Paul J. Cohen, then a 29-year-old mathematician at Stanford University, found a surprising answer to one of the great problems of modern set theory: Is there an order of infinity higher than the number of integers but lower than the number of points on a line? To make clear exactly what Cohen proved, something must first be said about those two lowest known levels of infinity.

It was Georg Ferdinand Ludwig Philipp Cantor who first discovered that beyond the infinity of the integers—an infinity to which he gave the name aleph-null—there are not only higher infinities but also an infinite number of them. Leading mathematicians were sharply divided in their reactions. Henri Poincaré called Cantorism a disease from which mathematics would have to recover, and Hermann Weyl spoke of Cantor's hierarchy of alephs as "fog on fog."

On the other hand, David Hilbert said, "From the paradise created for us by Cantor, no one will drive us out," and Bertrand Russell once praised Cantor's achievement as "probably the greatest of which the age can boast." Today only mathematicians of the intuitionist school and a few philosophers are still uneasy about the alephs. Most mathematicians long ago lost their fear of them, and the proofs by which Cantor established his "terrible dynasties" (as they have been called by the world-renowned Argentine writer Jorge Luis Borges) are now universally honored as being among the most brilliant and beautiful in the history of mathematics.

Any infinite set of things that can be counted 1, 2, 3, . . . has the cardinal number $aleph_0$ (aleph-null), the bottom rung of Cantor's aleph ladder. Of course, it is not possible actually to count such a set; one merely shows how it can be put into one-to-one correspondence with the counting numbers. Consider, for example, the infinite set of primes.

It is easily put in one-to-one correspondence with the positive integers:

$$
\begin{array}{cccccc}
1 & 2 & 3 & 4 & 5 & 6 \ldots \\
\downarrow & \downarrow & \downarrow & \downarrow & \downarrow & \downarrow \\
2 & 3 & 5 & 7 & 11 & 13 \ldots
\end{array}
$$

The set of primes is therefore an aleph-null set. It is said to be "countable" or "denumerable." Here we encounter a basic paradox of all infinite sets. Unlike finite sets, they can be put in one-to-one correspondence with a *part* of themselves or, more technically, with one of their "proper subsets." Although the primes are only a small portion of the positive integers, as a completed set they have the same aleph number. Similarly, the integers are only a small portion of the rational numbers (the integers plus all integral fractions), but the rationals form an aleph-null set too.

There are all kinds of ways in which this can be proved by arranging the rationals in a countable order. The most familiar way is to attach them, as fractions, to an infinite square array of lattice points and then count the points by following a zigzag path, or a spiral path if the lattice includes the negative rationals. Here is another method of ordering and counting the positive rationals that was proposed by the American logician Charles Sanders Peirce.

Start with the fractions 0/1 and 1/0. (The second fraction is meaningless, but that can be ignored.) Sum the two numerators and then the two denominators to get the new fraction 1/1, and place it between the previous pair: 0/1, 1/1, 1/0. Repeat this procedure with each pair of adjacent fractions to obtain two new fractions that go between them:

$$
\frac{0}{1} \quad \frac{1}{2} \quad \frac{1}{1} \quad \frac{2}{1} \quad \frac{1}{0} \, .
$$

The five fractions grow, by the same procedure, to nine:

$$
\frac{0}{1} \quad \frac{1}{3} \quad \frac{1}{2} \quad \frac{2}{3} \quad \frac{1}{1} \quad \frac{3}{2} \quad \frac{2}{1} \quad \frac{3}{1} \quad \frac{1}{0} \, .
$$

In this continued series every rational number will appear once and only once, and always in its simplest fractional form. There is no need, as there is in other methods of ordering the rationals, to eliminate fractions, such as 10/20, that are equivalent to simpler fractions also on the list, because no reducible fraction ever appears. If at each step you fill the cracks, so to speak, from left to right, you can count the fractions simply by taking them in their order of appearance.

This series, as Peirce said, has many curious properties. At each new step the digits above the lines, taken from left to right, begin by repeating the top digits of the previous step: 01, 011, 0112, and so on. And at each step the digits below the lines are the same as those above the lines but in reverse order. As a consequence, any two fractions equally distant from the central 1/1 are reciprocals of each other. Note also that for any adjacent pair, a/b, c/d, we can write such equalities as $bc - ad = 1$, and $c/d - a/b = 1/bd$. The series is closely related to what are called Farey numbers (after the English geologist John Farey, who first analyzed them), about which there is now a considerable literature.

It is easy to show that there is a set with a higher infinite number of elements than aleph-null. To explain one of the best of such proofs, a deck of cards is useful. First consider a finite set of three objects, say a key, a watch, and a ring. Each subset of this set is symbolized by a row of three cards (see Figure 25.1), a face-up card (white) indicates that the object above it is in the subset, a face-down card (black) indicates that it is not. The first subset consists of the original set itself. The next three rows indicate subsets that contain only two of the objects. They are followed by the three subsets of single objects and finally by the empty (or null) subset that contains none of the objects. For any set of n elements the number of subsets is 2^n. (It is easy to see why. Each element can be either included or not, so for one element there are two subsets, for two elements there are $2 \times 2 = 4$ subsets, for three elements there are $2 \times 2 \times 2 = 8$ subsets, and so on.) Note that this formula applies even to the empty set, since $2^0 = 1$ and the empty set has the empty set as its sole subset.

This procedure is applied to an infinite but countable (aleph-null) set of elements at the left in Figure 25.2. Can the subsets of this infinite set be put into one-to-one correspondence with the counting integers? Assume that they can. Symbolize each subset with a row of cards, as before, only now each row continues endlessly to the right. Imagine these infinite rows listed in any order whatever and numbered 1, 2, 3, . . . from the top down.

If we continue forming such rows, will the list eventually catch all the subsets? No—because there is an infinite number of ways to produce a subset that cannot be on the list. The simplest way is to consider

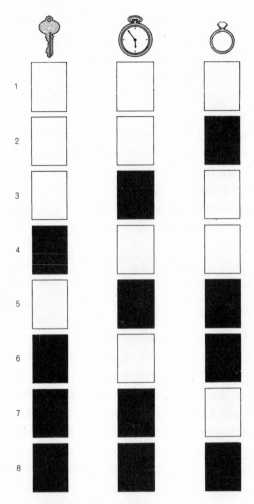

Figure 25.1. Subsets of a set of three elements

the diagonal set of cards indicated by the arrow and then suppose every card along this diagonal is turned over (that is, every face-down card is turned up, every face-up card is turned down). The new diagonal set cannot be the first subset because its first card differs from the first card of subset 1. It cannot be the second subset because its second card differs from the second card of subset 2. In general it cannot be the nth subset because its nth card differs from the nth card of subset n. Since we have produced a subset that cannot be on the list, even when the list is infinite, we are forced to conclude that the original assumption is false. The set of all subsets of an aleph-null set is a set with the cardi-

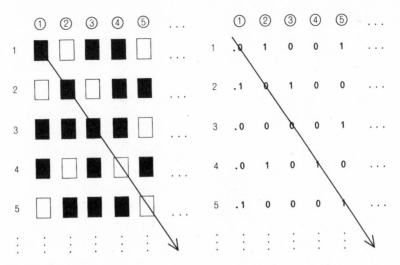

Figure 25.2. A countable infinity has an uncountable infinity of subsets *(left)* that correspond to the real numbers *(right)*.

nal number 2 raised to the power of aleph-null. This proof shows that such a set cannot be matched one-to-one with the counting integers. It is a higher aleph, an "uncountable" infinity.

Cantor's famous diagonal proof, in the form just given, conceals a startling bonus. It proves that the set of real numbers (the rationals plus the irrationals) is also uncountable. Consider a line segment, its ends numbered 0 and 1. Every rational fraction from 0 to 1 corresponds to a point on this line. Between any two rational points there is an infinity of other rational points; nevertheless, even after all rational points are identified, there remains an infinity of unidentified points—points that correspond to the unrepeating decimal fractions attached to such algebraic irrationals as the square root of 2 and to such transcendental irrationals as pi and *e.* Every point on the line segment, rational or irrational, can be represented by an endless decimal fraction. But these fractions need not be decimal; they can also be written in binary notation. Thus every point on the line segment can be represented by an endless pattern of 1's and 0's, and every possible endless pattern of 1's and 0's corresponds to exactly one point on the line segment. See Addendum.

Now, suppose each face-up card at the left in Figure 25.2 is replaced by 1 and each face-down card is replaced by 0, as shown at the right in the illustration. Put a binary point in front of each row and we have an infinite list of binary fractions between 0 and 1. But the diagonal set of symbols, after each 1 is changed to 0 and each 0 to 1, is not on the list, and the real numbers and points on the line are uncountable. By careful dealing with the duplications Cantor showed that the three sets—the subsets of aleph-null, the real numbers, and the totality of points on a line segment—have the same number of elements. Cantor called this cardinal number C, the "power of the continuum." He believed it was also $\aleph_1$ (aleph-one), the first infinity greater than aleph-null.

By a variety of simple, elegant proofs Cantor showed that C was the number of such infinite sets as the transcendental irrationals (the algebraic irrationals, he proved, form a countable set), the number of points on a line of infinite length, the number of points on any plane figure or on the infinite plane, and the number of points in any solid figure or in all of 3-space. Going into higher dimensions does not increase the number of points. The points on a line segment one inch long can be matched one-to-one with the points in any higher-dimensional solid, or with the points in the entire space of any higher dimension.

The distinction between aleph-null and aleph-one (we accept, for the moment, Cantor's identification of aleph-one with C) is important in geometry whenever infinite sets of figures are encountered. Imagine an infinite plane tessellated with hexagons. Is the total number of vertices aleph-one or aleph-null? The answer is aleph-null; they are easily counted along a spiral path (see Figure 25.3). On the other hand, the number of different circles of one-inch radius that can be placed on a sheet of typewriter paper is aleph-one because inside any small square near the center of the sheet there are aleph-one points, each the center of a different circle with a one-inch radius.

Consider in turn each of the five symbols J. B. Rhine uses on his "ESP" test cards (see Figure 25.4). Can it be drawn an aleph-one number of times on a sheet of paper, assuming that the symbol is drawn with ideal lines of no thickness and that there is no overlap or intersection of any lines? (The drawn symbols need not be the same size, but all must be similar in shape.) It turns out that all except one can be drawn an aleph-one number of times. Can the reader show which symbol is the exception?

Richard Schlegel, a physicist, attempted to relate the two alephs to cosmology by calling attention to a seeming contradiction in the steady-state theory. According to that theory, the number of atoms in the cosmos at the present time is aleph-null. (The cosmos is regarded as

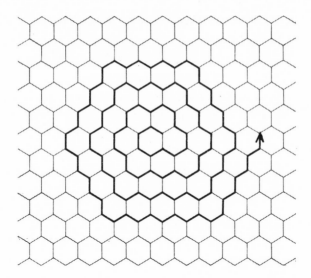

Figure 25.3. Spiral counts the vertices of a hexagonal tessellation

Figure 25.4. Five "ESP" symbols

infinite even though an "optical horizon" puts a limit on what can be seen.) Moreover, atoms are steadily increasing in number as the universe expands so that the average density remains constant. Infinite space can easily accommodate any finite number of doublings of the quantity of atoms, for aleph-null multiplied by two is aleph-null again. (If you have an aleph-null number of eggs in aleph-null boxes, one egg per box, you can accommodate another aleph-null set of eggs simply by shifting the egg in box 1 to box 2, the egg in box 2 to box 4, and so on, each egg going to a box whose number is twice the number of egg's previous box. This empties all the odd-numbered boxes, which can then be filled with another aleph-null set of eggs.) In the steady-state theory the universe extends into the indefinite past so that there would seem to be aleph-null periods of doubling already completed. This would give us 2 raised to the power of aleph-null atoms. As we have seen, this produces an aleph-one set. For example, even if there were only two atoms at an infinitely remote time in the past, after aleph-null doublings, they would have grown to an aleph-one set. But the cosmos cannot contain an aleph-one set of atoms. Any collection of distinct physical entities (as opposed to the ideal entities of mathematics) is countable and therefore, *at the most,* aleph-null.

In his paper, "The Problem of Infinite Matter in Steady-State Cosmology," Schlegel found the way out. Instead of regarding the past as a completed aleph-null set of finite time intervals (to be sure, ideal instants in time form an aleph-one continuum, but Schlegel is concerned with those finite time intervals during which doublings of atoms occur), we can view both the past and the future as infinite in the inferior sense of "becoming" rather than completed. Whatever date is suggested for the origin of the universe (remember, we are dealing with the steady-state model, not with a Big Bang or oscillating theory), we can always set an earlier date. In a sense there is a "beginning," but we can push it as far back as we please. There is also an "end," but we can push it as far forward as we please. As we go back in time, continually halving the number of atoms, we never halve them more than a finite number of times, with the result that their number never shrinks to less than aleph-null. As we go forward in time, doubling the number of atoms, we never double more than a finite number of times: therefore the set of atoms never grows larger than aleph-null. In either direction the leap is never made to a completed aleph-null set of time intervals. As a result the set of atoms never leaps to aleph-one and the disturbing contradiction does not arise.

Cantor was convinced that his endless hierarchy of alephs, each obtained by raising 2 to the power of the preceding aleph, represented all the alephs there are. There are none in between. Nor is there an Ultimate Aleph, such as certain Hegelian philosophers of the time identified with the Absolute. The endless hierarchy of infinities itself, Cantor argued, is a better symbol of the Absolute.

All his life Cantor tried to prove that there is no aleph between alephnull and C, the power of the continuum, but he never found a proof. In 1938 Kurt Gödel showed that Cantor's conjecture, which became known as the Continuum Hypothesis, could be assumed to be true, and that this could not conflict with the axioms of set theory.

What Cohen proved in 1963 was that the opposite could also be assumed. One can posit that C is *not* aleph-one; that there is at least one aleph between aleph-null and C, even though no one has the slightest notion of how to specify a set (for example, a certain subset of the transcendental numbers) that would have such a cardinal number. This too is consistent with set theory. Cantor's hypothesis is undecidable. Like the parallel postulate of Euclidean geometry, it is an independent axiom that can be affirmed or denied. Just as the two assumptions about Euclid's parallel axiom divided geometry into Euclidean and non-Euclidean, so the two assumptions about Cantor's hypothesis now divide the theory of infinite sets into Cantorian and non-Cantorian. It is even worse than that. The non-Cantorian side opens up the possibility of an infinity of systems of set theory, all as consistent as standard theory now is and all differing with respect to assumptions about the power of the continuum.

Of course Cohen did no more than show that the continuum hypothesis was undecidable within standard set theory, even when the theory is strengthened by the axiom of choice. Many mathematicians hope and believe that some day a "self-evident" axiom, not equivalent to an affirmation or denial of the continuum hypothesis, will be found, and that when this axiom is added to set theory, the continuum hypothesis will be decided. (By "self-evident" they mean an axiom which all mathematicians will agree is "true.") Indeed, Gödel expected and Cohen expects this to happen and are convinced that the continuum hypothesis is in fact false, in contrast to Cantor, who believed and hoped it was true. So far, however, these remain only pious Platonic hopes. What is undeniable is that set theory has been struck a gigantic cleaver blow, and exactly what will come of the pieces no one can say.

Addendum

In giving a binary version of Cantor's famous diagonal proof that the real numbers are uncountable, I deliberately avoided complicating it by considering the fact that every integral fraction between 0 and 1 can be represented as an infinite binary fraction in two ways. For example, ¼ is .01 followed by aleph-null zeroes and also .001 followed by aleph-null ones. This raises the possibility that the list of real binary fractions might be ordered in such a way that complementing the diagonal would produce a number on the list. The constructed number would, of course, have a *pattern* not on the list, but could not this be a pattern which expressed, in a different way, an integral fraction on the list?

The answer is no. The proof assumes that all possible infinite binary patterns are listed, therefore every integral fraction appears *twice* on the list, once in each of its two binary forms. It follows that the constructed diagonal number cannot match either form of any integral fraction on the list.

In every base notation there are two ways to express an integral fraction by an aleph-null string of digits. Thus in decimal notation ¼ = .2500000 . . . = .2499999. . . . Although it is not necessary for the validity of the diagonal proof in decimal notation, it is customary to avoid ambiguity by specifying that each integral fraction be listed only in the form that terminates with an endless sequence of nines, then the diagonal number is constructed by changing each digit on the diagonal to a different digit other than nine or zero.

Until I discussed Cantor's diagonal proof in *Scientific American,* I had not realized how strongly the opposition to this proof has persisted; not so much among mathematicians as among engineers and scientists. I received many letters attacking the proof. William Dilworth, an electrical engineer, sent me a clipping from the *LaGrange Citizen,* LaGrange, IL, January 20, 1966, in which he is interviewed at some length about his rejection of Cantorian "numerology." Dilworth first delivered his attack on the diagonal proof at the International Conference on General Semantics, New York, 1963.

One of the most distinguished of modern scientists to reject Cantorian set theory was the physicist P. W. Bridgman. He published a paper about it in 1934, and in his *Reflections of a Physicist* (Philosophical Library, 1955) he devotes pages 99–104 to an uncompromising attack on transfinite numbers and the diagonal proof. "I personally cannot see an iota of appeal in this proof," he writes, "but it appears to me to be a perfect nonsequitur—my mind will not do the things that it is obviously expected to do if it is indeed a proof."

The heart of Bridgman's attack is a point of view widely held by philosophers of the pragmatic and operationalist schools. Infinite numbers, it is argued, do not "exist" apart from human behavior. Indeed, all numbers are merely names for something that a person *does,* rather than names of "things." Because one can count 25 apples but cannot count an infinity of apples, "it does not make sense to speak of infinite numbers as 'existing' in the Platonic sense, and still less does it make sense to speak of infinite numbers of different orders of infinity, as does Cantor."

"An infinite number," declares Bridgman, "is a certain aspect of what one does when he embarks on carrying out a process . . . an infinite number is an aspect of a *program* of action."

The answer to this is that Cantor *did* specify precisely what one must "do" to define a transfinite number. The fact that one cannot carry out an infinite procedure no more diminishes the reality or usefulness of Cantor's alephs than the fact that one cannot fully compute the value of pi diminishes the reality or usefulness of pi. It is not, as Bridgman maintained, a question of whether one accepts or rejects the Platonic notion of numbers as "things." For an enlightened pragmatist, who wishes to ground all abstractions in human behavior, Cantorian set theory should be no less meaningful or potentially useful than any other precisely defined abstract system such as, say, group theory or a non-Euclidean geometry.

Several readers attacked Schlegel's claim that Cantor's alephs expose a contradiction in a steady-state theory of the universe. They focused on his argument that after a countable infinity of atom doublings the cosmos would contain an uncountable infinity of atoms. For details on this objection see Rudy Rucker's *Infinity and the Mind* (The Mathematical Association of America, 1982), pages 241–42.

For an account of crank objections to Cantor's alephs, see "Cantor's Diagonal Process," in Underwood Dudley's *Mathematical Cranks* (The Mathematical Association of America, 1992).

Answers

Which of the five ESP symbols cannot be drawn an aleph-one number of times on a sheet of paper, assuming ideal lines that do not overlap or intersect, and replicas that may vary in size but must be similar in the strict geometric sense?

Only the plus symbol cannot be aleph-one replicated. Figure 25.5 shows how each of the other four can be drawn an aleph-one number of times. In each case points on line segment *AB* form an aleph-one continuum. Clearly a set of nested or side-by-side figures can be drawn so that a different replica passes through each of these points, thus putting the continuum of points into one-to-one correspondence with a set of nonintersecting replicas. There is no comparable way to place replicas of the plus symbol so that they fit snugly against each other. The centers of any pair of crosses must be a finite distance apart (although this distance can be made as small as one pleases), forming a countable (aleph-null) set of points. The reader may enjoy devising a

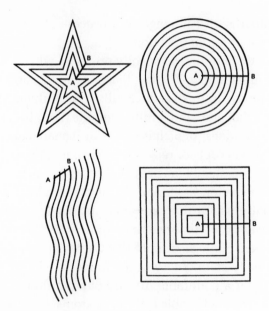

Figure 25.5. Proof for "ESP"-symbol problem

formal proof that aleph-one plus symbols cannot be drawn on a page. The problem is similar to one involving alphabet letters that can be found in Leo Zippin's *Uses of Infinity* (Random House, 1962), page 57. So far as I know, no one has yet specified precisely what conditions must be met for a linear figure to be aleph-one replicable. Some figures are aleph-one replicable by translation or rotation, some by shrinkage, some by translation plus shrinkage, some by rotation plus shrinkage. I rashly reported in my column that all figures topologically equivalent to a line segment or a simple closed curve were aleph-one replicable, but Robert Mack, then a high school student in Concord, MA, found a simple counterexample. Consider two unit squares, joined like a vertical domino, then eliminate two unit segments so that the remaining segments form the numeral 5. It is not aleph-one replicable.

Bibliography

Farey numbers

A. H. Beiler, *Recreations in the Theory of Numbers,* Dover, 1964, Chapter 16.

G. S. Cunningham, "Farey Sequences," *Enrichment Mathematics for High School,* National Council of Teachers of Mathematics, 1963, Chapter 1.

R. Honsberger, *Ingenuity in Mathematics,* The Mathematical Association of America New Mathematical Library, 1970, Chapter 5.

Transfinite numbers

J. Breuer, *Introduction to the Theory of Sets,* Prentice-Hall, 1958.

E. V. Huntington, *The Continuum,* Dover, 1955.

M. M. Zuckerman, *Sets and Transfinite Numbers,* Macmillan, 1974.

Cohen's proof

P. J. Cohen, *Set Theory and the Continuum Hypothesis,* W. A. Benjamin, 1966.

P. J. Cohen and R. Hersh, "Non-Cantorian Set Theory," *Scientific American,* December 1967, pp. 104–16.

D. S. Scott, "A Proof of the Independence of the Continuum Hypothesis." *Mathematical Systems Theory,* Vol. 1, 1967, pp. 89–111.

R. M. Smullyan, "The Continuum Problem," *The Encyclopedia of Philosophy,* Macmillan, 1967.

R. M. Smullyan, "The Continuum Hypothesis," *The Mathematical Sciences,* MIT Press, 1969, pp. 252–71.

Alephs and cosmology

R. Schlegel, "The Problem of Infinite Matter in Steady-State Cosmology," *Philosophy of Science,* Vol. 32, January 1965, pp. 21–31.

R. Schlegel. *Completeness in Science,* Appleton-Century-Crofts, 1967, pp. 138–49.

S. Ulam, "Combinatorial Analysis in Infinite Sets and Some Physical Theories," *SIAM Review,* Vol. 6, October 1964, pp. 343–55.

Chapter 26   **Supertasks**

Points
Have no parts or joints.
How then can they combine
To form a line?

                —J. A. Lindon

Every finite set of n elements has 2^n subsets if one includes the
original set and the null, or empty, set. For example, a set of three elements, *ABC*, has $2^3 = 8$ subsets: *ABC, AB, BC, AC, A, B, C,* and the null
set. As the philosopher Charles Sanders Peirce once observed (*Collected Papers* Vol. 4, p. 181), the null set "has obvious logical pecu-
liarities." You can't make any false statement about its members
because it has no members. Put another way, if you say anything logically contradictory about its members, you state a truth, because the solution set for the contradictory statement is the null set. Put
colloquially, you are saying something true about nothing.

In modern set theory it is convenient to think of the null set as an
"existing set" even though it has no members. It can also be said to have
2^n subsets because $2^0 = 1$, and the null set has one subset, namely itself.
And it is a subset of every set. If set *A* is included in set *B*, it means that
every member of set *A* is a member of set *B*. Therefore, if the null set is
to be treated as a legitimate set, all its members (namely none) must be
in set *B*. To prove it by contradiction, assume the null set is *not* included in set *B*. Then there must be at least one member of the null set
that is not a member of *B*, but this is impossible because the null set has
no members.

The n elements of any finite set obviously cannot be put into one-to-
one correspondence with its subsets because there are always more
than n subsets. Is this also true of infinite sets? The answer is yes, and
the general proof is one of the most beautiful in all set theory.

It is an indirect proof, a reductio ad absurdum. Assume that all elements of *N*, a set with any number or members, finite or infinite, are
matched one-to-one with all of *N*'s subsets. Each matching defines a
coloring of the elements:

1. An element is paired with a subset that includes that element. Let us call all such elements blue.
2. An element is paired with a subset that does *not* include that element. We call all such elements red.

The red elements form a subset of our initial set N. Can this subset be matched to a blue element? No, because every blue element is in its matching subset, therefore the red subset would have to include a blue element. Can the red subset be paired with a red element? No, because the red element would then be included in its subset and would therefore be blue. Since the red subset cannot be matched to either a red or blue element of N, we have constructed a subset of N that is not paired with any element of N. No set, even if infinite, can be put into one-to-one correspondence with its subsets. If n is a transfinite number, then 2^n—by definition it is the number of subsets of n—must be a higher order of infinity than n.

Georg Cantor, the founder of set theory, used the term aleph-null for the lowest transfinite number. It is the cardinal number of the set of all integers, and for that reason is often called a "countable infinity." Any set that can be matched one-to-one with the counting numbers, such as the set of integral fractions, is said to be a countable or aleph-null set. Cantor showed that when 2 is raised to the power of aleph-null—giving the number of subsets of the integers—the result is equal to the cardinal number of the set of all real numbers (rational or irrational), called the "power of the continuum," or C. It is the cardinal number of all points on a line. The line may be a segment of any finite length, a ray with a beginning but no end, or a line going to infinity in both directions. Figure 26.1 shows three intuitively obvious geometrical proofs that all three kinds of line have the same number of points. The slant

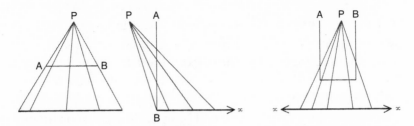

Figure 26.1. The number of points on a line segment AB is the same as on a longer line segment *(left)*, a ray *(center)*, and a line *(right)*.

lines projected from point P indicate how all points on the line segment AB can be put into one-to-one correspondence with all points on the longer segment, on a ray, and on an endless line.

The red-blue proof outlined above (Cantor published it in 1890) of course generates an infinite hierarchy of transfinite numbers. The ladder starts with the set of counting numbers, aleph-null, next comes C, then all the subsets of C, then all the subsets of all the subsets of C, and so on. The ladder can also be expressed like this:

$$\text{aleph-null, } C, 2^C, 2^{2^C}, 2^{2^{2^C}}, \ldots.$$

Cantor called C "aleph-one" because he believed that no transfinite number existed between aleph-null and C. And he called the next number aleph-two, the next aleph-three, and so on. For many years he tried unsuccessfully to prove that C was the next higher transfinite number after aleph-null, a conjecture that came to be called the Continuum Hypothesis. We now know, thanks to proofs by Kurt Gödel and Paul Cohen, that the conjecture is undecidable within standard set theory, even when strengthened by the axiom of choice. We can assume without contradiction that Cantor's alephs catch all transfinite numbers, or we can assume, also without contradiction, a non-Cantorian set theory in which there is an infinity of transfinite numbers between any two adjacent entries in Cantor's ladder. (See the previous chapter for a brief, informal account of this.)

Cantor also tried to prove that the number of points on a square is the next higher transfinite cardinal after C. In 1877 he astounded himself by finding an ingenious way to match all the points of a square to all the points of a line segment. Imagine a square one mile on a side, and a line segment one inch long (see Figure 26.2). On the line segment every point from 0 to 1 is labeled with an infinite decimal fraction: The point corresponding to the fractional part of pi is .14159 . . . , the point corresponding to 1/3 is .33333 . . . , and so on. Every point is represented by a unique string of aleph-null digits, and every possible aleph-null string of digits represents a unique point on the line segment. (A slight difficulty arises from the fact that a fraction such as .5000 . . . is the same as .4999 . . . , but it is easily overcome by dodges we need not go into here.)

Now consider the square mile. Using a Cartesian coordinate system, every point on the square has unique x and y coordinates, each of which can be represented by an endless decimal fraction. The illustra-

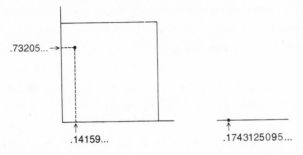

Figure 26.2. Points in square mile and on line segment

tion shows a point whose x coordinate is the fractional part of pi and whose y coordinate is the fractional part of the square root of 3, or .73205. . . . Starting with the x coordinate, alternate the digits of the two numbers: .1743125095. . . . The result is an endless decimal labeling a unique point on the line segment. Clearly this can be done with every point on the square. It is equally obvious that the mapping procedure can be reversed: we can select any point on the line segment and, by taking alternate digits of its infinite decimal, can split it into two endless decimals that as coordinates label a unique point on the square. (Here we must recognize and overcome the subtle fact that, for example, the following three distinct points on the segment—.449999 . . . , .459090 . . . , and .540909 . . .—all map the same point [½, ½] in the square.) In this way the points of any square can be put into one-to-one correspondence with the points on any line segment; therefore the two sets are equivalent and each has the cardinal number C.

The proof extends easily to a cube (by interlacing three coordinates) or to a hypercube of n dimensions (by interlacing n coordinates). Other proofs show that C also numbers the points in an infinite space of any finite number of dimensions, even an infinite space of aleph-null dimensions.

Cantor hoped that his transfinite numbers would distinguish the different orders of space but, as we have seen, he himself proved that this was not the case. Mathematicians later showed that it is the topological way the points of space go together that distinguishes one space from another. The matchings in the previous paragraphs are not continuous; that is, points close together on, for instance, the line are not necessarily close together on the square, and vice versa. Put another way, you cannot continuously deform a line to make it a square, or a square to make it a cube, or a cube to make a hypercube, and so on.

Supertasks

Is there a set in mathematics that corresponds to 2^C? Of course we know it is the number of all subsets of the real numbers, but does it apply to any familiar set in mathematics? Yes, it is the set of all real functions of x, even the set of all real one-valued functions. This is the same as the number of all possible permutations of the points on a line. Geometrically it is all the curves (including discontinuous ones) that can be drawn on a plane or even a small finite portion of a plane the size, say, of a postage stamp. As for 2 to the power of 2^C, no one has yet found a set, aside from the subsets of 2^C, equal to it. Only aleph-null, C, and 2^C seem to have an application outside the higher reaches of set theory. As George Gamow once said, "We find ourselves here in a position exactly opposite to that of . . . the Hottentot who had many sons but could not count beyond three." There is an endless ladder of transfinite numbers, but most mathematicians have only three "sons" to count with them. This has not prevented philosophers from trying to find metaphysical interpretations for the transfinite numbers. Cantor himself, a deeply religious man, wrote at length on such matters. In the United States, Josiah Royce was the philosopher who made the most extensive use of Cantor's alephs, particularly in his work *The World and the Individual.*

The fact that there is no highest or final integer is involved in a variety of bewildering paradoxes. Known as supertasks, they have been much debated by philosophers of science since they were first suggested by the mathematician Hermann Weyl. For instance, imagine a lamp (called the Thomson lamp after James F. Thomson, who first wrote about it) that is turned off and on by a push-button switch. Starting at zero time, the lamp is on for 1/2 minute, then it is off for 1/4 minute, then on for 1/8 minute, and so on. Since the sum of this halving series, 1/2 + 1/4 + 1/8 + . . . +, is 1, at the end of one minute the switch will have been moved aleph-null times. Will the lamp be on or off?

Everyone agrees that a Thomson lamp cannot be constructed. Is such a lamp logically conceivable or is it nonsense to discuss it in the abstract? One of Zeno's celebrated paradoxes concerns a constant-speed runner who goes half of a certain distance in 1/2 minute, a fourth of the distance in the next 1/4 minute, an eighth of the distance in the next 1/8 minute, and so on. At the end of one minute he has had no difficulty reaching the last point of the distance. Why, then, cannot we say that at the end of one minute the switch of the Thomson lamp has made its last move? The answer is that the lamp must then be on or off and this is the same as saying that there is a last integer that is either even or odd. Since the integers have no last digit, the lamp's operation seems logically absurd.

Another supertask concerns an "infinity machine" that calculates and prints the value of pi. Each digit is printed in half the time it takes to print the preceding one. Moreover, the digits are printed on an idealized tape of finite length, each digit having half the width of the one before it. Both the time and the width series converge to the same limit, so that in theory one might expect the pi machine, in a finite time, to print all the digits of pi on a piece of tape. But pi has no final digit to print, and so again the supertask seems self-contradictory.

One final example: Max Black (1909–1988) imagined a machine that transfers a marble from tray A to tray B in one minute and then rests for a minute as a second machine returns the marble to A. In the next half-minute the first machine moves the marble back to B; then it rests for a half-minute as the other machine returns it to A. This continues, in a halving time series, until the machines' movements become, as Black put it, a "grey blur." At the end of four minutes each machine has made aleph-null transfers. Where is the marble? Once more, the fact that there is no last integer to be odd or even seems to rule out the possibility, even in principle, of such a supertask. (The basic articles on supertasks, by Thomson, Black, and others, are reprinted in Wesley C. Salmon's 1970 paperback anthology *Zeno's Paradoxes*.)

One is tempted to say that the basic difference between supertasks and Zeno's runner is that the runner moves continuously whereas the supertasks are performed in discrete steps that form an aleph-null set. The situation is more complicated than that. Adolph Grünbaum, in *Modern Science and Zeno's Paradoxes*, argues convincingly that Zeno's runner could also complete his run by what Grünbaum calls a "staccato" motion of aleph-null steps. The staccato runner goes the first half of his distance in 1/4 minute, rests 1/4 minute, goes half of the remaining distance in 1/8 minute, rests 1/8 minute, and so on. When he is running, he moves twice as fast as his "legato" counterpart, but his overall average speed is the same, and it is always less than the velocity of light. Since the pauses of the staccato runner converge to zero, at the end of one minute he too will have reached his final point just as an ideal bouncing ball comes to rest after an infinity of discrete bounces. Grünbaum finds no logical objection to the staccato run, even though it cannot be carried out in practice. His attitudes toward the supertasks are complex and controversial. He regards infinity machines of certain designs as being logically impossible and yet in most cases, with suitable qualifications, he defends them as logically consistent variants of the staccato run.

These questions are related to an old argument to the effect that Cantor was mistaken in his claim that aleph-null and *C* are different orders of infinity. The proof is displayed in Figure 26.3. The left side is an endless list of integers in serial order. Each is matched with a number on the right that is formed by reversing the order of the digits and putting a decimal in front of them. Since the list on the left can go to infinity, it should eventually include every possible sequence of digits. If it does, the numbers on the right will also catch every possible sequence and therefore will represent all real numbers between 0 and 1. The real numbers form a set of size *C*. Since this set can be put in one-to-one correspondence with the integers, an aleph-null set, the two sets appear to be equivalent.

INTEGERS	DECIMAL FRACTIONS
1	.1
2	.2
3	.3
.	.
.	.
.	.
10	.01
11	.11
12	.21
.	.
.	.
100	.001
101	.101
.	.
.	.
1234	.4321
.	.
.	.
.	.

Figure 26.3. Fallacious proof concerning two alephs

I would be ashamed to give this proof were it not for the fact that every year or so I receive it from a correspondent who has rediscovered it and convinced himself that he has demolished Cantorian set theory. Readers should have little difficulty seeing what is wrong.

Addendum

Among physicists, no one objected more violently to Cantorian set theory than Percy W. Bridgman. In *Reflections of a Physicist* (1955) he says he "cannot see an iota of appeal" in Cantor's proof that the real numbers form a set of higher infinity than the integers. Nor can he find paradox in any of Zeno's arguments because he is unable to think of a line as a set of points (see the *Clerihew* by Lindon that I used as an epigraph) or a time interval as a set of instants.

"A point is a curious thing," he wrote in *The Way Things Are* (1959), "and I do not believe that its nature is appreciated, even by many mathematicians. A line is not composed of points in any real sense. . . . We do not construct the line out of points, but, given the line, we may construct points on it. 'All the points on the line' has the same sort of meaning that the 'entire line' has. . . . We *create* the points on a line just as we create the numbers, and we identify the points by the numerical values of the coordinates."

Merwin J. Lyng, in *The Mathematics Teacher* (April 1968, p. 393), gives an amusing variation of Black's moving-marble supertask. A box has a hole at each end: Inside the box a rabbit sticks his head out of hole *A*, then a minute later out of hole *B*, then a half-minute later out of hole *A*, and so on. His students concluded that after two minutes the head is sticking out of both holes, "but practically the problem is not possible unless we split hares."

For what it is worth, I agree with those who believe that paradoxes such as the staccato run can be stated without contradiction in the language of set theory, but as soon as any element is added to the task that involves a highest integer, you add something not permitted, therefore you add only nonsense. There is nothing wrong in the abstract about an ideal bouncing ball coming to rest, or a staccato moving point reaching a goal, but nothing meaningful is added if you assume that at each bounce the ball changes color, alternating red and blue; then ask what color it is when it stops bouncing, or if the staccato runner opens and shuts his mouth at each step and you ask if it is open or closed at the finish.

A number of readers called my attention to errors in this chapter, as I first wrote it as a column, but I wish particularly to thank Leonard Gillman, of the University of Texas at Austin for reviewing the column and suggesting numerous revisions that have greatly simplified and improved the text.

Answers

The fundamental error in the false proof that the counting numbers can be matched one-to-one with the real numbers is that, no matter how long the list of integers on the left (and their mirror reversals on the right), no number with aleph-null digits will ever appear on each side. As a consequence no irrational decimal fraction will be listed on the right. The mirror reversals of the counting numbers, with a decimal point in front of each, form no more than a subset of the integral fractions between 0 and 1. Not even 1/3 appears in this subset because its decimal form requires aleph-null digits. In brief, all that is proved is the well-known fact that the counting numbers can be matched one-to-one with a subset of integral fractions.

The false proof reminds me of a quatrain I once perpetrated:

> Pi vs *e*
>
> Pi goes on and on and on . . .
> And *e* is just as cursed.
> I wonder: Which is larger
> When their digits are reversed?

Bibliography

P. Benacerraf, "Tasks, Super-Tasks, and Modern Eleatics," *Journal of Philosophy,* Vol. 59, 1962, pp. 765–84.

J. Benardete, *Infinity, an Essay in Metaphysics,* Clarendon Press, 1964.

P. Clark and S. Read, "Hypertasks," *Synthese,* Vol. 61, 1984, pp. 387–90.

J. Earman and J. D. Norton, "Infinite Pains: The Trouble with Supertasks," *Benacerraf and His Critics,* A. Morton and S. P. Stich (eds.), Blackwell, 1966, pp. 231–61.

C. D. Gruender, "The Achilles Paradox and Transfinite Numbers," *British Journal for the Philosophy of Science,* Vol. 17, November 1966, pp. 219–31.

A. Grünbaum, *Modern Science and Zeno's Paradoxes,* Wesleyan University Press, 1967.

A. Grünbaum, "Are 'Infinity Machines' Paradoxical?" *Science,* Vol. 159, January 26, 1968, pp. 396–406.

A. Grünbaum, "Can an Infinitude of Operations Be Performed in a Finite Time?" *British Journal for the Philosophy of Science,* Vol. 20, October 1969, pp. 203–18.

C. W. Kilmister, "Zeno, Aristotle, Weyl and Shuard: Two-and-a-half millenia of worries over number," *The Mathematical Gazette,* Vol. 64, October 1980, pp. 149–58.

J. P. Laraudogoittia, "A Beautiful Supertask," *Mind,* Vol. 105, 1996, pp. 81–83.

J. P. Laraudogoittia, "Classical Particle Dynamics, Indeterminism, and a Supertask," *British Journal for the Philosophy of Science,* Vol. 48, 1997, pp. 49–54.

J. P. Laroudogoittia, "Some Relativistic and Higher Order Supertasks," *Philosophy of Science,* Vol. 65, 1998, pp. 502–17.

W. C. Salmon (ed.), *Zeno's Paradoxes,* Bobbs-Merrill, 1970.

J. P. Van Bendegem, "Ross' Paradox is an Impossible Super-Task," *British Journal for the Philosophy of Science,* Vol. 45, 1994, pp. 743–48.

For when there are no words [accompanying music] it is very difficult to recognize the meaning of the harmony and rhythm, or to see that any worthy object is imitated by them.   —PLATO, *Laws,* Book II

Plato and Aristotle agreed that in some fashion all the fine arts, including music, "imitate" nature, and from their day until the late 18th century "imitation" was a central concept in western aesthetics. It is obvious how representational painting and sculpture "represent," and how fiction and the stage copy life, but in what sense does music imitate?

By the mid-18th century philosophers and critics were still arguing over exactly how the arts imitate and whether the term is relevant to music. The rhythms of music may be said to imitate such natural rhythms as heartbeats, walking, running, flapping wings, waving fins, water waves, the periodic motions of heavenly bodies, and so on, but this does not explain why we enjoy music more than, say, the sound of cicadas or the ticking of clocks. Musical pleasure derives mainly from tone patterns, and nature, though noisy, is singularly devoid of tones. Occasionally wind blows over some object to produce a tone, cats howl, birds warble, bowstrings twang. A Greek legend tells how Hermes invented the lyre: he found a turtle shell with tendons attached to it that produced musical tones when they were plucked.

Above all, human beings sing. Musical instruments may be said to imitate song, but what does singing imitate? A sad, happy, angry, or serene song somehow resembles sadness, joy, anger, or serenity, but if a melody has no words and invokes no special mood, what does it copy? It is easy to understand Plato's mystification.

There is one exception: the kind of imitation that plays a role in "program music." A lyre is severely limited in the natural sounds it can copy, but such limitations do not apply to symphonic or electronic music. Program music has no difficulty featuring the sounds of thunder, wind, rain, fire, ocean waves, and brook murmurings; bird calls (cuckoos and crowing cocks have been particularly popular), frog croaks, the gaits of animals (the thundering hoofbeats in Wagner's *Ride of the Valkyries*), the flights of bumblebees; the rolling of trains, the clang of hammers; the battle sounds of marching soldiers, clashing armies, roaring cannons, and exploding bombs. *Slaughter on Tenth Avenue* includes a pistol shot and the wail of a police-car siren. In Bach's *Saint Matthew Passion* we hear the earthquake and the ripping of the temple veil. In the *Alpine Symphony* by Richard Strauss, cowbells are imitated by the shaking of cowbells. Strauss insisted he could tell that a certain female character in Felix Mottl's *Don Juan* had red hair, and he once said that someday music would be able to distinguish the clattering of spoons from that of forks.

Such imitative noises are surely a trivial aspect of music even when it accompanies opera, ballet, or the cinema; besides, such sounds play no role whatsoever in "absolute music," music not intended to "mean" anything. A Platonist might argue that abstract music imitates emotions, or beauty, or the divine harmony of God or the gods, but on more mundane levels music is the least imitative of the arts. Even nonobjective paintings resemble certain patterns of nature, but nonobjective music resembles nothing except itself.

Since the turn of the century most critics have agreed that "imitation" has been given so many meanings (almost all are found in Plato) that it has become a useless synonym for "resemblance." When it is made precise with reference to literature or the visual arts, its meaning is obvious and trivial. When it is applied to music, its meaning is too fuzzy to be helpful. In this chapter we take a look at a surprising discovery by Richard F. Voss, a physicist from Minnesota who joined the Thomas J. Watson Research Center of the International Business Machines Corporation after obtaining his Ph.D. at the University of California at Berkeley under the guidance of John Clarke. This work is not likely to restore "imitation" to the lexicon of musical criticism, but it does suggest a curious way in which good music may mirror a subtle statistical property of the world.

The key concepts behind Voss's discovery are what mathematicians and physicists call the spectral density (or power spectrum) of a fluctuating quantity, and its "autocorrelation." These deep notions are technical and hard to understand. Benoît Mandelbrot, who is also at the Watson Research Center, and whose work makes extensive use of spectral densities and autocorrelation functions, has suggested a way of avoiding them here. Let the tape of a sound be played faster or slower than normal. One expects the character of the sound to change considerably. A violin, for example, no longer sounds like a violin. There is a special class of sounds, however, that behave quite differently. If you play a recording of such a sound at a different speed, you have only to adjust the volume to make it sound exactly as before. Mandelbrot calls such sounds "scaling noises."

By far the simplest example of a scaling noise is what in electronics and information theory is called white noise (or "Johnson noise"). To be white is to be colorless. White noise is a colorless hiss that is just as dull whether you play it faster or slower. Its autocorrelation function, which measures how its fluctuations at any moment are related to previous fluctuations, is zero except at the origin, where of course it must be 1. The most commonly encountered white noise is the thermal noise produced by the random motions of electrons through an electrical resistance. It causes most of the static in a radio or amplifier and the "snow" on radar and television screens when there is no input.

With randomizers such as dice or spinners it is easy to generate white noise that can then be used for composing a random "white tune," one with no correlation between any two notes. Our scale will be one octave of seven white keys on a piano: do, re, me, fa, so, la, ti. Fa is our middle frequency. Now construct a spinner such as the one shown at the left in Figure 27.1. Divide the circle into seven sectors and label them with the notes. It matters not at all what arc lengths are assigned to these sectors; they can be completely arbitrary. On the spinner shown, some order has been imposed by giving fa the longest arc (the highest probability of being chosen) and assigning decreasing proba-

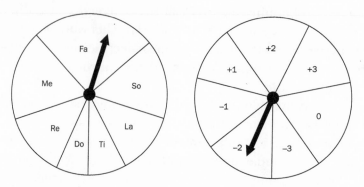

Figure 27.1. Spinners for white music *(left)* and brown music *(right)*

bilities to pairs of notes that are equal distances above and below fa. This has the effect of clustering the tones around fa.

To produce a "white melody" simply spin the spinner as often as you like, recording each chosen note. Since no tone is related in any way to the sequence of notes that precedes it, the result is a totally uncorrelated sequence. If you like, you can divide the circle into more parts and let the spinner select notes that range over the entire piano keyboard, black keys as well as white.

To make your white melody more sophisticated, use another spinner, its circle divided into four parts (with any proportions you like), and labeled 1, 1/2, 1/4, and 1/8 so that you can assign a full, a half, a quarter, or an eighth of a beat to each tone. After the composition is completed, tap it out on the piano. The music will sound just like what it is: random music of the dull kind that a two-year-old or a monkey might produce by hitting keys with one finger. Similar white music can be based on random number tables or the digits in an irrational number.

A more complicated kind of scaling noise is one that is sometimes called Brownian noise because it is characteristic of Brownian motion, the random movements of small particles suspended in a liquid and buffeted by the thermal agitation of molecules. Each particle executes a three-dimensional "random walk," the positions in which form a highly correlated sequence. The particle, so to speak, always "remembers" where it has been.

When tones fluctuate in this fashion, let us follow Voss and call it Brownian music or brown music. We can produce it easily with a spinner and a circle divided into seven parts as before, but now we label the regions, as shown at the right in Figure 27.1, to represent intervals between successive tones. These step sizes and their probabilities can be whatever we like. On the spinner shown, plus means a step up the scale of one, two, or three notes and minus means a step down of the same intervals.

Start the melody on the piano's middle C, then use the spinner to generate a linear random walk up and down the keyboard. The tune will wander here and there, and it will eventually wander off the keyboard. If we treat the ends of the keyboard as "absorbing barriers," the tune ends when we encounter one of them. We need not go into the ways in which we can treat the barriers as reflecting barriers, allowing the tune to bounce back, or as elastic barriers. To make the barriers

Fractal Music

elastic we must add rules so that the farther the tone gets from middle C, the greater is the likelihood it will step back toward C, like a marble wobbling from side to side as it rolls down a curved trough.

As before, we can make our brown music more sophisticated by varying the tone durations. If we like, we can do this in a brown way by using another spinner to give not the duration but the increase or decrease of the duration—another random walk but one along a different street. The result is a tune that sounds quite different from a white tune because it is strongly correlated, but a tune that still has little aesthetic appeal. It simply wanders up and down like a drunk weaving through an alley, never producing anything that resembles good music.

If we want to mediate between the extremes of white and brown, we can do it in two essentially different ways. The way chosen by previous composers of "stochastic music" is to adopt transition rules. These are rules that select each note on the basis of the last three or four. For example, one can analyze Bach's music and determine how often a certain note follows, say, a certain triplet of preceding notes. The random selection of each note is then weighted with probabilities derived from a statistical analysis of all Bach quadruplets. If there are certain transitions that never appear in Bach's music, we add rejection rules to prevent the undesirable transitions. The result is stochastic music that resembles Bach but only superficially. It sounds Bachlike in the short run but random in the long run. Consider the melody over periods of four or five notes and the tones are strongly correlated. Compare a run of five notes with another five-note run later on and you are back to white noise. One run has no correlation with the other. Almost all stochastic music produced so far has been of this sort. It sounds musical if you listen to any small part but random and uninteresting when you try to grasp the pattern as a whole.

Voss's insight was to compromise between white and brown input by selecting a scaling noise exactly halfway between. In spectral terminology it is called $1/f$ noise. (White noise has a spectral density of $1/f^0$; brownian noise has a spectral density of $1/f^2$. In "one-over-f" noise the exponent of f is 1 or very close to 1.) Tunes based on $1/f$ noise are moderately correlated, not just over short runs but throughout runs of any size. It turns out that almost every listener agrees that such music is much more pleasing than white or brown music.

In electronics $1/f$ noise is well known but poorly understood. It is sometimes called flicker noise. Mandelbrot, whose book *The Fractal*

Geometry of Nature (W. H. Freeman, 1982) has already become a modern classic, was the first to recognize how widespread $1/f$ noise is in nature, outside of physics, and how often one encounters other scaling fluctuations. For example, he discovered that the record of the annual flood levels of the Nile is a $1/f$ fluctuation. He also investigated how the curves that graph such fluctuations are related to "fractals," a term he invented. A scaling fractal can be defined roughly as any geometrical pattern (other than Euclidean lines, planes, and surfaces) with the remarkable property that no matter how closely you inspect it, it always looks the same, just as a slowed or speeded scaling noise always sounds the same. Mandelbrot coined the term fractal because he assigns to each of the curves a fractional dimension greater than its topological dimension.

Among the fractals that exhibit strong regularity the best-known are the Peano curves that completely fill a finite region and the beautiful snowflake curve discovered by the Swedish mathematician Helge von Koch in 1904. The Koch snowflake appears in Figure 27.2 as the boundary of the dark "sea" that surrounds the central motif. (For details on the snowflake's construction, and a discussion of fractals in general, see Chapter 3 of my *Penrose Tiles to Trapdoor Ciphers* [W. H. Freeman, 1989]).

The most interesting part of Figure 27.2 is the fractal curve that forms the central design. It was discovered by Mandelbrot and published for the first time as the cover of *Scientific American's* April 1978 issue. If you trace the boundary between the black and white regions from the tip of the point of the star at the lower left to the tip of the point of the star at the lower right, you will find this boundary to be a single curve. It is the third stage in the construction of a new Peano curve. At the limit this lovely curve will completely fill a region bounded by the traditional snowflake! Thus Mandelbrot's curve brings together two path-breaking fractals: the oldest of them all, Giuseppe Peano's 1890 curve, and Koch's later snowflake!

The secret of the curve's construction is the use of line segments of two unequal lengths and oriented in 12 different directions. The curve is much less regular than previous Peano curves and therefore closer to the modeling of natural phenomena, the central theme of Mandelbrot's book. Such natural forms as the gnarled branches of a tree or the shapes of flickering flames can be seen in the pattern.

At the left in Figure 27.3 is the first step of the construction. A

Figure 27.2. Benoît Mandelbrot's Peano–snowflake as it appeared on the cover of *Scientific American* (April 1978). The curve was drawn by a program written by Sigmund Handelman and Mark Laff.

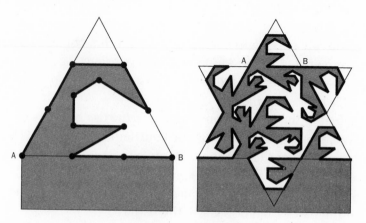

Figure 27.3. The first two steps in constructing Mandelbrot's Peano–snowflake curve

crooked line of nine segments is drawn on and within an equilateral triangle. Four of the segments are then divided into two equal parts, creating a line from *A* to *B* that consists of 13 long and short segments. The second step replaces each of these 13 segments with a smaller replica of the crooked line. These replicas (necessarily of unequal size) are oriented as is shown inside the star at the right in the illustration. A third repetition of the procedure generates the curve in Figure 27.2. (It belongs to a family of curves arising from William Gosper's discovery of the "flow-snake," a fractal pictured in Chapter 3 of my above cited book.) When the construction is repeated to infinity, the limit is a Peano curve that totally fills a region bordered by the Koch snowflake. The Peano curve has the usual dimension of 2, but its border, a scaling fractal of infinite length, has (as is explained in Mandelbrot's book) a fractal dimension of log 4/log 3, or 1.2618. . . .

Unlike these striking artificial curves, the fractals that occur in nature—coastlines, rivers, trees, star clustering, clouds, and so on—are so irregular that their self-similarity (scaling) must be treated statistically. Consider the profile of the mountain range in Figure 27.4, reproduced from Mandelbrot's book. This is not a photograph, but a computer-generated mountain scene based on a modified Brownian noise. Any

Figure 27.4. A modified Brownian landscape generated by a computer program

vertical cross section of the topography has a profile that models a random walk. The white patches, representing water or snow in the hollows below a certain altitude, were added to enhance the relief.

The profile at the top of the mountain range is a scaling fractal. This means that if you enlarge any small portion of it, it will have the same statistical character as the line you now see. If it were a true fractal, this property would continue forever as smaller and smaller segments are enlarged, but of course such a curve can neither be drawn nor appear in nature. A coastline, for example, may be self-similar when viewed from a height of several miles down to several feet, but below that the fractal property is lost. Even the Brownian motion of a particle is limited by the size of its microsteps.

Since mountain ranges approximate random walks, one can create "mountain music" by photographing a mountain range and translating its fluctuating heights to tones that fluctuate in time. Villa Lobos actually did this using mountain skylines around Rio de Janeiro. If we view nature statically, frozen in time, we can find thousands of natural curves that can be used in this way to produce stochastic music. Such music is usually too brown, too correlated, however, to be interesting. Like natural white noise, natural brown noise may do well enough, perhaps, for the patterns of abstract art but not so well as a basis for music.

When we analyze the dynamic world, made up of quantities constantly changing in time, we find a wealth of fractal like fluctuations that have $1/f$ spectral densities. In his book Mandelbrot cites a few: variations in sunspots, the wobbling of the earth's axis, undersea currents, membrane currents in the nervous system of animals, the fluctuating levels of rivers, and so on. Uncertainties in time measured by an atomic clock are $1/f$: the error is 10^{-12} regardless of whether one is measuring an error on a second, minute, or hour. Scientists tend to overlook $1/f$ noises because there are no good theories to account for them, but there is scarcely an aspect of nature in which they cannot be found.

T. Musha, a physicist at the Tokyo Institute of Technology, discovered that traffic flow past a certain spot on a Japanese expressway exhibited $1/f$ fluctuation. In a more startling experiment, Musha rotated a radar beam emanating from a coastal location to get a maximum variety of landscape on the radar screen. When he rotated the beam once, variations in the distances of all objects scanned by the beam produced a

Brownian spectrum. But when he rotated it twice and then subtracted one curve from the other the resulting curve—representing all the changes of the scene—was close to $1/f$.

We are now approaching an understanding of Voss's daring conjecture. The changing landscape of the world (or, to put it another way, the changing content of our total experience) seems to cluster around $1/f$ noise. It is certainly not entirely uncorrelated, like white noise, nor is it as strongly correlated as brown noise. From the cradle to the grave our brain is processing the fluctuating data that comes to it from its sensors. If we measure this noise at the peripheries of the nervous system (under the skin of the fingers), it tends, Mandelbrot says, to be white. The closer one gets to the brain, however, the closer the electrical fluctuations approach $1/f$. The nervous system seems to act like a complex filtering device, screening out irrelevant elements and processing only the patterns of change that are useful for intelligent behavior.

On the canvas of a painting, colors and shapes are static, reflecting the world's static patterns. Is it possible, Mandelbrot asked himself many years ago, that even completely nonobjective art, when it is pleasing, reflects fractal patterns of nature? He is fond of abstract art and maintains that there is a sharp distinction between such art that has a fractal base and such art that does not, and that the former type is widely considered the more beautiful. Perhaps this is why photographers with a keen sense of aesthetics find it easy to take pictures, particularly photomicrographs, of natural patterns that are almost indistinguishable from abstract expressionist art.

Motion can be added to visual art, of course, in the form of the motion picture, the stage, kinetic art, and dance, but in music we have meaningless, nonrepresentational tones that fluctuate to create a pattern that can be appreciated only over a period of time. Is it possible, Voss asked himself, that the pleasures of music are partly related to scaling noise of $1/f$ spectral density? That is, is this music "imitating" the $1/f$ quality of our flickering experience?

That may or may not be true, but there is no doubt that music of almost every variety does exhibit $1/f$ fluctuations in its changes of pitch as well as in the changing loudness of its tones. Voss found this to be true of classical music, jazz, and rock. He suspects it is true of all music. He was therefore not surprised that when he used a $1/f$ flicker noise from a transistor to generate a random tune, it turned out to be more pleasing than tunes based on white and brown noise sources.

Figure 27.5, supplied by Voss, shows typical patterns of white, $1/f$, and brown when noise values (vertical) are plotted against time (horizontal). These patterns were obtained by a computer program that simulates the generation of the three kinds of sequences by tossing dice. The white noise is based on the sum obtained by repeated tosses of 10 dice. These sums range from 10 to 60, but the probabilities naturally force a clustering around the median. The Brownian noise was gener-

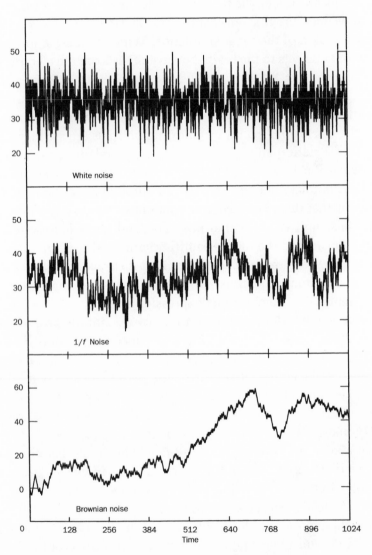

Figure 27.5. Typical patterns of white, $1/f$, and Brownian noise

ated by tossing a single die and going up one step on the scale if the number was even and down a step if the number was odd.

The 1/*f* noise was also generated by simulating the tossing of 10 dice. Although 1/*f* noise is extremely common in nature, it was assumed until recently that it is unusually cumbersome to simulate 1/*f* noise by randomizers or computers. Previous composers of stochastic music probably did not even know about 1/*f* noise, but if they did, they would have had considerable difficulty generating it. As this article was being prepared Voss was asked if he could devise a simple procedure by which readers could produce their own 1/*f* tunes. He gave some thought to the problem and to his surprise hit on a clever way of simplifying existing 1/*f* computer algorithms that does the trick beautifully.

The method is best explained by considering a sequence of eight notes chosen from a scale of 16 tones. We use three dice of three colors: red, green, and blue. Their possible sums range from 3 to 18. Select 16 adjacent notes on a piano, black keys as well as white if you like, and number them 3 through 18.

Write down the first eight numbers, 0 through 7, in binary notation, and assign a die color to each column as is shown in Figure 27.6. The first note of our tune is obtained by tossing all three dice and picking the tone that corresponds to the sum. Note that in going from 000 to 001 only the red digit changes. Leave the green and blue dice undisturbed, still showing the numbers of the previous toss. Pick up only the red die

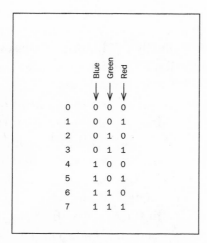

Figure 27.6. Binary chart for Voss's 1/*f* dice algorithm

Fractal Music

and toss it. The new sum of all three dice gives the second note of your tune. In the next transition, from 001 to 010, both the red and green digits change. Pick up the red and green dice, leaving the blue one undisturbed, and toss the pair. The sum of all three dice gives the third tone. The fourth note is found by shaking only the red die; the fifth, by shaking all three. The procedure, in short, is to shake only those dice that correspond to digit changes.

It is not hard to see how this algorithm produces a sequence halfway between white and brown. The least significant digits, those to the right, change often. The more significant digits, those to the left, are more stable. As a result, dice corresponding to them make a constant contribution to the sum over long periods of time. The resulting sequence is not precisely $1/f$ but is so close to it that it is impossible to distinguish melodies formed in this way from tunes generated by natural $1/f$ noise. Four dice can be used the same way for a $1/f$ sequence of 16 notes chosen from a scale of 21 tones. With 10 dice you can generate a melody of 2^{10}, or 1,024, notes from a scale of 55 tones. Similar algorithms can of course be implemented with generalized dice (octahedrons, dodecahedrons, and so on), spinners, or even tossed coins.

With the same dice simulation program Voss has supplied three typical melodies based on white, brown, and $1/f$ noise. The computer printouts of the melodies are shown in 27.7–27.9. In each case Voss varied both the melody and the tone duration with the same kind of noise. Above each tune are shown the noise patterns that were used.

Over a period of two years, tunes of the three kinds were played at various universities and research laboratories, for many hundreds of people. Most listeners found the white music too random, the brown too correlated, and the $1/f$ "just about right." Indeed, it takes only a glance at the music itself to see how the $1/f$ property mediates between the two extremes. Voss's earlier $1/f$ music was based on natural $1/f$ noise, usually electronic, even though one of his best compositions derives from the record of the annual flood levels of the Nile. He has made no attempt to impose constant rhythms. When he applied $1/f$ noise to a pentatonic (five-tone) scale and also varied the rhythm with $1/f$ noise, the music strongly resembled Oriental music. He has not tried to improve his $1/f$ music by adding transition or rejection rules. It is his belief that stochastic music with such rules will be greatly improved if the underlying choices are based on $1/f$ noise rather than the white noise so far used.

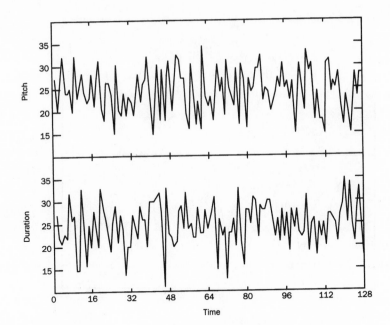

Figure 27.7. White music

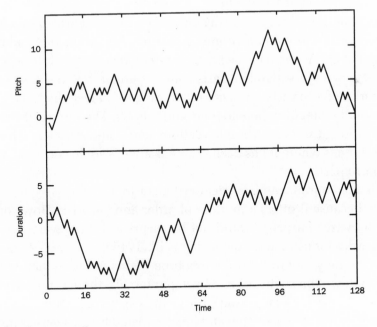

Figure 27.8. Brown music

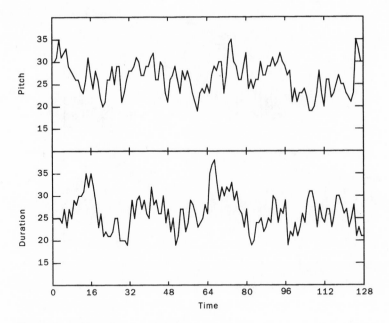

Figure 27.9. 1/*f* music

Note that 1/*f* music is halfway between white and brown in a fractal sense, not in the manner of music that has transition rules added to white music. As we have seen, such music reverts to white when we compare widely separated parts. But 1/*f* music has the fractal self-similarity of a coastline or a mountain range. Analyze the fluctuations on a small scale, from note to note, and it is 1/*f*. The same is true if you break a long tune into 10-note sections and compare them. The tune never forgets where it has been. There is always some correlation with its entire past.

It is commonplace in musical criticism to say that we enjoy good music because it offers a mixture of order and surprise. How could it be otherwise? Surprise would not be surprise if there were not sufficient order for us to anticipate what is likely to come next. If we guess too accurately, say in listening to a tune that is no more than walking up and down the keyboard in one-step intervals, there is no surprise at all. Good music, like a person's life or the pageant of history, is a wondrous mixture of expectation and unanticipated turns. There is nothing new about this insight, but what Voss has done is to suggest a mathematical measure for the mixture.

I cannot resist mentioning three curious ways of transforming a melody to a different one with the same $1/f$ spectral density for both tone patterns and durations. One is to write the melody backward, another is to turn it upside down, and the third is to do both. These transformations are easily accomplished on a player piano by reversing and/or inverting the paper roll. If a record or tape is played backward, unpleasant effects result from a reversal of the dying-away quality of tones. (Piano music sounds like organ music.) Reversal or inversion naturally destroys the composer's transition patterns, and that is probably what makes the music sound so much worse than it does when it is played normally. Since Voss composed his tunes without regard for short-range transition rules, however, the tunes all sound the same when they are played in either direction.

Canons for two voices were sometimes deliberately written, particularly in the 15th century, so that one melody is the other backward, and composers often reversed short sequences for contrapuntal effects in longer works. Figure 27.10 shows a famous canon that Mozart wrote as a joke. In this instance the second melody is almost the same as the one you see taken backward and upside down. Thus if the sheet is placed flat on a table, with one singer on one side and the other singer on the other, the singers can read from the same sheet as they harmonize!

No one pretends, of course, that stochastic $1/f$ music, even with added transition and rejection rules, can compete with the music of good composers. We know that certain frequency ratios, such as the three-to-two ratio of a perfect fifth, are more pleasing than others, either when the two tones are played simultaneously or in sequence. But just what composers do when they weave their beautiful patterns of meaningless sounds remains a mystery that even they do not understand.

It is here that Plato and Aristotle seem to disagree. Plato viewed all the fine arts with suspicion. They are, he said (or at least his Socrates said), imitations of imitations. Each time something is copied something is lost. A picture of a bed is not as good as a real bed, and a real bed is not as good as the universal, perfect idea of bedness. Plato was less concerned with the sheer delight of art than with its effects on character, and for that reason his *Republic* and *Laws* recommend strong state censorship of all the fine arts.

Aristotle, on the other hand, recognized that the fine arts are of value to a state primarily because they give pleasure and that this pleasure springs from the fact that artists do much more than make poor copies.

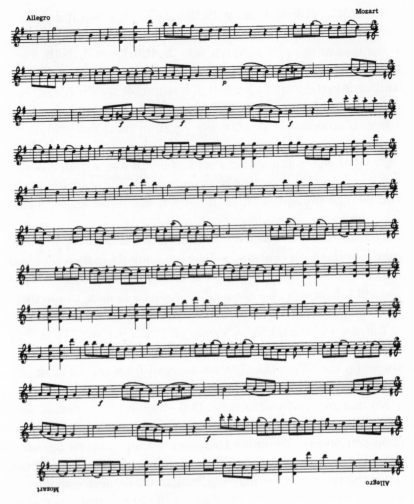

Figure 27.10. Mozart's palindromic and invertible canon

> They said, "You have a blue guitar,
> You do not play things as they are."
> The man replied, "Things as they are
> Are changed upon the blue guitar."

Wallace Stevens intended his blue guitar to stand for all the arts, but music, more than any other art and regardless of what imitative aspects it may have, involves the making of something utterly new. You may occasionally encounter natural scenes that remind you of a painting, or episodes in life that make you think of a novel or a play. You will never come on anything in nature that sounds like a symphony. As to whether mathematicians will someday write computer programs that will create good music—even a simple, memorable tune—time alone will tell.

Addendum

Irving Godt, who teaches music history at the Indiana University of Pennsylvania, straightened me out on the so-called Mozart canon with the following letter. It appeared in *Scientific American* (July 1978):

> A few musical errors slipped past Martin Gardner's critical eye when he took up "Mozart's palindromic and invertible canon" in his report on fractal curves and "one-over-*f*" fluctuations.
>
> Mozart scholars now agree that the canon is almost certainly not by Mozart, even though publishers have issued it under his name. For more than 40 years the compilers of the authoritative Köchel catalogue of Mozart's compositions have relegated it to the appendix of doubtful attributions, where along with three other pieces of a similar character, it bears the catalogue number K. Anh. C 10.16. We have no evidence that the piece goes back any further than the last century.
>
> The piece is not for two singers but for two violins. Singers cannot produce the simultaneous notes of the chords in the second measure (and elsewhere), and the ranges of the parts are quite impractical. To perform the piece the two players begin from opposite ends of the sheet of music and arrive at a result that falls far below the standard of Mozart's authentic canons and other *jeux d'esprit*. The two parts combine for long stretches of parallel octaves, they rarely achieve even the most rudimentary rhythmic or directional independence, and their harmony consists of little more than the most elementary writing in parallel thirds. This little counterfeit is not nearly as interesting as Mr. Gardner's columns.

John G. Fletcher wrote to suggest that because $1/f$ music lies between white and brown music it should be called tan music. The term "pink" has also been suggested, and actually used by some writers. *Fate* magazine (October 1978) ran a full-page advertisement for an LP record album produced by "Master Wilburn Burchette," of Spring Valley, CA, titled *Mind Storm*. The ad calls it "fantastic new deep-hypnotic music that uses a phenomenon known in acoustical science as 'pink sound' to open the mind to thrilling psychic revelations! This astonishing new music acts something like a crystal ball reflecting back the images projected by the mind. . . . Your spirit will soar as this incredible record album carries you to new heights of psychic awareness!"

Frank Greenberg called my attention to some "mountain music" composed by Sergei Prokofiev for Sergei Eisenstein's film *Alexander Nevsky* in 1938. "Eisenstein provided Prokofiev with still shots of individual scenes of the movie as it was being filmed. Prokofiev then took these scenes and used the silhouette of the landscape and human figures as a pattern for the position of the notes on the staff. He then orchestrated around these notes."

Bibliography

T. Johnson, "Self-Similar Melodies," Two-Eighteen Press, 1996.

S. K. Langer, "On Significance in Music," *Philosophy in a New Key*, Harvard University Press, 1957.

T. Musha, "$1/f$ Fluctuations in Biological Systems," *Sixth International Symposium on Noise in Physical Systems,*" National Bureau of Standards, 1981, pp. 143–46.

A. T. Scarpelli, "$1/f$ Random Tones: Making Music with Fractals," *Personal Computing*, Vol. 3, 1979, pp. 17–27.

M. Schroeder, "Noises: White, Pink, Brown, and Black," *Fractals, Chaos, Power Laws*, W. H. Freeman, 1991, Chapter 5.

D. E. Thomsen, "Making Music Fractally," *Science News*, Vol. 117, 1980, pp. 187ff.

B. J. West and M. Shlesinger, "The Noise in Natural Phenomena," *American Scientist*, Vol. 78, 1978, pp. 40–45.

R. F. Voss and J. Clarke, "$1/f$ Noise in Music and Speech," *Nature*, Vol. 258, 1975, pp. 317–18.

R. F. Voss and J. Clarke, "$1/f$ Noise in Music," *The Journal of the Acoustical Society of America*, Vol. 63, 1978, pp. 258–63.

A. van der Ziel., *Noise: Sources, Characterization, Measurement*, Prentice-Hall, 1970.

Chapter 28   *Surreal Numbers*

Some said "John, print it";
  others said, "Not so."
Some said "It might do good";
  others said, "No."

—JOHN BUNYAN, *Apology for His Book*

John Horton Conway, the almost legendary mathematician of the University of Cambridge (now at Princeton University), quotes the above lines at the end of the preface to his book, *On Numbers and Games* (Academic Press, 1976), or *ONAG*, as he and his friends call it. It is hard to imagine a mathematician who would say not so or no. The book is vintage Conway: profound, pathbreaking, disturbing, original, dazzling, witty, and splattered with outrageous Carrollian wordplay. Mathematicians, from logicians and set theorists to the humblest amateurs, will be kept busy for decades rediscovering what Conway has left out or forgotten and exploring the strange new territories opened by his work.

The sketch of Conway reproduced below could be titled "John 'Horned' (Horton) Conway." The infinitely regressing, interlocking horns form, at the limit, what topologists call a "wild" structure; this one is termed the Alexander horned sphere. Although it is equivalent to the simply connected surface of a ball, it bounds a region that is not simply connected. A loop of elastic cord circling the base of a horn cannot be removed from the structure even in an infinity of steps.

Conway is the inventor of the computer game Life, discussed in detail in Chapter 31 of this text. By carefully choosing a few ridiculously simple transition rules Conway created a cellular automaton structure of extraordinary depth and variety. Soon after, by invoking the simplest possible distinction—a binary division between two sets—and adding a few simple rules, definitions, and conventions, he constructed a rich field of numbers and an equally rich associated structure of two-person games.

The story of how Conway's numbers are created on successive "days," starting with the zeroth day, is told in Donald E. Knuth's nov-

John "Horned" (Horton) Conway,
as sketched by a colleague on
computer-printout paper

elette *Surreal Numbers* (Addison-Wesley, 1974). Because I discuss
Knuth's book in chapter 48, I shall say no more about it here except to
remind readers that the construction of the numbers is based on one
rule: If we are given a left set L and a right set R and no member of L is
equal to or greater than any member of R, then there is a number $\{L\,|\,R\}$
that is the "simplest number" (in a sense defined by Conway) in between.

By starting with literally nothing at all (the empty set) on the left and the right, { | }, one obtains a definition of zero. Everything else follows by the technique of plugging newly created numbers back into the left–right arrangement. The expression {0|0} is not a number, but {0| }, with the null set on the right, defines 1, { |0} defines –1, and so on.

Proceeding inductively, Conway is able to define all integers, all integral fractions, all irrationals, all of Georg Cantor's transfinite numbers, a set of infinitesimals (they are the reciprocals of Cantor's numbers, not the infinitesimals of nonstandard analysis), and infinite classes of weird numbers never before seen by man, such as

$$\sqrt[3]{(\omega + 1)} - \frac{\pi}{\omega}$$

where ω is omega, Cantor's first infinite ordinal.

Conway's games are constructed in a similar but more general way. The fundamental rule is: If L and R are any two sets of games, there is a game $\{L|R\}$. Some games correspond to numbers and some do not, but all of them (like the numbers) rest on nothing. "We remind the reader again," Conway writes, "that since ultimately we are reduced to questions about members of the empty set, no one of our inductions will require a 'basis.' "

In a "game" in Conway's system two players, Left and Right, alternate moves. (Left and Right designate players, such as Black and White or Arthur and Bertha, not who goes first or second.) Every game begins with a first position, or state. At this state and at each subsequent state a player has a choice of "options," or moves. Each choice completely determines the next state. In standard play the first person unable to make a legal move loses. This is a reasonable convention, Conway writes, "since we normally consider ourselves as losing when we cannot find any good move, we should obviously lose when we cannot find any move at all!" In "misère" play, which is usually much more difficult to analyze, the person who cannot move is the winner. Every game can be diagrammed as a rooted tree, its branches signifying each player's options at each successive state. On Conway's trees Left's options go up and to the left and Right's go up and to the right.

Games may be "impartial," as in Nim, which means that any legal move can be made by the player whose turn it is to move. If a game is not impartial, as in chess (where each player must move only his own pieces), Conway calls it a partizan game. His net thus catches both an enormous variety of familiar games, from Nim to chess, and an infinity of games never before imagined. Although his theory applies to games with an infinity of states or to games with an infinity of options or to both, he is concerned mainly with games that end after a finite number of moves. "Left and Right," he explains, "are both busy men, with heavy political responsibilities."

Conway illustrates the lower levels of his theory with positions taken from a partizan domino-placing game. (Conway calls it Domineering.) The board is a rectangular checkerboard of arbitrary size and shape. Players alternately place a domino to cover two adjacent squares, but Left must place his pieces vertically and Right must place his horizontally. The first player who is unable to move loses.

An isolated empty square

allows no move by either player. "No move allowed" corresponds to the empty set, so that in Conway's notation this simplest of all games is assigned the value { | } = 0, the simplest of all numbers. Conway calls it Endgame. Its tree diagram, shown at the right of the square, is merely the root node with no branches. Because neither side can move, the second player, regardless of whether he is Left or Right, is the winner. "I courteously offer you the first move in this game," writes Conway. Since you cannot move, he wins.

A vertical strip of two (or three) cells

offers no move to Right but allows one move for Left. Left's move leads to a position of value 0, so that the value of this region is {0 | } = 1. It is the simplest of all positive games, and it corresponds to the simplest positive number. Positive games are wins for Left regardless of who starts. The region's tree diagram is shown at the right above.

A horizontal strip of two (or three) cells

allows one move for Right but no move for Left. The value of the region is { | 0} = –1. It is the simplest of all negative games and corresponds to the simplest negative number. Negative games are wins for Right regardless of who starts.

A vertical strip of four (or five) cells

has a value of 2. Right has no moves. Left can, if he likes, take the two middle cells in order to leave a zero position, but his "best" move is to take two end cells, because that leaves him an additional move. If this region is the entire board, then of course either play wins, but if it is an isolated region in a larger board or in one of many boards in a "compound game," it may be important to make the move that maximizes the number of additional moves it leaves for the player. For this reason the tree shows only Left's best line of play. The value of the game is $\{1,0|\ \} = \{1|\ \} = 2$. A horizontal strip of four cells has a value of –2. If only one player can move in a region and he can fit n of his dominoes into it but no more, then clearly the region has the value $+n$ if the player is Left and $-n$ if the player is Right.

Things get more interesting if both players can move in a region, because then one player may have ways of blocking his opponent. Consider the following region:

Left can place a domino that blocks any move by Right, thus leaving a zero position and winning. Right cannot similarly block Left because Right's only move leaves a position of value 1. In Conway's notation the value of this position is $\{0, -1|1\} = \{0|1\}$, an expression that defines 1/2. The position therefore counts as half a move in favor of Left. By turning the L region on its side one finds that the position is $\{-1|0,1\} = \{-1|0\} = -1/2$, or half a move in favor of Right.

More complicated fractions arise in Conway's theory. For example, has a value of {1/2 | 1} = 3/4 of a move for Left, since 3/4 is the simplest number between 1/2 and 1, the values of the best options for Left and Right. In a game called partizan Hackenbush, Conway gives an example of a position in which Left is exactly 5/64 move ahead!

The values of some game positions are not numbers at all. The simplest example is illustrated in Domineering by this region:

Both Left and Right have opening moves only, so that the first person to play wins regardless of whether he is Left or Right. Since each player can reduce the value to 0, the value of the position is {0 | 0}. This is not a number. Conway symbolizes it with * and calls it "star." Another example would be a Nim heap containing a single counter. It is the simplest "fuzzy" game. Fuzzy values correspond to positions in which either player can win if he moves first.

The value of a compound game is simply the sum of the values of its component games. This statement applies also to the value of a position in a game in progress that has been divided by play into a set of subgames. For example, Figure 28.1 shows a position in a game of Domineering played on a standard chessboard. The values of the isolated regions are indicated. The position seems to be well balanced, but the regions have a sum of 1¼, which means that Left is one and one-fourth of a move ahead and therefore can win regardless of who moves next. This outcome would be tedious to decide by drawing a complete tree, but Conway's theory gives it quickly and automatically.

A game is not considered "solved" until its outcome (assuming that both players make their best moves) is known (that is, whether the game's value is zero, positive, negative, or fuzzy) and a successful strategy is found for the player who has the win. This stricture applies only to games that must end, but such games may offer infinite options, as in the game Conway calls "My Dad Has More Money than Yours." Players alternately name a sum of money for just two moves and the highest sum wins. Although the tree, Conway admits, is complicated, the outcome is clearly a second-player win.

Are these not trivial beginnings? Yes, but they provide a secure foundation on which Conway, by plugging newly created games back into his left–right scheme, carefully builds a vast and fantastic edifice. I

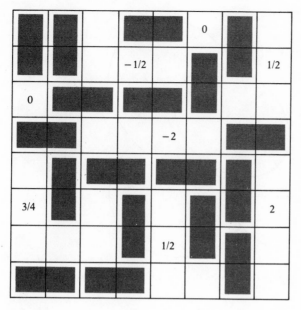

Figure 28.1. A Domineering position that will result in a win for Left

shall not proceed further with it here; instead I shall describe a few unusual games that Conway analyzes in the light of his theory. In all these
games we assume standard play in that the first person who is unable
to move loses.

1. Col (named for its inventor, Colin Vout). A map is drawn on brown
paper. *L* has black paint, *R* has white. They alternately color a region
with the proviso that regions sharing a border segment not be the same
color. It is useful to regard all regions bordering a white one as being
tinted white and all regions bordering a black one as being tinted black.
A region acquiring both tints drops out of the map as being an un-
paintable one.

Conway analyzes Col on the map's dual graph (see Figure 28.2),
defining what he calls "explosive nodes" and marking them with light-
ning bolts. Of course, the game can be played on white paper with pencils of any two colors. Vout has reported that in the set of all
topologically distinct connected maps of one region through five regions 9 are first-player wins and 21 are second-player wins. The game
in general is unsolved.

2. Snort (after Simon Norton). This is the same as Col except that
neighboring regions must be the same color. It too is unsolved. Conway

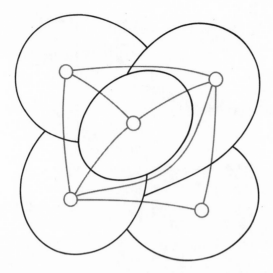

Figure 28.2. A five-region map, with its dual graph (*gray*), for playing Col or Snort

suspects it has a richer theory than Col. His most valuable tip: If you can color a region adjacent to every region of your opponent's color, do so.

3. Silver Dollar Game without the Dollar. The board is a horizontal strip of cells extending any length to the right (see Figure 28.3(A)). Pennies are arbitrarily placed on certain cells, one to a cell, to provide the initial position. Players alternately move a coin to the left to any empty cell. No jumps are allowed. Eventually all the coins jam at the left, and the person who cannot move loses.

The game is simply Nim in one of its endless disguises. (I assume the

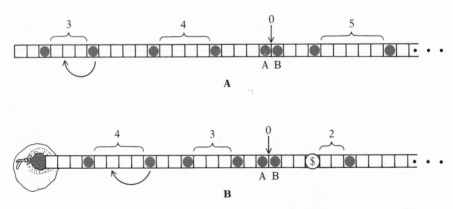

Figure 28.3. *(A)* The silver-coin game without the silver dollar. *(B)* The same game with the dollar.

INFINITY

reader is familiar with Nim and knows how to determine the winning strategy. If not, he can consult any number of books, including Conway's book or my *Scientific American Book of Mathematical Puzzles & Diversions* [Simon and Schuster, 1959]). Corresponding to the Nim heaps (or rows) of counters are the vacant cells between pennies, starting with the vacancy at the extreme right and including only alternate vacancies. In the illustration the Nim heaps are indicated by brackets and one arrow. The heaps are 3, 4, 0, and 5, so that the game is equivalent to playing Nim with rows of three, four, and five counters.

Rational play is exactly as in Nim: Move to reduce the Nim sum of the heaps to 0, a game with a second-player win. The one trivial difference is that here a heap can increase in size. If, however, you have the win and your opponent makes such a move, you immediately restore the heap to its previous size by moving the coin that is just to the right of the heap.

If in the illustrated position it is your move, you are sure to win if you make the move indicated by the curved arrow. If your opponent responds by moving counter *A* two cells to the left, the move raises the empty heap to 2. Your response is to move *B* two cells to the left, thus returning the heap to 0.

4. Silver Dollar Game with the Dollar. This is the same as the preceding game except that one of the coins (any one) is a silver dollar and the cell farthest to the left is a money bag (see Figure 28.3(B)). A coin farthest to the left can move into the bag. When the dollar is bagged, the game ends and the next player wins by taking the bag.

This game too is Nim disguised. Count the bag as being empty if the coin at its right is a penny and full if it is the dollar, and play Nim as before. If you have the win, your opponent will be forced to drop the dollar in the bag. If the winner is deemed to be the player who bags the dollar, count the bag as being full if the coin at its right is the coin just to the left of the dollar and count it as being empty otherwise. The position shown corresponds to Nim with heaps of 4, 3, 0, and 2. The first player wins in both versions only if he makes the move indicated by the curved arrow.

5. Rims. The initial position of this pleasant way of playing a variant of Nim consists of two or more groups of spots. The move is to draw a simple closed loop through any positive number of spots in one group. The loop must not cross itself or cross or touch any other loop. A game is shown in Figure 28.4.

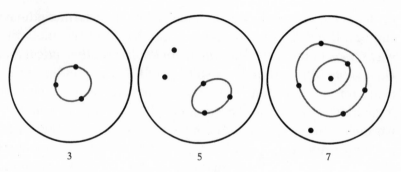

Figure 28.4. A position in Rims

Conway shows that Rims is the same as Nim played with the added rule that you are allowed, if you like, to take from the center of a row, leaving two new rows instead of one or more rows. Although the number of heaps can increase, the winning strategy is standard Nim strategy. If each loop is confined to one or two spots, the game is equivalent to the familiar game of Kayles. Conway calls it Rayles.

6. Prim and Dim. Let me introduce these games by explaining Prime Nim, a simpler game not discussed by Conway. It was first analyzed by Claude E. Shannon. Prime Nim is played the same way that Nim is except players must diminish heaps only by prime numbers, including 1 as a prime. "The game is actually a bit of a mathematical hoax," wrote Shannon (in a private communication), "since at first sight it seems to involve deep additive properties of the prime numbers. Actually the only property involved is that all multiples of the first nonprime are nonprime."

The first nonprime is 4. The strategy therefore is merely to regard each heap as being equal to the remainder when its number is divided by 4 and then to play standard Nim strategy with these modulo-4 numbers. If 1 is not counted as being prime, the strategy is less simple in standard play and is so complicated in misère play that I do not believe it has ever been solved.

Prim, suggested by Allan Tritter, requires that players take from each heap a number prime to the heap's number. In other words, the two numbers must not be equal or have a common divisor other than 1. Dim requires that a player remove a *divisor* of n (including 1 and n as divisors) from a heap of size n. Conway gives solutions to both games as well as variants in which taking 1 from a heap of 1 is allowed in Prim and taking n from a heap of n is not allowed in Dim.

7. Cutcake. This is a partizan game invented by Conway. It is played with a set of rectangular cakes, each scored into unit squares like a waffle. Left's move is to break a piece of cake into two parts along any horizontal lattice line, Right's is to break a piece along any vertical line. The game has a surprisingly simple theory.

Figure 28.5 shows a 4 × 7 cake. In Conway's notation its value is 0, which means it is a second-player win regardless of who goes first. It looks as if the vertical breaker, who has twice as many opening moves as his opponent, would have the advantage, but he does not if he goes first. Assume that the vertical breaker goes first and breaks along the line indicated by the arrow. What is the second player's winning response?

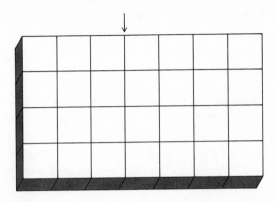

Figure 28.5. A format for Cutcake, with the first move being a vertical break at the arrow

I have given only a few examples of Conway's exotic nomenclature. Games can be short, small, all small, tame, restive, restless, divine, extraverted, and introverted. There are ups, downs, remote stars, semistars, and superstars. There are atomic weights and sets with such names as On, No, Ug, and Oz. Conway has a temperature theory, with thermographs on which hot positions are cooled by pouring cold water on them. He has a Mach principle for the small world: the atomic weight of a short all-small game is at least 1 if and only if the game exceeds the remote stars!

Conway's theorem 99 conveys some notion of the book's whimsical flavor. It tells us (I paraphrase to remove a slight error that Conway discovered too late to correct) that any short all-small game of atomic weight zero is dominated by some superstar. Only a feeling of incompleteness, Conway adds, prompts him to give a final theorem. Theorem 100 is: "This is the last theorem in this book."

Surreal Numbers

Addendum

Since this chapter first ran as a *Scientific American* column in 1976, Conway has written other books and papers, and articles about him have appeared, some of which are listed in the bibliography. Especially notable is the two-volume *Winning Ways,* on which he collaborated with Elwyn Berlekamp and Richard Guy. It has already become a classic of recreational mathematics and has made important technical contributions to game theory and combinatorics. For Conway's work on Penrose tiling, see Chapters 1 and 2 of my *Perrose Tiles to Trapdoor Ciphers.* For his work on sporadic groups and knot theory, see my *Last Recreations,* Chapters 9 and 5 respectively.

Conway spoke on his game theories at a conference on recreational mathematics at Miami University in September 1976. His lecture, reprinted as "A Gamut of Game Theories" in *Mathematics Magazine* (see bibliography), concludes:

> The theories can be applied to hundreds and thousands of games—really lovely little things; you can invent more and more and more of them. It's especially delightful when you find a game that somebody's already considered and possibly not made much headway with, and you find you can just turn on one of these automatic theories and work out the value of something and say, "Ah! Right is 47/64ths of a move ahead, and so she wins."

Martin Kruskal, now a mathematician at Rutgers University, became fascinated by Conway's surreal numbers. For many years he has been working on their clarification and elaboration, and their potential applications to other fields of mathematics. For an introduction to this exciting work, see the last two entries, by Polly Shulman and Robert Matthews, in the chapter's bibliography under the subhead of "Surreal Numbers." Kruskal is working on what he calls "surreal calculus," in which Conway's numbers take care of the infinitesimals so essential in the calculus of Newton and Leibniz.

Answers

The problem was to find the winning response to a first-player move in Cutcake. The cake is a 4×7 rectangle. If the first player breaks the cake vertically into a 4×4 square and a 4×3 rectangle, the unique winning reply is to break the 4×3 piece into two 2×3 rectangles.

Bibliography

On Games

E. Berlekamp, J. Conway, and R. Guy, *Winning Ways* (two volumes), Academic Press, 1982.

J. H. Conway, "All Numbers Great and Small," University of Calgary Mathematical Research Paper No. 149, 1972.

J. H. Conway, "All Games Bright and Beautiful," University of Calgary Mathematical Research Paper No. 295, 1975. Reprinted in *The American Mathematical Monthly,* Vol. 84, 1977, pp. 417–34.

J. H. Conway, *On Numbers and Games,* Academic Press, 1976.

J. H. Conway, "A Gamut of Game Theories," *Mathematics Magazine,* Vol. 51, 1978, pp. 5–12.

J. H. Conway and R. Guy, *The Book of Numbers,* Springer-Verlag, 1996.

S. W. Golomb, "A Mathematical Investigation of Games of 'Take Away,'" *Journal of Combinatorial Theory,* Vol. 1, 1966, pp. 443–58.

R. K. Guy and C. A. B. Smith, "The G-Values of Various Games," *Proceedings of the Cambridge Philosophical Society,* Vol. 52, Part 3, 1956, pp. 514–26.

On Surreal Numbers

H. Gonshor, *An Introduction to the Theory of Surreal Numbers,* London Mathematical Society Lecture Note Series 110, 1986.

D. E. Knuth, *Surreal Numbers,* Addison-Wesley, 1974.

R. Matthews, "The Man Who Played God with Infinity," *New Scientist,* September 2, 1995, pp. 36–40.

P. Shulman, "Infinity Plus One and Other Surreal Numbers," *Discover,* December 1995, pp. 97–105.

About Conway

D. Albers, "John Horton Conway—Talking a Good Game," *Mathematical Horizons,* Spring 1994, pp. 6–9.

M. Alpert, "Profile: Not Just Fun and Games," *Scientific American,* April 1999, pp. 40–42.

M. Browne, "Intellectual Duel," *The New York Times,* August 30, 1988.

R. K. Guy, "John Horton Conway: Mathematical Magus," *Two-Year College Mathematics Journal,* Vol. 13, 1982, pp. 290–99.

G. Kolata, "At Home in the Elusive World of Mathematics," *The New York Times,* October 12, 1993.

C. Moseley, "The Things that Glitter," *Princeton Alumni Weekly*, December 23, 1992, pp. 10–13.

C. Seife, "Mathematician," *The Sciences*, May/June 1994, pp. 12–15.

G. Taubes, "John Horton Conway: A Mathematical Madness in Cambridge," *Discover*, August 1984, pp. 40–50.

M. Thompson and M. Niemann, "Mathematical Mania," *Science Today*, Spring 1987, pp. 8–9.

VIII

Combinatorics

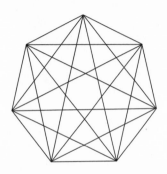

Chapter 29   *Hexaflexagons*

Flexagons are paper polygons, folded from straight or crooked strips of paper, which have the fascinating property of changing their faces when they are "flexed." Had it not been for the trivial circumstance that British and American notebook paper are not the same size, flexagons might still be undiscovered, and a number of top-flight mathematicians would have been denied the pleasure of analyzing their curious structures.

It all began in the fall of 1939. Arthur H. Stone, then a 23-year-old graduate student from England, in residence at Princeton University on a mathematics fellowship, had just trimmed an inch from his American notebook sheets to make them fit his English binder. For amusement he began to fold the trimmed-off strips of paper in various ways, and one of the figures he made turned out to be particularly intriguing. He had folded the strip diagonally at three places and joined the ends so that it made a hexagon (see Figure 29.1). When he pinched two adjacent triangles together and pushed the opposite corner of the hexagon toward the center, the hexagon would open out again, like a budding flower, and show a completely new face. If, for instance, the top and bottom faces of the original hexagon were painted different colors, the new face would come up blank and one of the colored faces would disappear!

This structure, the first flexagon to be discovered, has three faces. Stone did some thinking about it overnight, and on the following day confirmed his belief (arrived at by pure cerebration) that a more complicated hexagonal model could be folded with six faces instead of only three. At this point Stone found the structure so interesting that he showed his paper models to friends in the graduate school. Soon "flexagons" were appearing in profusion at the lunch and dinner tables.

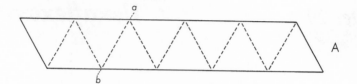

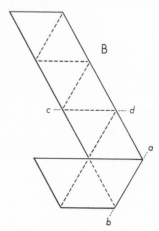

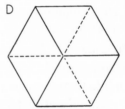

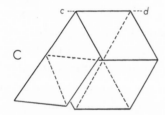

Figure 29.1. Trihexaflexagon is constructed by cutting a strip of paper so that it may be marked off in 10 equilateral triangles *(A)*.

The strip is folded backward along the line *ab* and turned over *(B)*.

It is then folded backward again along the line *cd* and the next to the last triangle is placed on top of the first *(C)*.

The last triangle is now folded backward and glued to the other side of the first *(D)*.

The figure may be flexed but it is not meant to be cut out. Fairly stiff paper at least an inch and a half wide is recommended.

A "Flexagon Committee" was organized to probe further into the mysteries of flexigation. The other members besides Stone were Bryant Tuckerman, a graduate student of mathematics; Richard P. Feynman, a graduate student in physics; and John W. Tukey, a young mathematics instructor.

The models were named hexaflexagons—"hexa" for their hexagonal form and "flexagon" for their ability to flex. Stone's first model is a tri-hexaflexagon ("tri" for the three different faces that can be brought into view); his elegant second structure is a hexahexaflexagon (for its six faces).

To make a hexahexaflexagon you start with a strip of paper which is divided into 19 equilateral triangles (see Figure 29.2). You number the

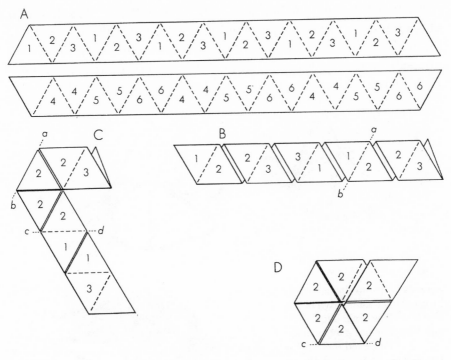

Figure 29.2. Hexahexaflexagon is constructed by cutting a strip of paper so that it may be marked off in 19 triangles *(A)*. The triangles on one side are numbered 1, 2, and 3; the triangles on the other, 4, 5, and 6. A similar pattern of colors or geometrical figures may also be used. The hexagon is then folded as shown. The figure can be flexed to show six different faces.

triangles on one side of the strip 1, 2, and 3, leaving the nineteenth triangle blank, as shown in the drawing. On the opposite side the triangles are numbered 4, 5, and 6, according to the scheme shown. Now you fold the strip so that the same underside numbers face each other—4 on 4, 5 on 5, 6 on 6, and so on. The resulting folded strip, illustrated in *(B)* in the series, is then folded back on the lines *ab* and *cd* (*C*), forming the hexagon *(D);* finally the blank triangle is turned under and pasted to the corresponding blank triangle on the other side of the strip. All this is easier to carry out with a marked strip of paper than it is to describe.

If you have made the folds properly, the triangles on one visible face of the hexagon are all numbered 1, and on the other face all are numbered 2. Your hexahexaflexagon is now ready for flexing. You pinch two adjacent triangles together (see Figure 29.3), bending the paper along the line between them, and push in the opposite corner; the fig-

Hexaflexagons

ure may then open up to face 3 or 5. By random flexing you should be able to find the other faces without much difficulty. Faces 4, 5, and 6 are a bit harder to uncover than 1, 2, and 3. At times you may find yourself trapped in an annoying cycle that keeps returning the same three faces over and over again.

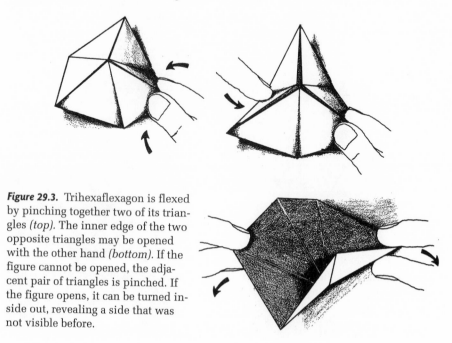

Figure 29.3. Trihexaflexagon is flexed by pinching together two of its triangles *(top)*. The inner edge of the two opposite triangles may be opened with the other hand *(bottom)*. If the figure cannot be opened, the adjacent pair of triangles is pinched. If the figure opens, it can be turned inside out, revealing a side that was not visible before.

Tuckerman quickly discovered that the simplest way to bring out all the faces of any flexagon was to keep flexing it at the same corner until it refused to open, then to shift to an adjacent corner. This procedure, known as the "Tuckerman traverse," will bring up the six faces of a hexahexaflexagon in a cycle of 12 flexes, but 1, 2, and 3 turn up three times as often as 4, 5, and 6. A convenient way to diagram a Tuckerman traverse is shown in Figure 29.4, the arrows indicating the order in which the faces are brought into view. This type of diagram can be applied usefully to the traversing of any type of flexagon. When the model is turned over, a Tuckerman traverse runs the same cycle in reverse order.

By lengthening the chain of triangles, the committee discovered, one can make flexagons with 9, 12, 15, or more faces: Tuckerman managed to make a workable model with 48! He also found that with a strip of paper cut in a zigzag pattern (that is, a strip with sawtooth rather than

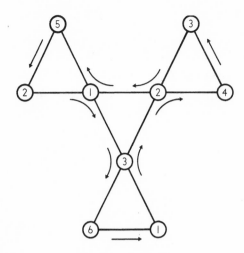

Figure 29.4. Diagram of a Tuckerman traverse on a hexahexaflexagon

,straight edges) it was possible to produce a tetrahexaflexagon (four faces) or a pentahexaflexagon. There are three different hexahexaflexagons—one folded from a straight strip, one from a chain bent into a hexagon, and one from a form that somewhat resembles a three-leaf clover. The decahexaflexagon (10 faces) has 82 different variations, all folded from weirdly bent strips. Flexagons can be formed with any desired number of faces, but beyond 10 the number of different species for each increases at an alarming rate. All even-numbered flexagons, by the way, are made of strips with two distinct sides, but those with an odd number of faces have only a single side, like a Möbius surface.

A complete mathematical theory of flexigation was worked out in 1940 by Tukey and Feynman. It shows, among other things, exactly how to construct a flexagon of any desired size or species. The theory has never been published, though portions of it have since been rediscovered by other mathematicians. Among the flexigators is Tuckerman's father, the distinguished physicist Louis B. Tuckerman, who was formerly with the National Bureau of Standards. Tuckerman senior devised a simple but efficient tree diagram for the theory.

Pearl Harbor called a halt to the committee's flexigation program, and war work soon scattered the four charter members to the winds. Stone became a lecturer in mathematics at the University of Manchester in England, and is now at the University of Rochester, NY. Feynman became a famous theoretical physicist at the California Institute of

Technology. Tukey, a professor of mathematics at Princeton, has made brilliant contributions to topology and to statistical theory which have brought him worldwide recognition. Tuckerman became a mathematician at IBM's research center in Yorktown Heights, NY.

One of these days the committee hopes to get together on a paper or two which will be the definitive exposition of flexagon theory. Until then the rest of us are free to flex our flexagons and see how much of the theory we can discover for ourselves.

Addendum

In constructing flexagons from paper strips it is a good plan to crease all the fold lines back and forth before folding the model. As a result, the flexagon flexes much more efficiently. Some readers made more durable models by cutting triangles from poster board or metal and joining them with small pieces of tape, or by gluing them to one. long piece of tape, leaving spaces between them for flexing. Louis Tuckerman keeps on hand a steel strip of such size that by wrapping paper tape of a certain width around it he can quickly produce a folded strip of the type shown in Figure 29.2(B). This saves considerable time in making flexagons from straight chains of triangles.

Readers passed on to me a large variety of ways in which flexagon faces could be decorated to make interesting puzzles or display striking visual effects. Each face of the hexahexa, for example, appears in at least two different forms, owing to a rotation of the component triangles relative to each other. Thus if we divide each face as shown in Figure 29.5, using different colors for the A, B, and C sections, the same face may appear with the A sections in the center as shown, or with the B or C sections in the center. Figure 29.6 shows how a geometrical pattern may be drawn on one face so as to appear in three different configurations.

Of the 18 possible faces that can result from a rotation of the triangles, three are impossible to achieve with a hexahexa of the type made from a straight strip. This suggested to one correspondent the plan of pasting parts of three different pictures on each face so that by flexing the model properly, each picture could presumably be brought together at the center while the other two would be fragmented around the rim. On the three inner hexagons that cannot be brought together, he pasted the parts of three pictures of comely, undraped young ladies to make

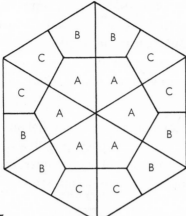

Figure 29.5.

Figure 29.6.

what he called a hexahexafrustragon. Another reader wrote that he achieved similar results by pasting together two adjacent triangular faces. This prevents one entire face from flexing into view, although the victim can see that it exists by peeking into the model's interior.

The statement that only 15 different patterns are possible on the straight-strip hexahexa must be qualified. An unsymmetrical coloring of the faces discloses the curious fact that three of these 15 patterns have mirror-image partners. If you number the inner corners of each pattern with digits from 1 to 6, writing them in clockwise order, you will find that three of these patterns turn up with the same digits in counterclockwise order. Bearing this asymmetry in mind, one can say that the six faces of this hexahexa exhibit a total of 18 different configurations. This was first called to my attention by Albert Nicholas, professor of education at Monmouth College, Monmouth, IL, where the making of flexagons became something of a craze in the early months of 1957.

I do not know who was the first to use a printed flexagon as an advertising premium or greeting card. The earliest sent to me was a trihexa distributed by the Rust Engineering Company of Pittsburgh to advertise their service award banquet in 1955. A handsome hexahexa, designed to display a variety of colored snowflake patterns, was used by *Scientific American* for their 1956 Christmas card.

For readers who would like to construct and analyze flexagons other than the two described in the chapter, here is a quick run-down on some low-order varieties.

1. The unahexa. A strip of three triangles can be folded flat and the opposite ends can be joined to make a Möbius strip with a triangular edge. Since it has one side only, made up of six triangles, one might call it a unahexaflexagon, though of course it isn't six-sided and it doesn't flex.
2. The duahexa. Simply a hexagon cut from a sheet of paper. It has two faces but doesn't flex.
3. The trihexa. This has only the one form described in this chapter.
4. The tetrahexa. This likewise has only one form. It is folded from the crooked strip shown in Figure 29.7(A).
5. The pentahexa. One form only. Folded from the strip in Figure 29.7 (B).
6. The hexahexa. There are three varieties, each with unique properties. One of them is described in this chapter. The other two are folded from the strips pictured in Figure 29.7(C).
7. The heptahexa. This can be folded from the three strips shown in Figure 29.7(D). The first strip can be folded in two different ways, making four varieties in all. The third form, folded from the overlapping figure-eight strip, is the first of what Louis Tuckerman calls the "street flexagons." Its faces can be numbered so that a Tuckerman traverse will bring uppermost the seven faces in serial order, like passing the street numbers on a row of houses.

The octahexa has 12 distinct varieties, the enneahexa has 27, and the decahexa, 82. The exact number of varieties of each order can be figured in more than one way depending on how you define a distinct variety. For example, all flexagons have an asymmetric structure which can be right-handed or left-handed, but mirror-image forms should hardly be classified as different varieties. For details on the number of nonequivalent hexaflexagons of each order, consult the paper by Oakley and Wisner listed in the bibliography.

Straight chains of triangles produce only hexaflexagons with orders that are multiples of three. One variety of a 12-faced hexa is particularly

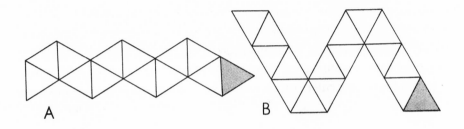

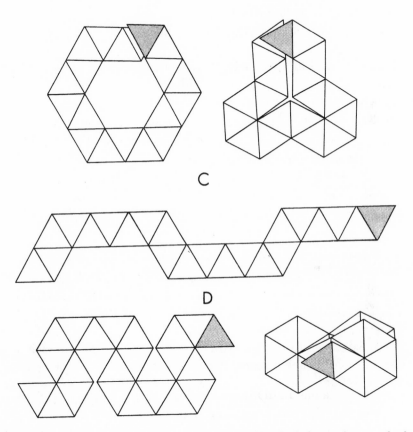

Figure 29.7. Crooked strips for folding hexaflexagons. The shaded triangles are tabs for pasting.

easy to fold. Start with a straight chain twice as long as the one used for the hexahexa. "Roll" it into the form shown in Figure 29.2(B). The strip is now the same length as the one used for the hexahexa. Fold this rolled strip exactly as if you were making a hexahexa. Result: a dodecahexaflexagon.

In experimenting with higher-order flexagons, a handy rule to bear in mind is that the sum of the number of leaves (thicknesses of paper) in two adjacent triangular sections always equals the number of faces. It is interesting to note also that if each face of a flexagon is given a number or symbol, and the symbol is marked on each triangular component, the order of symbols on the unfolded strip always exhibits a threefold symmetry. For example, the strip for the hexahexa in Figure 29.2 bears the following top and bottom pattern of digits:

$$123123 \quad 123123 \quad 123123$$
$$445566 \quad 445566 \quad 445566$$

A triple division similar to this is characteristic of all hexahexa-flexagons, although on models of odd order one of the three divisions is always inverted.

Of the hundreds of letters received about flexagons, the following two were the most amusing. They appeared in the March and May issues of *Scientific American,* 1957.

SIRS:

I was quite taken with the article entitled "Flexagons" in your December issue. It took us only six or seven hours to paste the hexahexa-flexagon together in the proper configuration. Since then it has been a source of continuing wonder.

But we have a problem. This morning one of our fellows was sitting flexing the hexahexaflexagon idly when the tip of his necktie became caught in one of the folds. With each successive flex, more of his tie vanished into the flexagon. With the sixth flexing he disappeared entirely.

We have been flexing the thing madly, and can find no trace of him, but we have located a sixteenth configuration of the hexahexaflexagon.

Here is our question: Does his widow draw workmen's compensation for the duration of his absence, or can we have him declared legally dead immediately? We await your advice.

NEIL UPTEGROVE

Allen B. Du Mont Laboratories, Inc.
Clifton, N.J.

SIRS:

The letter in the March issue of your magazine complaining of the disappearance of a fellow from the Allen B. Du Mont Laboratories "down" a hexahexaflexagon, has solved a mystery for us.

One day, while idly flexing our latest hexahexaflexagon, we were confounded to find that it was producing a strip of multicolored material. Further flexing of the hexahexaflexagon finally disgorged a gum-chewing stranger.

Unfortunately he was in a weak state and, owing to an apparent loss of memory, unable to give any account of how he came to be with us. His health has now been restored on our national diet of porridge, haggis and whisky, and he has become quite a pet around the department, answering to the name of Eccles.

Our problem is, should we now return him and, if so, by what method? Unfortunately Eccles now cringes at the very sight of a hexahexaflexagon and absolutely refuses to "flex."

ROBERT M. HILL

The Royal College of Science and Technology
Glasgow, Scotland

Although this chapter first ran as an article in *Scientific American* (December 1956), it became the first of my monthly columns to be headed "Mathematical Games." I had learned about hexaflexagons from Royal Vale Heath, a New York City broker who was also an amateur magician and mathematician. His little book *Mathemagic* (1933) was one of the earliest books about magic tricks that operate with mathematical principles.

I cannot now recall how Heath discovered flexagons, but he was able to steer me to the four Princeton University inventors. To gather material for my article, I made a trip to Princeton where I interviewed Tukey and Tuckerman. (Stone was then in England; Feynman was at Cal Tech.)

After my piece appeared, people all over Manhattan were flexing flexagons. Gerry Piel, publisher of *Scientific American,* called me into his office to ask if there was enough similar material to make a regular column. I assured him there was and immediately made the rounds of Manhattan's used bookstores to buy as many books on recreational math (there were not many) as I could find. Once the column got underway, I began to receive fresh ideas from mathematicians and writing the column became an easy and enjoyable task that lasted more than a quarter-century.

Although flexagons, as far as anyone knows, are completely free of applications (except of course as playthings), mathematicians continue to be intrigued by their whimsical properties. In *The Second "Scientific American" Book of Mathematical Puzzles and Diversions* I introduced their square-shaped cousins, the tetraflexagons. Both types have been discussed in many subsequent articles, but no one has yet written a definitive work on flexagon theory. Frank Bernhart, who taught mathematics at the Rochester Institute of Technology, knows more about flexagons than anyone. Some day, let us hope, he will find a publisher for a monograph about them.

Bibliography

D. A. Engel, "Hybrid Flexagons," *Journal of Recreational Mathematics*, Vol. 2, January 1969, pp. 35–41.

D. A. Engel, "Hexaflexagon + HHFG = Slipagon," *Journal of Recreational Mathematics*, No. 3, 1995, pp. 161–66.

M. Gilpin, "Symmetries of the Trihexaflexagon," *Mathematics Magazine*, Vol. 49, September 1976, pp. 189–92.

P. Hilton, J. Pedersen, and H. Walker, "The Faces of the Tri-Hexaflexagon," *Mathematics Magazine*, Vol. 70, October 1997, pp. 243–51.

P. Jackson, *Flexagons*, British Origami Society, 1978.

D. Johnson, *Mathemagic with Flexagons*, Activity Resources, 1974.

M. Jones, *The Mysterious Flexagons*, Crown, 1966. A book for children printed on heavy paper so the child can cut strips for folding.

M. Joseph, "Hexahexaflexagrams," *Mathematics Teacher*, Vol. 44 April 1951, pp. 247–48. Tells how to make a straight chain hexahexaflexagon.

P. Liebeck, "The Construction of Flexagons," *Mathematical Gazette*, Vol. 48, December 1964, pp. 397–402.

J. Madachy, *Mathematics on Vacation*, Scribner's, 1966, pp. 62–84.

F. G. Maunsell, "The Flexagon and the Hexahexaflexagram," *Mathematical Gazette*, Vol. 38, 1954, pp. 213–14. Describes the trihexa and hexahexa. See also Vol. 41, 1957, pp. 55–56, for a note on this article by Joan Crampin. It describes hexaflexagons folded from straight strips (orders 3, 6, 9, 12,. . .).

T. B. McLean, "V-Flexing the Hexaflexagon," *American Mathematical Monthly*, Vol. 86, June/July 1979, pp. 457–66.

S. Morris, "Hexplay," *Omni*, October 1984, pp. 89ff.

C. O. Oakley and R. J. Wisner, "Flexagons," *American Mathematical Monthly*, Vol. 64, March 1957, pp. 143–54.

T. O'Reilly, "Classifying and Counting Hexaflexagrams," *Journal of Recreational Mathematics*, Vol. 8, No. 3, 1975–76, pp. 182–87.

A. Panov, "To Flexland With Mr. Flexman," *Quantum*, March/April 1992, pp. 64–65.

A. Panov and A. Keltnin, "Flexland Revisited," *Quantum*, July/August, 1993, pp. 64–65.

W. R. Ransom, "A Six-Sided Hexagon," *School Science and Mathematics*, Vol. 52, 1952, p. 94. Tells how to make a trihexaflexagon.

W. Ransom, "Protean Shapes with Flexagons," *Recreational Mathematics Magazine,* No. 13, February 1963, pp. 35–37.

S. Scott, "How to Construct Hexaflexagons," *Recreational Mathematics Magazine,* No. 12, December 1962, pp. 43–49.

V. Silva, "Flexagons," *Eureka,* No. 52, March 1993, pp. 30–44.

R. F. Wheeler, "The Flexagon Family," *Mathematical Gazette,* Vol. 42 February 1958, pp. 1–6. A fairly complete analysis of straight and crooked strip forms.

Chapter 30   *The Soma Cube*

". . . no time, no leisure . . . not a moment to sit down and think—or if ever by some un- lucky chance such a crevice of time should yawn in the solid substance of their dis- tractions, there is always soma, delicious soma . . ." —Aldous Huxley, *Brave New World*

*T*he Chinese puzzle game called tangrams, believed to be thou- sands of years old, employs a square of thin material that is dissected into seven pieces. The game is to rearrange those pieces to form other figures. From time to time efforts have been made to devise a suitable analog in three dimensions. None, in my opinion, has been as suc- cessful as the Soma cube, invented by Piet Hein, the Danish writer whose mathematical games, Hex and Tac Tix, are discussed in the first *Scientific American Book of Mathematical Puzzles.* (In Denmark, Piet Hein was best known for his books of epigrammatic poems written under the pseudonym Kumbel.)

Piet Hein conceived of the Soma cube during a lecture on quantum physics by Werner Heisenberg. While the noted German physicist was speaking of a space sliced into cubes, Piet Hein's supple imagination caught a fleeting glimpse of the following curious geometrical theo- rem. If you take all the irregular shapes that can be formed by combin- ing no more than four cubes, all the same size and joined at their faces, these shapes can be put together to form a larger cube.

Let us make this clearer. The simplest irregular shape—"irregular" in the sense that it has a concavity or corner nook in it somewhere—is produced by joining three cubes as shown in Figure 30.1, piece 1. It is the only such shape possible with three cubes. (Of course no irregular shape is possible with one or two cubes.) Turning to four cubes, we find that there are six different ways to form irregular shapes by joining the cubes face to face. These are pieces 2 to 7 in the illustration. To iden- tify the seven pieces Piet Hein labels them with numerals. No two shapes are alike, although 5 and 6 are mirror images of each other. Piet Hein points out that two cubes can be joined only along a single coor- dinate, three cubes can add a second coordinate perpendicular to the

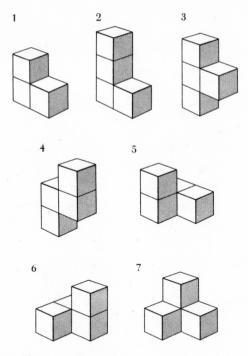

Figure 30.1. The seven Soma pieces

first, and four cubes are necessary to supply the third coordinate perpendicular to the other two. Since we cannot enter the fourth dimension to join cubes along a fourth coordinate supplied by five-cube shapes, it is reasonable to limit our set of Soma pieces to seven. It is an unexpected fact that these elementary combinations of identical cubes can be joined to form a cube again.

As Heisenberg talked on, Piet Hein swiftly convinced himself by doodling on a sheet of paper that the seven pieces, containing 27 small cubes, would form a 3 × 3 × 3 cube. After the lecture he glued 27 cubes into the shapes of the seven components and quickly confirmed his insight. A set of the pieces was marketed under the trade name Soma, and the puzzle has since become a popular one in the Scandinavian countries.

To make a Soma cube—and the reader is urged to do so, for it provides a game that will keep every member of the family entranced for hours—you have only to obtain a supply of children's blocks. The seven pieces are easily constructed by spreading rubber cement on the appropriate faces, letting them dry, then sticking them together.

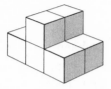

Figure 30.2. A form made up of two Soma pieces

As a first lesson in the art of Soma, see if you can combine any two pieces to form the stepped structure in Figure 30.2. Having mastered this trivial problem, try assembling all seven pieces into a cube. It is one of the easiest of all Soma constructions. More than 230 essentially dif-

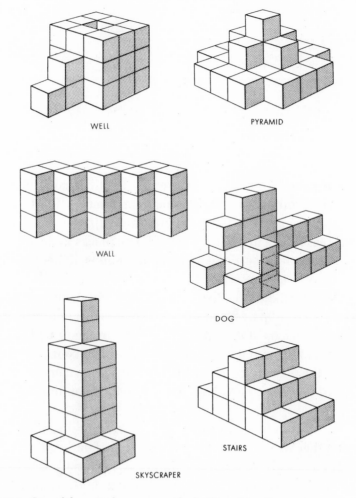

WELL

PYRAMID

WALL

DOG

SKYSCRAPER

STAIRS

Figure 30.3. One of these 12 forms cannot be built up from Soma pieces.

COMBINATORICS

ferent solutions (not counting rotations and reflections) have been tabulated by Richard K. Guy of the University of Malaya, in Singapore, but the exact number of such solutions has not yet been determined. A good strategy to adopt on this as well as other Soma figures is to set the more irregular shapes (5, 6, and 7) in place first, because the other pieces adjust more readily to remaining gaps in a structure. Piece 1 in particular is best saved until last.

After solving the cube, try your hand at the more difficult seven-piece structures in Figure 30.3. Instead of using a time-consuming trial and error technique, it is much more satisfying to analyze the con-

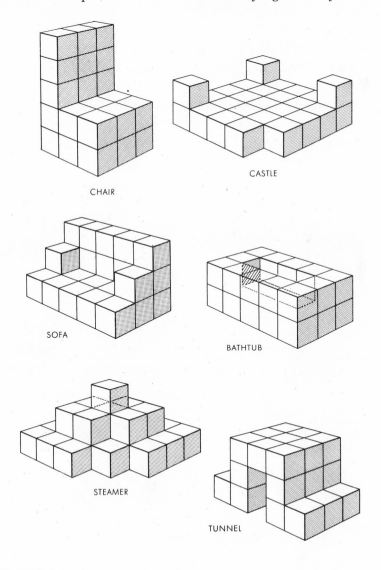

CHAIR

CASTLE

SOFA

BATHTUB

STEAMER

TUNNEL

structions and cut down your building time by geometrical insights. For example, it is obvious that pieces 5, 6, and 7 cannot form the steps to the well. Group competition can be introduced by giving each player a Soma set and seeing who can build a given figure in the shortest length of time. To avoid misinterpretations of these structures it should be said that the far sides of the pyramid and steamer are exactly like the near sides; both the hole in the well and the interior of the bathtub have a volume of three cubes; there are no holes or projecting pieces on the hidden sides of the skyscraper; and the column that forms the back of the dog's head consists of four cubes, the bottom one of which is hidden from view.

After working with the pieces for several days, many people find that the shapes become so familiar that they can solve Soma problems in their heads. Tests made by European psychologists have shown that ability to solve Soma problems is roughly correlated with general intelligence, but with peculiar discrepancies at both ends of the I.Q. curve. Some geniuses are very poor at Soma and some morons seem specially gifted with the kind of spatial imagination that Soma exercises. Everyone who takes such a test wants to keep playing with the pieces after the test is over.

Like the two-dimensional polyominoes, Soma constructions lend themselves to fascinating theorems and impossibility proofs of combinatorial geometry. Consider the structure in the left illustration of Figure 30.4. No one had succeeded in building it, and eventually a formal impossibility proof was devised. Here is the clever proof, discovered by Solomon W. Golomb, mathematician at the University of Southern California.

We begin by looking down on the structure as shown in the right illustration and coloring the columns in checkerboard fashion. Each column is two cubes deep except for the center column, which consists

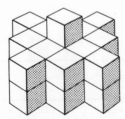 

Figure 30.4. An impossible Soma form   A means of labeling the form

COMBINATORICS

of three cubes. This gives us a total of eight white cubes and 19 black, quite an astounding disparity.

The next step is to examine each of the seven components, testing it in all possible orientations to ascertain the maximum number of black cubes it can possess if placed within the checkerboard structure. The chart in Figure 30.5 displays this maximum number for each piece. As you see, the total is 18 black to nine white, just one short of the 19–8 split demanded. If we shift the top black block to the top of one of the columns of white blocks, then the black–white ratio changes to the required 18–9, and the structure becomes possible to build.

SOMA PIECE	MAXIMUM BLACK CUBES	MINIMUM WHITE CUBES
1.	2	1
2.	3	1
3.	3	1
4.	2	2
5.	3	1
6.	3	1
7.	2	2
	18	9

Figure 30.5. Table for the impossibility proof

I must confess that one of the structures in Figure 30.3 is impossible to make. It should take the average reader many days, however, to discover which one it is. Methods for building the other figures will not be given in the answer section (it is only a matter of time until you succeed with any one of them), but I shall identify the figure that cannot be made.

The number of pleasing structures that can be built with the seven Soma pieces seems to be as unlimited as the number of plane figures that can be made with the seven tangram shapes. It is interesting to note that if piece 1 is put aside, the remaining six pieces will form a shape exactly like 1 but twice as high.

Addendum

When I wrote the column about Soma, I supposed that few readers would go to the trouble of actually making a set. I was wrong. Thousands of readers sent sketches of new Soma figures and many complained that their leisure time had been obliterated since they were bitten by the Soma bug. Teachers made Soma sets for their classes. Psy-

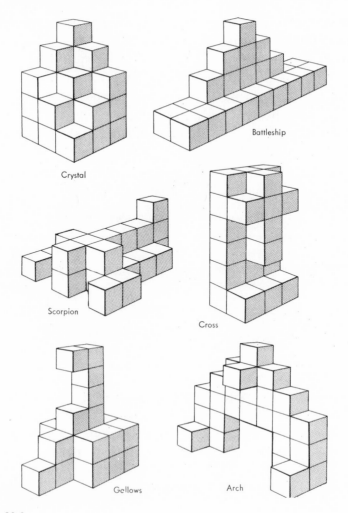

Crystal

Battleship

Scorpion

Cross

Gallows

Arch

Figure 30.6.

chologists added Soma to their psychological tests. Somaddicts made sets for friends in hospitals and gave them as Christmas gifts. A dozen firms inquired about manufacturing rights. Gem Color Company, 200 Fifth Avenue, New York, NY, marketed a wooden set—the only set authorized by Piet Hein.

From the hundreds of new Soma figures received from readers, I have selected the twelve that appear in Figure 30.6. Some of these figures were discovered by more than one reader. All are possible to construct.

The charm of Soma derives in part, I think, from the fact that only

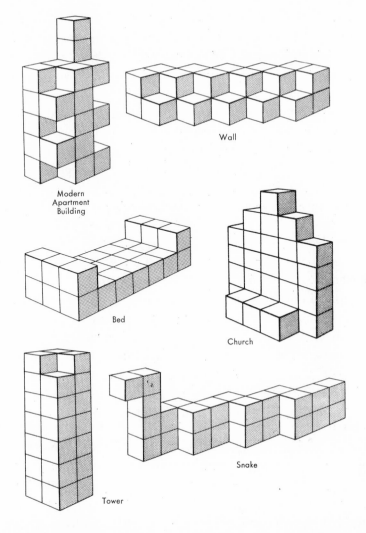

Wall

Modern
Apartment
Building

Bed

Church

Snake

Tower

seven pieces are used; one is not overwhelmed by complexity. All sorts of variant sets, with a larger number of pieces, suggest themselves, and I have received many letters describing them.

Theodore Katsanis of Seattle, in a letter dated December 23, 1957 (before the article on Soma appeared), proposed a set consisting of the eight different pieces that can be formed with four cubes. This set includes six of the Soma pieces plus a straight chain of four cubes and a 2 × 2 square. Katsanis called them "quadracubes"; other readers later suggested "tetracubes." The eight pieces will not, of course, form a cube; but they do fit neatly together to make a 2 × 4 × 4 rectangular solid. This is a model, twice as high, of the square tetracube. It is pos-

The Soma Cube

sible to form similar models of each of the other seven pieces. Katsanis also found that the eight pieces can be divided into two sets of four, each set making a 2 × 2 × 4 rectangular solid. These two solids can then be put together in different ways to make double-sized models of six of the eight pieces.

In a previous column I described the 12 pentominoes: flat shapes formed by connecting unit squares in all possible ways. Mrs. R. M. Robinson, wife of a mathematics professor at the University of California in Berkeley, discovered that if the pentominoes are given a third dimension, one unit thick, the twelve pieces will form a 3 × 4 × 5 rectangular solid. This was independently discovered by several others, including Charles W. Stephenson, M.D., of South Hero, VT. Dr. Stephenson also found ways of putting together the three-dimensional pentominoes to make rectangular solids of 2 × 5 × 6 and 2 × 3 × 10.

The next step in complexity is to the 29 pieces formed by putting five cubes together in all possible ways. Katsanis, in the same letter mentioned above, suggested this and called the pieces "pentacubes." Six pairs of pentacubes are mirror-image forms. If we use only one of each pair, the number of pentacubes drops to 23. Both 29 and 23 are primes, therefore no rectangular solids are possible with either set. Katsanis proposed a triplication problem: choose one of the 29 pieces, then use 27 of the remaining 28 to form a model of the selected piece, three times as high.

A handsome set of pentacubes was shipped to me in 1960 by David Klarner of Napa, CA. I dumped them out of the wooden box in which they were packed and have not yet succeeded in putting them back in. Klarner has spent considerable time developing unusual pentacube figures, and I have spent considerable time trying to build some of them. He writes that there are 166 *hexacubes* (pieces formed by joining six-unit cubes), of which he was kind enough *not* to send a set.

The seven Soma pieces are a subset of what are now called polycubes—polyhedrons formed by joining unit cubes by their faces. Since I introduced the Soma cube in a 1958 column it has been made and sold by numerous toy companies around the world. Here Parker Brothers sold the cube along with an instruction booklet written by Piet Hein. The firm also distributed three issues of *Soma Addict,* a newsletter edited by game agent Thomas Atwater.

Several computer programs verified that there are 240 ways, not counting rotations and reflections, to make the Soma cube. John Conway produced what he called the Somap. You'll find a picture of it in *Winning Ways,* Vol. 2, by Elwyn Berlekamp, Conway, and Richard Guy, pages 802–3. This amazing graph shows how you can start with any of 239 solutions to the cube, then transform it to any other solution by moving no more than two or three pieces. There is one solution unobtainable in this way.

J. Edward Hanrahan wrote to me about a Soma task he invented. The challenge is to form a $4 \times 4 \times 2$ structure so that its five "holes" on the top layer have the shape of each of the 12 pentominoes. The problem is solvable for each pentomino except the straight one which obviously can't fit into the structure.

Puzzle collector Jerry Slocum owns dozens of different dissections of a 3^3 cube into seven or fewer pieces, most of them marketed after my column on the Soma cube appeared. In the chapter on polycubes in *Knotted Doughnuts and Other Mathematical Entertainments,* I describe the Diabolical cube, sold in Victorian England. The earliest dissection of a 3^3 cube known, its six polycube pieces form the cube in 13 ways. I also describe the Miksinski cube, another six-piece dissection—one that has only two solutions.

Many later 3^3 dissections limit the number of solutions by coloring or decorating the unit cube faces in various ways. For example, the unit cubes are either black or white and the task is to make a cube that is checkered throughout or just on its six faces. Other dissections put one through six spots on the unit cubes so that the assembled cube will resemble a die. The unit cubes can be given different colors. The task is to form a cube with a specified pattern of colors on each face. Another marketed cube had digits 1 through 9 on the unit cubes, and the problem was to make a cube with each face a magic square. A puzzle company in Madison, WI, advertised a game using a set of nine polycubes. Players selected any seven at random by rolling a pair of dice, then tried to make a 3^3 cube with the pieces. In 1969 six different dissections of the cube, each made with five, six, or seven polycubes, were sold under the name Impuzzibles.

It's easy to cut a 3^3 cube into six or fewer polycubes that will make a cube in just one way, but with seven unmarked pieces, as I said earlier, it is not so easy. Two readers sent such dissections which they believed solved this problem, but I was unable to verify their claims. Rhoma, a slant version of Soma produced by a shear distortion that changes each unit cube as well as the large cube to a rhomboid shape, went on sale here. A more radically squashed Soma was sold in Japan. With such distortions the solution becomes unique.

John Brewer, of Lawrence, KS, in a little magazine he used to publish called *Hedge Apple and Devil's Claw* (Autumn 1995 issue), introduced the useful device of giving each Soma piece a different color. Solutions could then be represented by showing three sides of the cube with its unit cubes properly colored. He sent me a complete Somap using such pictures for each solution. His article also tells of his failed effort to locate Marguerite Wilson, the first to publish a complete set of solutions to the Soma cube.

Alan Guth, the M.I.T. physicist famous for his conjecture that a moment after the big bang the universe rapidly inflated, was quoted as follows in *Discover* (December 1997):

> My all-time favorite puzzle was a game called Soma, which I think was first marketed when I was in college. A set consisted of seven odd-shaped pieces that could be put together to make a cube, or a large variety of other three-dimensional shapes. Playing with two sets at once was even more fun. The instruction book claims that there are a certain number of different ways of making the cube, and I remember writing a computer program to verify this number. They were right, but I recall that they counted each of the 24 different orientations of the cube as a different "way" of making it.

Answer

The only structure in Figure 30.3 that is impossible to construct with the seven Soma pieces is the skyscraper.

Bibliography

J. Brunrell et al, "The Computerized Soma Cube," *Symmetry Unifying Human Understanding,* I. Hargittai (ed.), Pergamon, 1987.

G. S. Carson, "Soma Cubes," *Mathematics Teacher,* November 1973, pp. 583–92.

S. Farhi, *Soma Cubes,* Pentacubes Puzzles Ltd, Auckland, Australia, 1979. This 15-page booklet depicts 111 Soma structures and their solutions.

D. A. Macdonald and Y. Gürsel, "Solving Soma Cube and Polyomino Problems Using a Microcomputer," *Byte,* November 1989, pp. 26–41.

D. Spector, "Soma: A Unique Object for Mathematical Study," in *Mathematics Teacher,* May 1982, pp. 404–7.

M. Wilson, *Soma Puzzle Solutions,* Creative Publications, 1973.

Chapter 31   **The Game
                                                  of Life**

Most of the work of John Horton Conway, a distinguished
Cambridge mathematician now at Princeton University, has been in
pure mathematics. For instance, in 1967 he discovered a new group—
some call it "Conway's constellation"—that includes all but two of the
then-known sporadic groups. (They are called "sporadic" because they
fail to fit any classification scheme.) It is a breakthrough that has had
exciting repercussions in both group theory and number theory. It ties
in closely with an earlier discovery by John Leech of an extremely
dense packing of unit spheres in a space of 24 dimensions where each
sphere touches 196,560 others. As Conway has remarked, "There is a
lot of room up there."

In addition to such serious work Conway also enjoys recreational
mathematics. Although he is highly productive in this field, he seldom publishes his discoveries. In this chapter we consider Conway's
most famous brainchild, a fantastic solitaire pastime he calls "Life."
Because of its analogies with the rise, fall, and alterations of a society of living organisms, it belongs to a growing class of what are
called "simulation games." To play Life without a computer you
need a fairly large checkerboard and a plentiful supply of flat counters of two colors. (Small checkers or poker chips do nicely.) An Oriental "Go" board can be used if you can find flat counters small
enough to fit within its cells. (Go stones are awkward to use because
they are not flat.) It is possible to work with pencil and graph paper
but it is much easier, particularly for beginners, to use counters and
a board.

The basic idea is to start with a simple configuration of counters (organisms), one to a cell, then observe how it changes as you apply Conway's "genetic laws" for births, deaths, and survivals. Conway chose

his rules carefully, after a long period of experimentation, to meet three desiderata:

1. There should be no initial pattern for which there is a simple proof that the population can grow without limit.
2. There should be initial patterns that *apparently* do grow without limit.
3. There should be simple initial patterns that grow and change for a considerable period of time before coming to an end in three possible ways: Fading away completely (from overcrowding or from becoming too sparse), settling into a stable configuration that remains unchanged thereafter, or entering an oscillating phase in which they repeat an endless cycle of two or more periods.

In brief, the rules should be such as to make the behavior of the population both interesting and unpredictable.

Conway's genetic laws are delightfully simple. First note that each cell of the checkerboard (assumed to be an infinite plane) has eight neighboring cells, four adjacent orthogonally, four adjacent diagonally. The rules are:

1. Survivals. Every counter with two or three neighboring counters survives for the next generation.
2. Deaths. Each counter with four or more neighbors dies (is removed) from overpopulation. Every counter with one neighbor or none dies from isolation.
3. Births. Each empty cell adjacent to exactly three neighbors—no more, no fewer—is a birth cell. A counter is placed on it at the next move.

It is important to understand that all births and deaths occur *simultaneously*. Together they constitute a single generation or, as we shall usually call it, a "tick" in the complete "life history" of the initial configuration. Conway recommends the following procedure for making the moves:

1. Start with a pattern consisting of black counters.
2. Locate all counters that will die. Identify them by putting a black counter on top of each.
3. Locate all vacant cells where births will occur. Put a white counter on each birth cell.
4. After the pattern has been checked and double-checked to make sure

no mistakes have been made, remove all the dead counters (piles of two) and replace all newborn white organisms with black counters.

You will now have the first generation in the life history of your initial pattern. The same procedure is repeated to produce subsequent generations. It should be clear why counters of two colors are needed. Because births and deaths occur simultaneously, newborn counters play no role in causing other deaths or births. It is essential, therefore, to be able to distinguish them from live counters of the previous generation while you check the pattern to be sure no errors have been made. Mistakes are very easy to make, particularly when first playing the game. After playing it for a while you will gradually make fewer mistakes, but even experienced players must exercise great care in checking every new generation before removing the dead counters and replacing newborn white counters with black.

You will find the population constantly undergoing unusual, sometimes beautiful, and always unexpected change. In a few cases the society eventually dies out (all counters vanishing), although this may not happen until after a great many generations. Most starting patterns either reach stable figures—Conway calls them "still lifes"—that cannot change or patterns that oscillate forever. Patterns with no initial symmetry tend to become symmetrical. Once this happens the symmetry cannot be lost, although it may increase in richness.

Conway originally conjectured that no pattern can grow without limit. Put another way, any configuration with a finite number of counters cannot grow beyond a finite upper limit to the number of counters on the field. Conway offered a prize of $50 to the first person who could prove or disprove the conjecture before the end of 1970. One way to disprove it would be to discover patterns that keep adding counters to the field: A "gun" (a configuration that repeatedly shoots out moving objects such as the "glider," to be explained below) or a "puffer train" (a configuration that moves but leaves behind a trail of "smoke").

Let us see what happens to a variety of simple patterns.

A single organism or any pair of counters, wherever placed, will obviously vanish on the first tick.

A beginning pattern of three counters also dies immediately unless at least one counter has two neighbors. Figure 31.1 shows the five connected triplets that do not fade on the first tick.

(Their orientation is of course irrelevant.) The first three (a, b, c) van-

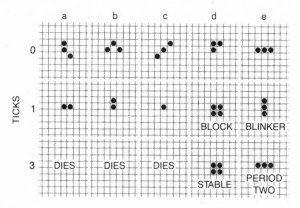

Figure 31.1. The fate of five triplets in "Life"

ish on the second tick. In connection with *c* it is worth noting that a single diagonal chain of counters, however long, loses its end counters on each tick until the chain finally disappears. The speed a chess king moves in any direction is called by Conway (for reasons to be made clear later) the "speed of light." We say, therefore, that a diagonal chain decays at each end with the speed of light.

Pattern *d* becomes a stable "block" (2 × 2 square) on the second tick. Pattern *e* is the simplest of what are called "flip-flops" (oscillating figures of period 2). It alternates between horizontal and vertical rows of three. Conway calls it a "blinker."

Figure 31.2 shows the life histories of the five tetrominoes (four rookwise-connected counters). The square *a* is, as we have seen, a still-life figure. Tetrominoes *b* and *c* reach a stable figure, called a "beehive," on the second tick. Beehives are frequently produced patterns. Tetromino *d* becomes a beehive on the third tick. Tetromino *e* is the most interesting of the lot. After nine ticks it becomes four isolated blinkers, a flip-flop called "traffic lights." It too is a common configuration. Figure 31.3 shows 12 common forms of still life.

The reader may enjoy experimenting with the 12 pentominoes (all possible patterns of five rookwise-connected counters) to see what happens to each. He will find that five vanish before the fifth tick, two quickly reach a stable loaf, and four in a short time become traffic lights. The only pentomino that does not end quickly (by vanishing, becoming stable, or oscillating) is the *R* pentomino (*a* in Figure 31.4). Conway has tracked it for 460 ticks. By then it has thrown off a number of gliders. Conway remarks: "It has left a lot of miscellaneous junk

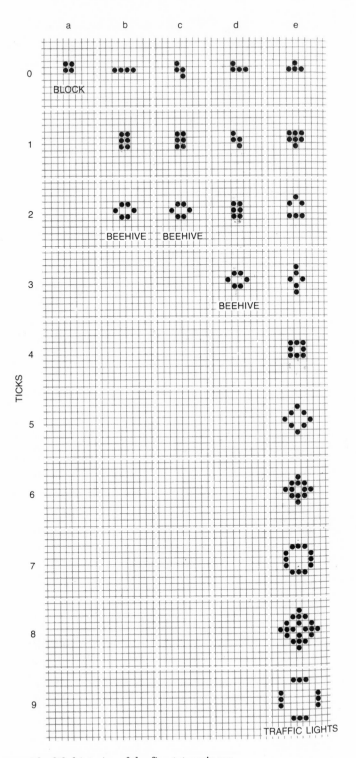

Figure 31.2. The life histories of the five tetrominoes

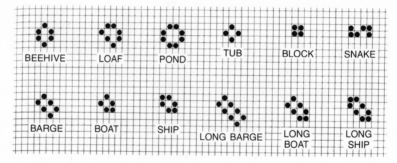

Figure 31.3. The commonest stable forms

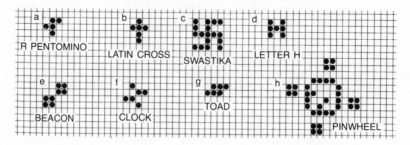

Figure 31.4. The R pentomino (a) and exercises for the reader

stagnating around, and has only a few small active regions, so it is not at all obvious that it will continue indefinitely." Its fate is revealed in the addendum to this chapter.

For such long-lived populations Conway sometimes uses a computer with a screen on which he can observe the changes. The program was written by M.J.T. Guy and S. R. Bourne. Without its help some discoveries about the game would have been difficult to make.

As easy exercises the reader is invited to discover the fate of the Latin cross (b in Figure 31.4), the swastika (c), the letter H (d), the beacon (e), the clock (f), the toad (g), and the pinwheel (h). The last three figures were discovered by Simon Norton. If the center counter of the H is moved up one cell to make an arch (Conway calls it "pi"), the change is unexpectedly drastic. The H quickly ends but pi has a long history. Not until after 173 ticks has it settled down to five blinkers, six blocks, and two ponds. Conway also has tracked the life histories of all the hexominoes and all but seven of the heptominoes. Some hexominoes enter the history of the R pentomino; for example, the pentomino becomes a hexomino on its first tick.

One of the most remarkable of Conway's discoveries is the five-counter glider shown in Figure 31.5. After two ticks it has shifted slightly and been reflected in a diagonal line. Geometers call this a "glide reflection"; hence the figure's name. After two more ticks the glider has righted itself and moved one cell diagonally down and to the right from its initial position. We mentioned earlier that the speed of a chess king is called the speed of light. Conway chose the phrase because it is the highest speed at which any kind of movement can occur on the board. No pattern can replicate itself rapidly enough to move at such speed. Conway has proved that the maximum speed diagonally is one-fourth the speed of light. Since the glider replicates itself in the same orientation after four ticks, and has traveled one cell diagonally, one says that it glides across the field at one-fourth the speed of light.

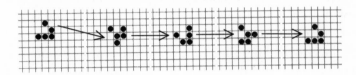

Figure 31.5. The "glider"

Movement of a finite figure horizontally or vertically into empty space, Conway has also shown, cannot exceed half the speed of light. Can any reader find a relatively simple figure that travels at such a speed? Remember, the speed is obtained by dividing the number of ticks required to replicate a figure by the number of cells it has shifted. If a figure replicates in four ticks in the same orientation after traveling two unit squares horizontally or vertically, its speed will be half that of light. Figures that move across the field by self-replication are extremely hard to find. Conway knows of four, including the glider, which he calls "spaceships" (the glider is a "featherweight spaceship"; the others have more counters). I will disclose their patterns in the Answer Section.

Figure 31.6 depicts three beautiful discoveries by Conway and his collaborators. The stable honey farm (*a* in Figure 31.6) results after 14 ticks from a horizontal row of seven counters. Since a 5 × 5 block in one move produces the fourth generation of this life history, it becomes a honey farm after 11 ticks. The "figure 8" (*b* in Figure 31.6), an oscillator found by Norton, both resembles an 8 and has a period of 8. The

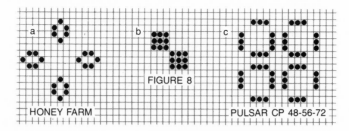

Figure 31.6. Three remarkable patterns, one stable and two oscillating

form *c*, in Figure 31.6 called "pulsar *CP* 48–56–72," is an oscillator with a life cycle of period 3. The state shown here has 48 counters, state two has 56, and state three has 72, after which the pulsar returns to 48 again. It is generated in 32 ticks by a heptomino consisting of a horizontal row of five counters with one counter directly below each end counter of the row.

Conway has tracked the life histories of a row of *n* counters through *n* = 20. We have already disclosed what happens through *n* = 4. Five counters result in traffic lights, six fade away, seven produce the honey farm, eight end with four beehives and four blocks, nine produce two sets of traffic lights, and 10 lead to the "pentadecathlon," with a life cycle of period 15. Eleven counters produce two blinkers, 12 end with two beehives, 13 with two blinkers, 14 and 15 vanish, 16 give "big traffic lights" (eight blinkers), 17 end with four blocks, 18 and 19 fade away, and 20 generate two blocks.

Conway also investigated rows formed by sets of *n* adjacent counters separated by one empty cell. When *n* = 5 the counters interact and become interesting. Infinite rows with *n* = 1 or *n* = 2 vanish in one tick, and if *n* = 3 they turn into blinkers. If *n* = 4 the row turns into a row of beehives.

The 5–5 row (two sets of five counters separated by a vacant cell) generates the pulsar *CP* 48–56–72 in 21 ticks. The 5–5–5 ends in 42 ticks with four blocks and two blinkers. The 5–5–5–5 ends in 95 ticks with four honey farms and four blinkers, 5–5–5–5–5 terminates with a spectacular display of eight gliders and eight blinkers after 66 ticks. Then the gliders crash in pairs to become eight blocks after 86 ticks. The form 5–5–5–5–5–5 ends with four blinkers after 99 ticks, and 5–5–5–5–5–5–5, Conway remarks, "is marvelous to sit watching on the computer screen." This ultimate destiny is given in the addendum.

COMBINATORICS

Addendum

My 1970 column on Conway's "Life" met with such an instant enthusiastic response among computer hackers around the world that their mania for exploring "Life" forms was estimated to have cost the nation millions of dollars in illicit computer time. One computer expert, whom I shall leave nameless, installed a secret switch under his desk. If one of his bosses entered the room he would press the button and switch his computer screen from its "Life" program to one of the company's projects.

The troublesome R pentomino becomes a 2-tick oscillator after 1,103 ticks. Six gliders have been produced and are traveling outward. The debris left at the center (see Figure 31.7) consists of four blinkers, one ship, one boat, one loaf, four beehives, and eight blocks. This was first established at Case Western Reserve University by Gary Filipski and Brad Morgan and later confirmed by scores of "Life" hackers here and abroad.

The fate of the 5–5–5–5–5–5–5 was first independently found by Robert T. Wainwright and a group of hackers at Honeywell's Computer Control Division, later by many others. The pattern stabilizes as a 2-tick oscillator after 323 ticks with four traffic lights, eight blinkers, eight loaves, eight beehives, and four blocks. Figure 31.8 reproduces a printout of the final steady state. Because symmetry cannot be lost in the history of any life form, the vertical and horizontal axes of the original symmetry are preserved in the final state. The maximum population (492 bits) is reached in generation 283, and the final population is 192.

When Conway visited me in 1970 and demonstrated his game of Life on my Go board, neither he nor I had any inkling of how famous the game would become, and what surprises about it would soon be discovered. It is far and away the most played, most studied of all cellular automata recreations.

Bill Gosper, when he worked in M.I.T.'s artificial intelligence laboratory run by Marvin Minsky, won Conway's $50 prize by discovering the glider gun. I can still recall my excitement when I received Gosper's telegram describing the gun. This is a Life pattern that shoots off a constant stream of gliders. The gun soon provided the base for proving that Life could function as a Turing machine capable, in principle, of performing all the calculations possible by a digital computer. The gun's reality was at once verified by Robert Wainwright who later

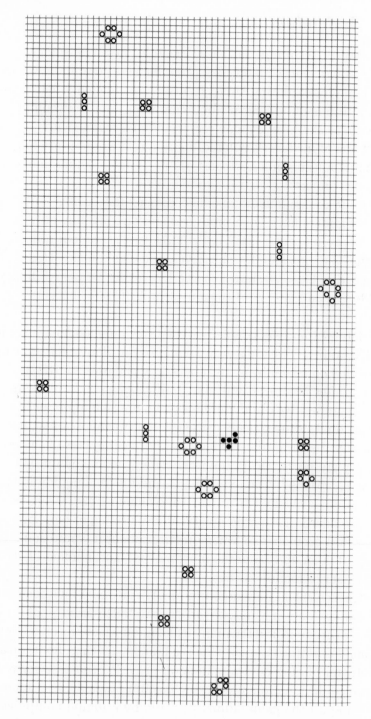

Figure 31.7. R pentomino's original *(black)* and final *(open dots)* state. (Six gliders are out of sight.)

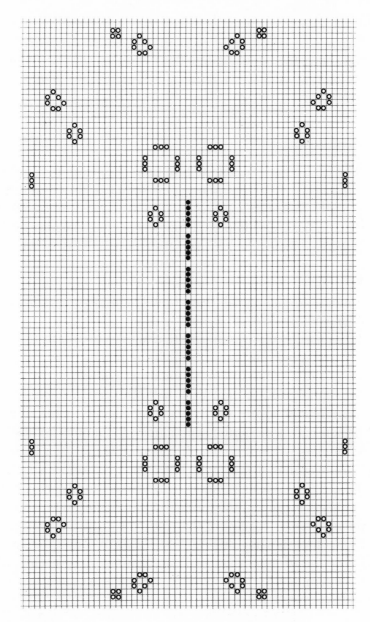

Figure 31.8. Initial pattern and final state of the 5–5–5–5–5–5–5 row

edited a periodical called *Lifeline* that reported ongoing research on Life.

For these and other amazing Life developments see the second and third chapters on Life in my *Wheels, Life, and Other Mathematical Amusements.*

Answers

The Latin cross dies on the fifth tick. The swastika vanishes on the sixth tick. The letter *H* also dies on the sixth tick. The next three figures are flip-flops: As Conway writes, "The toad pants, the clock ticks, and the beacon flashes, with period 2 in every case." The pinwheel's interior rotates 90 degrees clockwise on each move, the rest of the pattern remaining stable. Periodic figures of this kind, in which a fixed outer border is required to move the interior, Conway calls "billiard-table configurations" to distinguish them from "naturally periodic" figures such as the toad, clock, and beacon.

The three unescorted ships (in addition to the glider, or featherweight spaceship) are shown in Figure 31.9. To be precise, each becomes a spaceship in 1 tick. (The patterns in Figure 31.9 never recur.) All three travel horizontally to the right with half the speed of light. As they move they throw off sparks that vanish immediately as the ships continue on their way. Unescorted spaceships cannot have bodies longer than six counters without giving birth to objects that later block their motion. Conway has discovered, however, that longer spaceships, which he calls "overweight" ones, can be escorted by two or more smaller ships that prevent the formation of blocking counters. Figure 31.10 shows a larger spaceship that can be escorted by two smaller ships. Except for this same ship, lengthened by two units, longer ships require a flotilla of more than two companions. A spaceship with a body of 100 counters, Conway found, can be escorted safely by a flotilla of 33 smaller ships.

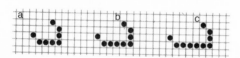

Figure 31.9. Lightweight *(left)*, middleweight *(center)*, and heavyweight *(right)* spaceships

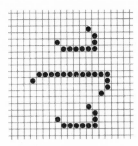

Figure 31.10. Overweight spaceship with two escorts

Bibliography

On cellular automata theory:

M. A. Arbib, *Theories of Abstract Automata,* Prentice-Hall, 1969.

A. W. Burks (ed), *Essays on Cellular Automata,* University of Illinois Press, 1970.

E. F. Codd, *Cellular Automata,* Academic Press, 1968.

M. L. Minsky, *Computation: Finite and Infinite Machines,* Prentice-Hall, 1967.

M. L. Minsky and S. Papert, *Perceptrons,* MIT Press, 1969.

J. von Neumann, *Theory of Self-Replicating Automata,* University of Illinois Press, 1966.

On the Game of Life:

P. Anthony, *Ox,* Avon, 1976. A novel involving Conway's Life and hexaflexagon theory.

J. Barry, "The Game of Life: Is It Just a Game?" *The London Times,* Sunday, June 13, 1971.

E. Berlekamp, J. Conway, and R. Guy, "What is Life?" *Winning Ways,* Vol. 2, Academic Press, 1982, Chapter 25.

R. A. Bosch, "Integer Programming and the Game of Life," *SIAM Review,* Vol. 41, September 1999, pp. 594–604.

K. Brecher, "Spirals: Magnificent Mystery," *Science Digest,* Spring 1980, p. 74ff.

D. J. Buckingham, "Some Facts of Life," *Byte,* Vol. 3, December 1978, pp. 54–67.

G. J. Chaitin, "Toward a Mathematical Definition of Life," Part 1, *ACM SICACT News,* Vol. 4, January 1970, pp. 12–18; Part 2, *IBM Research Report RC 6919,* December 1977.

M. Dresden and D. Wang, "Life Games and Statistical Methods," *Proceedings of the National Academy of Science,* Vol. 72, March 1975, pp. 956–68.

S. Evans, "APL Makes Life Easy (and Vice Versa)," *Byte,* Vol. 5, October 1980, pp. 192–93.

F. Hapgood, "Let There Be Life," *Omni,* April 1987, p. 43ff.

C. Helmers, "Life Line," *Byte,* Vol. 1, September 1975, pp. 72–80; Part 2, October 1975, pp. 34–42; Part 3, December 1975, pp. 48–55; Part 4, January 1976, pp. 32–41.

P. Macaluso, "Life After Death," *Byte,* Vol. 6, July 1981, pp. 326–33.

J. K. Millen, "One-Dimensional Life," *Byte,* Vol. 3, December 1978, pp. 68–74.

J. Millium, J. Reardon, and P. Smart, "Life with Your Computer," *Byte,* Vol. 3, December 1978, pp. 45–50.

M. Miyamoto, "An Equilibrium State for a One-Dimensional Life Game," *Journal of Mathematics of Kyoto University,* Vol. 19, 1979, pp. 525–40.

M. D. Niemiec, "Life Algorithms," *Byte,* Vol. 4, January 1979, pp. 90–97.

W. Pardner, "Life is Simple with APL," *PC Tech Journal,* Vol. 2, September 1984, pp. 129–47.

H. A. Peelle, "The Game of Life," *Recreational Computing,* Vol. 7, May/June 1979, pp. 16–27.

D. T. Piele, "Population Explosion: An Activity Lesson," *Mathematics Teacher,* October 1974, pp. 496–502.

W. Poundstone, *The Recursive Universe,* Morrow, 1984.

L. E. Schulman and P. E. Seiden, "Statistical Mechanics of a Dynamical System Based on Conway's Game of Life," *IBM Research Report RC 6802,* October 1977.

R. Soderstrom, "Life Can Be Easy," *Byte,* Vol. 4, April 1979, pp. 166–69.

R. Wainwright (ed.), *Lifeline,* a newsletter on Life. Issues 1 through 11, March 1971 through September 1973.

M. Zeleny, "Self-Organization of Living Systems," *International Journal of General Systems,* Vol. 4, 1977, pp. 13–28.

Chapter 32   *Paper Folding*

*The easiest way to refold a road map is dif-
ferently.*   —Jones's Rule of the Road

One of the most unusual and frustrating unsolved problems in
modern combinatorial theory, proposed many years ago by Stanislaw
M. Ulam, is the problem of determining the number of different ways
to fold a rectangular "map." The map is precreased along vertical and
horizontal lines to form a matrix of identical rectangles. The folds are
confined to the creases, and the final result must be a packet with any
rectangle on top and all the others under it. Since there are various
ways to define what is meant by a "different" fold, we make the definition precise by assuming that the cells of the unfolded map are numbered consecutively, left to right and top to bottom. We wish to know
how many permutations of these n cells, reading from the top of the
packet down, can be achieved by folding. Cells are numbered the same
on both sides, so that it does not matter which side of a cell is "up" in
the final packet. Either end of the packet can be its "top," and as a result every fold will produce two permutations, one the reverse of the
other. The shape of each rectangle is irrelevant because no fold can rotate a cell 90 degrees. We can therefore assume without altering the
problem that all the cells are identical squares.

The simplest case is the $1 \times n$ rectangle, or a single strip of n squares.
It is often referred to as the problem of folding a strip of stamps along
their perforated edges until all the stamps are under one stamp. Even
this special case is still unsolved in the sense that no nonrecursive formula has been found for the number of possible permutations of n
stamps. Recursive procedures (procedures that allow calculating the
number of folds for n stamps provided that the number for $n - 1$ stamps
is known) are nonetheless known. The total number of permutations of
n objects is $n!$ (that is, factorial n, or $n \times (n - 1) \times (n - 2) \cdots \times 1$). All $n!$
permutations can be folded with a strip of two or three stamps, but for

four stamps only 16 of the 4! = 24 permutations are obtainable (see Figure 32.1). For five stamps the number of folds jumps to 50 and for six stamps it is 144. John E. Koehler wrote a computer program, reported in a 1968 paper, with which he went as high as $n = 16$, for which 16,861,984 folds are possible. W. F. Lunnon, in another 1968 paper, carried his results to $n = 24$ and, in a later paper, to $n = 28$. Koehler showed in his article that the number of possible stamp folds is the same as the number of ways of joining n dots on a circle by chords of two alternating colors in such a way that no chords of the same color intersect.

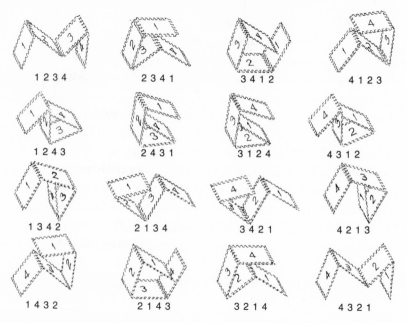

Figure 32.1. The 16 ways to fold a four-stamp strip

The simplest rectangle that is not a strip is the trivial 2×2 square. It is easy to find that only eight of the 4! = 24 permutations can be folded, half of which (as explained above) are reversals of the other half. The 2×3 rectangle is no longer trivial, because now it becomes possible to tuck one or more cells into open pockets. This greatly confuses matters. As far as I know, nothing has been published on the nonstrip rectangles. I was able to fold 60 of the 6! = 720 permutations (10 folds for each cell on top), but it is possible I missed a few.

An amusing pastime is to find six-letter words that can be put on the

2×3 map (lettering from left to right and from the top down) so that the map can be folded into a packet that spells, from the top down, an anagram of the original word. Each cell should be labeled the same on both sides to make it easier to identify in the packet. For example, it is not hard to fold ILL-FED to spell FILLED and SQUIRE to spell RISQUE. On the other hand, OSBERG (an anagram for the last name of the Argentine writer Jorge Luis Borges that appears on page 361 of Vladimir Nabokov's novel *Ada*) cannot be folded to BORGES, nor can BORGES be folded to OSBERG. Can the reader give a simple proof of both impossibilities?

The 2×4 rectangle is the basis of two map-fold puzzles by Henry Ernest Dudeney (see page 130 of his *536 Puzzles & Curious Problems,* Scribner's, 1967). Dudeney asserts there are 40 ways to fold this rectangle into a packet with cell No. 1 on top, and although he speaks tantalizingly of a "little law" he discovered for identifying certain possible folds, he offers no hint as to its nature. I have no notion how many of the $8! = 40{,}320$ permutations can be folded.

When one considers the 3×3, the smallest nontrivial square, the problem becomes fantastically complex. As far as I know, the number of possible folds (of the $9! = 362{,}880$ permutations) has not been calculated, although many paperfold puzzles have exploited this square. One was an advertising premium, printed in 1942 by a company in Mt. Vernon, NY, that is shown in Figure 32.2.

On one side of the sheet are pictures of Hitler and Mussolini. In the

Figure 32.2. A map-fold problem from World War II

top left corner is a window with open spaces die-cut between the bars. The illustration incorrectly shows a picture of Tojo and a second window below him. Those two cells should have been blank. Tojo and the second window are on the *underside* of the two cells. The task is to imprison two of the three dictators by folding the square along its grid lines into a 1 × 1 packet so that on each side of the packet is a window through which you see the face of a dictator. The fold is not difficult, though it does require a final tuck.

A much tougher puzzle using a square of the same size is the creation of Robert Edward Neale, a Protestant minister, retired professor of psychiatry and religion from the Union Theological Seminary, and the author of the influential book *In Praise of Play* (Harper & Row, 1969). Neale is a man of many avocations. One of them is origami, the Oriental art of paper folding, a field in which he is recognized as one of the country's most creative experts. Magic is another of Neale's side interests; his famous trick of the bunny in the top hat, done with a folded dollar bill, is a favorite among magicians. The hat is held upside down. When its sides are squeezed, a rabbit's head pops up.

Figure 32.3 shows Neale's hitherto unpublished Beelzebub puzzle. Start by cutting a square from a sheet of paper or thin cardboard, crease it to make nine cells, then letter the cells (the same letter on opposite sides of each cell) as indicated. First try to fold the square into a packet that spells (from the top down) these eight pseudonyms of the fallen angel who, in Milton's *Paradise Lost,* is second in rank to Satan himself: Bel Zeebub, Bub Blezee, Ube Blezbe, Bub Zelbee, Bub Beelze, Zee Bubble, Buz Lebeeb, Zel Beebub. If you can master these names, you are ready to tackle the really fiendish one: Beelzebub, the true name of "the prince of the devils" (Matthew 12:24). Its extremely difficult fold will be explained in the Answer Section. No one who succeeds in fold-

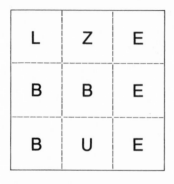

Figure 32.3. Robert Edward Neale's Beelzebub puzzle

ing all nine names will wonder why the general map-folding problem is still unsolved.

Neale has invented a variety of remarkable paper-fold puzzles, but there is space for only two more. One is in effect a nonrectangular "map" with a crosscut at the center (see Figure 32.4). The numbers may represent six colors: all the 1-cells are one color, the 2-cells are a second color, and so on. Here opposite sides of each cell are different. After numbering or coloring as shown at the top in the illustration, turn the sheet over (turn it *sideways,* exchanging left and right sides) and then number or color the back as shown at the bottom. The sheet must now be folded to form a curious species of tetraflexagon.

To fold the tetraflexagon, position the sheet as shown at the top in the

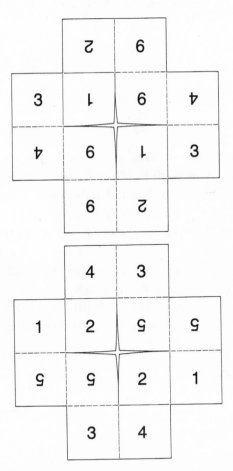

Figure 32.4. Front *(top)* and back *(bottom)* of the unfolded tetraflexagon

illustration. (It helps if you first press the creases so that the solid lines are what origamians call "mountain folds" and the dotted lines are "valley folds.") Reach underneath and seize from below the two free corners of the 1-cells, holding the corner of the upper cell between the tip of your left thumb and index finger and the corner of the lower cell between the tip of your right thumb and finger. A beautiful maneuver can now be executed, one that is easy to do when you get the knack even though it is difficult to describe. Pull the corners simultaneously down and away from each other, turning each 1-cell over so that it becomes a 5-cell as you look down at the sheet. The remaining cells will come together to form two opentop boxes with a 6-cell at the bottom of each box (see Figure 32.5).

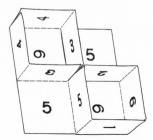

Figure 32.5. First step in folding the tetraflexagon

Shift your grip to the two inside corners of the 5-cells—corners diagonally opposite the corners you were holding. Push down on these corners, at the same time pulling them apart. The boxes will collapse so that the sheet becomes a flat 2 × 2 tetraflexagon with four 1-cells on top and four 2-cells on the underside (see Figure 32.6). If the collapsing is not properly done, you will find a 4-cell in place of a 1-cell, and/or a 3-cell in place of a 2-cell. In either case simply tuck the wrong square out of sight, replacing it with the correct one.

The tetraflexagon is flexed by folding it in half (the two sides going back), then opening it at the center crease to discover a new "face," all

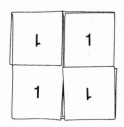

Figure 32.6. Final step in folding the tetraflexagon

COMBINATORICS

of whose cells have the same number (or color). It is easy to flex and find faces 1, 2, 3, and 4. It is not so easy to find faces 5 and 6.

One of Neale's most elegant puzzles is his "Sheep and Goats," which begins with a strip of four squares and a tab for later gluing (see Figure 32.7). Precrease the sheet (folding it both ways) along all dotted lines. Then color half of each square *(dark grey in illustration)* black—on both sides, as if the ink had soaked right through the paper.

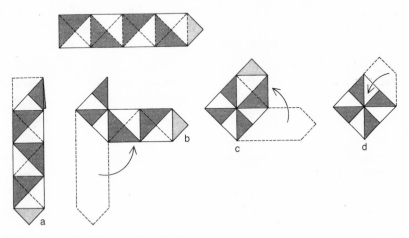

Figure 32.7. Neale's "Sheep and Goats" problem

The strip is folded as shown in steps *(a)*, *(b)*, *(c)*, and *(d)*. The first fold is back and down. The next three folds are valley folds, first to the right, then up, then left. After the last fold slide the tab *under* the double-leaved black triangle at the top left of the square. Glue the tab to the bottom leaf of the triangle. You should now have a square with four black and four white triangles on each side. These are the sheep and goats.

The problem is: By folding only along precreased lines, change the paper to a square of the same size that is all white on one side and all black on the other. In other words, separate the sheep from the goats. It is not easy, but it is a delight to make the moves rapidly once you master the steps—which I shall diagram in the Answer Section, along with the answer to the tetraflexagon puzzle. (To make the manipulations smoother, it is a good plan to trim a tiny sliver from all single edges after the square has been folded and glued.)

Anyone interested in learning some of Neale's more traditional origami figures will find six of his best (including his Thurber dog) in

Samuel Randlett's *The Best of Origami* (E. P. Dutton, 1963). Some of his dollar-bill folds (including the jumping frog) are in *Folding Money: Volume II,* edited by Randlett (Magic, Inc., 1968).

Addendum

The "little law" that Henry Ernest Dudeney hinted about in connection with his map-fold problem has probably been rediscovered. Mark B. Wells of the Los Alamos Scientific Laboratory used a computer to confirm that the 2×3 map has 10 folds for each cell on top. The program also found that the order-3 square has 152 folds for each cell on top. In his 1971 paper Lunnon proved that for any rectangular map every cyclic permutation of every possible fold is also a possible fold. Thus it is necessary to determine only the folds for one cell on top because the cyclic permutations of these folds give all the other folds. For example, since 123654789 is a possible fold, so also are 236547891, 365478912, and so on. It is a strange law because the folds for cyclic permutations differ wildly. It is not yet known whether the law applies to all polyomino-shaped maps or to maps with equilateral triangles as cells.

In his 1971 paper Lunnon used an ingenious diagram based on two perpendicular slices through the center of the final packet. He was able to write a simple backtrack program for $x \times y$ maps, extend the problem to higher dimensions, and discover several remarkable theorems. For example, the edges of one cross section always diagram x linear maps of y cells each, and the edges of the other cross section diagram y linear maps of x cells each.

The 2×3, 2×4, 2×5, 2×6, and 3×4 maps have respectively 60, 320, 1,980, 10,512, and 15,552 folds. The order-3 square has 1,368 folds, the order-4 has 300,608, the order-5 has 186,086,600. In all cases the number of folds is the same for each cell on top, as required by cyclic law. The order-2 cube, folded through the fourth dimension, has 96 folds. The order-3 cube has 85,109,616. Many other results are tabulated by Lunnon in his 1971 paper, but a nonrecursive formula for even planar maps remains elusive.

The linear map-fold function, as Lunnon calls it, is the limit approached by the ratio between adjacent values of the number of possible folds for a $1 \times n$ strip. It is very close to 3.5. In his unpublished 1981 paper Lunnon narrows the upper and lower bounds to 3.3868 and 3.9821.

In 1981 Harmony Books in the United States, and Pan Books in England, brought out a large paperback book called *Folding Frenzy*. It contains six 3 × 3 squares, with red and green patterns on both sides, and five pages partially die-cut. Without removing any pages, there are nine puzzles to solve by folding the squares. The puzzles are credited to Jeremy Cox.

In describing one of his map-fold puzzles (*Modern Puzzles,* No. 214) Dudeney mentioned a curious property of map folds that is not at all obvious until you think about it carefully. It applies not only to rectangular maps, but also to maps in the shape of any polyomino; that is, a shape formed by joining unit squares at their edges. Assume that any such map is red on one side, white on the other. No matter how it is folded into a 1 × 1 packet, the colors on the tops of each cell will alternate regardless of which side of the packet is up. If the cells of the map are colored like a checkerboard, with each cell the same color on both sides, the final packet (after any sort of folding) will have leaves that alternate colors. If the checkerboard coloring is such that each cell is red on one side, white on the other, all cells in the folded packet will have their red sides facing one way, their white sides facing the other way.

It occurred to me in 1971 that the parity principles involved here could be the basis for a variety of magic tricks. One appeared under the title "Paradox Papers" in Karl Fulves' magic periodical, *The Pallbearers Review.* It goes like this: Fold a sheet of paper twice in each direction so that the creases make 16 cells. It is a good plan to fold the paper each way along every crease to make refolding easier later on.

Assume in your mind that the cells are checkerboard colored black and red, with red at the top left corner. Five red playing cards are taken from a deck and someone selects one of them. With a red pencil jot the names of the five cards in five cells, using abbreviations such as 4D and QH. Tell your audience that you are taking cells at random, but actually you must put the name of the chosen card on one of the "black" cells, and the other four names on "red" cells.

Have another card chosen, this time from a set of five black cards. Turn the sheet over, side for side, and jot the names of the five black cards on cells, again apparently at random. Use a black pencil. Put the chosen card on a "red" cell, the others on "black" cells.

Ask someone to fold the sheet any way he likes to make a 1 × 1 packet. With a pair of shears, trim around the four sides of the packet. Deal the 16 pieces on the table. Five names will be seen, all the same color except for one—the chosen card of the other color. Turn over the 16 pieces. The same will be true of the other sides.

Gene Nielsen, in the May 1972 issue of the same journal, suggested the following variant. Pencil X's and O's on all the cells, alternating them checkerboard fashion. Turn over the sheet horizontally, and put exactly the same pattern on the other side. Spectators will not realize that each cell has an X on one side, an O on the other. Someone folds the sheet randomly into a packet. Pretend you are using PK to influence the folding so that it will produce a startling result. Trim the sides of the packet and spread the pieces on the table. All X's face one way, all O's face the other way.

Swami, a magic periodical published in Calcutta by Sam Dalal, printed my "Paper Fold Prediction" in its July 1973 issue. Start by numbering the cells of a 3 × 3 sheet from 1 through 9, taking the cells in the usual way from left to right and top down. Put the digits on one side of the paper only. After someone folds the sheet randomly, trim the sides of the packet and spread the pieces. Add all the numbers showing. Reverse the pieces and add the digits on the other sides. The two sums will be different. Explain that by randomly folding the sheet, the nine digits are randomly split into two sets. Clearly there is no way to know in advance what either sum will be when the pieces are spread.

Repeat the same procedure, but this time use a 4 × 4 square with cells numbered 1 through 16. The sheet is randomly folded and the edges are trimmed. Before spreading the pieces, hold them to your forehead and announce that the sum will be 68. Put down the packet, either side uppermost, and spread the pieces. The numbers showing will total 68. Discard the pieces before anyone discovers that the sum on the reverse sides also is 68.

The trick works because if the original square has an odd number of cells, the sums on the two sides will not be equal. (On the 3 × 3 they will be 20 and 25.) However, if the square has an even number of cells, the sum is a constant equal to $(n^2 + n)/4$ where n is the highest number. You can now repeat the trick with a 5 × 5 square, but instead of predicting a sum, predict that the *difference* between the sums on the two sides of the pieces will be 13.

The principle applies to cells numbered with other sequences. For example, hand a wall calendar to someone and ask him to tear out the page for the month of his birth. He then cuts from the page any 4 × 4 square of numbers. The sheet is folded, the packet is trimmed, the pieces are spread, and the visible numbers are added. The sum will be equal to four times the sum of the sheet's lowest and highest numbers. You can predict this as soon as you see the square that has been cut, or you can divine the number later by ESP.

Some other suggestions. Allow a spectator to write any digit he likes in each cell of a sheet of any size, writing left to right and top down. As he writes, keep a running total in your head by subtracting the second number from the first, adding the third, subtracting the fourth, and so on. The running total is likely to fluctuate between plus and minus. The number you end with, whether plus or minus, will be the difference between the two sums after the sheet is folded, trimmed, and the pieces are spread.

Magic squares lend themselves to prediction tricks of a similar nature. For example, suppose a 4 × 4 map bears the numbers of a magic square. After folding, trim only on two opposite sides of the packet. This will produce four strips. Have someone select one of the four. The other three are destroyed. You can predict the sum of the numbers on the selected strip because it will be the magic square's constant. Of course you do not tell the audience that the numbers form a magic square.

Many of these tricks adapt easily to nonsquare sheets, such as a 3 × 4. See my *Mental Magic* (Sterling, 1999), page 54, for a trick using a 3 × 4 sheet bearing the letters A through G.

In 1984 C. Malorana, of Washington, DC, printed what he called "The Delayed Justice Puzzle." It is the same as the 3 × 3 fold puzzle I described except that the faces of the three dictators are replaced by three faces of Nixon. "Ol' Tricky Dick escaped justice once," the instructions read, "but this time I think we'll get him! Try and fold the paper (only along the dashed lines of course) so that Ol' Tricky is behind the bars on each side of the packet."

Answers

A simple proof that on the 2 × 3 rectangle OSBERG cannot be folded to spell BORGES (or vice versa) is to note that in each case the fold requires that two pairs of cells touching only at their corners would have to be brought together in the final packet. It is evident that no fold can put a pair of such cells together.

The square puzzle with the faces and prison windows is solved from the starting position shown. Fold the top row back and down, the left column toward you and right, the bottom row back and up. Fold the right packet of three cells back and tuck it into the pocket. A face is now behind bars on each side of the final packet. The central face of the square cannot be put behind bars because its cell is diagonally adjacent to each of the window cells.

Space prevents my giving solutions for the eight pseudonyms of Beelzebub, but Beelzebub itself can be obtained as follows. Starting with the layout shown, fold the bottom row toward you and up to cover *BBE*. Fold the left column toward you and right to cover *ZU*. Fold the top row toward you and down, but reverse the crease between *L* and *Z* so that *LZ* goes between *B* and *B* on the left and the upper *E* goes on top of the lower *E*. You now have a rectangle of two squares. On the left, from the top down, the cells are *BLZBUB,* on the right *EEE*. The final move is difficult. Fold the right panel (*EEE*) toward you and left. The three *E*'s are tucked so that the middle *E* goes between *Z* and *B,* and the other two *E*'s together go between *B* and *L.* Once you grasp what is required it is easier to combine this awkward move with the previous one. The result is a tightly locked packet that spells Beelzebub. The solution is unique. If the cells of the original "map" are numbered 1 through 9, the final packet is 463129785.

To find the 5-face of the tetraflexagon, start with face 1 on the top and 2 on the bottom. Mountain-fold in half vertically, left and right panels going back, so that if you were to open the flexagon at the center crease you would see the 4-face. Instead of opening it, however, move left the lower inside square packet (with 4 and 3 on its outsides) and move right the upper square packet (also with 4 and 3 on its outsides). Insert your fingers and open the flexagon into a cubical tube open at the top and bottom. Collapse the tube the other way. This creates a new tetraflexagon structure that can be flexed to show faces 1, 3, and 5.

A similar maneuver creates a structure that shows faces 2, 4, and 6. Go back to the original structure that shows faces 1, 2, 3, and 4 and repeat the same moves as before except that you begin with the 2-face uppermost and the 1-face on the underside.

Figure 32.8 shows how to separate the sheep from the goats:

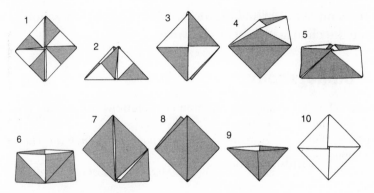

Figure 32.8. Solution to the "Sheep and Goats" problem

1. Start with the two-color square folded as shown.
2. Fold in half along the horizontal diagonal by folding the bottom corner up to make a "hat" with a white triangle at the lower left-hand corner.
3. Open the hat's base and continue opening until you can flatten the hat to make the small square shown.
4. Insert a left finger into the pocket on the right of the upper face of drawing *(3)* Pull upward and flatten as shown.
5. Turn the paper over sideways and repeat the previous move on the other side. The result is a rectangle with a white triangle in the upper right-hand corner.
6. Open the rectangle into a cubical tube open at the top and bottom. Collapse the tube the other way to make a rectangle again, except that now it is colored as shown.
7. Insert your right thumb into the pocket on the left of drawing *(6)* lift up the flap and flatten it as shown.
8. Turn the paper over sideways and do the same on the other side. You should now have a small square, black on both sides.
9. Reach into the square from above, open it, and flatten to make an inverted hat, black on both sides.
10. Open the hat by separating its bottom points and flatten the large square that results. It will be the same size as the square you started with, but now it is all white on one side and all black on the other—all sheep and all goats.

Repeating the same sequence of moves will mix the sheep and the goats again. With practice the folds can be done so rapidly that you can hold the square out of sight under a table for just a few moments and produce the change almost as if by magic.

Bibliography

S. Baker, "Postage Stamps," "Computer Solutions to Three Topological-Combinatorial Problems," 1976, unpublished.

J. E. Koehler, "Folding a Strip of Stamps," *Journal of Combinatorial Theory,* Vol. 5, September 1968, pp. 135–52.

W. F. Lunnon, "A Map-Folding Problem," *Mathematics of Computation,* Vol. 22, January 1968, pp. 193–99.

W. F. Lunnon, "Multi-Dimensional Stamp Folding," *The Computer Journal,* Vol. 14, February 1971, pp. 75–79.

W. F. Lunnon, "The Map-Folding Problem Revisited," unpublished.

W. F. Lunnon, "Bounds for the Map-Folding Function," Cardiff, August, 1981, unpublished.

R. Neale, "The Devil's Fold," *Games,* November/December 1979, p. 33.

A. Sainte-Lague, *Avec des nombres et des lignes* (With Numbers and Lines), Paris: Libraire Vuibert, 1937; third edition 1946. The linear $(1 \times n)$ postage stamp problem is discussed in Part 2, Chapter 2, pages 147–62. I am indebted to Victor Meally for telling me of this earliest known analysis of the problem.

J. Touchard, "Contributions à l'étude du problème des timbres poste," *Canadian Journal of Mathematics,* Vol. 2, 1950, pp. 385–98.

Chapter 33

Ramsey Theory

Prove that at a gathering of any six people, some three of them are either mutual acquaintances or complete strangers to each other. —PROBLEM E 1321, *The American Mathematical Monthly,* June–July, 1958

*T*his chapter was originally written in 1977 to honor the appearance of *The Journal of Graph Theory,* a periodical devoted to one of the fastest growing branches of modern mathematics. Frank Harary, the founding editor, is the author of the world's most widely used introduction to the subject. The current managing editor is Fan Chung.

Graph theory studies sets of points joined by lines. Two articles in the first issue of the new journal dealt with Ramsey graph theory, a topic that has a large overlap with recreational mathematics. Although a few papers on Ramsey theory, by the Hungarian mathematician Paul Erdös and others, appeared in the 1930s, it was not until the late 1950s that work began in earnest on the search for what are now called Ramsey numbers. One of the great stimulants to this search was the innocent-seeming puzzle quoted above. It was making the mathematical-folklore rounds as a graph problem at least as early as 1950, and at Harary's suggestion it was included in the William Lowell Putnam Mathematical Competition of 1953.

It is easy to transform this puzzle into a graph problem. Six points represent the six people. Join every pair of points with a line, using a red pencil, say, to indicate two people who know each other and a blue pencil for two strangers. The problem now is to prove that no matter how the lines are colored, you cannot avoid producing either a red triangle (joining three mutual acquaintances) or a blue triangle (joining three strangers).

Ramsey theory, which deals with such problems, is named for an extraordinary University of Cambridge mathematician, Frank Plumpton Ramsey. Ramsey was only 26 when he died in 1930, a few days after an abdominal operation for jaundice. His father, A. S. Ramsey, was president of Magdalene College, Cambridge, and his younger brother Michael was archbishop of Canterbury from 1961 to 1974. Economists know him for his remarkable contributions to economic theory. Logicians know him for his simplification of Bertrand Russell's ramified theory of types (it is said that Ramsey Ramseyfied the ramified theory) and for his division of logic paradoxes into logical and semantical classes. Philosophers of science know him for his subjective interpretation of probability in terms of beliefs and for his invention of the "Ramsey sentence," a symbolic device that greatly clarifies the nature of the "theoretical language" of science.

In 1928 Ramsey read to the London Mathematical Society a now classic paper, "On a Problem of Formal Logic." (It is reprinted in *The Foundations of Mathematics,* a posthumous collection of Ramsey's essays edited by his friend R. B. Braithwaite.) In this paper Ramsey proved a deep result about sets that is now known as Ramsey's theorem. He proved it first for infinite sets, observing this to be easier than his next proof, for finite sets. Like so many theorems about sets, it turned out to have a large variety of unexpected applications to combinatorial problems. The theorem in its full generality is too complicated to explain here, but for our purposes it will be sufficient to see how it applies to graph-coloring theory.

When all pairs of n points are joined by lines, the graph is called a complete graph on n points and is symbolized by K_n. Since we are concerned only with topological properties, it does not matter how the points are placed or the lines are drawn. Figure 33.1 shows the usual ways of depicting complete graphs on two through six points. The lines identify every subset of n that has exactly two members.

Suppose we arbitrarily color the lines of a K_n graph red or blue. We might color the lines all red or all blue or any mixture in between. This is called a two-coloring of the graph. The coloring is of course a simple way to divide all the two-member subsets of n into two mutually exclusive classes. Similarly, a three-coloring of the lines divides them into three classes. In general, an r-coloring divides the pairs of points into r mutually exclusive classes.

A "subgraph" of a complete graph is any kind of graph contained in the complete graph in the sense that all the points and lines of the subgraph are in the larger graph. It is easy to see that any complete graph is a subgraph of any complete graph on more points. Many simple graphs have names. Figure 33.2 shows four families: paths, cycles, stars, and wheels. Note that the wheel on four points is another way of

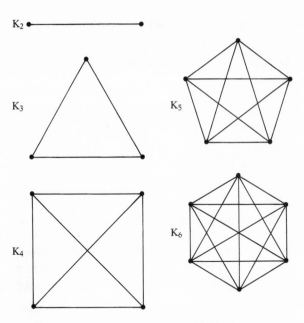

Figure 33.1. Complete graphs on two through six points

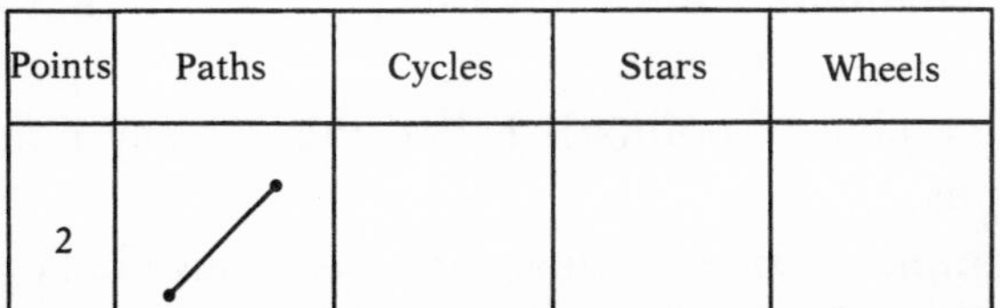

Points	Paths	Cycles	Stars	Wheels
2				
3				
4				
5				
6				

Figure 33.2. Four important families of simple graphs

Ramsey Theory

drawing K_4. It is often called a tetrahedron because it is a planar projection of the tetrahedron's skeleton.

Consider now the following problem involving six pencils of different colors. To each color we assign any kind of graph we like. For example:

1. Red: a pentagon (five-point cycle).
2. Orange: a tetrahedron.
3. Yellow: a seven-point star.
4. Green: a 13-point path.
5. Blue: an eight-point wheel.
6. Purple: a bow tie (two triangles sharing just one point).

We now ask a curious question. Are there complete graphs that, if their lines are arbitrarily six-colored, are certain to contain as a subgraph at least one of the six graphs listed above? In other words, no matter how we color one of these complete graphs with the six pencils are we certain to get either a red pentagon or an orange tetrahedron or a yellow seven-point star, and so on? Ramsey's theorem proves that beyond a certain size all complete graphs have this property. Let's call the smallest graph of this infinite set the Ramsey graph for the specified set of subgraphs. Its number of points is called the Ramsey number for that set of subgraphs.

Every Ramsey graph provides both a game and a puzzle. For our example, the game is as follows. Two players take turns picking up any one of the six pencils and coloring a line of the Ramsey graph. The first person to complete the coloring of one of the specified subgraphs is the loser. Since it is a Ramsey graph, the game cannot be a draw. Moreover, it is the smallest complete graph on which a draw is not possible.

The related puzzle involves a complete graph with one fewer point than the Ramsey graph. This obviously is the largest complete graph on which the game can be a draw. Such a graph is called the critical coloring for the specified set of subgraphs. The puzzle consists in finding a coloring for the critical graph in which none of the subgraphs appears.

I have no idea what the Ramsey number is for the six subgraphs given. Its complete graph would be so large (containing hundreds of points) that playing a game on it would be out of the question, and the associated puzzle is far too difficult to be within the range of a feasible computer search. Nevertheless, Ramsey games and puzzles with smaller complete graphs and with pencils of just two colors can be quite entertaining.

The best-known Ramsey game is called Sim after mathematician Gustavus Simmons who was the first to propose it. Sim is played on the complete graph with six points (K_6), which models the problem about the party of six people. It is not hard to prove that 6 is the Ramsey number for the following two subgraphs:

1. Red: triangle (K_3).
2. Blue: triangle (K_3).

In "classical" Ramsey theory it is customary to use solitary numbers for complete graphs, and so we can express the above result with this compact notation: $R(3,3) = 6$. The R stands for Ramsey number, the first 3 for a triangle of one color (say, red) and the second 3 for a triangle of another color (say, blue). In other words, the smallest complete graph that forces a "monochromatic" (all red or all blue) triangle when the graph is two-colored is 6. Thus if two players alternately color the K_6 red and blue, one player is certain to lose by completing a triangle of his color. The corresponding and easy puzzle is to two-color the critical graph, K_5, so that no monochromatic triangle appears.

It turns out that when K_6 is two-colored, at least two monochromatic triangles are forced. (If there are exactly two and they are of opposite color, they form a bow tie.) This raises an interesting question. If a complete graph on n points is two-colored, how many monochromatic triangles are forced? A. W. Goodman was the first to answer this in a 1959 paper, "On Sets of Acquaintances and Strangers at Any Party." Goodman's formula is best broken into three parts: If n has the form $2u$, the number of forced monochromatic triangles is $\frac{1}{3}u(u-1)(u-2)$. If n is $4u + 1$, the number is $\frac{2}{3}2u(u-1)(4u+1)$. If n is $4u+3$, it is $\frac{1}{3}2u(u+1)(4u-1)$. Thus for complete graphs of 6 through 12 points the numbers of forced one-color triangles are 2, 4, 8, 12, 20, 28, and 40.

Random two-coloring will usually produce more monochromatic triangles than the number forced. When the coloring of a Ramsey graph contains exactly the forced number of triangles and no more, it is called extremal. Is there always an extremal coloring in which the forced triangles are all the same color? (Such colorings have been called blue-empty, meaning that the number of blue triangles is reduced to zero.) In 1961 Léopold Sauvé showed that the answer is no for all odd n, except for $n = 7$. This suggests a new class of puzzles. For example, draw the complete graph on seven points. Can you two-color it so that there are no blue triangles and no more than four red triangles? It is not easy.

Very little is known about "classical" Ramsey numbers. They are the number of points in the smallest complete graph that forces a given set of smaller complete graphs. There is no known practical procedure for finding classical Ramsey numbers. An algorithm is known: One simply explores all possible colorings of complete graphs, going up the ladder until the Ramsey graph is found. This task grows so exponentially in difficulty and at such a rapid rate, however, that it quickly becomes computationally infeasible. Even less is known about who wins—the first player or the second—if a Ramsey game is played rationally. Sim has been solved (it is a second-player win), but almost nothing is known about Ramsey games involving larger complete graphs.

So far we have considered only the kind of Ramsey game that Harary calls an avoidance game. As he has pointed out, at least three other kinds of game are possible. For example, in an "achievement" game (along the lines of Sim) the first player to complete a monochromatic triangle wins. In the other two games the play continues until all the lines are colored, and then either the player who has the most triangles of his color or the player who has the fewest wins. These last two games are the most difficult to analyze, and the achievement game is the easiest. In what follows "Ramsey game" denotes the avoidance game.

Apart from $R(3,3) = 6$, the basis of Sim, only the following seven other nontrivial classical Ramsey numbers are known for two-colorings:

1. $R(3,4) = 9$. If K_9 is two-colored, it forces a red triangle (K_3) or a blue tetrahedron (K_4). No one knows who wins if this is played as a Ramsey game, or who wins on any $R(n,n)$ game with n larger than 3.
2. $R(3,5) = 14$.
3. $R(4,4) = 18$. If K_{18} is two-colored, a monochromatic tetrahedron (K_4) is forced. This is not a bad Ramsey game, although the difficulty of identifying tetrahedrons makes it hard to play. The graph and its coloring correspond to the fact that at a party of 18 people there is either a set of four acquaintances or four total strangers.
4. $R(3,6) = 18$. At the same party there is either a set of three acquaintances or six total strangers. In coloring terms, if a complete graph of 18 points is two-colored it forces either a red triangle or a blue complete graph of six points.

5. $R(3,7) = 23$.
6. $R(3,8) = 28$.
7. $R(3,9) = 36$.
8. $R(4,5) = 25$.
9. $R(6,7) = 298$.

As of March 1996, here are known bounds for eight other Ramsey numbers:

$R(3,10) = 40–43$.
$R(4,6) = 35–41$.
$R(4,7) = 49–61$.
$R(5,5) = 43–49$.
$R(5,6) = 58–87$.
$R(5,7) = 80–143$.
$R(6,6) = 102–165$.
$R(7,7) = 205–540$.

What is the smallest number of people that must include either a set of five acquaintances or a set of five strangers? This is equivalent to asking for the smallest complete graph that cannot be two-colored without producing a monochromatic complete graph with five points, which is the same as asking for the Ramsey number of $R(5,5)$. So great is the jump in complexity from $R(4,4)$ to $R(5,5)$ that Stefan Burr, a leading expert on Ramsey theory who now teaches computer science at City College of the CUNY in New York City, thinks it possible that the number will never be known. Even $R(4,5)$, he believes, is so difficult to analyze that it is possible its value also may never be found.

Only one other classical Ramsey number is known, and it is for three colors. $R(3) = 3$ is trivial because if you one-color a triangle, you are sure to get a one-color triangle. We have seen that $R(3,3)$ equals 6. $R(3,3,3)$ equals 17. This means that if K_{17} is three-colored, it forces a monochromatic triangle. Actually it forces more than one, but the exact number is not known.

$R(3,3,3) = 17$ was first proved in 1955. The Ramsey game for this graph uses pencils of three different colors. Players alternately color a line, using any color they want to, until a player loses by completing a monochromatic triangle. Who wins if both players make their best possible moves? No one knows. The corresponding Ramsey puzzle is to three-color K_{16}, the critical graph, so that no monochromatic triangle appears.

What about $R(3,3,3,3)$, the minimum complete graph that forces a one-color triangle when it is four-colored? It is unknown, although an upper bound of 64 was proved by Jon Folkman, a brilliant combinatorialist who committed suicide in 1964 at the age of 31, following an operation for a massive brain tumor. The best lower bound, 51, was established by Fan Chung, who gave the proof in her Ph.D. thesis.

Classical Ramsey theory generalizes in many fascinating ways. We have already considered the most obvious way: the seeking of what are called generalized Ramsey numbers for r-colorings of complete graphs that force graphs other than complete ones. Václav Chvátal and Harary were the pioneers in this territory, and Burr has been mining it for many years. Consider the problem of finding Ramsey numbers for minimum complete graphs that force a monochromatic star of n points. Harary and Chvátal were the first to solve it for two-coloring. In 1973 Burr and J. A. Roberts solved it for any number of colors.

Another generalized Ramsey problem is to find Ramsey numbers for two-colorings of K_n that force a specified number of monochromatic "disjoint" triangles. (Triangles are disjoint if they have no common point.) In 1975 Burr, Erdös, and J. H. Spencer showed the number to be $5d$, where d is the number of disjoint triangles and is greater than two. The problem is unsolved for more than two colors.

The general case of wheels is not even solved for two colors. The Ramsey number for the wheel of four points, the tetrahedron, is, as we have seen, 18. The wheel of five points (a wheel with a hub and four spokes) was shown to have a Ramsey number of 15 by Tim Moon, a Nigerian mathematician. The six-point wheel is unsolved, although its Ramsey number is known to have bounds of 17 to 20 inclusive, and there are rumors of an unpublished proof that the value is 17.

Figure 33.3, a valuable chart supplied by Burr and published for the first time in my 1977 column, lists the 113 graphs with no more than six lines and no isolated points, all of which have known generalized Ramsey numbers. Note that some of these graphs are not connected. In such cases the entire pattern, either all red or all blue, is forced by the complete graph with the Ramsey number indicated.

Every item on Burr's chart is the basis of a Ramsey game and puzzle, although it turns out that the puzzles—finding critical colorings for the critical graphs—are much easier than finding critical colorings for classical Ramsey numbers. Note that the chart gives six variations of Sim. A two-coloring of K_6 not only forces a monochromatic triangle but also forces a square, a four-point star (sometimes called a "claw"), a five-point path, a pair of disjoint paths of two and three points (both the same color), a square with a tail and the simple "tree" that is 15 on Burr's chart. The triangle with a tail (8), the five-point star (12), the Latin cross (27), and the fish (51) might be worth looking into as Ramsey games on K_7.

Ronald L. Graham of Bell Laboratories, one of the nation's top combinatorialists, has made many significant contributions to generalized Ramsey theory. It would be hard to find a creative mathematician who less resembles the motion-picture stereotype. In his youth Graham and two friends were professional trampoline performers who worked for a circus under the name of the Bouncing Baers. He is also one of the country's best jugglers and former president of the International Juggler's Association. The ceiling of his office is covered with a large net that he can lower and attach to his waist, so that when he is practicing with six or seven balls, any missed ball obligingly rolls back to him.

In 1968 Graham found an ingenious solution for a problem of the Ramsey type posed by Erdös and András Hajnal. What is the smallest graph of any kind, not containing K_6, that forces a monochromatic triangle when it is two-colored? Graham's unique solution is the eight-point graph shown in Figure 33.4. The proof is a straightforward reductio ad absurdum. It begins with the assumption that a two-coloring that avoids monochromatic triangles is possible and then shows that this forces such a triangle. At least two lines from the top point must be, say, gray, and the graph's symmetry allows us to make the two outside lines gray with no loss of generality. The end points of these two lines must then be joined by a colored line (shown dotted) to prevent the formation of a gray triangle. Readers may enjoy trying to complete the argument.

What about similar problems when the excluded subgraph is a complete graph other than K_6? The question is meaningless for K_3 because K_3 is itself a triangle. K_5 is unsolved. The best-known solution is a 16-point graph discovered by two Bulgarian mathematicians. K_4 is even further from being solved. Folkman, in a paper published posthumously, proved that such a Ramsey graph exists, but his construction used more than $2 \uparrow \uparrow \uparrow 2^{901}$ points. This is such a monstrous number that there is no way to express it without using a special arrow notation. The notation is introduced by Donald E. Knuth in his article "Mathematics and Computer Science: Coping with Finiteness" in *Sci-*

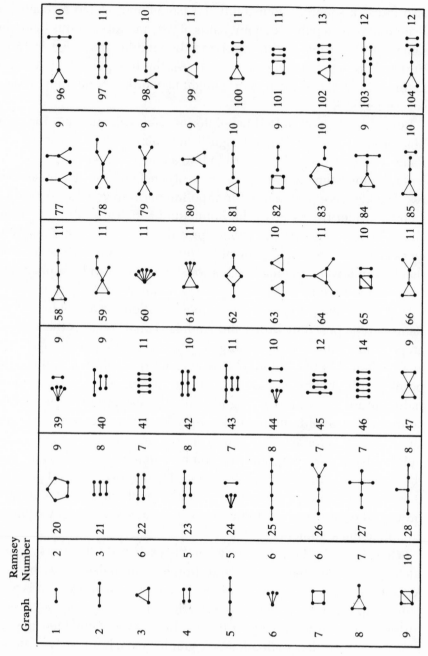

Figure 33.3. Simple graphs for which the generalized Ramsey number is known

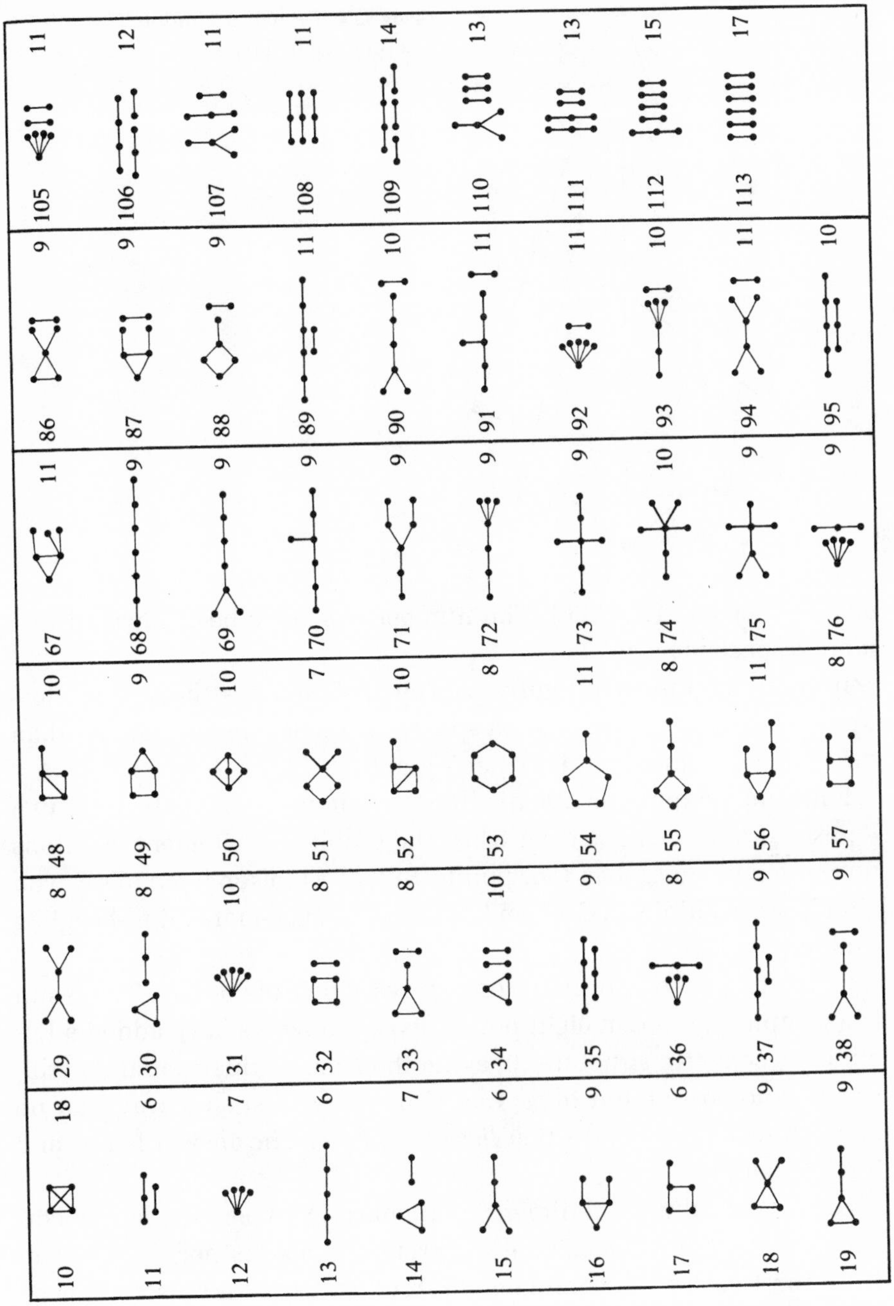
10 18
11 6
12 7
13 6
14 7
15 6
16 9
17 6
18 9
19 9
29 8
30 8
31 10
32 8
33 8
34 10
35 9
36 8
37 9
38 9
48 10
49 9
50 10
51 7
52 10
53 8
54 11
55 8
56 11
57 8
67 11
68 9
69 9
70 9
71 9
72 9
73 9
74 10
75 9
76 9
86 9
87 9
88 9
89 11
90 10
91 11
92 11
93 10
94 11
95 10
105 11
106 12
107 11
108 11
109 14
110 13
111 13
112 15
113 17

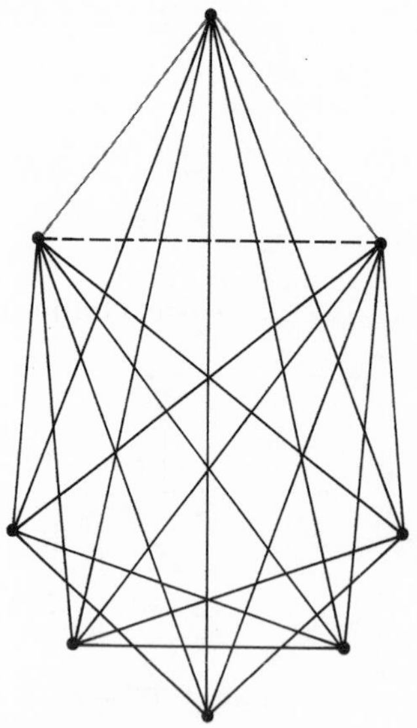

Figure 33.4. Graham's solution to a problem by Erdös

ence (December 17, 1976). The number has since been lowered to a more respectable size.

Imagine the universe tightly packed with spheres the size of electrons. The total number of such spheres is inconceivably smaller than the number occurring in Folkman's graph.

Folkman's graph dramatically illustrates how enormously difficult a Ramsey problem can be even when the problem's statement mentions no graph with more than four points. But, as Al Jolson liked to say, you ain't heard nothin' yet. Graham has found an even more mind-boggling example.

Consider a cube with lines joining every pair of corners. The result is a complete graph on eight points, except now we have added a Euclidean geometric structure. Imagine the lines of this spatial K_8 arbitrarily colored red and blue. Can it be done in such a way that no monochromatic K_4 results that *lies on a plane*? The answer is yes, and it is not hard to do.

Let us generalize to *n*-dimensional cubes. A hypercube has 2^n corners. On the four-dimensional hypercube, it also is possible to two-color the lines of the complete graph of 2^4, or 16, points so that no one-color complete planar graph of four points results. The same can be done with the 2^5 hypercube of 32 points. This suggests the following Euclidean Ramsey problem: What is the smallest dimension of a hypercube such that if the lines joining all pairs of corners are two-colored, a planar K_4 of one color will be forced? Ramsey's theorem guarantees that the question has an answer only if the forced K_4 is not confined to a plane.

The existence of an answer when the forced monochromatic K_4 is planar was first proved by Graham and Bruce L. Rothschild in a far-reaching generalization of Ramsey's theorem that they found in 1970. Finding the actual number, however, is something else. Graham has established an upper bound, but it is a bound so vast that it holds the record for the largest number ever used in a serious mathematical proof.

To convey at least a vague notion of the size of Graham's number we must first attempt to explain Knuth's arrow notation. The number written $3 \uparrow 3$ is $3 \times 3 \times 3 = 3^3 = 27$. The number $3 \uparrow \uparrow 3$ denotes the expression $3 \uparrow (3 \uparrow 3)$. Since $3 \uparrow 3$ equals 27, we can write $3 \uparrow \uparrow 3$ as $3 \uparrow 27$ or 3^{27}. As a slanting tower of exponents it is

$$3^{3^3}.$$

The tower is only three levels high, but written as an ordinary number it is 7,625,597,484,987. This is a big leap from 27, but it is still such a small number that we can actually print it.

When the huge number $3 \uparrow \uparrow \uparrow 3 = 3 \uparrow \uparrow (3 \uparrow \uparrow 3) = 3 \uparrow \uparrow 3^{27}$ is written as a tower of 3's, it reaches a height of 7,625,597,484,987 levels. Both the tower and the number it represents are now too big to be printed without special notation.

Consider $3 \uparrow \uparrow \uparrow \uparrow 3 = 3 \uparrow \uparrow \uparrow (3 \uparrow \uparrow \uparrow 3)$. Inside the parentheses is the gigantic number obtained by the preceding calculation. It is no longer possible to indicate in any simple way the height of the tower of 3's that expresses $3 \uparrow \uparrow \uparrow \uparrow 3$. The height is another universe away from $3 \uparrow \uparrow \uparrow 3$. If we break $3 \uparrow \uparrow \uparrow \uparrow 3$ down to a series of the double-arrow operations, it is $3 \uparrow \uparrow (3 \uparrow \uparrow (3 \uparrow \uparrow \cdots \uparrow \uparrow (3 \uparrow \uparrow 3) \cdots))$, where the number of steps to be iterated is $3 \uparrow \uparrow \uparrow 3$. As Knuth says, the dots "suppress a lot of detail." $3 \uparrow \uparrow \uparrow \uparrow 3$ is unimaginably larger than $3 \uparrow \uparrow \uparrow 3$, but it is still small as finite numbers go, since most finite numbers are very much larger.

We are now in a position to indicate Graham's number. It is represented in Figure 33.5. At the top is $3 \uparrow \uparrow \uparrow \uparrow 3$. This gives the number of arrows in the number just below it. That number in turn gives the number of arrows below it. This continues for 2^6, or 64, layers. It is the bottom number that Graham has proved to be an upper bound for the hypercube problem.

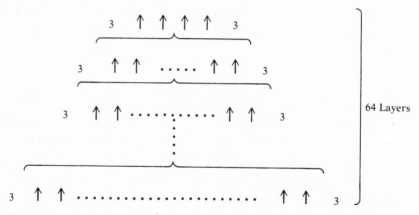

Figure 33.5. Graham's upper bound for the solution to a Euclidean Ramsey problem

Now hold on to your hat. Ramsey-theory experts believe the actual Ramsey number for this problem is probably 6. As Stanislaw M. Ulam said many times in his lectures, "The infinite we shall do right away. The finite may take a little longer."

Addendum

Since this chapter first appeared in *Scientific American* in 1977, such rapid progress has been made in Ramsey theory that a bibliography of papers would run to more than a thousand titles and any effort to summarize here even the main results would be impossible. Fortunately, two excellent books are now available as introductions to the general field (of which Ramsey graph theory is only a part): Ronald Graham's *Rudiments of Ramsey Theory* and *Ramsey Theory*, a book Graham coauthored with two colleagues.

Ramsey theory now includes, among many other things, problems involving the partitioning of lines in any graph or the partitioning of points in any space. Euclidean Ramsey theory, pioneered in the 1970s by Graham, Erdös, and others, concerns the k-coloring of all points in a given Euclidean space and determining what patterns are forced. For example, no matter how the points on the plane are two-colored, the coloring will force the vertices of a monochromatic triangle of any specified size and shape except the equilateral triangle. (The plane can be colored with stripes of two alternating colors of such widths that no

450   COMBINATORICS

equilateral triangle, say of side 1, can have all its corners the same color.)

If all the points in Euclidean three-dimensional space are two-colored, will it force an equilateral triangle of any size? Yes. Consider the four points of any regular tetrahedron. No matter how the points are two-colored, at least three must be the same color, and of course those three will form a monochromatic equilateral triangle. For other theorems of this type, much harder to prove, see the 1973 paper by Erdös, Graham, and others.

During the past decade Frank Harary and his associates have been analyzing Ramsey games. Only a small fraction of their results have been published, but Harary is planning a large book on what he calls achievement and avoidance games of the Ramsey type.

In Stefan Burr's chart (Figure 33.3) showing all graphs with six or fewer lines, you will see that in the interval of integers 2 through 18, only 4 and 16 are missing as generalized Ramsey numbers. There are three graphs of seven lines each that have generalized Ramsey numbers of 16, but there is no graph that has a generalized Ramsey number of 4. Is every positive integer except 4 (ignoring 1, which is meaningless) a generalized Ramsey number? In 1970 Harary showed that the answer is yes. The impossibility of 4 is easily seen by inspecting Burr's chart. The tetrahedron (complete graph for four points) cannot be the smallest graph that forces a subgraph because every possible subgraph of a tetrahedron has a Ramsey number higher or lower than 4.

What is the smallest complete graph which, if two-colored, will force a monochromatic complete graph for five points? In other words, what is the generalized Ramsey number $R(5,5)$? In 1975 Harary offered $100 for the first solution, but the prize remains unclaimed.

"Ramsey theory is only in its infancy," writes Harary in his tribute to Ramsey in the special 1983 issue of *The Journal of Graph Theory* cited in the bibliography and adds: "There is no way that Frank Ramsey could have foreseen the theory his work has inspired." D. H. Mellor, in another tribute to Ramsey in the same issue, has this memorable sentence: "Ramsey's enduring fame in mathematics . . . rests on a theorem he didn't need, proved in the course of trying to do something we now know can't be done!"

The discovery that Ramsey Number $R(4,5)$ equals 25 was first announced in 1993 by the *Rochester Democrat and Chronicle* because one of the discoverers, Stanislaw Radziszowski, was a professor at the Rochester Institute of Technology. Someone using the initials B.V.B. reported this to Ann Landers who published the letter in her syndicated advice column on June 22, 1993. B.V.B. put it this way:

Two professors, one from Rochester, the other from Australia, have worked for three years, used 110 computers, and communicated 10,000 miles by electronic mail, and finally have learned the answer to a question that has baffled scientists for 63 years. The question is this: If you are having a party and want to invite at least four people who know each other and five who don't, how many people should you invite? The answer is 25. Mathematicians and scientists in countries worldwide have sent messages of congratulations.

I don't want to take anything away from this spectacular achievement, but it seems to me that the time and money spent on this project could have been better used had they put it toward finding ways to get food to the millions of starving children in war-torn countries around the world.

Miss Landers replied as follows:

Dear B.V.B.: There has to be more to this "discovery" than you recounted. The principle must be one that can be applied to solve important scientific problems. If anyone in my reading audience can provide an explanation in language a lay person can understand, I will print it. Meanwhile I am "Baffled in Chicago."

I sent Miss Landers this letter:

You asked (June 22) for an explanation of the proof that 25 people are required at a party if at least four are mutually acquainted, and at least five are mutual strangers. This is a theorem in what is called Ramsey theory. It can be modeled by points on paper; each point representing a person. Every point is joined to every other point by a line.

If a line is colored red it means that its two points are persons who know each other. If a line is blue, it means its two persons are strangers. We now can ask: what is the smallest number of points needed so that no matter how the lines are bicolored, there is sure to be either four points mutually joined by red lines, or five mutually joined by blue lines? The answer, 25, was not proved until recently. For the problem's background, see the chapter on Ramsey theory in my book *Penrose Tiles to Trapdoor Ciphers* (W. H. Freeman, 1989).

As far as I know, the letter was not printed. Perhaps Miss Landers remained baffled as to what earthly use this result could possibly have.

Radziszowski and his associate Brendon McKay published their proof in a paper titled "$R(4,5) = 25$" in *The Journal of Graph Theory* (Vol. 19, No. 3, 1995, pp. 309–21).

Answers

Figure 33.6 (note the color symmetry) shows how to two-color the complete graph on seven points so that it is blue-empty (gray lines) and contains four (the minimum) red triangles (black lines). If you enjoyed working on this problem, you might like to tackle K_8 by two-coloring the complete graph on eight points so that it is blue-empty and has eight (the minimum) red triangles.

Garry Lorden, in his 1962 paper, "Blue-Empty Chromatic Graphs," showed that problems of this type are uninteresting. If the number of points n is even, Goodman had earlier shown that the graph can be made blue-empty extremal simply by coloring red two complete subgraphs of $n/2$ points. If n is odd, K_7 is (as I said earlier) the only complete graph that can be made blue-empty extremal by two-coloring. When two-coloring complete graphs where n is odd and not 7, how do you create a blue-empty graph that is not extremal but has the smallest number of red triangles? Lorden showed that this is easily done by partitioning the graph into two complete red subgraphs, one of $(n + 1)/2$ points, the other of $(n - 1)/2$ points. Some extra red lines may be required, but their addition is trivial.

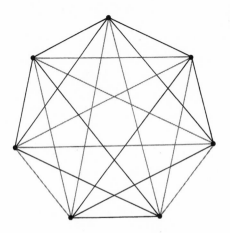

Figure 33.6. Solution to a Ramsey graph puzzle

Bibliography

M. Albertson, "People Who Know People," *Mathematics Magazine,* Vol. 67, October 1994, pp. 278–81.

S. Burr, "Generalized Ramsey Theory for Graphs—A Survey," *Graphs and Combinatorics,* R. Bari and F. Harary (eds.), Springer-Verlag, 1974, pp. 58–75.

B. Cipra, "A Visit to Asymptopia," *Science,* Vol. 267, February 17, 1995, pp. 264–65.

P. Erdös, et al., "Euclidean Ramsey Theorems I," *Journal of Combinatorial Theory,* Ser. A, Vol. 14, 1973, pp. 341–63.

A. W. Goodman, "On Sets of Acquaintances and Strangers at Any Party," *The American Mathematical Monthly,* 66, 1959, pp. 778–83.

R. Graham, "On Edgewise 2-colored Graphs with Monochromatic Triangles and Containing No Complete Hexagon," *Journal of Combinatorial Theory,* Vol. 4, 1968, p. 300.

R. Graham, B. Rothschild, and J. Spencer, *Ramsey Theory,* Wiley, 1980.

R. Graham, *Rudiments of Ramsey Theory, The American Mathematical Society,* 1981.

R. Graham and J. Spencer, "Ramsey Theory," *Scientific American,* July 1990, pp. 112–17.

F. Harary, *Graph Theory,* Addison-Wesley, 1969.

G. Lorden, "Blue-Empty Chromatic Graphs," *The American Mathematical Monthly,* 69, 1962, pp. 114–20.

J. Nesetril and V. Rodl (eds.), *Mathematics of Ramsey Theory,* Springer, 1990.

I. Peterson, "Party Numbers," *Science News,* Vol. 144, July 17, 1993, pp. 46–47.

W. Slany, "Graph Ramsey Games," *DBAI Technical Report,* Technische Universitat Wien, Vienna, Austria, November 5, 1999, pp. 1–45. The paper lists 69 references.

The Journal of Graph Theory, 7, Spring 1983. This issue, dedicated to Frank Ramsey, contains 19 papers about Ramsey and Ramsey theory.

Chapter 34

Bulgarian Solitaire and Other Seemingly Endless Tasks

With useless endeavor,
Forever, forever,
Is Sisyphus rolling
His stone up the mountain!

   —Henry Wadsworth Longfellow,
     The Masque of Pandora

Suppose you have a basket containing 100 eggs and also a supply of egg cartons. Your task is to put all the eggs into the cartons. A step (or move) consists of putting one egg into a carton or taking one egg from a carton and returning it to the basket. Your procedure is this: After each two successive packings of an egg you move an egg from a carton back to the basket. Although this is clearly an inefficient way to pack the eggs, it is obvious that eventually all of them will get packed.

Now assume the basket can hold any finite number of eggs. The task is unbounded if you are allowed to start with as many eggs as you like. Once the initial number of eggs is specified, however, a finite upper bound is set on the number of steps needed to complete the job.

If the rules allow transferring any number of eggs back to the basket any time you like, the situation changes radically. There is no longer an upper bound on the steps needed to finish the job even if the basket initially holds as few as two eggs. Depending on the rules, the task of packing a finite number of eggs can be one that must end, one that cannot end, or one that you can choose to make either finite or infinite in duration.

We now consider several entertaining mathematical tasks with the following characteristic. It seems intuitively true that you should be able to delay completing the task forever, when actually there is no way to avoid finishing it in a finite number of moves.

Our first example is from a paper by the philosopher-writer-logician Raymond M. Smullyan. Imagine you have an infinite supply of pool balls, each bearing a positive integer, and for every integer there is an infinite number of balls. You also have a box that contains a finite quantity of numbered balls. Your goal is to empty the box. Each step consists of removing a ball and replacing it with any finite number of balls of lower rank. The 1 balls are the only exceptions. Since no ball has a rank lower than 1, there are no replacements for a 1 ball.

It is easy to empty the box in a finite number of steps. Simply replace each ball higher than 1 with a 1 ball until only 1 balls remain, then take out the 1 balls one at a time. The rules allow you, however, to replace a ball with a rank above 1 with *any* finite number of balls of lower rank. For instance, you may remove a ball of rank 1,000 and replace it with a billion balls of rank 999, with 10 billion of rank 998, with a billion billion of rank 987, and so on. In this way the number of balls in the box may increase beyond imagining at each step. Can you not prolong the emptying of the box forever? Incredible as it may seem at first, there is no way to avoid completing the task.

Note that the number of steps needed to empty the box is unbounded in a much stronger way than it is in the egg game. Not only is there no bound on the number of eggs you begin with but also each time you remove a ball with a rank above 1 there is no bound to the number of balls you may use to replace it. To borrow a phrase from John Horton Conway, the procedure is "unboundedly unbounded." At every stage of the game, as long as the box contains a single ball other than a 1 ball, it is impossible to predict how many steps it will take to empty the box of all but 1 balls. (If all the balls are of rank 1, the box will of course empty in as many steps as there are 1 balls.) Nevertheless, no matter how clever you are in replacing balls, the box eventually must empty after a finite number of moves. Of course, we have to assume that although you need not be immortal, you will live long enough to finish the task.

Smullyan presents this surprising result in a paper, "Trees and Ball Games," in *Annals of the New York Academy of Sciences* (Vol. 321, 1979, pp. 86–90). Several proofs are given, including a simple argument by induction. I cannot improve on Smullyan's phrasing:

> If all balls in the box are of rank 1, then we obviously have a losing game. Suppose the highest rank of any ball in the box is 2. Then we have at the outset a finite number of 2's and a finite number of 1's. We can't keep throwing away 1's forever; hence we must sooner or later throw out one of our 2's. Then we have one less 2 in the box (but possibly many more 1's than we started with). Again, we can't keep throwing out 1's forever, and so we must sooner or later throw out another 2. We

see that after a finite number of steps we must throw away our last 2, and then we are back to the situation in which we have only 1's. We already know this to be a losing situation. This proves that the process must terminate if the highest rank present is 2. Now, what if the highest rank is 3? We can't keep throwing away just balls of rank ≤ 2 forever (we just proved that!); hence we must sooner or later throw out a 3. Then again we must sooner or later throw out another 3, and so we must eventually throw out our last 3. This then reduces the problem to the preceding case when the highest rank present is 2, which we have already solved.

Smullyan also proves that the game ends by modeling it with a tree graph. A "tree" is a set of line segments each of which joins two points, and in such a way that every point is connected by a unique path of segments leading to a point called the tree's root. The first step of a ball game, filling the box, is modeled by representing each ball as a point, numbered like the ball and joined by a line to the tree's root. When a ball is replaced by other balls of lower rank, its number is erased and the new balls indicated by a higher level of numbered points are joined to the spot where the ball was removed. In this way the tree grows steadily upward, its "end points" (points that are not the root and are attached to just one segment) always representing the balls in the box at that stage of the game.

Smullyan proves that if this tree ever becomes infinite (has an infinity of points), it must have at least one infinite branch stretching upward forever. This, however, is clearly impossible because the numbers along any branch steadily decrease and therefore must eventually terminate in 1. Since the tree is finite, the game it models must end. As in the ball version, there is no way to predict how many steps are needed to complete the tree. At that stage, when the game becomes bounded, all the end points are labeled 1. The number of these 1 points may, of course, exceed the number of electrons in the universe, or any larger number. Nevertheless, the game is not Sisyphean. It is certain to end after a finite number of moves.

Smullyan's basic theorem, which he was the first to model as a ball game, derives from theorems involving the ordering of sets that go back to Georg Cantor's work on the transfinite ordinal numbers. It is closely related to a deep theorem about infinite sets of finite trees that was first proved by Joseph B. Kruskal and later in a simpler way by C. St. J. A. Nash-Williams. More recently Nachum Dershowitz and Zohar Manna have used similar arguments to show that certain computer programs, which involve "unboundedly unbounded" operations, must eventually come to a halt.

A special case of Smullyan's ball game is modeled by numbering a finite tree upward from the root as in Figure 34.1, left. We are allowed to chop off any end point, along with its attached segment, then add to the tree as many new branches as we like, and wherever we like, provided all the new points are of lower rank than the one removed. For example, the figure at the right in the illustration shows a possible new growth after a 4 point has been chopped off. In spite of the fact that after each chop the tree may grow billions on billions of new branches, after a finite number of chops the tree will be chopped down. Unlike the more general ball game, we cannot remove any point we like, only the end points, but because each removed point is replaced by points of lower rank, Smullyan's ball theorem applies. The tree may grow inconceivably bushier after each chop, but there is a sense in which it always gets closer to the ground until eventually it vanishes.

A more complicated way of chopping down a tree was proposed by Laurie Kirby and Jeff Paris in *The Bulletin of the London Mathematical Society* (Vol. 14, Part 4, No. 49, July 1982, pp. 285–93). They call their tree graph a hydra. Its end points are the hydra's heads, and Hercules wants to destroy the monster by total decapitation. When a head is severed, its attached segment goes with it. Unfortunately after the first chop the hydra acquires one or more new heads by growing a new branch from a point (call it k) that is one step below the lost segment. This new branch is an exact replica of the part of the hydra that extends up from k. The figure at the top right in Figure 34.2 shows the hydra after Hercules has chopped off the head indicated by the sword in the figure at the top left.

The situation for Hercules becomes increasingly desperate because when he makes his second chop, *two* replicas grow just below the severed segment (Figure 34.2, *bottom left*). And *three* replicas grow after the third chop (Figure 34.2, *bottom right*), and so on. In general, n replicas sprout at each nth chop. There is no way of labeling the hydra's points to make this growth correspond to Smullyan's ball game; nevertheless, Kirby and Paris are able to show, utilizing an argument based on a remarkable number theorem found by the British logician R. L. Goodstein, that no matter what sequence Hercules follows in cutting off heads, the hydra is eventually reduced to a set of heads (there may be millions of them even if the starting form of the beast is simple) that are

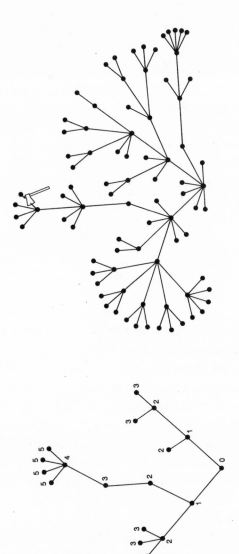

Figure 34.1.

all joined directly to the root. They are then eliminated one by one until the hydra expires from lack of heads.

A useful way to approach the hydra game is to think of the tree as modeling a set of nested boxes. Each box contains all the boxes reached by moving upward on the tree, and it is labeled with the maximum number of levels of nesting that it contains. Thus in the first figure of the hydra the root is a box of rank 4. Immediately above it on the left

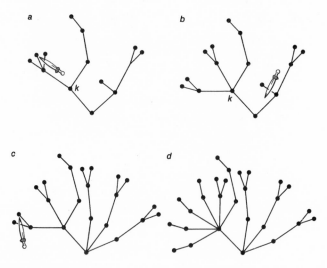

Figure 34.2.

is a 3 box and on the right is a 2 box, and so on. All end points are empty boxes of rank 0. Each time a 0 box (hydra head) is removed the box immediately below gets duplicated (along with all its contents), but each of the duplicates as well as the original box now contains one fewer empty box. Eventually you are forced to start reducing ranks of boxes, like the ranks of balls in the ball game. An inductive argument similar to Smullyan's will show that ultimately all boxes become empty, after which they are removed one at a time.

I owe this approach to Dershowitz, who pointed out that it is not even necessary for the hydra to limit its growth to a consecutively increasing number of new branches. After each chop as many finite duplicates as you like may be allowed to sprout. It may take Hercules much longer to slay the monster, but there is no way he can permanently avoid doing so if he keeps hacking away. Note that the hydra never gets taller as it widens. Some of the more complicated growth programs considered by Dershowitz and Manna graph as trees that can grow taller as well as wider, and such trees are even harder to prove terminating.

Our next example of a task that looks as if it could go on forever when it really cannot is known as the 18-point problem. You begin with a line segment. Place a point anywhere you like on it. Now place a second point so that each of the two points is within a different half of the line segment. (The halves are taken to be "closed intervals," which means that the end points are not considered "inside" the interval.) Place a third point so that each of the three is in a different third of the line. At this stage it becomes clear that the first two points cannot be just anywhere. They cannot, for example, be close together in the middle of the line or close together at one end. They must be carefully placed so that when the third point is added, each will be in a different third of the line. You proceed in this way, placing every nth point so that the first n points always occupy different $1/n$th parts of the line. If you choose locations carefully, how many spots can you put on the line?

Intuitively it seems as if the number should be endless. A line segment obviously can be divided into as many equal parts as you like and each may contain a point. The catch is that the points must be serially numbered to meet the task's conditions. It turns out, astonishingly, that you cannot get beyond 17 points! Regardless of how clever you are at placing 17 points, the 18th will violate the rules and the game ends. In fact, it is not even easy to place 10 points. Figure 34.3 shows one way to place six.

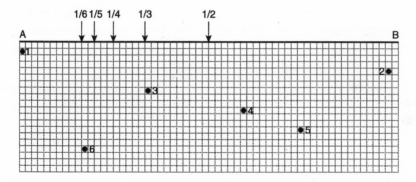

Figure 34.3.

This unusual problem first appeared in *One Hundred Problems in Elementary Mathematics* (problems 6 and 7) by the Polish mathematician Hugo Steinhaus. (Basic Books published a translation in 1964, and there is now a Dover soft-cover reprint.) Steinhaus gives a 14-point solution, and he states in a footnote that M. Warmus has proved 17 is the limit. The first published proof, by Elwyn R. Berlekamp and Ronald L. Graham, is in their paper "Irregularities in the Distributions of Finite Sequences" (*Journal of Number Theory*, Vol. 2, No. 2, May 1970, pp. 152–61).

Warmus, a Warsaw mathematician, did not publish his shorter proof until six years later in the same journal (Vol. 8, No. 3, August 1976, pp. 260–63). He gives a 17-point solution, and he adds that there are 768 patterns for such a solution, or 1,536 if you count their reversals as being different.

Our last example of a task that ends suddenly in a counterintuitive way is one you will enjoy modeling with a deck of playing cards. Its origin is unknown, but Graham, who told me about it, says that European mathematicians call it Bulgarian solitaire for reasons he has not been able to discover. Partial sums of the series $1 + 2 + 3 + \cdots$ are known as triangular numbers because they correspond to triangular arrays such as the 10 bowling pins or the 15 pool balls. The task involves any triangular number of playing cards. The largest number you can get from a standard deck is 45, the sum of the first nine counting numbers.

Form a pile of 45 cards, then divide it into as many piles as you like, with an arbitrary number of cards in each pile. You may leave it as a single pile of 45, or cut it into two, three, or more piles, cutting anywhere you want, including 44 cuts to make 45 piles of one card each. Now keep repeating the following procedure. Take one card from each pile and place all the removed cards on the table to make a new pile. The piles need not be in a row. Just put them anywhere. Repeat the procedure to form another pile, and keep doing it.

As the structure of the piles keeps changing in irregular ways it seems unlikely you will reach a state where there will be just one pile with one card, one pile with two cards, one with three, and so on to one with nine cards. If you should reach this improbable state, without getting trapped in loops that keep returning the game to a previous state, the game must end, because now the state cannot change. Repeating the procedure leaves the cards in exactly the same consecutive state as before. It turns out, surprisingly, that regardless of the initial state of the game, you are sure to reach the consecutive state in a finite number of moves.

Bulgarian solitare is a way of modeling some problems in partition theory that are far from trivial. The partitions of a counting number n are all the ways a positive integer can be expressed as the sum of positive integers without regard to their order. For example, the triangular number 3 has three partitions: $1 + 2$, $1 + 1$, and 3. When you divide a packet of cards into an arbitrary number of piles, any number to a pile, you are forming a partition of the packet. Bulgarian solitaire is a way of changing one partition to another by subtracting 1 from each number in the partition, then adding a number equal to the number of subtracted 1's. It is not obvious this procedure always gives rise to a chain of partitions, without duplicates, that ends with the consecutive partition. I am told it was first proved in 1981 by Jørgen Brandt, a Danish mathematician, but I do not know his proof or whether it has been published.

Bulgarian solitaire for any triangular number of cards can be diagrammed as a tree with the consecutive partition labeling its root and all other partitions represented by the tree's points. The picture at the left in Figure 34.4 shows the simple tree for the three-card game. In the picture at the right of the illustration is the less trivial tree for the 11 partitions of six cards. The theorem that any game ends with the consecutive partition is equivalent to the theorem that all the partitions of a triangular number will graph as a connected tree, with each partition one step above its successor in the game and the consecutive partition at the tree's root.

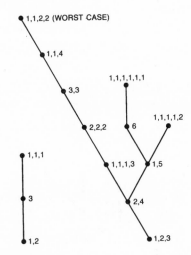

Figure 34.4.

Note that the highest point on the six-card tree is six steps from the root. This partition, 1,1,2,2, is the "worst" starting case. It is easy to see that the game must end in no more than six steps from any starting partition. It has been conjectured that any game must end in no more than $k(k - 1)$ steps, where k is any positive integer in the formula for triangular numbers $\frac{1}{2}k(k + 1)$. The computer scientist Donald E. Knuth asked his Stanford University students to test the conjecture by computer. They confirmed it for $k = 10$ or less, so that the conjecture is almost certainly true, but so far a proof has been elusive.

Figure 34.5 shows the tree for Bulgarian solitaire with 10 cards ($k = 4$). There are now three worst cases at the top, each 12 steps from the root. Note also that the tree has 14 end points. We can call them Eden partitions because unless you start with them they never arise in a game. They are all those partitions whose number of parts exceeds the highest number of parts by 2 or more.

The picture at the left in Figure 34.6 shows the standard way of using dots to diagram partition 1,1,2,3,3, at the tree's top. If this pattern is ro-

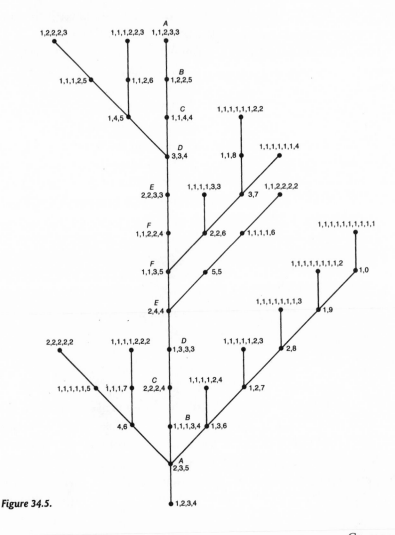

Figure 34.5.

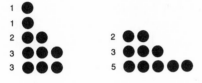

Figure 34.6.

tated and mirror-reflected, it becomes the pattern at the right in the illustration. Its rows now give the partition 2,3,5. Each partition is called the conjugate of the other. The relation is obviously symmetrical. A partition unchanged by conjugation is said to be self-conjugate. On the 10 tree there are just two such partitions, the root and 1,1,1,2,5. When the remaining partitions are paired as conjugates, an amazing pattern appears along the trunk. The partitions pair as is shown by the letters. This symmetry holds along the main trunk of all Bulgarian trees so far investigated.

If the symmetry holds for all such trees, we have a simple way to determine the worst case at the top. It is the conjugate of the partition (there is always only one) just above the root. An even faster way to find the trunk's top is to prefix 1 to the root and diminish its last number by 1.

The Bulgarian operation can be diagrammed by removing the leftmost column of its flush-left dot pattern, turning the column 90 degrees, and adding it as a new row. Only diagrams of the 1, 2, 3, 4, . . . form are unaltered by this. If you could show that no sequence of operations on any partition other than the consecutive one would return a diagram to its original state, you would have proved that all Bulgarian games graph as trees and therefore must end when their root is reached.

If the game is played with 55 cards ($k = 10$), there are 451,276 ways to partition them, so that drawing a tree would be difficult. Even the 15-card tree, with 176 points, calls for computer aid. How are these numbers calculated? Well, it is a long and fascinating story. Let us say partitions are *ordered,* so that 3, for example, would have four ordered partitions (usually called "compositions"): $1 + 2$, $2 + 1$, $1 + 1 + 1$, and 3. It turns out that the formula for the total number of compositions is simply 2^{n-1}. But when the partitions are unordered, as they are in the solitaire card game, the situation is unbelievably disheveled. Although there are many recursive procedures for counting unordered partitions, using at each step the number of known partitions for all smaller numbers, an exact asymptotic formula was finally obtained. The big breakthrough was made by the British mathematician G. H. Hardy, working with his Indian friend Srinivasa Ramanujan. Their not quite exact formula was perfected by Hans A. Rademacher in 1937. The Hardy-Ramanujan-Rademacher formula is a horribly shaggy infinite series that involves (among other things) pi, square roots, complex roots, and derivatives of hyperbolic functions! George E. Andrews, in his standard textbook on partition theory, calls it an "unbelievable identity" and "one of the crowning achievements" in the history of his subject.

The sequence of partitions for $n = 1$, $n = 2$, $n = 3$, $n = 4$, $n = 5$, and $n = 6$ is 1,2,3,5,7,11, and so you might expect the next partition to be the next prime, 13. Alas, it is 15. Maybe all partitions are odd. No, the next partition is 22. One of the deep unsolved problems in partition theory is whether, as n increases, the even and odd partitions approach equality in number.

If you think partition theory is little more than a mathematical pastime, let me close by saying that a way of diagramming sets of partitions, using number arrays known as the Young tableaux, has become enormously useful in particle physics. But that's another ball game.

Addendum

Many readers sent proofs of the conjecture that Bulgarian solitaire must end in $k(k-1)$ steps, and the proof was later given in several articles listed in the bibliography. Ethan Akin and Morton Davis began their 1983 paper as follows:

> Blast Martin Gardner! There you are, minding your own business, and *Scientific American* comes along like a virus. All else forgotten, you must struggle with infection by one of his fascinating problems. In the August 1983 issue he introduced us to Bulgarian solitaire.

Bibliography

E. Akin and M. Davis, "Bulgarian Solitaire," *American Mathematical Monthly*, Vol. 92, April 1985, pp. 237–50.

G. E. Andrews, *Number Theory: The Theory of Partitions*, Addison-Wesley Publishing Co., 1976.

T. Bending, "Bulgarian Solitaire," *Eureka*, No. 50, April 1990, pp. 12–19.

J. Brandt, "Cycles of Partitions," *Proceedings of the American Mathematical Society*, Vol. 85, July 1982, pp. 483–86.

M. Carver, "Hercules Hammers Hydra Herd," *Discover*, November 1987, pp. 94–95, 104.

N. Dershowitz and Z. Manna, "Proving Termination with Multiset Orderings," *Communications of the ACM,* Vol. 22, No. 8, August 1979, pp. 465–76.

G. Etienne, "Tableaux de Young et Solitaire Bulgare," *The Journal of Combinatorial Theory,* Ser. A, Vol. 58, November 1991, pp. 181–97.

R. W. Hamming, "The Tennis Ball Paradox," *Mathematics Magazine,* Vol. 62, October 1989, pp. 268–73.

K. Igusa, "Solution of the Bulgarian Solitaire Conjecture," *Mathematics Magazine,* Vol. 58, November 1985, pp. 259–71.

L. Kirby and J. Paris, "Accessible Independence Results for Peano Arithmetic," *The Bulletin of the London Mathematical Society,* Vol. 14, No. 49, Part 4, July 1983, pp. 285–93.

C. Si. J. A. Nash-Williams, "On Well-quasi-ordering Finite Trees," *Proceedings of the Cambridge Philosophical Society,* Vol. 59, Part 4, October 1963, pp. 833–35.

A. Nicholson, "Bulgarian Solitaire," *Mathematics Teacher,* Vol. 86, January 1993, pp. 84–86.

J. Propp, "Some Variants of Ferrier Diagrams," *Journal of Combinatorial Theory,* Ser. A, Vol. 52, September 1989, pp. 98–128.

R. M. Smullyan, "Trees and Ball Games," *Annals of the New York Academy of Sciences,* Vol. 321, 1979, pp. 86–90.

IX

Games and Decision Theory

Chapter 35

A Matchbox Game-Learning Machine

*I knew little of chess, but as only a few
pieces were on the board, it was obvious that
the game was near its close. . . . [Moxon's]
face was ghastly white, and his eyes glittered
like diamonds. Of his antagonist I had only
a back view, but that was sufficient; I should
not have cared to see his face.*

*T*he quotation is from Ambrose Bierce's classic robot story,
"Moxon's Master" (reprinted in Groff Conklin's science-fiction anthology, *Thinking Machines*). The inventor Moxon has constructed a chess-playing robot. Moxon wins a game. The robot strangles him.

Bierce's story reflects a fear that someday computers will develop a
will of their own. Let it not be thought that this worries only those who
do not understand computers. Before his death Norbert Wiener anticipated with increasing apprehension the day when complex government decisions would be turned over to sophisticated game-theory
machines. Before we know it, Wiener warned, the machines may shove
us over the brink into a suicidal war.

The greatest threat of unpredictable behavior comes from the learning machines: computers that improve with experience. Such machines
do not do what they have been told to do but what they have *learned*
to do. They quickly reach a point at which the programmer no longer
knows what kinds of circuits his machine contains. Inside most of
these computers are randomizing devices. If the device is based on the
random decay of atoms in a sample radioactive material, the machine's
behavior is not (most physicists believe) predictable even in principle.

Much of the research on learning machines has to do with computers that steadily improve their ability to play games. Some of the work
is secret—war is a game. The first significant machine of this type was
an IBM 704 computer programed by Arthur L. Samuel of the IBM research department at Poughkeepsie, NY. In 1959 Samuel set up the
computer so that it not only played a fair game of checkers but also was
capable of looking over its past games and modifying its strategy in the
light of this experience. At first Samuel found it easy to beat his ma-

chine. Instead of strangling him, the machine improved rapidly, soon reaching the point at which it could clobber its inventor in every game. So far as I know no similar program has yet been designed for chess, although there have been many ingenious programs for nonlearning chess machines.

In 1960 the Russian chess grandmaster Mikhail Botvinnik was quoted as saying that the day would come when a computer would play master chess. "This is of course nonsense," wrote the American chess expert Edward Lasker in an article on chess machines in the Fall 1961 issue of a magazine called *The American Chess Quarterly.* But it was Lasker who was talking nonsense. A chess computer has three enormous advantages over a human opponent: (1) it never makes a careless mistake; (2) it can analyze moves ahead at a speed much faster than a human player can; (3) it can improve its skill without limit. There is every reason to expect that a chess-learning machine, after playing thousands of games with experts, will someday develop the skill of a master. It is even possible to program a chess machine to play continuously and furiously against itself. Its speed would enable it to acquire in a short time an experience far beyond that of any human player.

It is not necessary for the reader who would like to experiment with game-learning machines to buy an electronic computer. It is only necessary to obtain a supply of empty matchboxes and colored beads. This method of building a simple learning machine is the happy invention of Donald Michie, a biologist at the University of Edinburgh. Writing on "Trial and Error" in *Penguin Science Survey 1961,* Vol. 2, Michie describes a ticktacktoe learning machine called MENACE (Matchbox Educable Naughts And Crosses Engine) that he constructed with 300 matchboxes.

MENACE is delightfully simple in operation. On each box is pasted a drawing of a possible ticktacktoe position. The machine always makes the first move, so only patterns that confront the machine on odd moves are required. Inside each box are small glass beads of various colors, each color indicating a possible machine play. A V-shaped cardboard fence is glued to the bottom of each box, so that when one shakes the box and tilts it, the beads roll into the V. Chance determines the color of the bead that rolls into the V's corner. First-move boxes contain four beads of each color, third-move boxes contain three beads of each color, fifth-move boxes have two beads of each color, seventh-move boxes have single beads of each color.

The robot's move is determined by shaking and tilting a box, opening the drawer, and noting the color of the "apical" bead (the bead in the V's apex). Boxes involved in a game are left open until the game ends. If the machine wins, it is rewarded by adding three beads of the apical color to each open box. If the game is a draw, the reward is one bead per box. If the machine loses, it is punished by extracting the apical bead from each open box. This system of reward and punishment closely parallels the way in which animals and even humans are taught and disciplined. It is obvious that the more games MENACE plays, the more it will tend to adopt winning lines of play and shun losing lines. This makes it a legitimate learning machine, although of an extremely simple sort. It does not make (as does Samuel's checker machine) any self-analysis of past plays that causes it to devise new strategies.

Michie's first tournament with MENACE consisted of 220 games over a two-day period. At first the machine was easily trounced. After 17 games the machine had abandoned all openings except the corner opening. After the twentieth game it was drawing consistently, so Michie began trying unsound variations in the hope of trapping it in a defeat. This paid off until the machine learned to cope with all such variations. When Michie withdrew from the contest after losing eight out of ten games, MENACE had become a master player.

Since few readers are likely to attempt building a learning machine that requires 300 matchboxes, I have designed hexapawn, a much simpler game that requires only 24 boxes. The game is easily analyzed—indeed, it is trivial—but the reader is urged *not* to analyze it. It is much more fun to build the machine, then learn to play the game while the machine is also learning.

Hexapawn is played on a 3 × 3 board, with three chess pawns on each side as shown in Figure 35.1. Dimes and pennies can be used instead of actual chess pieces. Only two types of move are allowed: (1) A pawn may advance straight forward one square to an empty square; (2) a pawn may capture an enemy pawn by moving one square diagonally, left or right, to a square occupied by the enemy. The captured piece is removed from the board. These are the same as pawn moves in chess, except that no double move, en passant capture, or promotion of pawns is permitted.

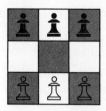

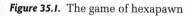
Figure 35.1. The game of hexapawn

The game is won in any of three ways:

1. By advancing a pawn to the third row.
2. By capturing all enemy pieces.
3. By achieving a position in which the enemy cannot move.

Players alternate moves, moving one piece at a time. A draw clearly is impossible, but it is not immediately apparent whether the first or second player has the advantage.

To construct HER (Hexapawn Educable Robot) you need 24 empty matchboxes and a supply of colored beads. Small candies that come in different colors—M&M's or Skittles—work nicely. Each matchbox bears one of the diagrams in Figure 35.2. The robot always makes the second move. Patterns marked "2" represent the two positions open to HER on the second move. You have a choice between a center or an end opening, but only the left end is considered because an opening on the right would obviously lead to identical (although mirror-reflected) lines of play. Patterns marked "4" show the eleven positions that can confront HER on the fourth (its second) move. Patterns marked "6" are the eleven positions that can face HER on the sixth (its last) move. (I have included mirror-image patterns in these positions to make the working easier; otherwise 19 boxes would suffice.)

Inside each box place a single bead to match the color of each arrow on the pattern. The robot is now ready for play. Every legal move is represented by an arrow; the robot can therefore make all possible moves and only legal moves. The robot has no strategy. In fact, it is an idiot.

The teaching procedure is as follows. Make your first move. Pick up the matchbox that shows the position on the board. Shake the matchbox, close your eyes, open the drawer, remove one bead. Close the drawer, put down the box, place the bead on top of the box. Open your eyes, note the color of the bead, find the matching arrow and move accordingly. Now it is your turn to move again. Continue this procedure until the game ends. If the robot wins, replace all the beads and play

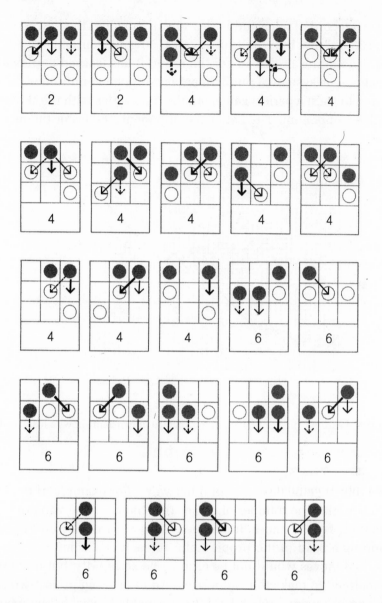

Figure 35.2. Labels for HER matchboxes. (The four different kinds of arrows represent four different colors.)

again. If it loses, punish it by confiscating only the bead that represents its *last* move. Replace the other beads and play again. If you should find an empty box (this rarely happens), it means the machine has no move that is not fatal and it resigns. In this case confiscate the bead of the preceding move.

A Matchbox Game-Learning Machine 475

Keep a record of wins and losses so you can chart the first 50 games. Figure 35.3 shows the results of a typical 50-game tournament. After 36 games (including 11 defeats for the robot) it has learned to play a perfect game. The system of punishment is designed to minimize the time required to learn a perfect game, but the time varies with the skill of the machine's opponent. The better the opponent, the faster the machine learns.

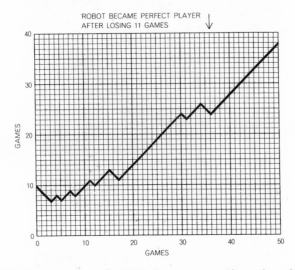

Figure 35.3. Learning curve for HER's first 50 games (downslant shows a loss; upslant, a win)

The robot can be designed in other ways. For example, if the intent is to maximize the number of games that the machine wins in a tournament of, say, 25 games, it may be best to reward (as well as punish) by adding a bead of the proper color to each box when the machine wins. Bad moves would not be eliminated so rapidly, but it would be less inclined to make the bad moves. An interesting project would be to construct a second robot, HIM (Hexapawn Instructable Matchboxes), designed with a different system of reward and punishment but equally incompetent at the start of a tournament. Both machines would have to be enlarged so they could make either first or second moves. A tournament could then be played between HIM and HER, alternating the first move, to see which machine would win the most games out of 50.

Similar robots are easily built for other games. Stuart C. Hight, director of research studies at the Bell Telephone Laboratories in Whip-

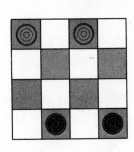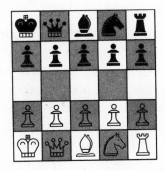

Figure 35.4. Matchbox machines can be built for minicheckers *(left)* but not for minichess *(right)*.

pany, NJ, once built a matchbox learning machine called NIMBLE (Nim Box Logic Engine) for playing Nim with three piles of three counters each. The robot plays either first or second and is rewarded or punished after each game. NIMBLE required only 18 matchboxes and played almost perfectly after 30 games. For an analysis of the game of Nim, see Chapter 15 of my *Scientific American Book of Mathematical Puzzles & Diversions.*

By reducing the size of the board the complexity of many familiar games can be minimized until they are within the scope of a matchbox robot. The game of Go, for example, can be played on the intersections of a 2 × 2 checkerboard. The smallest nontrivial board for checkers is shown in Figure 35.4. It should not be difficult to build a matchbox machine that would learn to play it. Readers disinclined to do this may enjoy analyzing the game. Does either side have a sure win or will two perfect players draw?

When chess is reduced to the smallest board on which all legal moves are still possible, as shown in Figure 35.4, the complexity is still far beyond the capacity of a matchbox machine. In fact, I have found it impossible to determine which player, if either, has the advantage. Minichess is recommended for computer experts who wish to program a simplified chess-learning machine and for all chess players who like to sneak in a quick game during a coffee break.

Addendum

Many readers who experimented with matchbox learning machines were kind enough to write to me about them. L. R. Tanner, at

Westminster College, Salt Lake City, UT, made good use of HER as a concession at a college carnival. The machine was designed to learn by rewards only, so that customers would always have a chance (though a decreasing one) of winning, and prizes to winners were increased in value as HER became more proficient.

Several readers built two matchbox machines to be pitted against each other. John Chambers, Toronto, called his pair THEM (Two-way Hexapawn Educable Machines). Kenneth W. Wiszowaty, science teacher at Phillip Rogers Elementary School, Chicago, sent me a report by his seventh-grade pupil, Andrea Weiland, on her two machines which played against each other until one of them learned to win every time. John House, Waterville, OH, called his second machine RAT (Relentless Autolearning Tyrant), and reported that after 18 games RAT conceded that HER would win all subsequent games.

Peter J. Sandiford, director of operations research for Trans-Canada Air Lines, Montreal, called his machines Mark I and Mark II. As expected, it took 18 games for Mark I to learn how to win every time and Mark II to learn how to fight the longest delaying action. Sandiford then devised a devilish plan. He arranged for two students, a boy and a girl from a local high school mathematics club, who knew nothing about the game, to play hexapawn against each other after reading a handout describing the rules. "Each contestant was alone in a room," writes Sandiford, "and indicated his moves to a referee. Unknown to the players the referees reported to a third room containing the jellybean computers and scorekeepers. The players thought they were playing each other by remote control, so to speak, whereas they were in fact playing independently against the computers. They played alternately black and white in successive games. With much confusion and muffled hilarity we in the middle tried to operate the computers, keep the games in phase, and keep the score."

The students were asked to make running comments on their own moves and those of their opponent. Some sample remarks:

"It's the safest thing to do without being captured; it's almost sure to win."

"He took me, but I took him too. If he does what I expect, he'll take my pawn, but in the next move I'll block him."

"Am I stupid!"

"Good move! I think I'm beat."

"I don't think he's really thinking. By now he shouldn't make any more careless mistakes."

"Good game. She's getting wise to my action now."

"Now that he's thinking, there's more competition."

"Very surprising move . . . couldn't he see I'd win if he moved forward?"

"My opponent played well. I guess I just got the knack of it first."

When the students were later brought face to face with the machines they had been playing, they could hardly believe, writes Sandiford, that they had not been competing against a real person.

Richard L. Sites, at M.I.T., wrote a FORTRAN program for an IBM 1620 so that it would learn to play Octapawn, a 4 × 4 version of hexapawn that begins with four white pawns on the first row and four black pawns on the fourth row. He reports that the first player has a sure win with a corner opening. At the time of his writing, his program had not yet explored center openings.

Judy Gomberg, Maplewood, NJ, after playing against a matchbox machine that she built, reported that she learned hexapawn faster than her machine because "every time it lost I took out a candy and ate it."

Robert A. Ellis, at the computing laboratory, Ballistics Research Laboratories, Aberdeen Proving Ground, MD, told me about a program he wrote for a digital computer which applied the matchbox-learning technique to a ticktacktoe-learning machine. The machine first plays a stupid game, choosing moves at random, and is easily trounced by human opponents. Then the machine is allowed to play 2,000 games against itself (which it does in two or three minutes), learning as it goes. After that, the machine plays an excellent strategy against human opponents.

My defense of Botvinnik's remark that computers will some day play master chess brought a number of irate letters from chess players. One grandmaster assured me that Botvinnik was speaking with tongue in cheek. The interested reader can judge for himself by reading a translation of Botvinnik's speech (which originally appeared in *Komsomolskaya Pravda,* January 3, 1961) in *The Best in Chess,* edited by I. A. Horowitz and Jack Straley Battell (New York: Dutton, 1965, pp. 63–69). "The time will come," Botvinnik concludes, "when mechanical chessplayers will be awarded the title of International Grandmaster . . . and it will be necessary to promote two world championships, one for humans, one for robots. The latter tournament, naturally, will not be between machines, but between their makers and program operators."

An excellent science-fiction story about just such a tournament, Fritz Leiber's "The 64-Square Madhouse," appeared in *If,* May 1962, and was reprinted in Leiber's *A Pail of Air* (New York: Ballantine, 1964). Lord Dunsany, by the way, has twice given memorable descriptions of chess games played against computers. In his short story "The Three Sailors' Gambit" (in *The Last Book of Wonder*) the machine is a magic crystal. In his novel *The Last Revolution* (a 1951 novel about the computer revolution that has never, unaccountably, been published in the United States) it is a learning computer. The description of the narrator's first game with the computer, in the second chapter, is surely one of the funniest accounts of a chess game ever written.

The hostile reaction of master chess players to the suggestion that computers will someday play master chess is easy to understand; it has been well analyzed by Paul Armer in a Rand report (p-2114-2, June 1962) on *Attitudes Toward Intelligent Machines.* The reaction of chess players is particularly amusing. One can make out a good case against computers writing top-quality music or poetry, or painting great art, but chess is not essentially different from ticktacktoe except in its enormous complexity, and learning to play it well is precisely the sort of thing computers can be expected to do best.

Master checker-playing machines will undoubtedly come first. Checkers is now so thoroughly explored that games between champions almost always end in draws, and in order to add interest to such games, the first three moves are now chosen by chance. Richard Bellman, writing "On the Application of Dynamic Programming to the Determination of Optimal Play in Chess and Checkers," *Proceedings of the National Academy of Sciences* (Vol. 53, February 1965, pp. 244–47), says that "it seems safe to predict that within ten years checkers will be a completely decidable game."

Chess is, of course, of a different order of complexity. One suspects it will be a long time before one can (so goes an old joke in modern dress) play the first move of a chess game against a computer and have the computer print, after a period of furious calculation, "I resign." In 1958 some responsible mathematicians predicted that within 10 years computers would be playing master chess, but this proved to be wildly overoptimistic. Tigran Petrosian, when he became world chess champion, was quoted in *The New York Times* (May 24, 1963) as expressing doubts that computers would play master chess within the next 15 or 20 years.

Hexapawn can be extended simply by making the board wider but keeping it three rows deep. John R. Brown, in his paper "Extenda-pawn—An Inductive Analysis" (*Mathematics Magazine,* Vol. 38, November 1965, pp. 286–99) gives a complete analysis of this game. If n is the number of columns, the game is a win for the first player if the final digit of n is 1, 4, 5, 7, or 8. Otherwise the second player has the win.

My hexapawn game was marketed in two forms. In 1969 IBM adapted it to a board that used a spinner to choose one of four colors and buttons to place on game positions to reward or punish the primitive "learning machine." Titled "Hexapawn: A Game You Play to Lose," it was distributed by IBM to high school science and mathematics classes, and to the general public. In 1970 Gabriella, a Farmingdale, NY, firm, produced and sold a similar product they called the Gabriella Computer Kit. See *Science News* (October 26, 1970) for an advertisement. The game was also described in the first chapter of *We Build Our Own Computers,* a projects handbook by A. B. Bolt and others (Cambridge University Press, 1966).

Computers that learn from experience may offer the most promising way to simulate the human mind. Based on what are called neural networks, they rely heavily on the parallel processing of data. The neural network literature is so vast that I made no attempt to cite references. Our brain is, of course, an enormously complicated, parallel-processing learning machine, the operation of which is still only dimly understood.

Computer chess has made rapid progress since this chapter was published as a *Scientific American* column in 1962. Programs now play on the master level, and on occasion have defeated grand masters. The best programs do not learn from experience, but owe their skill to the fantastic speed with which they explore long sequences of moves. It is only a matter of time before chess programs achieve grand master ratings.

A proof of first player win on the 3×3 Go field is given in the 1972 paper by Edward Thorp and William Walden. As far as I know, my 5×5 minichess has not been solved in the sense that it is not known whether the first or second player has the win when the game is played rationally, or whether the game is a draw. It is possible that many chess computer programs would be able to crack it if someone only bothered to give them the game as a problem.

Another curious form of minichess was created by a young woman named Chi Chi Hackenberg, whose invention was featured in *Eye* (November 1968, pp. 93–94). It uses a field of 4 × 8, or half the standard chessboard.

At the start, all eight white pieces are on the first row, in their usual formation. The second row has five white pawns, three pawns missing on the king's bishop and knight columns and the queen's knight column. The 13 black pieces are in mirror-image formation on the other two rows, making six vacant cells on the board. There are two new rules. White is not allowed to move a pawn on its opening move, and pawns are permitted to move vertically backward either one step or to capture diagonally backward.

Hackenberg says that she thinks White's best opening is to take the rook's pawn with its king's knight, threatening checkmate with KB to N2. However, Black can mount a counteroffensive by taking the white queen's pawn with its queen's bishop's pawn for the first in a series of checks that include a capture of White's queen. This seems so disastrous for White that the opening suggested by Hackenberg may not be the best after all. Or can White, after the checks subside, mount a viable counterattack? As far as I know, this miniversion of chess also has not been exhaustively analyzed.

Richard Bellman's prediction that checkers would be solved by the end of 1975—that is, it would become known if the first or second player had the win or the game is a draw—proved wide of the mark. As I write now (in 2000), checkers is far from decided, although checker programs today play on the same level as the best players.

Mikhail Botvinnik accurately predicted that computers would soon reach grand master level and that tournaments would be held between computer programs. Such contests are now annual events. In 1997 Deep Blue, IBM's chess program, defeated Gary Kasparov, the world's chess champion, in a six-game contest in Manhattan. Kasparov won only one game, tied three, and lost two.

There is now an enormous literature on the designing of computers that learn from experience and on the construction of better and better chess programs. I have made no attempt to introduce this literature in my brief bibliography.

Answers

The checker game on the 4 × 4 board is a draw if both sides play as well as possible. As shown in Figure 35.5, Black has a choice of three openings: (1) C5, (2) C6, (3) D6.

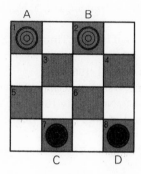

A   B

C   D

Figure 35.5. Checker game is drawn if played rationally

The first opening results in an immediate loss of the game when White replies A3. The second opening leads to a draw regardless of how White replies. The third opening is Black's strongest. It leads to a win if White replies A3 or B3. But White can reply B4 and draw.

With respect to the 3 × 3 simplified Go game, also mentioned as suitable for a matchbox learning machine, I am assured by Jay Eliasberg, vice-president of the American Go Association, that the first player has a sure win if he plays on the center point of the board and rationally thereafter.

The 4 × 4 checker game is trivial, but when the board is enlarged to 5 × 5 the result is both challenging and surprising. Robert L. Caswell, a chemist with the United States Department of Agriculture, wrote to me about this minigame, which he said had earlier been proposed to him. The game begins with three white checkers on the first row, three black checkers on the fifth row. All standard rules apply with Black moving first. One might guess the game to be drawn if played rationally, but the absence of "double corners" where kings can move back and forth makes this unlikely. Caswell discovered that not only does one side have a sure win but, if the loser plays well, the final win is spectacular. Rather than spoil the fun, I leave it to the reader to analyze the game and decide which player can always win.

Bibliography

W. R. Ashby, "Can a Mechanical Chess-player Outplay Its Designer?" *British Journal for the Philosophy of Science,* Vol. 3, No. 44, (1952).

A. Bernstein and M. de V. Roberts, "Computer v. Chess-player," *Scientific American,* June 1958, pp. 96–105.

A. Newell, J. C. Shaw, and H. A. Simon, "Chess-playing Programs and the Problem of Complexity," *IBM Journal of Research and Development,* Vol. 2, No. 4, October 1958, pp. 320–35.

A. Newell and H. A. Simon, "Computer Simulation of Human Thinking," *Science,* Vol. 134, 1961, pp. 2011–17.

N. Nilsson, *Learning Machines,* New York: McGraw-Hill, 1965.

A. L. Samuel, "Some Studies in Machine Learning, Using the Game of Checkers." *IBM Journal of Research and Development,* Vol. 3, No. 3, July 1959, pp. 210–29.

C. E. Shannon, "Game Playing Machines," *Journal of the Franklin Institute,* Vol. 260, No. 6, December 1955, pp. 447–53.

J. M. Smith and D. Michie, "Machines That Play Games," *The New Scientist,* No. 260, November 9, 1961, pp. 367–69.

Chapter 36

Sprouts and Brussels Sprouts

I made sprouts fontaneously. . . .
—James Joyce, *Finnegans Wake,* p. 542

"A friend of mine, a classics student at Cambridge, introduced me recently to a game called 'Sprouts' which became a craze at Cambridge last term. The game has a curious topological flavor."

So began a letter I received in 1967 from David Hartshorne, then a mathematics student at the University of Leeds. Soon other British readers were writing to me about this amusing pencil-and-paper game that had sprouted suddenly on the Cambridge grounds.

I am pleased to report that I successfully traced the origin of this game to its source: the joint creative efforts of John Horton Conway, then a teacher of mathematics at Sidney Sussex College, Cambridge, and Michael Stewart Paterson, then a graduate student working at Cambridge on abstract computer programming theory.

The game begins with *n* spots on a sheet of paper. Even with as few as three spots, Sprouts is more difficult to analyze than ticktacktoe, so that it is best for beginners to play with no more than three or four initial spots. A move consists of drawing a line that joins one spot to another or to itself and then placing a new spot anywhere along the line. These restrictions must be observed:

1. The line may have any shape but it must not cross itself, cross a previously drawn line, or pass through a previously made spot.
2. No spot may have more than three lines emanating from it.

Players take turns drawing curves. In normal Sprouts, the recommended form of play, the winner is the last person able to play. As in Nim and other games of the "take-away" type, the game can also be played in "misère" form, a French term that applies to a variety of card games in the whist family in which one tries to *avoid* taking tricks. In misère Sprouts the first person unable to play is the winner.

The typical three-spot normal game shown in Figure 36.1 was won on the seventh move by the first player. It is easy to see how the game got its name, for it sprouts into fantastic patterns as the game progresses. The most delightful feature is that it is not merely a combinatorial game, as so many connect-the-dots games are, but one that actually exploits the topological properties of the plane. In more technical language, it makes use of the Jordan-curve theorem, which asserts that simple closed curves divide the plane into outside and inside regions.

One might guess at first that a Sprouts game could keep sprouting for-

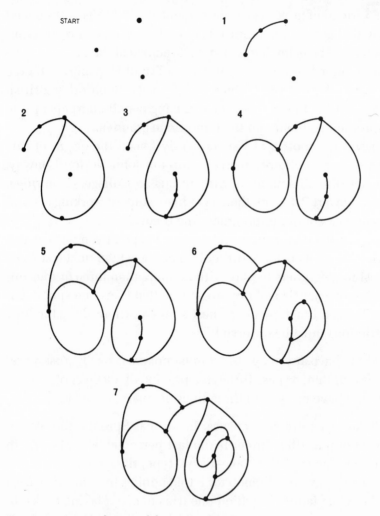

Figure 36.1. A typical game of three-spot Sprouts

ever, but Conway offers a simple proof that it must end in at most $3n - 1$ moves. Each spot has three "lives"—the three lines that may meet at that point. A spot that acquires three lines is called a "dead spot" because no more lines can be drawn to it. A game that begins with n spots has a starting life of $3n$. Each move kills two lives, at the beginning and at the end of the curve, but adds a new spot with a life of 1. Each move therefore decreases the total life of the game by 1. A game obviously cannot continue when only one life remains, since it requires at least two lives to make a move. Accordingly no game can last beyond $3n - 1$ moves. It is also easy to show that every game must last at least $2n$ moves. The three-spot game starts with nine lives, must end on or before the eighth move, and must last at least six moves.

The one-spot game is trivial. The first player has only one possible move: connecting the spot to itself. The second player wins in the normal game (loses in misère) by joining the two spots, either inside or outside the closed curve. These two second moves are equivalent, as far as playing the game is concerned, because before they are made there is nothing to distinguish the inside from the outside of the closed curve. Think of the game as being played on the surface of a sphere. If we puncture the surface by a hole inside a closed curve, we can stretch the surface into a plane so that all points previously outside the curve become inside, and vice versa. This topological equivalence of inside and outside is important to bear in mind because it greatly simplifies the analysis of games that begin with more than two spots.

With two initial spots, Sprouts immediately takes on interest. The first player seems to have a choice of five opening moves (see Figure 36.2), but the second and third openings are equivalent for reasons of symmetry. The same holds true of the fourth and fifth, and in light of the inside–outside equivalence just explained, all four of these moves can be considered identical. Only two topologically distinct moves, therefore, require exploring. It is not difficult to diagram a complete

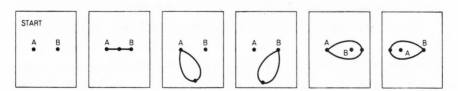

Figure 36.2. Initial spots *(A and B)* and first player's possible opening moves in two-spot game

tree chart of all possible moves, inspection of which shows that in both normal and misère forms of the two-spot game the second player can always win.

Conway found that the first player can always win the normal three-spot game and the second player can always win the misère version. Denis P. Mollison has shown that the first player has the win in normal four- and five-spot games. In response to a 10-shilling bet made with Conway that he could not complete his analysis within a month, Mollison produced a 49-page proof that the second player wins the normal form of the six-spot game. The second player wins the misère four-spot game.

Although no strategy for perfect play has been formulated, one can often see toward the end of a game how to draw closed curves that will divide the plane into regions in such a way as to lead to a win. It is the possibility of this kind of planning that makes Sprouts an intellectual challenge and enables a player to improve his skill at the game. But Sprouts is filled with unexpected growth patterns, and there seems to be no general strategy that one can adopt to make sure of winning.

Sprouts was invented on the afternoon of Tuesday, February 21, 1967, when Conway and Paterson had finished having tea in the mathematics department's common room and were doodling on paper in an effort to devise a new pencil-and-paper game. Conway had been working on a game invented by Paterson that originally involved the folding of attached stamps, and Paterson had put it into pencil-and-paper form. They were thinking of various ways of modifying the rules when Paterson remarked, "Why not put a new dot on the line?"

"As soon as this rule was adopted," Conway has written me, "all the other rules were discarded, the starting position was simplified to just n points, and Sprouts sprouted." The importance of adding the new spot was so great that all parties concerned agree that credit for the game should be on a basis of 3/5 to Paterson and 2/5 to Conway. "And there are complicated rules," Conway adds, "by which we intend to share any monies which might accrue from the game."

"The day after Sprouts sprouted," Conway continues, "it seemed that everyone was playing it. At coffee or tea times there were little groups of people peering over ridiculous to fantastic Sprout positions. Some people were already attacking Sprouts on toruses, Klein bottles, and the like, while at least one man was thinking of higher-dimensional versions. The secretarial staff was not immune; one found the remains of

Sprout games in the most unlikely places. Whenever I try to acquaint somebody new to the game nowadays, it seems he's already heard of it by some devious route. Even my three- and four-year-old daughters play it, though I can usually beat them."

The name "Sprouts" was given the game by Conway. An alternative name, "measles," was proposed by a graduate student because the game is catching and it breaks out in spots, but Sprouts was the name by which it quickly became known. Conway later invented a superficially similar game that he calls "Brussels Sprouts" to suggest that it is a joke. I shall describe this game but leave to the reader the fun of discovering why it is a joke before the explanation is given in the answer section.

Brussels Sprouts begins with n crosses instead of spots. A move consists of extending any arm of any cross into a curve that ends at the free arm of any other cross or the same cross; then a crossbar is drawn anywhere along the curve to create a new cross. Two arms of the new cross will, of course, be dead, since no arm may be used twice. As in Sprouts, no curve may cross itself or cross a previously drawn curve, nor may it go through a previously made cross. As in Sprouts, the winner of the normal game is the last person to play and the winner of the misère game is the first person who cannot play.

After playing Sprouts, Brussels Sprouts seems at first to be a more complicated and more sophisticated version. Since each move kills two crossarms and adds two live crossarms, presumably a game might never end. Nevertheless, all games do end and there is a concealed joke that the reader will discover if he succeeds in analyzing the game. To make the rules clear, a typical normal game of two-cross Brussels Sprouts is shown that ends with victory for the second player on the eighth move (see Figure 36.3).

A letter from Conway reports several important breakthroughs in Sproutology. They involve a concept he calls the "order of moribundity" of a terminal position, and the classification of "zero order" positions into five basic types: louse, beetle, cockroach, earwig, and scorpion (see Figure 36.4). The larger insects and arachnids can be infested with lice, sometimes in nested form, and Conway draws one pattern he says is "merely an inside-out earwig inside an inside-out louse." Certain patterns, he points out, are much lousier than others. And there is the FTOZOM (fundamental theorem of zero-order moribundity), which is quite deep. Sproutology is sprouting so rapidly that I shall have to postpone my next report on it for some time.

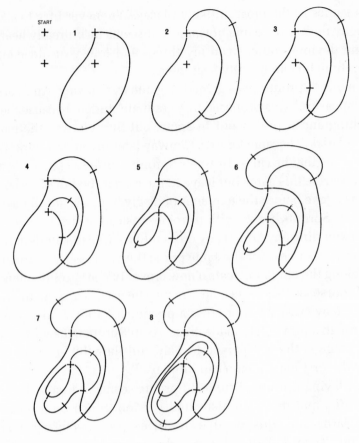

Figure 36.3. Typical game of two-cross Brussels Sprouts

Figure 36.4.

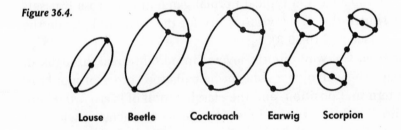

Louse   Beetle   Cockroach   Earwig   Scorpion

Addendum

Sprouts made an instant hit with *Scientific American* readers, many of whom suggested generalizations and variations of the game. Ralph J. Ryan III proposed replacing each spot with a tiny arrow, extending from one side of the line and allowing new lines to be drawn only to the arrow's point. Gilbert W. Kessler combined spots and cross-bars in a game he called "succotash." George P. Richardson investigated Sprouts on the torus and other surfaces. Eric L. Gans considered a generalization of Brussels Sprouts (called "Belgian Sprouts") in which spots are replaced by "stars"—n crossbars crossing at the same point. Vladimir Ygnetovich suggested the rule that a player, on each turn, has a choice of adding one, two, or no spots to his line.

Several readers questioned the assertion that every game of normal Sprouts must last at least $2n$ moves. They sent what they believed to be counterexamples, but in each case failed to notice that every isolated spot permits two additional moves.

Since this chapter appeared in *Scientific American* in 1967, David Applegate, Guy Jacobson, and Danny Sleator wrote the first computer program for analyzing Sprouts (see bibliography). They found that in normal Sprouts the first player wins when n (the number of spots) is 3, 4, 5, 9, 10, and 11. Beyond $n = 11$ their program was unable to cope with sprouting complexity. They did not go beyond $n = 9$ for misère Sprouts. The program proved a first player win when $n = 1$, 5, and 6.

The authors conjecture that in normal Sprouts the first player wins if $n = 3$, 4, or 5 (modulo 6) and wins in misère Sprouts if $n = 0$ or 1 (modulo 5). They add that Sleator believes both conjectures, but Applegate disbelieves both. Details about their program can be obtained from Applegate or Jacobson at Bell Labs, in Murray Hill, NJ, or Sleator at Carnegie Mellon University.

Answers

Why is the game of Brussels Sprouts, which appears to be a more sophisticated version of Sprouts, considered a joke by its inventor, John Horton Conway? The answer is that it is impossible to play Brussels Sprouts either well or poorly because every game must end in exactly $5n - 2$ moves, where n is the number of initial crosses. If played in standard form (the last to play is the winner), the game is always won by the first player if it starts with an odd number of crosses, by the second player if it starts with an even number. (The reverse is true, of course, in misère play.) After introducing someone to Sprouts, which is a genuine contest, one can switch to the fake game of Brussels Sprouts, make bets, and always know in advance who will win every game. I leave to the reader the task of proving that each game must end in $5n - 2$ moves.

Bibliography

P. Anthony, *Macroscopes,* Avon, 1969. In this novel by a popular science-fiction writer, Sprouts is discussed in the first chapter, and games are described and illustrated in several later chapters.

D. Applegate, G. Jacobson, and D. Sleator, "Computer Analysis of Sprouts," *The Mathematician and Pied Puzzler,* E. Berlekamp and T. Rodgers (eds.), A.K. Peters, 1999, pp. 199–201.

M. Baxter, "Unfair Games," *Eureka,* No. 50, April 1990, pp. 60–65.

E. Berkekamp, J. Conway, and R. Guy, "Sprouts and Brussels Sprouts," *Winning Ways,* Vol. 2, Academic Press, 1982, pp. 564–70.

M. Cooper, "Graph Theory and the Game of Sprouts," *American Mathematical Monthly,* Vol. 100, May 1993, pp. 478–82.

H. D'Alarcao and T. E. Moore, "Euler's Formula and a Game of Conway's," *Journal of Recreational Mathematics,* Vol. 9, No. 4, 1976–1977, pp. 249–51. The authors prove that Brussels Sprouts must end after $5n - 2$ moves.

T. K. Lam, "Connected Sprouts," *American Mathematical Monthly,* Vol. 101, February 1997, pp. 116–19.

Chapter 37

Harary's Generalized Ticktacktoe

*T*he world's simplest, oldest, and most popular pencil-and-paper game is still ticktacktoe, and combinatorial mathematicians, often with the aid of computers, continue to explore unusual variations and generalizations of it. In one variant that goes back to ancient times the two players are each given three counters, and they take turns first placing them on the 3 × 3 board and then moving them from cell to cell until one player gets his three counters in a row. Moving-counter ticktacktoe is the basis for a number of modern commercial games, such as John Scarne's Teeko and a game called Touché, in which concealed magnets cause counters to flip over and become opponent pieces.

Standard ticktacktoe can obviously be generalized to larger fields. For example, the old Japanese game of Go-Moku ("five stones") is essentially five-in-a-row ticktacktoe played on a Go board. Another way to generalize the game is to play it on "boards" of three or more dimensions.

In March, 1977, Frank Harary devised a delightful new way to generalize ticktacktoe. Harary was then a mathematician at the University of Michigan. He is now the Distinguished Professor of Computer Science at New Mexico State University, in Las Cruces. He has been called Mr. Graph Theory because of his tireless, pioneering work in this rapidly growing field that is partly combinatorial and partly topological. Harary is the founder of the *Journal of Combinatorial Theory* and the *Journal of Graph Theory,* and the author of *Graph Theory,* considered the world over to be the definitive textbook on the subject. His papers on graph theory, written alone or in collaboration with others, number more than 500. Harary ticktacktoe, as I originally called his generalization of the game, opens up numerous fascinating areas of recreational mathematics. Acting on his emphatic request, I now call it animal ticktacktoe for reasons we shall see below.

We begin by observing that standard ticktacktoe can be viewed as a two-color geometric-graph game of the type Harary calls an achievement game. Replace the nine cells of the ticktacktoe board with nine points joined by lines, as is shown in Figure 37.1. The players are each assigned a color, and they take turns coloring points on the graph. The first player to complete a straight line of three points in his color wins. This game is clearly isomorphic with standard ticktacktoe. It is well known to end in a draw if both players make the best possible moves.

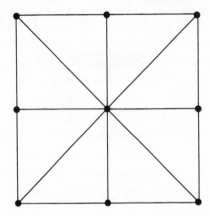

Figure 37.1. Ticktacktoe as a two-coloring game

Let us now ask: What is the smallest square on which the first player can force a win by coloring a straight (non-diagonal) three-point path? It is easy to show that it is a square of side four. Harary calls this side length the board number b of the game. It is closely related to the Ramsey number of generalized Ramsey graph theory, a number that plays an important part in the Ramsey games. (Ramsey theory is a field in which Harary has made notable contributions. It was in a 1972 survey paper on Ramsey theory that Harary first proposed making a general study of games played on graphs by coloring the graph edges.) Once we have determined the value of b we can ask a second question. In how few moves can the first player win? A little doodling shows that on a board of side four the first player can force a win in only three moves. Harary calls this the move number m of the game.

In ticktacktoe a player wins by taking cells that form a straight, order-3 polyomino that is either edge- or corner-connected. (The corner-connected figure corresponds to taking three cells on a diagonal.) Poly-ominoes of orders 1 through 5 are depicted in Figures 37.2 and 37.3. The polyomino terminology was coined by Solomon W. Golomb, who was the first to make a detailed study of these figures. Harary prefers to follow the usage of a number of early papers on the subject and call them "animals." I shall follow that practice here.

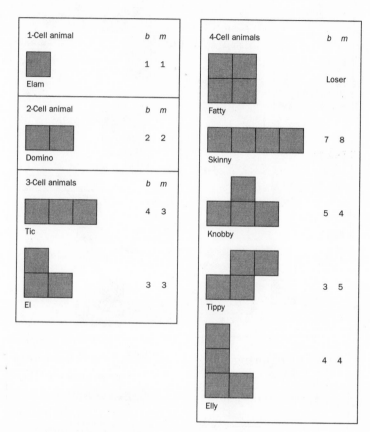

Figure 37.2. Animals of one through four cells

We are now prepared to explain Harary's generalization. Choose an animal of any order (number of square cells) and declare its formation to be the objective of a ticktacktoelike game. As in ticktacktoe we shall play not by coloring spots on a graph but by marking cells on square matrixes with noughts and crosses in the usual manner or by coloring cells red and green as one colors edges in a Ramsey graph game. Each player tries to label or color cells that will form the desired animal. The

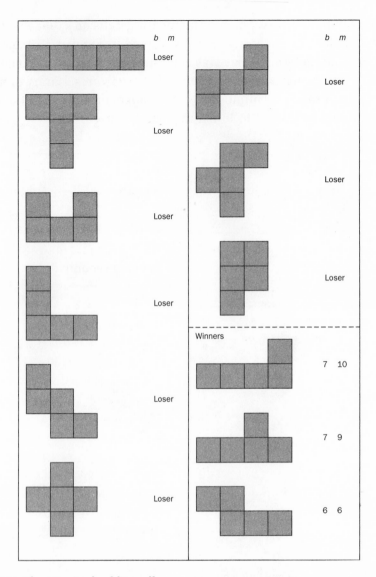

Figure 37.3. The 12 animals of five cells

animal will be accepted in any orientation and, if it is asymmetrical, in either of its mirror-image forms.

Our first task is to determine the animal's board number, that is, the length of the side of the smallest square on which the first player can, by playing the best possible strategy, force a win. If such a number exists, the animal is called a winner, and it will be a winner on all larger square fields. If there is no board number, the animal is called a loser.

If the animal chosen as the objective of a game is a loser, the second player can always force a draw, but he can never force a win. The clever proof of this fact is well known and applies to most ticktacktoelike games. Assume that the second player has a winning strategy. The first player can "steal" the strategy by first making an irrelevant opening move (which can never be a liability) and thereafter playing the winning strategy. This finding contradicts the assumption that the second player has a winning strategy, and so that assumption must be false. Hence the second player can never force a win. If the animal is a winner and b is known, we next seek m, the minimum number of moves in which the game can be won.

For the 1-cell animal (the monomino), which is trivially a winner, b and m are both equal to 1. When, as in this case, m is equal to the number of cells in the animal, Harary calls the game economical, because a player can win it without having to take any cell that is not part of the animal. The game in which the objective is the only 2-cell animal (the domino) is almost as trivial. It is also economical, with b and m both equal to 2. The games played with the two 3-cell animals (the trominoes) are slightly more difficult to analyze, but the reader can easily demonstrate that both are economical: for the L-shaped 3-cell animal b and m are both equal to 3, and for the straight 3-cell animal b equals 4 and m equals 3. This last game is identical with standard ticktacktoe except that corner-connected, or diagonal, rows of three cells are not counted as wins.

It is when we turn to the 4-cell animals (the tetrominoes) that the project really becomes interesting. Harary has given each of the five order-4 animals names, as is shown in Figure 37.2. Readers may enjoy proving that the b and m numbers given in the illustration are correct. Note that Fatty (the square tetromino) has no such numbers and so is labeled a loser. It was Andreas R. Blass, then one of Harary's colleagues at Michigan, who proved that the first player cannot force Fatty on a field of any size, even on the infinite lattice. Blass's result was the first surprise of the investigation into animal ticktacktoe. From this finding it follows at once that any larger animal containing a 2×2 square also is a loser: the second player simply plays to prevent Fatty's formation. More generally, any animal that contains a loser of a lower order is itself a loser. Harary calls a loser that contains no loser of lower order a basic loser. Fatty is the smallest basic loser.

The proof that Fatty is a minimal loser is so simple and elegant that it can be explained quickly. Imagine the infinite plane tiled with dominoes in the manner shown at the top of Figure 37.4. If Fatty is drawn anywhere on this tiling, it must contain a domino. Hence the second player's strategy is simply to respond to each of his opponent's moves by taking the other cell of the same domino. As a result the first player will never be able to complete a domino, and so he will never be able to complete a Fatty. If an animal is a loser on the infinite board, it is a loser on all finite boards. Therefore Fatty is always a loser regardless of the board size.

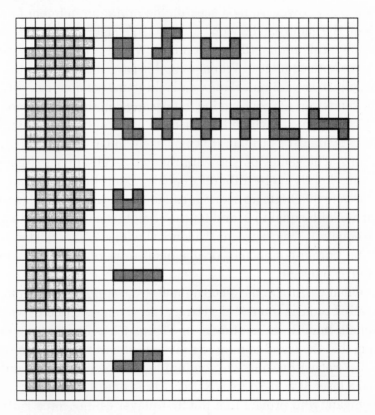

Figure 37.4. Tiling patterns *(left)* for the 12 basic losers *(right)*

Early in 1978 Harary and his colleagues, working with only the top four domino tilings shown in Figure 37.4 established that all but three of the 12 5-cell animals are losers. Among the nine losers only the one containing Fatty is not a basic loser. Turning to the 35 6-cell animals, all but four contain basic losers of lower order. Of the remaining four possible winners three can be proved losers with one of the five tilings

Games and Decision Theory

shown in the illustration. The animals that can be proved basic losers with each tiling pattern are shown alongside the pattern. In every case the proof is the same: it is impossible to draw the loser on the associated tiling pattern (which is assumed to be infinite) without including a domino; therefore the second player can always prevent the first player from forming the animal by following the strategy already described for blocking Fatty. There are a total of 12 basic losers of order 6 or lower.

It is worth noting how the tiling proof that the straight animal of five cells is a loser (another proof that was first found by Blass) bears on the game of Go-Moku. If the game is limited to an objective of five adjacent cells in a horizontal or vertical line (eliminating wins by diagonal lines), the second player can always force a draw. When diagonal wins are allowed, the game is believed to be a first-player win, although as far as I know that has not yet been proved even for fields larger than the Go board.

The only 6-cell animal that may be a winner is the one I named Snaky:

Although they have not yet been able to prove this animal is a winner, they conjecture its board number, if any, is no larger than 15 and its move number is no larger than 13. This assertion is the outstanding unsolved problem in animal ticktacktoe theory. Perhaps a reader can prove Snaky is a loser or conversely show how the first player can force the animal on a square field and determine its board and move numbers.

All the 107 order-7 animals are known to be losers because each contains a basic loser. Therefore since every higher-order animal must contain an order-7 animal, it can be said with confidence that there are no winners beyond order 6. If Snaky is a winner, as Harary and his former doctoral student Geoffrey Exoo conjecture, there are, by coincidence, exactly a dozen winners—half of them economical—and a dozen basic losers.

Any 4- or 5-cell animal can be the basis of a pleasant pencil-and-paper game or a board game. If both players know the full analysis, then depending on the animal chosen either the first player will win or the second player will force a draw. As in ticktacktoe between inexpert players, however, if this knowledge is lacking, the game can be entertaining. If the animal chosen as the objective of the game is a winner, the game is best played on a board of side b or $b - 1$. (Remember that a square of side $b - 1$ is the largest board on which the first player cannot force a win.)

All the variations and generalizations of animal ticktacktoe that have been considered so far are, as Harary once put it, "Ramseyish." For example, one can play the misère, or reverse, form of any game—in Harary's terminology an avoidance game—in which a player wins by forcing his opponent to color the chosen animal.

Avoidance games are unusually difficult to analyze. The second player trivially wins if the animal to be avoided is the monomino. If the domino is to be avoided, the second player obviously wins on the 2 × 2 square and almost as obviously on the 2 × 3 rectangle.

On a square board of any size the first player can be forced to complete the L-shaped 3-cell animal. Obviously the length of the square's side must be at least 3 for the game to be meaningful. If the length of the side is odd, the second player will win if he follows each of his opponent's moves by taking the cell symmetrically opposite the move with respect to the center of the board. If the first player avoids taking the center, he will be forced to take it on his last move and so will lose. If he takes it earlier in the game without losing, the second player should follow with any safe move. If the first player then takes the cell that is symmetrically opposite the second player's move with respect to the center, the second player should again make a harmless move, and so on; otherwise he should revert to his former strategy. If the length of the square's side is even, this type of symmetrical play leads to a draw, but the second player can still win by applying more complicated tactics.

On square boards the straight 3-cell animal cannot be forced on the first player. The proof of this fact is a bit difficult, even for the 3 × 3 square, but as a result no larger animal containing the straight 3-cell species can be forced on any square board. (The situation is analogous to that of basic losers in animal-achievement games.) Hence among the 4-cell animals only Fatty and Tippy remain as possible nondraws. Fatty can be shown to be a draw on any square board, but Tippy can be forced on the first player on all square boards of odd side. The complete analysis of all animal-avoidance games is still in the early stages and appears to present difficult problems.

Harary has proposed many other nontrivial variants of the basic animal games. For example, the objective of a game can be two or more different animals. In this case the first player can try to form one animal and the second player, the other, or both players can try to form either one. In addition, achievement and avoidance can be combined in the same game, and nonrectangular boards can be used. It is possible to include three or more players in any game, but this twist introduces coalition play and leads to enormous complexities. The rules can also be revised to accept corner-connected animals or animals that are both edge- and corner-connected. At the limit, of course, one could make any pattern whatsoever the objective of a ticktacktoelike game, but such broad generalizations usually lead to games that are too complicated to be interesting.

Another way of generalizing these games is to play them with polyiamonds (identical edge-joined equilateral triangles) or polyhexes (identical edge-joined regular hexagons) respectively on a regular triangular field or a regular hexagonal field. One could also investigate games played with these animals on less regular fields. An initial investigation of triangular forms, by Harary and Heiko Harborth, is listed in the bibliography.

The games played with square animals can obviously be extended to boards of three or more dimensions. For example, the 3-space analogue of the polyomino is the polycube: n unit cubes joined along faces. Given a polycube, one could seek b and m numbers based on the smallest cubical lattice within which the first player can force a win and try to find all the polycubes that are basic losers. This generalization is almost totally unexplored, but see the bibliography for a paper on the topic by Harary and Michael Weisbach.

As I have mentioned, Blass, now at Pennsylvania University, is one of Harary's main collaborators. The others include Exoo, A. Kabell, and Heiko Harborth, who is investigating games with the triangular and hexagonal cousins of the square animals. Harary is still planning a book on achievement and avoidance games in which all these generalizations of ticktacktoe and many other closely related games will be explored.

Addendum

In giving the proof that a second player cannot have the win in most ticktacktoelike games, I said that if the first player always wins on a board of a certain size, he also wins on any larger board. This is true of the square boards with which Harary was concerned, but is not necessarily true when such games are played on arbitrary graphs. A. K. Austin and C. J. Knight, mathematicians at the University of Sheffield, in England, sent the following counterexample.

Consider the graph at the left of Figure 37.5, on which three-in-a-row wins. The first player wins by taking A. The second player has a choice of taking a point in either the small or the large triangle. Whichever he chooses, the first player takes a corner point in the other triangle. The opponent must block the threatened win, then a play in the remaining corner of the same triangle forces a win.

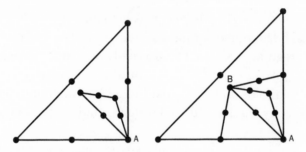

Figure 37.5. First player wins on graph at left, but second player can force a draw on enlarged graph at right.

Now enlarge the "board" by adding two points as shown on the right in Figure 37.5. The second player can draw by playing at B. If the first player does not start with A, the second player draws by taking A.

Bibliography

M. Gardner, "Ticktacktoe Games," *Wheels, Life, and Other Mathematical Amusements*, W. H. Freeman, 1983, Chapter 5.

S. W. Golomb, *Polyominoes*, Scribner's, 1965; Revised edition, Princeton University Press, 1994.

F. Harary, *Graph Theory*, Addison-Wesley, 1969.

F. Harary, "Achieving the Skinny Animal," *Eureka*, No. 42, Summer 1982, pp. 8–14; Errata, No. 43, Easter, 1983, pp. 5–6.

F. Harary, "Is Snaky a Winner?" *Combinatorics*, Vol. 2, April 1993, pp. 79–82. Harary offered $100 to the first person who could prove on or before December 31, 1995, whether Snaky is a winner. The question remains open.

F. Harary, *Achievement and Avoidance Games*, in preparation.

F. Harary and H. Harborth, "Extremal Animals," *Journal of Combinatorics, Information, and System Sciences*, 1, 1976, pp. 1–8.

F. Harary and H. Harborth, "Achievement and Avoidance Games with Triangular Animals," *Journal of Recreational Mathematics*, 18, 1985–86, pp. 110–15.

F. Harary and M. Weisbach, "Polycube Achievement Games," *Journal of Recreational Mathematics*, 15, 1982–83, pp. 241–46.

Chapter 38   *The New Eleusis*

> *I shall always consider the best guesser
> the best Prophet.*   —CICERO, *De Divinatione*

> *Don't never prophesy—onless ye know.*
> —JAMES RUSSELL LOWELL, *The Biglow Papers*

*I*n June 1959, I had the privilege of introducing in *Scientific American* a remarkable simulation game called Eleusis. The game, which is played with an ordinary deck of cards, is named for the ancient Eleusinian mysteries, religious rites in which initiates learned a cult's secret rules. Hundreds of ingenious simulation games have been developed for modeling various aspects of life, but Eleusis is of special interest to mathematicians and scientists because it provides a model of induction, the process at the very heart of the scientific method. Since then Eleusis has evolved into a game so much more exciting to play than the original version that I feel I owe it to readers to bring them up to date. I will begin, however, with some history.

Eleusis was invented in 1956 by Robert Abbott of New York, who at the time was an undergraduate at the University of Colorado. He had been studying that sudden insight into the solution of a problem that psychologists sometimes call the "Aha" reaction. Great turning points in science often hinge on these mysterious intuitive leaps. Eleusis turned out to be a fascinating simulation of this facet of science, even though Abbott did not invent it with this in mind. In 1963 Abbott's complete rules for the game appeared in his book, *Abbott's New Card Games* (hardcover, Stein & Day; paperback, Funk & Wagnalls).

Martin D. Kruskal, a distinguished mathematical physicist at Princeton University, became interested in the game and made several important improvements. In 1962 he published his rules in a monograph titled *Delphi: A Game of Inductive Reasoning.* Many college professors around the country used Eleusis and Delphi to explain scientific method to students and to model the Aha process. Artificial intelligence scientists wrote computer programs for the game. At the System Development Corporation in Santa Monica, research was done on Eleusis under the direction of J. Robert Newman. Litton Industries based a full-page advertisement on Eleusis. Descriptions of the game appeared in European books and periodicals. Abbott began receiving letters from all over the world with suggestions on how to make Eleusis a more playable game.

In 1973 Abbott discussed the game with John Jaworski, a young British mathematician who had been working on a computer version of Eleusis for teaching induction. Then Abbott embarked on a three-year program to reshape Eleusis, incorporating all the good suggestions he could. The new game is not only more exciting, its metaphorical level has been broadened as well. With the introduction of the roles of Prophet and False Prophet the game now simulates the search for any kind of truth. Here, then, based on a communication from Abbott, are the rules of New Eleusis as it is now played by aficionados.

At least four players are required. As many as eight can play, but beyond that the game becomes too long and chaotic.

Two standard decks, shuffled together are used. (Occasionally a round will continue long enough to require a third deck.) A full game consists of one or more rounds (hands of play) with a different player dealing each round. The dealer may be called by such titles as God, Nature, Tao, Brahma, the Oracle (as in Delphi), or just Dealer.

The dealer's first task is to make up a "secret rule." This is simply a rule that defines what cards can be legally played during a player's turn. In order to do well, players must figure out what the rule is. The faster a player discovers the rule, the higher his score will be.

One of the cleverest features of Eleusis is the scoring (described below), which makes it advantageous to the dealer to invent a rule that is neither too easy to guess nor too hard. Without this feature dealers would be tempted to formulate such complex rules that no one would guess them, and the game would become dull and frustrating.

An example of a rule that is too simple is: "Play a card of a color different from the color of the last card played." The alternation of colors would be immediately obvious. A better rule is: "Play so that primes and nonprimes alternate." For mathematicians, however, this might be too simple. For anyone else it might be too difficult. An example of a rule that is too complicated is: "Multiply the values of the last 3 cards played and divide by 4. If the remainder is 0, play a red card or a card with a value higher than 6. If the remainder is 1, play a black card or a picture card. If the remainder is 2, play an even card or a card with a value lower than 6. If the remainder is 3, play an odd card or a 10." No one will guess such a rule, and the dealer's score will be low.

Here are three examples of good rules for games with inexperienced players:

1. If the last legally played card was odd, play a black card. Otherwise play a red one.
2. If the last legally played card was black, play a card of equal or higher value. If the last card played was red, play a card of equal or lower value. (The values of the jack, queen, king, and ace are respectively 11, 12, 13, and 1.)
3. The card played must be either of the same suit or the same value as the last card legally played.

The secret rules must deal only with the sequence of legally played cards. Of course, advanced players may use rules that refer to the entire pattern of legal and illegal cards on the table, but such rules are much harder to guess and are not allowed in standard play. Under no circumstances should the secret rule depend on circumstances external to the cards. Examples of such improper rules are those that depend on the sex of the last player, the time of day, whether God scratches his (or her) ear, and so on.

The secret rule must be written down in unambiguous language, on a sheet of paper that is put aside for future confirmation. As Kruskal proposed, the dealer may give a truthful hint before the play begins. For example, he may say "Suits are irrelevant to the rule," or "The rule depends on the two previously played cards."

After the secret rule has been recorded, the dealer shuffles the double deck and deals 14 cards to each player and none to himself. He places a single card called the "starter" at the extreme left of the playing surface, as is indicated in Figure 38.1. To determine who plays first the dealer counts clockwise around the circle of players, starting with the player on his left and excluding himself. He counts until he reaches the number on the starter card. The player indicated at that number begins the play that then continues clockwise around the circle.

A play consists of placing one or more cards on the table. To play a single card the player takes a card from his hand and shows it to everyone. If according to the rule the card is playable, the dealer says "Right." The card is then placed to the right of the starter card, on the "main line" of correctly played cards extending horizontally to the right.

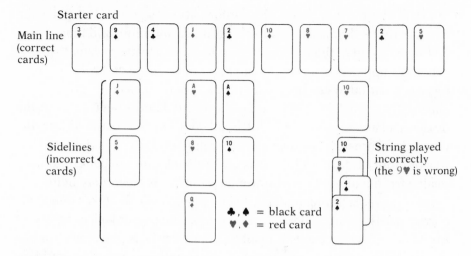

Figure 38.1. A typical round of Eleusis at an early stage

If the card fails to meet the rule, the dealer says "Wrong." In this case the card is placed directly below the last card played. Vertical columns of incorrect cards are called "sidelines." (Kruskal introduced both the layout and the terminology of the main line and sidelines.) Thus consecutive incorrect plays extend the same sideline downward. If a player displays a wrong card, the dealer gives him two more cards as a penalty, thereby increasing his hand.

If a player thinks he has discovered the secret rule, he may play a "string" of 2, 3, or 4 cards at once. To play a string he overlaps the cards slightly to preserve their order and shows them to everyone. If all the cards in the string conform to the rule, the dealer says "Right." Then all the cards are placed on the main line with no overlapping, as if they were correctly played single cards.

If one or more cards in a string are wrong, the dealer declares the entire string wrong. He does not indicate which cards do not conform to the rule. The wrong cards are left overlapping to keep their identity as a string and the entire string goes below the last card played. The player is then dealt twice as many cards as there are in the string.

The layout shown in Figure 38.1 demonstrates all the rules of Eleusis mentioned so far. The dealer's secret rule for this layout is the first of the three given above.

Players improve their score by getting rid of as many cards as possible, and of course they can do this best if they guess the secret rule. At the start of a round there is little information to go on, and plays are necessarily random. As the round continues and more and more information is added to the layout, the rule becomes steadily easier to guess.

It may happen that a player thinks he knows the secret rule but finds he has no card that can be legally played. He then has the option of declaring "No play." In this case he shows his hand to everyone. If the dealer declares him right and his hand contains four cards or less, the cards are returned to the deck and the round ends. If he is right and has five or more cards, then his cards are put back into the deck, and he is dealt a fresh hand with four fewer cards than he previously held.

If the player is wrong in declaring no play, the dealer takes one of his correct cards and puts it on the main line. The player keeps the rest of his hand and, as a penalty, is dealt five more cards. A player who thinks he has no correct play but has not figured out the secret rule should realize that the odds are against his using the no play option successfully. He would do better to play a card at random.

When a player thinks he knows the secret rule, he has the opportunity to prove it and increase his score. He does so by declaring himself a Prophet. The Prophet immediately takes over the dealer's duties, calling plays right or wrong and dealing penalty cards when the others play. He can declare himself a Prophet only if all the following conditions prevail:

1. He has just played (correctly or incorrectly), and the next player has not played.
2. There is not already a Prophet.
3. At least two other players besides himself and the dealer are still in the round.
4. He has not been a Prophet before in this round.

When a player declares himself a Prophet, he puts a marker on the last card he played. A chess king or queen may be used. The Prophet keeps his hand but plays no more cards unless he is overthrown. The play continues to pass clockwise around the player's circle, skipping the Prophet.

Each time a player plays a card or string, the Prophet calls the play right or wrong. The dealer then either validates or invalidates the Prophet's statement by saying "Correct" or "Incorrect." If the Prophet is correct, the card or string is placed on the layout—on the main line if right or on a sideline if wrong—and the Prophet gives the player whatever penalty cards are required.

If the dealer says "Incorrect," the Prophet is instantly overthrown. He is declared a False Prophet. The dealer removes the False Prophet's marker and gives him five cards to add to his hand. He is not allowed to become a Prophet again during the same round, although any other player may do so. The religious symbolism is obvious, but as Abbott points out, there is also an amusing analogy here with science: "The Prophet is the scientist who publishes. The False Prophet is the scientist who publishes too early." It is the fun of becoming a Prophet and of overthrowing a False Prophet that is the most exciting feature of New Eleusis.

After a Prophet's downfall the dealer takes over his former duties. He completes the play that overthrew the Prophet, placing the card or string in its proper place on the layout. If the play is wrong, however, no penalty cards are given. The purpose of this exemption is to encourage players to make unusual plays—even deliberately wrong ones—in the hope of overthrowing the Prophet. In Karl Popper's language, it encourages scientists to think of ways of "falsifying" a colleague's doubtful theory.

If there is a Prophet and a player believes he has no card to play, things get a bit complicated. This seldom happens, and so you can skip this part of the rules now and refer to it only when the need arises. There are four possibilities once the player declares no play:

1. Prophet says, "Right"; dealer says, "Correct." The Prophet simply follows the procedure described earlier.
2. Prophet says, "Right"; dealer says, "Incorrect." The Prophet is immediately overthrown. The dealer takes over and handles everything as usual, except that the player is not given any penalty cards.
3. Prophet says, "Wrong"; dealer says, "Incorrect." In other words, the player is right. The Prophet is overthrown, and the dealer handles the play as usual.
4. Prophet says, "Wrong"; dealer says, "Correct." In this case the Prophet now must pick one correct card from the player's hand and put it on the main line. If he does this correctly, he deals the player the five penalty cards and the game goes on. It is possible, however, for the Prophet to make a mistake at this point and pick an incorrect

card. If that happens, the Prophet is overthrown. The wrong card goes back into the player's hand and the dealer takes over with the usual procedure, except that the player is not given penalty cards.

After 30 cards have been played and there is no Prophet in the game, players are expelled from the round when they make a wrong play, that is, if they play a wrong card or make a wrong declaration of no play. An expelled player is given the usual penalty cards for his final play and then drops out of the round, retaining his hand for scoring.

If there is a Prophet, expulsions are delayed until at least 20 cards have been laid down after the Prophet's marker. Chess pawns are used as markers so that it is obvious when expulsion is possible. As long as there is no Prophet, a white pawn goes on every tenth card placed on the layout. If there is a Prophet, a black pawn goes on every tenth card laid down after the Prophet's marker. When a Prophet is overthrown, the black pawns and the Prophet's marker are removed.

A round can therefore go in and out of the phase when expulsions are possible. For example, if there are 35 cards on the layout and no Prophet, Smith is expelled when he plays incorrectly. Next Jones plays correctly and declares herself a Prophet. If Brown then plays incorrectly, she is not expelled because 20 cards have not yet been laid down after the Prophet's marker.

A round can end in two ways: (1) when a player runs out of cards or (2) when all players (excluding a Prophet, if there is one) have been expelled.

The scoring in Eleusis is as follows:

1. The greatest number of cards held by anyone (including the Prophet) is called the "high count." Each player (including the Prophet) subtracts the number of cards in his hand from the high count. The difference is his score. If he has no cards, he gets a bonus of four points.
2. The Prophet, if there is one, also gets a bonus. It is the number of main-line cards that follow his marker plus twice the number of side-line cards that follow his marker, that is, a point for each correct card since he became a Prophet and two points for each wrong card.
3. The dealer's score equals the highest score of any player. There is one exception: If there is a Prophet, count the number of cards (right and wrong) that precede the Prophet's marker and double this number; if the result is smaller than the highest score, the dealer's score is that smaller number.

If there is time for another round, a new dealer is chosen. In principle the game ends after every player has been dealer, but this could take most of a day. To end the game before everyone has dealt, each player adds his scores for all the rounds played plus 10 more points if he has not been a dealer. This compensates for the fact that dealers tend to have higher-than-average scores.

The layout in Figure 38.2 shows the end of a round with five players. Smith was the dealer. The round ended when Jones got rid of her cards. Brown was the Prophet and ended with 9 cards. Robinson was expelled when he incorrectly played the 10 of spades; he had 14 cards. Adams had 17 cards at the end of the game.

The high count is 17. Therefore Adams' score is 17 minus 17, or 0. Robinson's score is 17 minus 14, or 3. Jones receives 17 minus 0, or 17, plus a 4-point bonus for having no cards, so that her score is 21. Brown gets 17 minus 9, or 8, plus the Prophet's bonus of 34 (12 main-line and 11 sideline cards following his marker), making a total score of 42. This is the highest score for the round. Twice the number of cards preceding the Prophet's marker is 50. Smith, the dealer, receives 42 because it is the smaller of the two numbers 42 and 50.

Readers are invited to look over this layout and see if they can guess the secret rule. The play has been standard, and so that the rule is confined strictly to the main-line sequence.

Some miscellaneous advice from Abbott should help inexperienced Eleusis players. Since layouts tend to be large, the best way to play the game is on the floor. Of course a large table can be used as well as miniature cards on a smaller table. If necessary, the main line can be broken on the right and continued below on the left.

Remember that in Eleusis the dealer maximizes his score by choosing a rule that is neither too easy nor too difficult. Naturally this depends both on how shrewdly the dealer estimates the ability of the players and how accurately he evaluates the complexity of his rule. Both estimates require considerable experience. Beginning players tend to underestimate the complexity of their rules.

For example, the rule used in the first layout is simple. Compare it with: "Play a red card, then a black card, then an odd card, then an even card, and repeat cyclically." This rule seems to be simpler, but in practice the shift from the red–black variable to the even–odd variable makes it difficult to discover. Abbott points out that in general restrictive rules that allow only about a fourth of the cards to be acceptable on

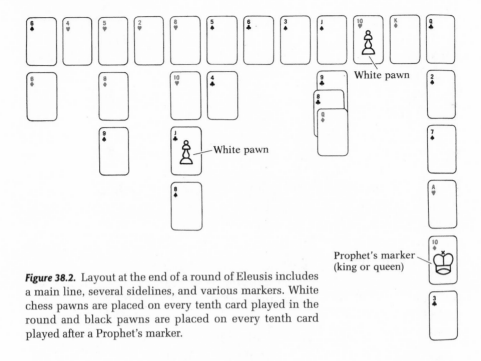

Figure 38.2. Layout at the end of a round of Eleusis includes a main line, several sidelines, and various markers. White chess pawns are placed on every tenth card played in the round and black pawns are placed on every tenth card played after a Prophet's marker.

any given play are easier to guess than less restrictive rules that allow half or more of the cards to be acceptable.

I shall not belabor the ways in which the game models a search for truth (scientific, mathematical, or metaphysical) since I discussed them in my first column on the game. I shall add only the fantasy that God or Nature may be playing thousands, perhaps a countless number, of simultaneous Eleusis games with intelligences on planets in the universe, always maximizing his or her pleasure by a choice of rules that the lesser minds will find not too easy and not too hard to discover if given enough time. The supply of cards is infinite, and when a player is expelled, there are always others to take his place.

Prophets and False Prophets come and go, and who knows when one round will end and another begin? Searching for any kind of truth is an exhilarating game. It is worth remembering that there would be no game at all unless the rules were hidden.

Addendum

Two unusual and excellent induction games have been invented since Eleusis and Delphi, both with strong analogies to scien-

GAMES AND DECISION THEORY

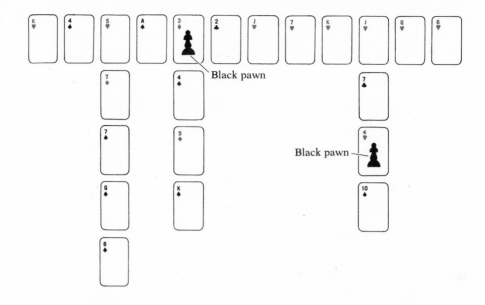

Black pawn

Black pawn

tific method. For Sid Sackson's board game Patterns, see Chapter 4 of my *Mathematical Circus* (Knopf, 1979). Pensari, using 32 special cards, was the brainchild of Robert Katz. His *Pensari Guide Book* (1986), which accompanies the cards, runs to 42 pages. Full-page advertisements for Pensari appeared in *Science News* in a number of 1987 issues.

Years ago, while reading *The Life and Letters of Thomas H. Huxley,* edited by his son Leonard (Vol. 1, Appleton, 1901, p. 262), I came across the following delightful paragraph. It is from a letter Huxley sent to Charles Kingsley in 1863.

> This universe is, I conceive, like to a great game being played out, and we poor mortals are allowed to take a hand. By great good fortune the wiser among us have made out some few of the rules of the game, as at present played. We call them "Laws of Nature," and honour them because we find that if we obey them we win something for our pains. The cards are our theories and hypotheses, the tricks our experimental verifications. But what sane man would endeavour to solve this problem: given the rules of a game and the winnings, to find whether the cards are made of pasteboard or goldleaf? Yet the problem of the metaphysicians is to my mind no saner.

In recent years Robert Abbott has been inventing bizarre mazes. Many of them can be found in his book *Mad Mazes,* published in 1990 by Bob Adams. "A Genius for Games," an article about Abbott by David Buxbaum, appeared in the *Mensa Bulletin,* May 1979.

Answers

The problem was to guess the secret rule that determined the final layout for a round in the card game Eleusis. The rule was: "If the last card is lower than the preceding legally played card, play a card higher than the last card, otherwise play a lower one. The first card played is correct unless it is equal to the starter card."

Bibliography

R. Abbott, *The New Eleusis,* Privately published, 1977.

T. Dietterich, master's thesis, "The Methodology of Knowledge Layers for Inducing Descriptions of Sequentially Ordered Events," Department of Computer Science, University of Illinois at Urbana, 1980.

M. D. Kruskal, *Delphi: A Game of Inductive Reasoning,* Plasma Physics Laboratory, Princeton University, 1962. A monograph of 16 pages.

H. C. Romesburg, "Simulating Scientific Inquiry with the Card Game Eleusis," *Science Education,* 3, 1979, pp. 599–608.

S. Sackson, "Eleusis: The Game with the Secret Rule," *Games,* May/June 1978, pp. 18–19.

R. W. Schmittberger, *New Rules for Classic Games,* Wiley, 1992.

X
Physics

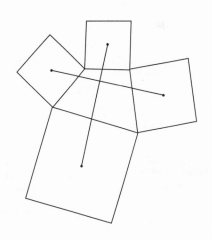

Chapter 39   *Time Travel*

"It's against reason," said Filby.
"What reason?" said the Time Traveller.

—H.G. WELLS, *The Time Machine*

H. G. Wells's short novel *The Time Machine,* an undisputed masterpiece of science fiction, was not the first story about a time machine. That distinction belongs to "The Clock That Went Backward," a pioneering but mediocre yarn by Edward Page Mitchell, an editor of the New York *Sun.* It was published anonymously in the *Sun* on September 18, 1881, seven years before young Wells (he was only 22) wrote the first version of his famous story.

Mitchell's tale was so quickly forgotten that science-fiction buffs did not even know of its existence until Sam Moskowitz reprinted it in his anthology of Mitchell's stories, *The Crystal Man* (1973). Nor did anyone pay much attention to Wells's fantasy when it was serialized in 1888 in *The Science Schools Journal* under the horrendous title "The Chronic Argonauts." Wells himself was so ashamed of this clumsily written tale that he broke it off after three installments and later destroyed all the copies he could find. A completely rewritten version, "The Time Traveller's Story," was serialized in *The New Review* beginning in 1894. When it came out as a book in 1895, it brought Wells instant recognition.

One of the many remarkable aspects of Wells's novella is the introduction in which the Time Traveller (his name is not revealed, but in Wells's first version he is called Dr. Nebo-gipfel) explains the theory behind his invention. Time is a fourth dimension. An instantaneous cube cannot exist. The cube we see is at each instant a cross section of a "fixed and unalterable" four-dimensional cube having length, breadth, thickness, and duration. "There is no difference between Time and any of the three dimensions of Space," says the Time Traveller, "except that our consciousness moves along it." If we could view a person from outside our space-time (the way human history is viewed by the Eternals in Isaac Asimov's *The End of Eternity* or by the Tralfamadorians in Kurt Vonnegut's *Slaughterhouse-Five*), we would see that person's past, present, and future all at once, just as in 3-space we see all parts of a wavy line that traces on a time chart the one-dimensional spatial movements of mercury in a barometer.

Reading these remarks today, one might suppose that Wells had been familiar with Hermann Minkowski's great work of tidying up Einstein's special theory of relativity. The line along which our consciousness crawls is, of course, our "world line": the line that traces our movements in 3-space on a four-dimensional Minkowski space-time graph. (*My World Line* is the title of George Gamow's autobiography.) But Wells's story appeared in its final form 10 years before Einstein published his first paper on relativity!

When Wells wrote his story, he regarded the Time Traveller's theories as little more than metaphysical hanky-panky designed to make his fantasy more plausible. A few decades later physicists were taking such hanky-panky with the utmost seriousness. The notion of an absolute cosmic time, with absolute simultaneity between distant events, was swept out of physics by Einstein's equations. Virtually all physicists now agree that if an astronaut were to travel to a distant star and back, moving at a velocity close to that of light, he could in theory travel thousands of years into the earth's future. Kurt Gödel constructed a rotating cosmological model in which one can, in principle, travel to any point in the world's past as well as future, although travel to the past is ruled out as physically impossible. In 1965 Richard P. Feynman received a Nobel prize for his space-time approach to quantum mechanics in which antiparticles are viewed as particles momentarily moving into the past.

Hundreds of science-fiction stories and in recent years many movies, have been based on time travel, many of them raising questions about time and causality that are as profound as they are sometimes funny. To give the most hackneyed example, suppose you traveled back to last month and shot yourself through the head. Not only do you know before making the trip that nothing like this happened but, assuming that somehow you could murder your earlier self, how could you exist next month to make the trip? Fredric Brown's "First Time Machine" opens with Dr. Grainger exhibiting his machine to three friends. One of them uses the device to go back 60 years and kill his hated grandfather when the man was a youth. The story ends 60 years later with Dr. Grainger showing his time machine to two friends.

It must not be thought that logical contradictions arise only when people travel in time. The transportation of *anything* can lead to paradox. There is a hint of this in Wells's story. When the Time Traveller sends a small model of his machine into the past or the future (he does not know which), his guests raise two objections. If the time machine went into the future, why do they not see it now, moving along its world line? If it went into the past, why did they not see it there before the Time Traveller brought it into the room?

One of the guests suggests that perhaps the model moves so fast in time it becomes invisible, like the spokes of a rotating wheel. But what if a time-traveling object stops moving? If you have no memory of a cube on the table Monday, how could you send it back to Monday's table on Tuesday? And if on Tuesday you go into the future, put the cube on the table Wednesday, then return to Tuesday, what happens on Wednesday if on Tuesday you destroy the cube?

Objects carried back and forth in time are sources of endless confusion in certain science-fiction tales. Sam Mines once summarized the plot of his own story, "Find the Sculptor," as follows: "A scientist builds a time machine, goes 500 years into the future. He finds a statue of himself commemorating the first time traveler. He brings it back to his own time and it is subsequently set up in his honor. You see the catch here? It had to be set up in his own time so that it would be there waiting for him when he went into the future to find it. He had to go into the future to bring it back so it could be set up in his own time. Somewhere a piece of the cycle is missing. When was the statue made?"

A splendid example of how paradox arises, even when nothing more than messages go back in time, is provided by the conjecture that tachyons, particles moving faster than light, might actually exist. Relativity theory leaves no escape from the fact that anything moving faster than light would move backward in time. This is what inspired A. H. Reginald Buller, a Canadian botanist, to write his often quoted limerick:

> *There was a young lady named Bright*
> *Who traveled much faster than light.*
> > *She started one day*
> > *In the relative way,*
> *And returned on the previous night.*

Tachyons, if they exist, clearly cannot be used for communication. G. A. Benford, D. L. Book, and W. A. Newcomb have chided physicists who are searching for tachyons for overlooking this. In "The Tachyonic Antitelephone," they point out that certain methods of looking for tachyons are based on interactions that make possible, in theory, communication by tachyons. Suppose physicist Jones on the earth is in communication by tachyonic antitelephone with physicist Alpha in another galaxy. They make the following agreement. When Alpha receives a message from Jones, he will reply immediately. Jones promises to send a message to Alpha at three o'clock earth time, if and only if he has not received a message from Alpha by one o'clock. Do you see the difficulty? Both messages go back in time. If Jones sends his message at three, Alpha's reply could reach him before one. "Then," as the authors put it, "the exchange of messages will take place if and only if it does not take place . . . a genuine . . . causal contradiction." Large sums of money have already gone down the drain, the authors believe, in efforts to detect tachyons by methods that imply tachyonic communication and are therefore doomed to failure.

Time dilation in relativity theory, time travel in Gödel's cosmos, and reversed time in Feynman's way of viewing antiparticles are so carefully hedged by other laws that contradictions cannot arise. In most time-travel stories the paradoxes are skirted by leaving out any incident that would generate a paradox. In some stories, however, logical contradictions explicitly arise. When they do, the author may leave them paradoxical to bend the reader's mind or may try to escape from paradox by making clever assumptions.

Before discussing ways of avoiding the paradoxes, brief mention should be made of what might be called pseudo-time-travel stories in which there is no possibility of contradiction. There can be no paradox, for example, if one simply observes the past but does not interact with it. The electronic machine in Eric Temple Bell's "Before the Dawn," which extracts motion pictures of the past from imprints left by light on ancient rocks, is as free of possible paradox as watching a video tape of an old television show. And paradox cannot arise if a person travels into the future by going into suspended animation, like Rip van Winkle, or Woody Allen in his motion picture *Sleeper,* or the sleepers in such novels as Edward Bellamy's *Looking Backward* or Wells's *When the Sleeper Wakes.* No paradox can arise if one dreams of the past (as in Mark Twain's *A Connecticut Yankee at King Arthur's Court* or in the 1986 motion picture *Peggy Sue Got Married*), or goes forward in a reincarnation, or lives for a while in a galaxy where change is so slow in relation to earth time that when he returns, centuries on the earth have gone by. But when someone actually travels to the past or the future, interacts with it, and returns, enormous difficulties arise.

In certain restricted situations paradox can be avoided by invoking Minkowski's "block universe," in which all history is frozen, as it were, by one monstrous space-time graph on which all world lines are eternal and unalterable. From this deterministic point of view one can allow certain kinds of time travel in either direction, although one must pay a heavy price for it. Hans Reichenbach, in a muddled discussion in *The Philosophy of Space and Time* (Dover, 1957, pp. 140–2), puts it this way: Is it possible for a person's world line to "loop" in the sense that it returns him to a spot in space-time, a spot very close to where he once had been and where some kind of interaction, such as speech, occurs between the two meeting selves? Reichenbach argues that this cannot be ruled out on logical grounds; it can only be ruled out on the ground that we would have to give up two axioms that are strongly confirmed by experience: (1) A person is a unique individual who maintains his identity as he ages, and (2) a person's world line is linearly ordered so that what he considers "now" is always a unique spot along the line. (Reichenbach does not mention it, but we would also have to abandon any notion of free will.) If we are willing to give up these things, says Reichenbach, we can imagine without paradox certain kinds of loops in a person's world line.

Reichenbach's example of a consistent loop is as follows. One day you meet a man who looks exactly like you but who is older. He tells you he is your older self who has traveled back in time. You think him insane and walk on. Years later you discover how to go back in time. You visit your younger self. You are compelled to tell him exactly what your older duplicate had told you when you were younger. Of course, he thinks that you are insane. You separate. Each of you leads a normal life until the day comes when your older self makes the trip back in time.

Hilary Putnam, in "It Ain't Necessarily So," argues in similar fashion that such world-line loops need not be contradictory. He draws a Feynman graph (see Figure 39.1) on which particle pair-production and pair-annihilation are replaced by person pair-production and pair-annihilation. The zigzag line is the world line of time traveler Smith. At time t_2 he goes back to t_1, converses with his younger self, then continues to lead a normal life. How would this be observed by someone

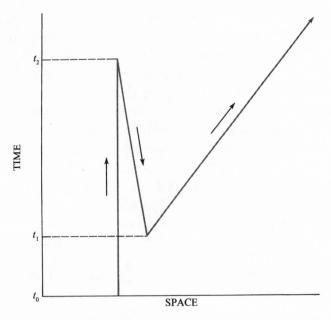

Figure 39.1. Feynman graph for a time traveler to the past

whose world line is normal? Simply put a ruler at the bottom of the chart, its edge parallel to the space axis, and move it slowly upward. At t_0 you see young Smith. At t_1 an older Smith suddenly materializes out of thin air in the same room along with an anti-Smith, who is seated in his time machine and living backward. (If he is smoking, you see his cigarette butt lengthen into a whole cigarette, and so on.) Perhaps the two forward Smiths converse. Finally, at t_2, young Smith, backward Smith, and the backward-moving time machine vanish. The older Smith and his older time machine continue on their way. The fact that we can draw a space-time diagram of these events, Putnam insists, is proof that they are logically consistent.

It is true that they are consistent, but note that Putnam's scenario, like Reichenbach's, involves such weak interaction between the Smiths that it evades the deeper contradictions that arise in time-travel fiction. What happens if the older Smith kills the younger Smith? Will Putnam kindly supply a Feynman graph?

There is only one good way out, and science-fiction scribblers have been using it for more than half a century. According to Sam Moskowitz, the device was first explicitly employed to resolve time-travel paradoxes by David R. Daniels in "Branches of Time," a tale that appeared in *Wonder Stories* in 1934. The basic idea is as simple as it is fantastic. Persons can travel to any point in the future of their universe, with no complications, but the moment they enter the past, the universe splits into two parallel worlds, each with its own time track. Along one track rolls the world as if no looping had occurred. Along the other track spins the newly created universe, its history permanently altered. When I say "newly created," I speak, of course, from the standpoint of the time traveler's consciousness. For an observer in, say, a fifth dimension the traveler's world line simply switches from one space-time continuum to another on a graph that depicts all the universes branching like a tree in a metauniverse.

Forking time paths appear in many plays, novels, and short stories by non-science-fiction writers. J. B. Priestley uses it in his popular play *Dangerous Corner,* as Lord Dunsany had done earlier in his play *If.* Mark Twain discusses it in *The Mysterious Stranger.* Jorge Luis Borges plays with it in his "Garden of Forking Paths." But it was the science-fiction writers who sharpened and elaborated the concept.

Let's see how it works. Suppose you go back to the time of Napoleon in Universe 1 and assassinate him. The world forks. You are now in Universe 2. If you like, you can return to the present of Universe 2, a universe in which Napoleon had been mysteriously murdered. How much would this world differ from the old one? Would you find a duplicate of yourself there? Maybe. Maybe not. Some stories assume that the slightest alteration of the past would introduce new causal chains that would have a multiplying effect and produce vast historical changes. Other tales assume that history is dominated by such powerful overall forces that even major alterations of the past would damp out and the future would soon be very much the same.

In Ray Bradbury's "A Sound of Thunder," Eckels travels back to an ancient geological epoch under elaborate precautions to prevent any serious alteration of the past. For example, he wears an oxygen mask to prevent his microbes from contaminating animal life. But Eckels violates a prohibition and accidentally steps on a living butterfly. When he returns to the present, he notices subtle changes in the office of the firm that arranged his trip. He is killed for having illegally altered the future.

Hundreds of other stories by fantasy and science-fiction writers have played variations on this theme. One of the saddest is Lord Dunsany's "Lost" (in *The Fourth Book of Jorkens,* 1948). A man travels to his past, by way of an Oriental charm, to right some old mistakes. Of course, this alters history. When he gets back to the present, he is missing his wife and home. "Lost! Lost!" he cries. "Don't go back down the years trying to alter anything. Don't even wish to. . . . And, mind you, the whole length of the Milky Way is more easily traveled than time, amongst whose terrible ages I am lost."

It is easy to see that in such a metacosmos of branching time paths, it is not possible to generate paradox. The future is no problem. If you travel to next week, you merely vanish for a week and reappear in the future a week younger than you would have been. But if you go back and murder yourself in your crib, the universe obligingly splits. Universe 1 goes on as before, with you vanishing from it when you grow up and make the trip back. Perhaps this happens repeatedly, each cycle creating two new worlds. Perhaps it happens only once. Who knows? In any case, Universe 2 with you and the dead baby in it rolls on. You are not annihilated by your deed, because now you are an alien from Universe 1 living in Universe 2.

In such a metacosmos it is easy (as many science-fiction writers have done) to fabricate duplicates of yourself. You can go back a year in Universe 1, live for a year with yourself in Universe 2, then again go back a year to visit two replicas of yourself in Universe 3. Clearly, by repeating such loops you can create as many replicas of yourself as you please. They are genuine replicas, not pseudo-replicas as in the scenarios by Reichenbach and Putnam. Each has his independent world line. History might become extremely chaotic, but there is one type of event that can never occur: a logically contradictory one.

This vision of a metacosmos containing branching worlds may seem crazy, but respectable physicists have taken it quite seriously. In Hugh Everett III's Ph.D. thesis " 'Relative State' Formulation of Quantum Mechanics" (*Reviews of Modern Physics* 29, July 1957, pp. 454–62) he outlines a metatheory in which the universe at every micromicroinstant branches into countless parallel worlds, each a possible combination of microevents that could occur as a result of microlevel uncertainty. The paper is followed by John A. Wheeler's favorable assessment in which he points out that classical physicists were almost as uncomfortable at first with the radical notions of general relativity.

"If there are infinite universes," wrote Fredric Brown in *What Mad Universe,* "then all possible combinations must exist. Then, somewhere, *everything must be true.* . . . There is a universe in which Huckleberry Finn is a real person, doing the exact things Mark Twain described him as doing. There are, in fact, an infinite number of universes in which a Huckleberry Finn is doing every possible variation of what Mark Twain *might* have described him as doing. . . . And infinite universes in which the states of existence are such that we would have no words or thoughts to describe them or to imagine them."

What if the universe never forks? Suppose there is only one world, this one, in which all world lines are linearly ordered and objects preserve their identity, come what may. Brown considers this possibility in his story "Experiment." Professor Johnson holds a brass cube in his hand. It is six minutes to three o'clock. At exactly three, he tells his colleagues, he will place the cube on his time machine's platform and send it five minutes into the past.

"Therefore," he remarks, "the cube should, at five minutes before three, vanish from my hand and appear on the platform, five minutes before I place it there."

"How can you place it there, then?" asked one of his colleagues.

"It will, as my hand approaches, vanish from the platform and appear in my hand to be placed there."

At five minutes to three the cube vanishes from Professor Johnson's hand and appears on the platform, having been sent back five minutes in time by his future action of placing the cube on the platform at three.

"See? Five minutes before I shall place it there, it *is* there!"

"But," says a frowning colleague, "what if, now that it has already appeared five minutes before you place it there, you should change your mind about doing so and *not* place it there at three o'clock? Wouldn't there be a paradox of some sort involved?"

Professor Johnson thinks this is an interesting idea. To see what happens, he does not put the cube on the platform at three.

There is no paradox. The cube remains. But the entire universe, including Professor Johnson, his colleagues, and the time machine, disappears.

Addendum

J. A. Lindon, a British writer of comic verse, sent me his sequel to the limerick about Miss Bright:

> *When they questioned her, answered Miss Bright,*
> *"I was there when I got home that night;*

So I slept with myself,
Like two shoes on a shelf,
Put-up relatives shouldn't be tight!"

Many readers called attention to two difficulties that could arise from time travel in either direction. If travelers stay at the same spot in space-time, relative to the universe, the earth would no longer be where it was. They might find themselves in empty space, or inside something solid. In the latter case, would the solid body prevent them from arriving? Would one or the other be shoved aside? Would there be an explosion?

The second difficulty is thermodynamic. After the time traveler departs, the universe will have lost a bit of mass-energy. When he arrives, the universe gains back the same amount. During the interval between leaving and arriving, the universe would seem to be violating the law of mass-energy conservation.

I mentioned briefly what is now called the "many-worlds interpretation" of QM (quantum mechanics). The best reference is a 1973 collection of papers on the topic, edited by Bryce DeWitt and Neill Graham. Assuming that the universe constantly splits into billions of parallel worlds, the interpretation provides an escape from the indeterminism of the Copenhagen interpretation of QM, as well as from the many paradoxes that plague it.

Some physicists who favor the many-worlds interpretation have argued that the countless duplicate selves and parallel worlds produced by the forking paths are not "real," but only artifacts of the theory. In this interpretation of the many-worlds interpretation, the theory collapses into no more than a bizarre way of saying the same things that are said in the Copenhagen interpretation. Everett himself, in his original 1957 thesis, added in proof this famous footnote:

In reply to a preprint of this article some correspondents have raised the question of the "transition from possible to actual," arguing that in "reality" there is—as our experience testifies—no such splitting of observer states, so that only one branch can ever actually exist. Since this point may occur to other readers the following is offered in explanation.

The whole issue of the transition from "possible" to "actual" is taken care of in the theory in a very simple way—there is no such transition, nor is such a transition necessary for the theory to be in accord with our experience. From the viewpoint of the theory *all* elements of a superposition (all "branches") are "actual," none any more "real" than the rest. It is unnecessary to suppose that all but one are somehow destroyed, since all the separate elements of a superposition individually obey the wave equation with complete indifference to the presence or absence ("actuality" or not) of any other elements. This total lack of effect of one branch on another also implies that no observer will ever be aware of any "splitting" process.

Arguments that the world picture presented by this theory is contradicted by experience, because we are unaware of any branching process, are like the criticism of the Copernican theory that the mobility of the earth as a real physical fact is incompatible with the common sense interpretation of nature because we feel no such motion. In both cases the argument fails when it is shown that the theory itself predicts that our experience will be what it in fact is. (In the Copernican case the addition of Newtonian physics was required to be able to show that the earth's inhabitants would be unaware of any motion of the earth.)

The many-worlds interpretation has been called a beautiful theory nobody can believe. Nevertheless, a number of top physicists have indeed accepted—some still do—its outrageous multiplicity of logically possible worlds. Here is DeWitt defending it in "Quantum Mechanics and Reality," a 1970 article reprinted in the collection he edited with Graham:

> The obstacle to taking such a lofty view of things, of course, is that it forces us to believe in the reality of all the simultaneous worlds . . . in each of which the measurement has yielded a different outcome. Nevertheless, this is precisely what [the inventors of the theory] would have us believe. . . . This universe is constantly splitting into a stupendous number of branches, all resulting from the measurement like interactions between its myriads of components. Moreover, every quantum transition taking place on every star, in every galaxy, in every remote corner of the universe is splitting our local world on earth into myriads of copies of itself.
>
> I still recall vividly the shock I experienced on first encountering this multiworld concept. The idea of 10^{100+} slightly imperfect copies of oneself all constantly splitting into further copies, which ultimately become unrecognizable, is not easy to reconcile with commonsense.

Although John Wheeler originally supported the many-worlds interpretation, he has since abandoned it. I quote from the first chapter of his *Frontiers of Time* (Center for Theoretical Physics, 1978):

Imaginative Everett's thesis is, and instructive, we agree. We once subscribed to it. In retrospect, however, it looks like the wrong track. First, this formulation of quantum mechanics denigrates the quantum. It denies from the start that the quantum character of nature is any clue to the plan of physics. Take this Hamiltonian for the world, that Hamiltonian, or any other Hamiltonian, this formulation says. I am a principle too lordly to care which, or why there should be any Hamiltonian at all. You give me whatever world you please, and in return I give you back many worlds. Don't look to me for help in understanding this universe.

Second, its infinitely many unobservable worlds make a heavy load of metaphysical baggage. They would seem to defy Mendeléev's demand of any proper scientific theory, that it should "expose itself to destruction."

Wigner, Weizsäcker, and Wheeler have made objections in more detail, but also in quite contrasting terms, to the relative-state or many-worlds interpretation of quantum mechanics. It is hard to name anyone who conceives of it as a way to uphold determinism.

In the paper titled "Rotating Cylinders and the Possibility of Global Causality Violation," physicist Frank Tipler raised the theoretical possibility of constructing a machine that would enable one to go forward or backward in time. (Tipler is one of the enthusiasts for the many-worlds interpretation, the coauthor of a controversial book (*The Anthropic Cosmological Principle,* Oxford University Press, 1986) and sole author of an even more controversial book, *The Physics of Immorality* (1994)). Taking off from Gödel's rotating cosmos and from recent work on the space-time pathologies surrounding black holes, Tipler imagines a massive cylinder, infinitely long, and rotating so rapidly that its surface moves faster than half the speed of light. Space-time near the cylinder would be so distorted that, according to Tipler's calculations, astronauts could orbit the cylinder, going with or against its spin, and travel into their past or future.

Tipler speculated on the possibility that such a machine could be built with a cylinder of finite length and mass, but later concluded that such a device was impossible to construct with any known forms of matter and force. Such doubts did not inhibit Poul Anderson from using Tipler's cylinder for time travel in his novel *The Avatar,* nor did it stop Robert Forward from writing "How to Build a Time Machine" (*Omni,* May 1980). "We already know the theory," *Omni* editors commented above Forward's backward article, "All that's needed is some advanced engineering."

I can understand the attraction of the many-world's interpretation of quantum mechanics provided it is taken as no more than another way of talking about quantum mechanics. If, however, the countless other universes are assumed to be as real as the one we are in,—real in the sense that the moon is real and unicorns are not—then the price one has to pay is horrendous. It is true that the many-world's interpretation simplifies concepts, but the multiplicity of other worlds is the greatest violation of Occam's razor in the history of theoretical physics. I applaud Wheeler for abandoning his early defense of the theory and admire Roger Penrose and other quantum experts for having the courage to reject the realistic interpretation of the many-world's interpretation as wild and temporary fashionable nonsense.

In 1995 England issued a stamp honoring H.G. Wells's *The Time Machine.*

I close with two pearls of wisdom from the stand-up comic "Professor" Irwin Corey: "The past is behind us" and "the future lies ahead."

Bibliography

B. Ash (ed.), "Time and the *N*th Dimension" and "Lost and Parallel Worlds," *The Visual Encyclopedia of Science Fiction,* Harmony, 1977.

G. A. Benford, D. L. Bock, and W. A. Newcomb, "The Tachyonic Antitelephone," *Physical Review D,* Vol. 2, July 15, 1970, pp. 263–65.

G. Benford, "Time Again," *Fantasy and Science Fiction,* July 1993, pp. 96–106.

S. K. Blau, "Would a Topology Change Allow Ms. Bright to Travel Backward in Time?" *American Journal of Physics,* Vol. 66, March 1998, pp. 179–85.

G. Conklin (ed.), *Science Fiction Adventures in Dimension,* Vanguard, 1953.

D. Deutsch and M. Lockwood, "The Quantum Physics of Time Travel," *Scientific American,* March 1994. (See also the letters section in September 1994.)

B. S. DeWitt and N. Graham (eds.), *The Many-Worlds Interpretation of Quantum Mechanics,* Princeton University Press, 1973.

J. Earman, "On Going Backward in Time," *Philosophy of Science,* Vol. 34, September 1967, pp. 211–22.

J. Earman, *Bangs, Crunches, Whimpers, and Shrieks,* Oxford University Press, 1995.

G. Feinberg, "Particles That Go Faster Than Light," *Scientific American,* February 1970, pp. 69–77.

D. Freedman, "Cosmic Time Travel," *Discover,* June 1989, pp. 58–63.

J. Richard Gott III, "Will We Travel Back (or Forward) in Time?" *Time,* April 10, 2000.

J. C. Graves and J. E. Roper, "Measuring Measuring Rods," *Philosophy of Science* 32, January 1965, pp. 39–56.

P. Horwich, "On Some Alleged Paradoxes of Time Travel," *Journal of Philosophy,* Vol. 72, 1975, pp. 432–44.

S. J. Jakiel and R. Levinthal, "The Laws of Time Travel," *Extrapolation,* Vol. 21, 1980, pp. 130–38.

D. Lewis, "The Paradoxes of Time Travel," *American Philosophical Quarterly,* Vol. 13, 1976, pp. 145–52.

L. Marder, *Time and the Space-Traveler,* Allen & Unwin, 1971.

H. Moravec, "Time Travel and Computing," *Extropy,* No. 9, Summer 1992, pp. 22–28.

P. J. Nahin, *Time Machines,* American Institute of Physics, 1993.

P. Nicholls (ed.), "Time Travel," 'Time Paradoxes," "Alternate Worlds," and "Parallel Worlds," *The Science Fiction Encyclopedia,* Doubleday, 1979.

B. Parker, *Cosmic Time Travel,* Plenum, 1991.

P. Parsons, "A Warped View of Time Travel," *Science,* Vol. 274, October 11, 1996, pp. 202–3.

J. B. Priestley, "Time and Fiction in Drama," *Man and Time,* Doubleday, 1964.

H. Putnam, "It Ain't Necessarily So," *The Journal of Philosophy,* Vol. 59, October 25, 1962, pp. 658–71; reprinted in Putnam's *Philosophical Papers,* Vol. 1, Cambridge University Press, 1975.

I. Redmount, "Wormholes, Time Travel, and Quantum Gravity," *New Scientist,* April 28, 1990, pp. 57–61.

L. S. Schulman, "Tachyon Paradoxes," *American Journal of Physics,* Vol. 39, May 1971, pp. 481–84.

J. J. C. Smart, "Is Time Travel Possible?" *The Journal of Philosophy,* Vol. 60, 1963, pp. 237–41.

N.J.J. Smith, "Bananas Enough for Time Travel," *British Journal of the Philosophy of Science,* Vol. 48, 1997, pp. 363–89.

P.V.D. Stern (ed.), *Travelers in Time,* Doubleday, 1947.

I. Stewart, "The Real Physics of Time Travel," *Analog Science Fiction and Fact,* January 1994, pp. 106–29.

F. J. Tipler, "Rotating Cylinders and the Possibility of Global Causality Violation," *Physical Review D,* Vol. 9, April 15, 1974, pp. 2203–6.

Chapter 40

Does Time Ever Stop?

It is impossible to meditate on time and the mystery of the creative passage of nature without an overwhelming emotion at the limitations of human intelligence.

—Alfred North Whitehead, *The Concept of Nature*

*T*here has been a great deal of interest among physicists as to whether there are events on the elementary-particle level that cannot be time-reversed, that is, events for which imagining a reversal in the direction of motion of all the particles involved is imagining an event that cannot happen in nature. Richard Feynman has suggested an approach to quantum mechanics in which antiparticles are viewed as particles momentarily traveling backward in time. Cosmologists have speculated about two universes for which all the events in one are reversed relative to the direction of time in the other: in each universe intelligent organisms would live normally from past to future, but if the organisms in one universe could in some way observe events in the other (which many physicists consider an impossibility), they would find those events going in the opposite direction. It has even been conjectured that if our universe stops expanding and starts to contract, there will be a time reversal, but it is far from clear what that would mean.

In this chapter I shall consider two bizarre questions about time. Indeed, these questions are of so little concern to scientists that only philosophers and writers of fantasy and science fiction have had much to say about them: Is it meaningful to speak of time stopping? Is it meaningful to speak of altering the past?

Neither question should be confused with the familiar subject of time's relativity. Newton believed the universe was pervaded by a single absolute time that could be symbolized by an imaginary clock off somewhere in space (perhaps outside the cosmos). By means of this clock the rates of all the events in the universe could be measured. The notion works well within a single inertial frame of reference such as the surface of the earth, but it does not work for inertial systems moving in relation to each other at high speeds. According to the theory of relativity, if a spaceship were to travel from our solar system to another solar system with a velocity close to that of light, events would proceed much slower on the spaceship than they would on the earth. In a sense, then, such a spaceship is traveling through time into the future. Passengers on the spaceship might experience a round-trip voyage as taking only a few years, but they would return to find that centuries of earth-years had elapsed.

The notion that different parts of the universe can change at different rates of time is much older than the theory of relativity. In the Scholastic theology of the Middle Ages angels were considered to be nonmaterial intelligences living by a time different from that of earthly creatures; God himself was thought to be entirely outside of time. In the first act of Lord Byron's play, *Cain, A Mystery,* the fallen angel Lucifer says:

> *With us acts are exempt from time*
>   *and we*
> *Can crowd eternity into an hour*
> *Or stretch an hour into eternity*
> *We breathe not by a mortal*
>   *measurement—*
> *But that's a mystery.*

In the 20th century hundreds of science-fiction stories played with the relativity of time in different inertial systems, but the view that time can speed up or slow down in different parts of our universe is central to many older tales. A popular medieval legend tells of a monk who is entranced for a minute or two by the song of a magical bird. When the bird stops singing, the monk discovers that several hundred years have passed. In a Moslem legend Mohammed is carried by a mare into the seventh heaven. After a long visit the prophet returns to the earth just in time to catch a jar of water the horse had kicked over before starting its ascent.

Washington Irving's "Rip Van Winkle" is this country's best-known story about someone who sleeps for what seems to him to be a normal time while two decades of earth-years rush by. King Arthur's daughter Gyneth slept for 500 years under a spell cast by Merlin. Every culture has similar sleeper legends. H. G. Wells used the device in *When the Sleeper Wakes,* and it is a common practice in science fiction to put astronauts into a cryogenic sleep so they can survive interstellar voyages that are longer than their normal life span. In Wells's short story "The

New Accelerator" a scientist discovers a way to speed up a person's biological time so that the world seems to come almost to a halt. This device too is frequently encountered in later science fiction.

The issue under consideration here, however, is not how time can vary but whether time can be said to stop entirely. It is clearly meaningful to speak of all motion ceasing in one part of the universe, whether or not such a part exists. In the theory of relativity the speed of light is an unattainable limit for any object with mass. If a spaceship could attain the speed of light (which the theory of relativity rules out because the mass of the ship would increase to infinity), then time on the spaceship would stop in the sense that all change on it would cease. In earth time it might take 100 years for the spaceship to reach a destination, but to astronauts on the spaceship the destination would be reached instantaneously. One can also imagine a piece of matter or even a human being reduced to such a low temperature (by some as yet unknown means) that even all subatomic motions would be halted. For that piece of matter, then, one could say that time had stopped. Actually it is hard to understand why the piece of matter would not vanish.

The idea of time stopping creates no problems for writers of fantasy who are not constrained by the real world. For example, in L. Frank Baum's "The Capture of Father Time," one of the stories in his *American Fairy Tales,* a small boy lassoes Time, and for a while everything except the movements of the boy and Father Time stops completely. In Chapter 22 of James Branch Cabell's *Jurgen: A Comedy of Justice,* outside time sleeps while Jurgen enjoys a pleasurable stay in Cocaigne with Queen Anaïtis. Later in the novel Jurgen stares into the eyes of the God of his grandmother and is absolutely motionless for 37 days. In Jorge Luis Borges' story "The Secret Miracle" a writer is executed by a firing squad. Between the command to fire and the writer's death God stops all time outside the writer's brain, giving him a year to complete his masterpiece.

Many similar examples from legend and literature show that the notion of time stopping in some part of the universe is not logically inconsistent. But what about the idea of time stopping throughout the universe? Does the notion that everything stops moving for a while and then starts again have any meaning?

If it is assumed that there is an outside observer—perhaps a god—watching the universe from a region of hypertime, then of course the notion of time stopping does have meaning, just as imagining a god in hyperspace gives meaning to the notion of everything in the universe turning upside down. The history of our universe may be like a three-dimensional motion picture a god is enjoying. When the god turns off the projector to do something else, a few millenniums may go by before he comes back and turns it on again. (After all, what are a few millenniums to a god?) For all we can know a billion centuries of hypertime may have elapsed between my typing the first and the second word of this sentence.

Suppose, however, all outside observers are ruled out and "universe" is taken to mean "everything there is." Is there still a way to give a meaning to the idea of all change stopping for a while? Although most philosophers and scientists would say there is not, a few have argued for the other side. For example, in "Time without Change" Sydney S. Shoemaker, a philosopher at Cornell University, makes an unusual argument in support of the possibility of change stopping.

Shoemaker is concerned not with the real world but with possible worlds designed to prove that the notion of time stopping everywhere can be given a reasonable meaning. He proposes several worlds of this kind, all of them based on the same idea. I shall describe only one such world here, in a slightly dramatized form.

Imagine a universe divided into regions A, B, and C. In normal times inhabitants of each region can observe the inhabitants of the other two and communicate with them. Every now and then, however, a mysterious purple glow permeates one of the regions. The glow always lasts for a week and is invariably followed by a year in which all change in the region ceases. In other words, for one year absolutely nothing happens there. Shoemaker calls the phenomenon a local freeze. Since no events take place, light cannot leave the region, and so the region seems to vanish for a year. When it returns to view, its inhabitants are unaware of any passage of time, but they learn from their neighbors that a year, as measured by clocks in the other two regions, has elapsed. To the inhabitants of the region that experienced the local freeze it seems that instantaneous changes have taken place in the other two regions. As Shoemaker puts it: "People and objects will appear to have moved in a discontinuous manner or to have vanished into thin air or to have materialized out of thin air; saplings will appear to have grown instantaneously into mature trees, and so on."

In the history of each of the three regions local freezes, invariably preceded by a week of purple light, have happened thousands of times. Now suppose that suddenly, for the first time in history, purple light appears simultaneously in regions *A, B,* and *C* and lasts for a week. Would it not be reasonable, Shoemaker asks, for scientists in the three regions to conclude that change had ceased for a year throughout the entire universe even though no minds were aware of it?

Shoemaker considers several objections to his thesis and counters all of them ingeniously. Interested readers can consult his paper and then read a technical analysis of it in the fifth chapter of G. Schlesinger's *Confirmation and Confirmability.* Schlesinger agrees with Shoemaker that an empirical, logically consistent meaning can be found for the sentence "A period of time *t* has passed during which absolutely nothing happened." Note that similar arguments about possible worlds can provide meanings for such notions as everything in a universe turning upside down, mirror-reversing, doubling in size, and so on.

The question of whether the past can be changed is even stranger than that of whether time can stop. Writers have often speculated about what might have happened if the past had taken a different turn. J. B. Priestley's play *Dangerous Corner* dealt with this question, and there have been innumerable "what if" stories in both science fiction and other kinds of literature. In all time-travel stories where someone enters the past, the past is necessarily altered. The only way the logical contradictions created by such a premise can be resolved is by positing a universe that splits into separate branches the instant the past is entered. In other words, while time in the old branch "gurgles on" (a phrase from Emily Dickinson) time in the new branch gurgles on in a different way toward a different future. When I speak of altering the past, however, I mean altering it throughout a single universe with no forking time paths. (Pseudoalterations of the past, such as the rewriting of history satirized by George Orwell in *1984,* obviously do not qualify.) Given this context, can an event, once it has happened, ever be made not to have happened?

The question is older than Aristotle, who in his *Ethics* (Book 6) writes: "It is to be noted that nothing that is past is an object of choice, for example, no one chooses to have sacked Troy; for no one *deliberates* about the past, but about what is future and capable of being otherwise, while what is past is not capable of not having taken place; hence Agathon is right in saying: 'For this alone is lacking even to God, to make undone things that have once been done.' "

Thomas Aquinas believed God to be outside of time and thus capable of seeing all his creation's past and future in one blinding instant. (Even though human beings have genuine power of choice, God knows how each one will choose; it is in this way that Aquinas sought to harmonize predestination and free will.) For Aquinas it was not possible for God to do absolutely impossible things, namely those that involve logical contradiction. For example, God could not make a creature that was both a human being and a horse (that is, a complete human being and a complete horse, rather than a mythical combination of parts such as a centaur), because that would involve the contradiction of assuming a creature to be simultaneously rational and nonrational.

Similarly, God cannot alter the past. That would be the same as asserting that the sack of Troy both took place and did not take place. Aquinas agreed with Aristotle that the past must forever be what it was, and it was this view that became the official position of medieval Scholasticism. It is not so much that God's omnipotence is limited by the law of contradiction but rather that the law is part of God's nature. "It is best to say," Aquinas wrote, "that what involves contradiction cannot be done rather than that God cannot do it." Modern philosophers would say it this way. God can't make a four-sided triangle, not because he can't make objects with four sides but because a triangle is *defined* as a three-sided polygon. The phrase "four-sided triangle" is therefore a nonsense phrase, one without meaning.

Edwyn Bevan, in a discussion of time in his book *Symbolism and Belief,* finds it odd that Aquinas would deny God the ability to alter the past and at the same time allow God to alter the future. In the 10th question of *Summa Theologica* (Ia. 10, article 5.3), Aquinas wrote: "God can cause an angel not to exist in the future, even if he cannot cause it not to exist while it exists, or not to have existed when it already has." For Aquinas to have suggested that for God the past is unalterable and the future is not unalterable, Bevan reasons, is surely to place God in some kind of time, thus contradicting the assertion that God is outside of time.

I know of no scientist or secular philosopher who has seriously believed the past could be altered, but a small minority of theologians have maintained that it could be. The greatest of them was Peter Damian, the zealous Italian reformer of the Roman Catholic church in the 11th century. In *On Divine Omnipotence,* his most controversial treatise, Damian argued that God is in no way bound by the law of con-

tradiction, that his omnipotence gives him the power to do all contradictory things including changing the past. Although Damian, who
started out as a hermit monk, argued his extreme views skillfully, he regarded all reasoning as superfluous, useful only for supporting revealed
theology. It appears that he, like Lewis Carroll's White Queen, would
have defended everyone's right to believe six impossible things before
breakfast. (Damian was also a great promoter of self-flagellation as a
form of penance, a practice that became such a fad during his lifetime
that some monks flogged themselves to death.)

One of my favorite Lord Dunsany stories is the best example I know
of from the literature of fantasy that illustrates Damian's belief in the
possibility of altering the past. It is titled "The King That Was Not," and
you will find it in Dunsany's early book of wonder tales *Time and the
Gods*. It begins as follows: "The land of Runazar hath no King nor ever
had one; and this is the law of the land of Runazar that, seeing that it
hath never had a King, it shall not have one for ever. Therefore in
Runazar the priests hold sway, who tell the people that never in
Runazar hath there been a King."

The start of the second paragraph is surprising: "Althazar, King of
Runazar. . . ." The story goes on to recount how Althazar ordered his
sculptors to carve marble statues of the gods. His command was
obeyed, but when the great statues were undraped, their faces were
very much like the face of the king. Althazar was pleased and rewarded
his sculptors handsomely with gold, but up in Pegāna (Dunsany's
Mount Olympus) the gods were outraged. One of them, Mung, leaned
forward to make his sign against Althazar, but the other gods stopped
him: "Slay him not, for it is not enough that Althazar shall die, who
hath made the faces of the gods to be like the faces of men, but he must
not even have ever been."

> "Spake we of Althazar, a King?"
> asked one of the gods.
> "Nay, we spake not."
> "Dreamed we of one Althazar?"
> "Nay, we dreamed not."

Below Pegāna, in the royal palace, Althazar suddenly passed out of
the memory of the gods and so "became no longer a thing that was or
had ever been." When the priests and the people entered the throne
room, they found only a robe and a crown. "The gods have cast away

the fragment of a garment," said the priests, "and lo! from the fingers of the gods hath slipped one little ring."

Addendum

When I wrote about time stopping I was using a colloquial expression to mean that change ceases. Because there are no moving "clocks" of any sort for measuring time, one can say in a loose sense that time stops. Of course time does not move or stop any more than length can extend or not extend. It is the universe that moves. You can refute the notion that time "flows" like a river simply by asking: "At what rate does it flow?" Shoemaker wanted to show in his paper not that time stops, then starts again, but that all change can stop and some sort of transcendental hypertime still persists. Change requires time, but perhaps, Shoemaker argued, time does not require change in our universe.

Harold A. Segal, in a letter in *The New York Times* (January 11, 1987) quoted a marvelous passage from Shakespeare's *As You Like It* (Act III, Scene 2) in which Rosalind explains how time can amble, trot, gallop, or stand still for different persons in different circumstances. It trots for the "young maid" between her engagement and marriage. It ambles for a priest who knows no Latin because he is free from the burden of "wasteful learning." It ambles for the rich man in good health who "lives merrily because he feels no pain." It gallops for the thief who awaits his hanging. For whom does it stand still? "With lawyers in the vacation; for they sleep between term and term, and then they perceive not how Time moves."

Isaac Asimov, in an editorial in *Asimov's Science-Fiction Magazine* (June 1986) explained why it would not be possible for a person to walk about and observe a world in which all change had stopped. To move, she would have to push aside molecules, and this would inject time into the outside world. She would be as frozen as the universe, even though dancing atoms in her brain might continue to let her think. Asimov could have added that she would not even be able to *see* the world because sight depends on photons speeding from the world into one's eyes.

Two readers, Edward Adams and Henry Lambert, independently wrote to say that the god Koschei, in *Jurgen,* could alter the past. At the end of the novel he eliminates all of Jurgen's adventures as never having happened. However, Jurgen recalls that Horvendile (the name Cabell often used for himself) once told him that he (Horvendile) and Koschei were one and the same!

Edward Fredkin is a computer scientist who likes to think of the universe as a vast cellular automaton run by an inconceivably complex algorithm that tells the universe how to jump constantly from one state to the next. Whoever or whatever is running the program could, of course, shut it down at any time, then later start it running again. We who are part of the program would have no awareness of such gaps in time.

On the unalterability of the past, readers reminded me of the stanza in Omar's *Rubaiyat* about the moving finger that having writ moves on, and all our piety and wit cannot call it back to cancel half a line. Or as Ogden Nash once put it:

> One thing about the past,
> It's likely to last.

I touched only briefly on the many science-fiction stories and novels that deal with time slowing down or halting. For references on some of the major tales see the section "When time stands still" on page 153 of *The Visual Encyclopedia of Science Fiction*. The most startling possibility, seriously advanced by some physicists, is that the universe comes to a complete stop billions of times every microsecond, then starts up again. Like a cellular automaton it jumps from state to state. Between the jumps, nothing changes. The universe simply does not exist. Time is quantized. An electron doesn't move smoothly from here to there. It moves in tiny jumps, occupying no space in between.

The fundamental unit of quantized time has been called the "chronon." Between chronons one can imagine one or more parallel universes operating within our space, but totally unknown to us. Think of a film with two unrelated motion pictures running on alternate frames. Between the frames of our universe, who knows what other exotic worlds are unrolling in the intervals between our chronons? Both motion pictures and cellular automata are deterministic, but in this vision of parallel universes running in the same space, there is no need to assume determinism. Chance and free will could still play creative roles in making the future of each universe unpredictable in principle.

Bibliography

E. Bevan, *Symbolism and Belief,* Kennikar Press, 1968.

E. C. Brewer, "Seven Sleepers" and "Sleepers," *The Reader's Handbook,* rev. ed., Lippincott, 1898.

C. D. Broad, "Time," *Encyclopedia of Religion and Ethics,* James Hastings (ed.), Scribner's, 1922.

K. Bulmer, "Time and Nth Dimensions," *The Visual Encyclopedia of Science Fiction,* Harmony Books, 1977, pp. 145–54.

M. Gardner, *The New Ambidextrous Universe,* W. H. Freeman, 1990.

P. Nicholls (ed.), "Time Travel," *The Science Fiction Encyclopedia.* Doubleday, 1979.

P. Nicholls (ed.), "Time Travel and Other Universes," *The Science in Science Fiction,* Knopf, 1983, Chapter 5.

G. Schlesinger, *Confirmation and Confirmability,* Oxford University Press, 1974.

S. S. Shoemaker, "Time Without Change," *The Journal of Philosophy,* Vol. 66, 1969, pp. 363–81.

P.V.D. Stern (ed.), *Travelers in Time: Strange Tales of Man's Journeying into the Past and the Future.* Doubleday, 1947.

Chapter 41

Induction and Probability

> The universe, so far as known to us,
> is so constituted, that whatever is
> true in any one case, is true in all
> cases of a certain description; the only
> difficulty is, to find what description.

—JOHN STUART MILL, *A System of Logic*

*I*magine that we are living on an intricately patterned carpet. It may or may not extend to infinity in all directions. Some parts of the pattern appear to be random, like an abstract expressionist painting; other parts are rigidly geometrical. A portion of the carpet may seem totally irregular, but when the same portion is viewed in a larger context, it becomes part of a subtle symmetry.

The task of describing the pattern is made difficult by the fact that the carpet is protected by a thick plastic sheet with a translucence that varies from place to place. In certain places we can see through the sheet and perceive the pattern; in others the sheet is opaque. The plastic sheet also varies in hardness. Here and there we can scrape it down so that the pattern is more clearly visible. In other places the sheet resists all efforts to make it less opaque. Light passing through the sheet is often refracted in bizarre ways, so that as more of the sheet is removed, the pattern is radically transformed. Everywhere there is a mysterious mixing of order and disorder. Faint lattices with beautiful symmetries appear to cover the entire rug, but how far they extend is anyone's guess. No one knows how thick the plastic sheet is. At no place has anyone scraped deep enough to reach the carpet's surface, if there is one.

Already the metaphor has been pushed too far. For one thing, the patterns of the real world, as distinct from this imaginary one, are constantly changing, like a carpet that is rolling up at one end while it is unraveling at the other end. Nevertheless, in a crude way the carpet can introduce some of the difficulties philosophers of science encounter in trying to understand why science works.

Induction is the procedure by which carpetologists, after examining parts of the carpet, try to guess what the unexamined parts look like.

Suppose the carpet is covered with billions of tiny triangles. Whenever a blue triangle is found, it has a small red dot in one corner. After finding thousands of blue triangles, all with red dots, the carpetologists conjecture that all blue triangles have red dots. Each new blue triangle with a red dot is a confirming instance of the law. Provided that no counterexample is found, the more confirming instances there are, the stronger is the carpetologists' belief that the law is true.

The leap from "some" blue triangles to "all" is, of course, a logical fallacy. There is no way to be absolutely certain, as one can be in working inside a deductive system, what any unexamined portion of the carpet looks like. On the other hand, induction obviously works, and philosophers justify it in other ways. John Stuart Mill did so by positing, in effect, that the carpet's pattern has regularities. He knew this reasoning was circular, since it is only by induction that carpetologists have learned that the carpet is patterned. Mill did not regard the circle as vicious, however, and many contemporary philosophers (R. B. Braithwaite and Max Black, to name two) agree. Bertrand Russell, in his last major work, tried to replace Mill's vague "nature is uniform" with something more precise. He proposed a set of five posits about the structure of the world that he believed were sufficient to justify induction.

Hans Reichenbach advanced the most familiar of several pragmatic justifications. If there is any way to guess what unexamined parts of the carpet look like, Reichenbach argued, it has to be by induction. If induction does not work, nothing else will, and so science might as well use the only tool it has. "This answer is not fallacious," wrote Russell, "but I cannot say that I find it very satisfying."

Rudolf Carnap agreed. His opinion was that all these ways of justifying induction are correct but trivial. If "justify" is meant in the sense that a mathematical theorem is justified, then David Hume was right: There is no justification. But if "justify" is taken in any of several weaker senses, then, of course, induction can be defended. A more interesting task, Carnap insisted, is to see whether it is possible to construct an inductive logic.

It was Carnap's great hope that such a logic could be constructed. He foresaw a future in which a scientist could express in a formalized language a certain hypothesis together with all the relevant evidence. Then by applying inductive logic, he could assign a probability value, called the degree of confirmation, to the hypothesis. There would be nothing final about that value. It would go up or down or stay the same as new evidence accumulated. Scientists already think in terms of such a logic, Carnap maintained, but only in a vague, informal way. As the tools of science become more powerful, however, and as our knowledge of probability becomes more precise, perhaps eventually we can create a calculus of induction that will be of practical value in the endless search for scientific laws.

In Carnap's *Logical Foundations of Probability* (University of Chicago Press, 1950) and also in his later writings, he tried to establish a base for such a logic. How successful he was is a matter of dispute. Some philosophers of science share his vision (John G. Kemeny for one) and have taken up the task where Carnap left off. Others, notably Karl Popper and Thomas S. Kuhn, regard the entire project as having been misconceived.

Carl G. Hempel, one of Carnap's admirers, has argued sensibly that before we try to assign quantitative values to confirmations, we should first make sure we know in a qualitative way what is meant by "confirming instance." It is here that we run into the worst kinds of difficulty.

Consider Hempel's notorious paradox of the raven. Let us approach it by way of 100 playing cards. Some of them have a picture of a raven drawn on the back. The hypothesis is: "All raven cards are black." You shuffle the deck and deal the cards face up. After turning 50 cards without finding a counterinstance, the hypothesis certainly becomes plausible. As more and more raven cards prove to be black, the degree of confirmation approaches certainty and may finally reach it.

Now consider another way of stating the same hypothesis: "All nonblack cards are not ravens." This statement is logically equivalent to the original one. If you test the new statement on another shuffled deck of 100 cards, holding them face up and turning them as you deal, clearly each time you deal a nonblack card and it proves to have no raven on the back, you confirm the guess that all nonblack cards are not ravens. Since this is logically equivalent to "All raven cards are black," you confirm that also. Indeed, if you deal all the cards without finding a red card with a raven, you will have completely confirmed the hypothesis that all raven cards are black.

Unfortunately, when this procedure is applied to the real world, it seems not to work. "All ravens are black" is logically the same as "All nonblack objects are not ravens." We look around and see a yellow object. Is it a raven? No, it is a buttercup. The flower surely confirms (albeit weakly) that all nonblack objects are not ravens, but it is hard to see how it has much relevance to "All ravens are black." If it does, it also confirms that all ravens are white or any color except yellow. To make things worse, "All ravens are black" is logically equivalent to "Any object is either black or not a raven." And that is confirmed by any black object whatever (raven or not) as well as by any nonraven (black or not). All of which seems absurd.

Nelson Goodman's "grue" paradox is equally notorious. An object is "grue" if it is green until, say, January 1, 2500, and blue thereafter. Is the law "All emeralds are grue" confirmed by observations of green emeralds? A prophet announces that the world will exist until January 1, 2500, when it will disappear with a bang. Every day the world lasts seems to confirm the prediction, yet no one supposes that it becomes more probable.

To make matters still worse, there are situations in which confirmations make a hypothesis less likely. Suppose you turn the cards of a shuffled deck looking for confirmations of the guess that no card has green spots. The first 10 cards are ordinary playing cards, then suddenly you find a card with blue spots. It is the eleventh confirming instance, but now your confidence in the guess is severely shaken. Paul Berent has pointed out several similar examples. A man 99 feet tall is discovered. He is a confirming instance of "All men are less than 100 feet tall," yet his discovery greatly *weakens* the hypothesis. Finding a normal-size man in an unlikely place (such as Saturn's moon Titan) is another example of a confirming instance that would weaken the same hypothesis.

Confirmations may even falsify a hypothesis. Ten cards with all values from the ace through the 10 are shuffled and dealt face down in a row. The guess is that no card with value n is in the nth position from the left. You turn the first nine cards. Each card confirms the hypothesis. But if none of the turned cards is the 10, the nine cards taken together refute the hypothesis.

Here is another example. Two piles of three cards each are on the table. One pile consists of the jack, queen, and king of hearts, the other consists of the jack, queen, and king of clubs. Each has been shuffled.

Smith draws a card from the heart pile, Jones takes a card from the club pile. The hypothesis is that the pair of selected cards consists of a king and queen. The probability of this is 2/9. Smith looks at his card and sees that it is a king. Without naming it, he announces that his card has confirmed the hypothesis. Why? Because knowing that his card is a king raises the probability of the hypothesis being true from 2/9 to 3/9 or 1/3. Jones now sees that he (Jones) has drawn a king, so he can make the same statement Smith made. Each card, taken in isolation, is a confirming instance. Yet, both cards taken together falsify the hypothesis.

Carnap was aware of such difficulties. He distinguished sharply between "degree of confirmation," a probability value based on the total relevant evidence, and what he called "relevance confirmation," which has to do with how new observations alter a confirmation estimate. Relevance confirmation cannot be given simple probability values. It is enormously complex, swarming with counterintuitive arguments. In Chapter 6 of Carnap's *Logical Foundations* he analyzes a group of closely related paradoxes of confirmation relevance that are easily modeled with cards.

For example, it is possible that data will confirm each of two hypotheses but disconfirm the two taken together. Consider a set of 10 cards, half with blue backs and half with green ones. The green-backed cards (with the hearts and spades designated H and S) are QH, $10H$, $9H$, KS, QS. The blue-backed cards are KH, JH, $10S$, $9S$, $8S$. The 10 cards are shuffled and dealt face down in a row.

Hypothesis A is that the property of being a face card (a king, a queen, or a jack) is more strongly associated with green backs than with blue. An investigation shows that this is true. Of the five cards with green backs, three are face cards versus only two face cards with blue backs. Hypothesis B is that the property of being a red card (hearts or diamonds) is also more strongly associated with green backs than with blue. A second investigation confirms this. Three green-backed cards are red, but there are only two red cards with blue backs. Intuitively one assumes that the property of being both red and a face card is more strongly associated with green backs than blue, but that is not the case. Only one red face card has a green back, whereas two red face cards have blue backs!

It is easy to think of ways, fanciful or realistic, in which similar situations can arise. A woman wants to marry a man who is both rich and kind. Some of the bachelors she knows have hair and some are bald. Being a statistician, she does some sampling. Project *A* establishes that 3/5 of the men with hair are rich but only 2/5 of the bald men are rich. Project *B* discloses that 3/5 of the men with hair are kind but only 2/5 of the bald men are kind. The woman might hastily conclude that she should marry a man with hair, but if the distribution of the attributes corresponds to that of the face cards and red cards mentioned in the preceding example, her chances of getting a rich, kind man are twice as great if she sets her sights on a bald man.

Another research project shows that 3/5 of a group of patients taking a certain pill are immune to colds for five years, compared with only 2/5 in the control group who were given a placebo. A second project shows that 3/5 of a group receiving the pill were immune to tooth cavities for five years, compared with 2/5 who got the placebo. The combined statistics could show that twice as many among those who got the placebo are free for five years from both colds and cavities, compared with those who got the pill.

A striking instance of how a hypothesis can be confirmed by two independent studies, yet disconfirmed by the total results, is provided by the following game. It can be modeled with cards, but to vary the equipment let's use 41 poker chips and four hats (see Figure 41.1). On table *A* is a black hat containing five colored chips and six white chips. Beside it is a gray hat containing three colored chips and four white chips. On table *B* is another pair of black and gray hats. In the black hat there are six colored chips and three white chips. In the gray hat there are nine colored chips and five white chips. The contents of the four hats are shown by the charts in the illustration.

You approach table *A* with the desire to draw a colored chip. Should you take a chip from the black hat or from the gray one? In the black hat five of the eleven chips are colored, so that the probability of getting a colored chip is 5/11. This is greater than 3/7, which is the probability of getting a colored chip if you take a chip from the gray hat. Clearly your best bet is to take a chip from the black hat.

The black hat is also your best choice on table *B*. Six of its nine chips are colored, giving a probability of 6/9, or 2/3, that you will get a colored chip. This exceeds the probability of 9/14 that you will get a colored chip if you choose to take a chip from the gray hat.

Now suppose that the chips from both black hats are combined in

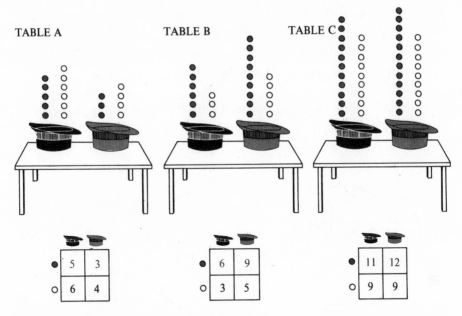

Figure 41.1. E. H. Simpson's reversal paradox

one black hat and that the same is done for the chips in the two gray hats (see Table C in Figure 41.1.) If you want to get a colored chip, surely you should take a chip from the black hat. The astonishing fact is that this is not true! Of the 20 chips now in the black hat, 11 are colored, giving a probability of 11/20 that you will get a colored chip. This is exceeded by a probability of 12/21 that you will get a colored chip if you take a chip from the gray hat.

The situation has been called Simpson's paradox by Colin R. Blyth, who found it in a 1951 paper by E. H. Simpson. The paradox has turned out to be older, but the name has persisted. Again, it is easy to see how the paradox could arise in actual research. Two independent investigations of a drug, for example, might suggest that it is more effective on men than it is on women, whereas the combined data would indicate the reverse.

One might imagine that such situations are too artificial to arise in statistical research. In an investigation to see if there was sex bias in the admissions of men and women to graduate studies at the University of California at Berkeley, however, Simpson's paradox actually turned up.

(See "Sex Bias in Graduate Admissions: Data from Berkeley," by P. J. Bickel, E. A. Hammel, and J. W. O'Connell.)

Blyth has invented another paradox that is even harder to believe than Simpson's. It can be modeled with three sets of cards or three unfair dice that are weighted to give the required probability distributions to their faces. We shall model it with the three spinners shown in Figure 41.2, because they are easy to construct by anyone who wants to verify the paradox empirically.

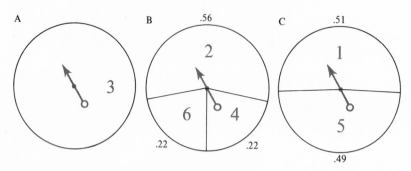

Figure 41.2. The three spinners for Colin R. Blyth's paradox

Spinner A, with an undivided dial, is the simplest. No matter where the arrow stops, it gives a value of 3. Spinner B gives values of 2, 4, or 6 with the respective probability distributions of .56, .22, and .22. Spinner C gives values of 1 or 5 with the probabilities of .51 and .49.

You pick a spinner; a friend picks another. Each of you flicks his arrow, and the highest number wins. If you can later change spinners on the basis of experience, which spinner should you choose? When the spins are compared in pairs, we find that A beats B with a probability of $1 \times .56 = .56$. A beats C with a probability of $1 \times .51 = .51$. B beats C with a probability of $(1 \times .22) + (.22 \times .51) + (.56 \times .51) = .6178$. Clearly A, which beats both of the others with a probability of more than 1/2, is the best choice. C is the worst because it is beaten with a probability of more than 1/2 by both of the others.

Now for the crunch. Suppose you play the game with two others and you have the first choice. The three spinners are flicked, and the high number wins. Calculating the probabilities reveals an extraordinary fact. A is the worst choice; C is the best! A wins with a probability of $.56 \times .51 = .2856$, or less than 1/3. B wins with a probability of $(.44 \times$

.51) + (.22 × .49) = .3322, or almost 1/3. C wins with a probability of .49 × .78 = .3822, or more than 1/3.

Consider the havoc this can wreak in statistical testing. Assume that drugs for a certain illness are rated in effectiveness with numbers 1 through 6. Drug *A* is uniformly effective at a value of 3 (spinner A). Studies show that drug *C* varies in effectiveness. Fifty-one percent of the time it has value 1, and 49 percent of the time it has value 5 (spinner C). If drugs *A* and *C* are the only two on the market and a doctor wants to maximize a patient's chance of recovery, he clearly chooses drug *A*.

What happens when drug *B*, with values and a probability distribution corresponding to spinner B, becomes available? The bewildered doctor, if he considers all three drugs, finds *C* preferable to *A*.

Blyth has an even more mind-blowing way of dramatizing the paradox. Every night, a statistician eats at a restaurant that offers apple pie and cherry pie. He rates his satisfaction with each kind of pie in values 1 through 6. The apple pie is uniformly 3 (spinner A); the cherry varies in the manner of spinner C. Naturally, the statistician always takes apple.

Occasionally the restaurant has blueberry pie. Its satisfaction varies in the manner of spinner B.

> Waitress: Shall I bring your apple pie?
> Statistician: No. Seeing that today you also have blueberry, I'll take the cherry.

The waitress would consider that a joke. Actually, the statistician is rationally maximizing his expectation of satisfaction. (An error. See Addendum.) Is there any paradox that points up more spectacularly the kinds of difficulty Carnap's followers must overcome in their efforts to advance his program?

Addendum

Many readers quite properly chided me for carelessness when I described Colin R. Blyth's paradox of the man and the three pies. It was I (not Blyth) who said that the man's decision was to maximize his "expectation of satisfaction." What he is maximizing is, in Blyth's words, "his best chance" of getting the most satisfying pie. It is a subtle but important difference. Both the dining statistician and the doctor have a choice between two intents: maximizing their average of satisfaction in the long run or maximizing their chance of getting the best pie or drug on a particular occasion.

To put it another way, Blyth's pie eater is minimizing his regret: the probability that he will see a better pie on the next table. His doctor counterpart, as Paul Chernick suggested, could be trying to avoid a malpractice suit that might result if a dissatisfied patient went to another doctor and got more effective treatment. "Is the case of a scientist closer to that of a player in the spinner game," asked George Mavrodes, "or is it closer to that of the statistical pie eater? . . . I do not know the answer to that question."

John F. Hamilton, Jr., revised the dialogue between the waitress and the statistician as follows:

Waitress: Which pie will be better tonight, *A* or *B*?
Statistician: The odds are on *A*.
Waitress: What about *A* and *C*?
Statistician: Again, *A* will probably win.
Waitress: I see, you mean *A* will probably be the best of all.
Statistician: No, actually *C* has the greatest chance of being the best.
Waitress: OK, cut the funny stuff. Which pie do you want to order, *A* or *C*?
Statistician: Neither. I'll have a slice of *B*, please.

The paradoxes of confirmation are not, of course, paradoxes in the sense of being contradictions, but paradoxes in the wider sense of being counterintuitive results that make nonsense of earlier naïve attempts, by John Stuart Mill and others, to define the meaning of "confirming instance." Philosophers who discuss the paradoxes are not ignorant of statistical theory. It is precisely because statistical theory demands so many careful distinctions that the task of formulating an inductive logic is so difficult.

Richard C. Jeffrey, in *The Logic of Decision Making* (University of Chicago Press, 2 ed., 1983), formulates an amusing variant of Goodman's "grue" paradox. We define a "goy" as a girl born before 2000 or a boy born after that date, and a "birl" as a boy born before 2000 or a girl born after that date. Until now, no goy has had a penis, and all birls have. Hence by induction, *(A)* the first goy born after 2000 will have no penis, and *(B)* the first birl born after that date will.

However, the first goy born after 2000 will be a boy, which contradicts *A*. Similarly, the first birl born after 2000 will be a girl, which contradicts *B*.

England's famous recent philosopher of science, Karl Popper, spent most of his life arguing against Rudolf Carnap that the so-called straight rule—that the more confirming evidence is found for a conjecture, the more the conjecture is probably true—has no role in deciding the worth of a theory. Instead, he insisted that science advances by having its conjectures pass tests for falsification. Carnap's commonsense view was that both confirmation and disconfirmation are essential to the scientific method.

Indeed, the falsification of a conjecture always adds confirmation to its opposite, and confirmation of a conjecture always tends to disconfirm its opposite. Consider the conjecture: There are life forms on Mars. If a Martian probe finds life forms it will confirm the conjecture and simultaneously falsify the conjecture that there are no life forms on Mars. And if the probe finds no life on Mars it will confirm the second conjecture and falsify the first. Moreover, falsification can be as fallible and uncertain as a confirmation. Popper got a lot of contorted debate out of his attack on what he called "induction." In my opinion his attack was extreme and is destined to fade from the history of the philosophy of science.

Bibliography

Confirmation Theory

P. Berent, "Disconfirmation by Positive Instances," *Philosophy of Science,* No. 39, December 1972, p. 522.

W. C. Salmon, "Confirmation," *Scientific American,* May 1973, pp. 75–83.

G. Schlesinger, *Confirmation and Confirmability,* Oxford University Press, 1974.

R. Swinburne, *An Introduction to Confirmation Theory,* Methuen, 1973.

Simpson's Paradox

E. F. Beckenbach, "Baseball Statistics," *Mathematics Teacher,* Vol. 72, May 1979, pp. 351–52.

P.J. Bickel, E. A. Hammel, and J. W. O'Connell, "Sex Bias in Graduate Admissions: Data from Berkeley," *Science,* Vol. 187, February 1985, pp. 398–404.

C. R. Blyth, "On Simpson's Paradox and the Sure-Thing Principle," *Journal of the American Statistical Association,* Vol. 67, June 1972, pp. 364–81.

R. Falk and M. Bar-Hillel, "Magic Possibilities of the Weighted Average," *Mathematics Magazine,* Vol. 53, March 1980, pp. 106–7.

T. R. Knapp, "Instances of Simpson's Paradox," *The College Mathematics Journal,* Vol. 16, June 1985, pp. 209–11.

C. Wagner, "Simpson's Paradox in Real Life," *American Statistician,* Vol. 36, February 1982, pp. 46–48.

Grue

F. Jackson, "Grue," *Journal of Philosophy,* Vol. 72, 1975, pp. 113–31.

D. Stalker, *Grue!,* Open Court, 1994. The book's bibliography lists 316 references.

R. Swinburne, "Grue," *Analysis,* Vol. 28, 1968, pp. 123–28.

T. E. Wilerson, "A Gruesome Note," *Mind,* Vol. 82, 1973, pp. 276–77.

Chapter 42   *Simplicity*

Fulfilling absolute decree
In casual simplicity.

       —EMILY DICKINSON

Ms. Dickinson's lines are about a small brown stone in a road, but if we view the stone as part of the universe, fulfilling nature's laws, all sorts of intricate and mysterious events are taking place within it on the microlevel. The concept of "simplicity," in both science and mathematics, raises a host of deep, complicated, still unanswered questions. Are the basic laws of nature few in number, or many, or perhaps infinite, as Stanislaw Ulam believed and others believe? Are the laws themselves simple or complex? Exactly what do we mean when we say one law or mathematical theorem is simpler than another? Is there any objective way to measure the simplicity of a law or a theory or a theorem?

Most biologists, especially those who are doing research on the brain, are impressed by the enormous complexity of living organisms. In contrast, although quantum theory has become more complicated with the discovery of hundreds of unexpected particles and interactions, most physicists retain a strong faith in the ultimate simplicity of basic physical laws.

This was especially true of Albert Einstein. "Our experience," he wrote, "justifies us in believing that nature is the realization of the simplest conceivable mathematical ideas." When he chose the tensor equations for his theory of gravitation, he picked the simplest set that would do the job, then published them with complete confidence that (as he once said to the mathematician John G. Kemeny) "God would not have passed up an opportunity to make nature that simple." It has even been argued that Einstein's great achievements were intellectual expressions of a psychological compulsion that Henry David Thoreau, in *Walden,* expressed as follows:

"Simplicity, simplicity, simplicity! I say, let your affairs be as two or three, and not a hundred or a thousand; instead of a million count half a dozen, and keep your accounts on your thumb nail."

In Peter Michelmore's biography of Einstein, he tells us that "Einstein's bedroom was monkish. There were no pictures on the wall, no carpet on the floor. . . . He shaved roughly with bar soap. He often went barefoot around the house. Only once every few months he would allow Elsa [his wife] to lop off swatches of his hair. . . . Most days he did not find underwear necessary. He also dispensed with pajamas and, later, with socks. 'What use are socks?' he asked. 'They only produce holes.' Elsa put her foot down when she saw him chopping off the sleeve of a new shirt from the elbow down. He explained that cuffs had to be buttoned or studded and washed frequently—all a waste of time."

"Every possession," Einstein said, "is a stone around the leg." The statement could have come straight out of *Walden*.

Yet nature seems to have a great many stones around her legs. Basic laws are simple only in first approximations; they become increasingly complex as they are refined to explain new observations. The guiding motto of the scientist, Alfred North Whitehead wrote, should be: "Seek simplicity and distrust it." Galileo picked the simplest workable equation for falling bodies, but it did not take into account the altitude of the body and had to be modified by the slightly more complicated equations of Newton. Newton too had great faith in simplicity. "Nature is pleased with simplicity," he wrote, echoing a passage in Aristotle, "and affects not the pomp of superfluous causes." Yet Newton's equations in turn were modified by Einstein, and there are physicists, such as Robert Dicke, who believe that Einstein's gravitational equations must be modified by still more complicated formulas.

It is dangerous to argue that because many basic laws are simple, the undiscovered laws also will be simple. Simple first approximations are obviously the easiest to discover first. Because the "aim of science is to seek the simplest explanations of complex facts," to quote Whitehead again (Chapter 7 of *The Concept of Nature*), we are apt to "fall into the error" of thinking that nature is fundamentally simple "because simplicity is the goal of our quest."

This we can say. Science sometimes simplifies things by producing theories that reduce to the same law phenomena previously considered unrelated—for example, the equivalence of inertia and gravity in general relativity. Science equally often discovers that behind apparently simple phenomena, such as the structure of matter, there lurks unsuspected complexity. Johannes Kepler struggled for years to defend the circular orbits of planets because the circle was the simplest closed curve. When he finally convinced himself that the orbits were ellipses, he wrote of the ellipse as "dung" he had to introduce to rid astronomy of vaster amounts of dung. It is a perceptive statement because it suggests that the introduction of more complexity on one level of a theory can introduce greater overall simplicity.

Nevertheless, at each step along the road simplicity seems to enter into a scientist's work in some mysterious way that makes the simplest workable hypothesis the best bet. "Simplest" is used here in a strictly objective sense, independent of human observation, even though no one yet knows how to define it. Naturally there are all sorts of ways one theory can be simpler than another in a pragmatic sense, but these ways are not relevant to the big question we are asking. As philosopher Nelson Goodman has put it, "If you want to go somewhere quickly, and several alternate routes are equally likely to be open, no one asks why you take the shortest." In other words, if two theories are not equivalent—lead to different predictions—and a scientist considers them equally likely to be true, he will test first the theory that he considers the "simplest" to test.

In this pragmatic sense, simplicity depends on a variety of factors: the kind of apparatus available, the extent of funding, the available time, the knowledge of the scientist and his assistants, and so on. Moreover, a theory may seem simple to one scientist because he understands the mathematics and complicated to another scientist less familiar with the math. A theory may have a simple mathematical form but predict complex phenomena that are difficult to test, or it may be a complicated theory that predicts simple results. As Charles Peirce pointed out, circumstances may be such that it is more economical to test first the *least* plausible of several hypotheses.

These subjective and pragmatic factors obviously play roles in research, but they fail to touch the heart of the mystery. The deep question is: Why, other things being equal, is the simplest hypothesis usually the most likely to be on the right track—that is, most likely to be confirmed by future research?

Consider the following "simple" instance of a scientific investigation. A physicist, searching for a functional relationship between two variables, records his observations as spots on a graph. Not only will he draw the simplest curve that comes close to the spots but also he allows simplicity to overrule the actual data. If the spots fall near a straight line, he will not draw a wavy curve that passes through every spot. He will assume that his observations are probably a bit off, pick a straight line that misses every spot, and guess that the function is a simple linear equation such as $x = 2y$ (see Figure 42.1). If this fails to predict new observations, he will try a curve of next-higher degree, say a hyperbola or a parabola. The point is that, other things being equal, the simpler curve has the higher probability of being right. A truly astonishing number of basic laws are expressed by low-order equations. Nature's preference for extrema (maxima and minima) is another familiar example of simplicity because in both cases they are the values when the function's derivative equals zero.

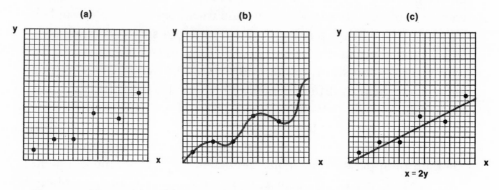

Figure 42.1. Observed data *(a)*, a possible function curve *(b)*, most likely function *(c)*

Simplicity sometimes overrules data in evaluating even the most complex, high-level theories, such as the theory of relativity or theories about elementary particles. If a theory is sufficiently simple and beautiful, and has great explanatory power, these facts often count more for it than early experiments, which seem to falsify the theory, count against it.

This raises some of the most perplexing questions in the philosophy of science. How can this particular kind of simplicity—the kind that adds to the probability that a law or theory is true—be defined? If it can be defined, can it be measured? Scientists tend to scorn both questions. They make intuitive judgments of simplicity without worrying about exactly what it is. Yet it is conceivable that someday a way to measure simplicity may have great practical value. Consider two theories that explain all known facts about fundamental particles. They are equal in their power to predict new observations, although the predictions are different. Both theories cannot be true. Both may be false. Each demands a different test and each test costs $1 million. If simplicity enters into the probability of a theory's being right, there is an obvious advantage in being able to measure simplicity so that the simplest theory can be tested first.

No one today knows how to measure this kind of simplicity or even how to define it. *Something* in the situation must be minimized, but what? It is no good to count the terms in a theory's mathematical formulation, because the number depends on the notation. The same formula may have 10 terms in one notation and three in another. Einstein's famous $E = mc^2$ looks simple only because each term is a shorthand symbol for concepts that can be written with formulas involving other concepts. This happens also in pure mathematics. The only way to express pi with integers is as the limit of an infinite series, but by writing π the entire series is squeezed into one symbol.

Minimizing the powers of terms also is misleading. For one thing, a linear equation such as $x = 2y$ graphs as a straight line only when the coordinates are Cartesian. With polar coordinates it graphs as a spiral. For another thing, minimizing powers is no help when equations are not polynomials. Even when they are polynomial, should one say that an equation such as $w = 13x + 23y + 132z$ is "simpler" than $x = y^2$?

In comparing the simplest geometric figures the notion of simplicity is annoyingly vague. In one of Johnny Hart's *B.C.* comic strips a caveman invents a square wagon wheel. Because it has too many corners and therefore too many bumps, he goes back to his drawing board and invents a "simpler" wheel in the shape of a triangle. Corners and bumps have been minimized, but the inventor is still further from the simplest wheel, the circle, which has no corners. Or should the circle be called the most complicated wheel because it is a "polygon" with an infinity of corners? The truth is that an equilateral triangle is simpler than a square in that it has fewer sides and corners. On the other hand, the square is simpler in that the formula for its area as a function of its side has fewer terms.

One of the most tempting of many proposed ways to measure the simplicity of a hypothesis is to count its number of primitive concepts. This, alas, is another blind alley. One can artificially reduce concepts by combining them. Nelson Goodman brings this out clearly in his famous "grue" paradox about which dozens of technical papers have been written. Consider a simple law: All emeralds are green. We now introduce the concept "grue." It is the property of being green if observed, say, before January 1, 2001, and being blue if observed thereafter. We state a second law: All emeralds are grue.

Both laws have the same number of concepts. Both have the same "empirical content" (they explain all observations). Both have equal predictive power. A single instance of a wrong color, when an emerald is examined at any future time, can falsify either hypothesis. Everyone prefers the first law because "green" is simpler than "grue"—it does not demand new theories to explain the sudden change of color of emeralds on January 1, 2001. Although Goodman has done more work than anyone on this narrow aspect of simplicity, he is still far from final results, to say nothing of the more difficult problem of measuring the overall simplicity of a law or theory. The concept of simplicity in science is far from simple! It may turn out that there is no single measure of simplicity but many different kinds, all of which enter into the complex final evaluation of a law or theory.

Surprisingly, even in pure mathematics similar difficulties arise. Mathematicians usually search for theorems in a manner not much different from the way physicists search for laws. They make empirical tests. In pencil doodling with convex quadrilaterals—a way of experimenting with physical models—a geometer may find that when he draws squares outwardly on a quadrilateral's sides and joins the centers of opposite squares, the two lines are equal and intersect at 90 degrees (see Figure 42.2). He tries it with quadrilaterals of different shapes, always getting the same results. Now he sniffs a theorem. Like

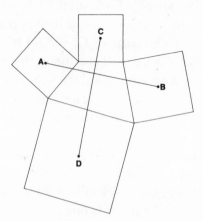

Figure 42.2. A "simple" geometrical theorem

a physicist, he picks the simplest hypothesis. He does not, for example, test first the conjecture that the two lines have a ratio of one to 1.00007 and intersect with angles of 89 and 91 degrees, even though this conjecture may equally well fit his crude measurements. He tests first the simpler guess that the lines are always perpendicular and equal. His "test," unlike the physicist's, is a search for a deductive proof that will establish the hypothesis with certainty.

Combinatorial theory is rich in similar instances where the simplest guess is usually the best bet. As in the physical world, however, there are surprises. Consider the following problem discovered by Leo Moser. Two or more spots are placed anywhere on a circle's circumference. Every pair is joined by a straight line. Given n spots, what is the maximum number of regions into which the circle can be divided? Figure 42.3 gives the answers for two, three, and four spots. The reader is asked to search for the answers for five and six spots and, if possible, find the general formula.

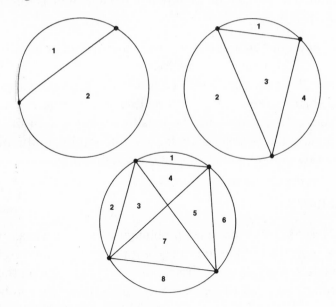

Figure 42.3. A combinatorial problem

Addendum

The beautiful theorem about the squares on the sides of an arbitrary convex quadrilateral is known as Von Aubel's theorem. Many readers, disappointed that I did not provide a proof, sent excellent proofs of their own. I lack space for any here, but you will find a simple vector proof in "Von Aubel's Quadrilateral Theorem," by Paul J. Kelly, *Mathematics Magazine,* January 1966, pages 35–37. A different proof based on symmetry operations is in *Geometric Transformations,* by I. M. Yaglom (Random House, 1962, pp. 95–96, problem 24b).

As Kelly points out, the theorem can be generalized in three ways that make it even more beautiful.

1. The quadrilateral need not be convex. The lines joining the centers of opposite squares may not intersect, but they remain equal and perpendicular.
2. Any three or even all four of the quadrilateral's corners may be collinear. In the first case the quadrilateral degenerates into a triangle with a "vertex" on one side; in the second case, into a straight line with two "vertices" on it.
3. One side of the quadrilateral may have zero length. This brings two corners together at a single point which may be treated as the center of a square of zero size.

The second and third generalizations were discovered by a reader, W. Nelson Goodwin, Jr., who drew the four examples shown in Figure 42.4. Note that the theorem continues to hold if opposite sides of a quadrilateral shrink to zero. The resulting line may be regarded as one of the lines connecting midpoints of opposite squares of zero size, and of course it equals and is perpendicular to a line joining midpoints of two squares drawn on opposite sides of the original line.

Answers

Leo Moser's circle-and-spots problem is an amusing example of how easily an empirical induction can go wrong in experimenting with pure mathematics. For one, two, three, four, and five spots on the circle, the maximum number of regions into which the circle can be divided by joining all pairs of spots with straight lines is 1, 2, 4, 8, 16. . . . One might conclude that this simple doubling series continues and that the maximum number of regions for n spots is 2^{n-1}. Unfortunately this formula fails for all subsequent numbers of spots. Figure 42.5 shows how six spots give a maximum of 31, not 32, regions. The correct formula is:

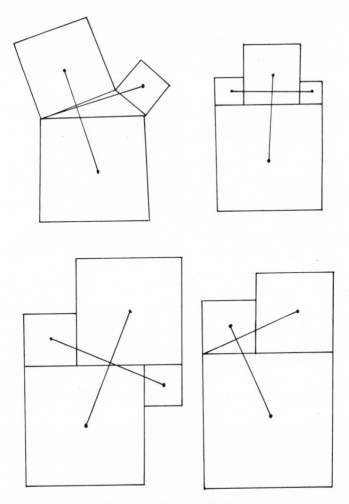

Figure 42.4. Curious generalizations of Von Aubel's theorem

$$n + \binom{n}{4} + \binom{n-1}{2}.$$

A parenthetical expression $\binom{m}{k}$ is the number of ways m objects can be combined, taken k at a time. (It equals $m!/k!(m-k)!$) Moser has pointed out that the formula gives the sum of the rows of numbers at the left of the diagonal line drawn on Pascal's triangle, as shown in the illustration.

Written out in full, the formula is:

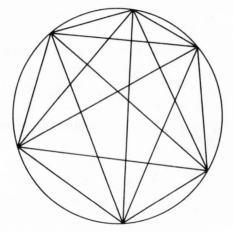

SPOTS									REGIONS
1				1					1
2			1		1				2
3		1		2		1			4
4	1		3		3		1		8
5	1	4		6		4	1		16
6	1	5	10		10	5	1		31
7	1	6	15	20		15	6	1	57

Figure 42.5. Solution to Leo Moser's spot problem

$$\frac{n^4 - 6n^3 + 23n^2 - 18n + 24}{24}.$$

When the positive integers are plugged into n, the formula generates the sequence: 1, 2, 4, 8, 16, 31, 57, 99, 163, 256, 386, 562. . . . The problem is a delightful illustration of Whitehead's advice to seek simplicity but distrust it.

I have been unable to determine where or when Moser first published this problem, but in a letter he says he thinks it was in *Mathematics Magazine* about 1950. It has since appeared in numerous books and periodicals, with varying methods of solution. A partial list of references is given in the bibliography for this chapter.

Almost no progress, if any, has been made in finding a way to measure the simplicity of a theory in a manner that would be useful to scientists. One recent suggestion, based on a technique for defining random numbers, is to translate a theory into a string of binary digits.

The theory's simplicity is then defined by the length of the shortest computer program that will print the string. This isn't much help. Quite apart from the difficulty of finding the shortest algorithm for a long binary expression, how is the string formed in the first place? How, for example, can you express superstring theory by a string of binary digits?

Scientists all agree that somehow the simpler of two theories, each with the same explanatory and predictive power, has the better chance of being fruitful, but no one knows why. Maybe it's because the ultimate laws of nature are simple, but who can be positive there really are ultimate laws? Some physicists suspect there may be infinite levels of complexity. At each level the laws may get progressively simpler until suddenly the experimenters open a trap door and another complicated subbasement is discovered.

Bibliography

S. Barker, "On Simplicity in Empirical Theories," *Philosophy of Science,* Vol. 28, 1961, pp. 162–71.

M. Bunge, *The Myth of Simplicity,* Prentice-Hall, 1963.

K. Friedman, "Empirical Simplicity and Testability," *British Journal for the Philosophy of Science,* Vol. 28, 1972, pp. 25–33.

N. Goodman, "The Test of Simplicity," *Science,* Vol. 128, October 31, 1958, pp. 1064–69.

N. Goodman "Science and Simplicity," *Philosophy of Science Today,* S. Morgenbesser (ed.), Basic Books, 1967.

C. G. Hempel, "Simplicity," *Philosophy of Natural Science,* Prentice-Hall, 1966, pp. 40–45.

M. Hesse "Simplicity," *The Encyclopedia of Philosophy,* Macmillan Company and the Free Press, 1967.

M. Hesse, "Ramifications of 'Grue,' " *The British Journal for the Philosophy of Science,* Vol. 20, May 1969, pp. 13–25.

W. Jefferson and J. Berger, "Occam's Razor and Bayesian Analysis," *American Scientist,* Vol. 80, January/February 1992, pp. 64–72.

J. G. Kemeny, "Two Measures of Simplicity," *Journal of Philosophy,* Vol. 52, 1955, pp. 722–33.

K. Popper, *The Logic of Scientific Discovery,* Basic Books, 1959, Chapter 7.

H. R. Post, "Simplicity in Scientific Theories," *British Journal of the Philosophy of Science,* Vol. 11, May 1960, pp. 32–41.

W. V. Quine and J. S. Ullian, "Hypothesis," *The Web of Belief,* Random House, 1970, Chapter V.

G. Schlesinger, "The Principle of Minimal Assumption," *British Journal of the Philosophy of Science,* Vol. 11, May 1960, pp. 55–59. See also *Discussion,* Vol. 11, February 1961, pp. 328–31.

G. Schlesinger, "Induction and Parsimony," *American Philosophical Quarterly,* Vol. 8, April 1971.

E. Sober, *Simplicity,* Clarendon Press, 1975.

R. Wright, "The Razor's Edge," *The Sciences,* July/August, 1987, pp. 2–5.

XI

Logic and Philosophy

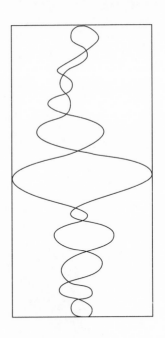

Chapter 43

The Unexpected Hanging

A new and powerful paradox has come to light." This is the opening sentence of a mind-twisting article by Michael Scriven that appeared in the July 1951 issue of the British philosophical journal *Mind*. Scriven, who bears the title of "professor of the logic of science" at the University of Indiana, is a man whose opinions on such matters are not to be taken lightly. That the paradox is indeed powerful has been amply confirmed by the fact that more than 20 articles about it have appeared in learned journals. The authors, many of whom are distinguished philosophers, disagree sharply in their attempts to resolve the paradox. Since no consensus has been reached, the paradox is still very much a controversial topic.

No one knows who first thought of it. According to the Harvard University logician W. V. Quine, who wrote one of the articles, the paradox was first circulated by word of mouth in the early 1940s. It often took the form of a puzzle about a man condemned to be hanged.

The man was sentenced on Saturday. "The hanging will take place at noon," said the judge to the prisoner, "on one of the seven days of next week. But you will not know which day it is until you are so informed on the morning of the day of the hanging."

The judge was known to be a man who always kept his word. The prisoner, accompanied by his lawyer, went back to his cell. As soon as the two men were alone the lawyer broke into a grin. "Don't you see?" he exclaimed. "The judge's sentence cannot possibly be carried out." (see Figure 43.1.)

"I don't see," said the prisoner.

"Let me explain. They obviously can't hang you next Saturday. Saturday is the last day of the week. On Friday afternoon you would still be alive and you would know with absolute certainty that the hanging

Figure 43.1. The prisoner eliminates all possible days.

would be on Saturday. You would know this *before* you were told so on Saturday morning. That would violate the judge's decree."

"True," said the prisoner.

"Saturday, then is positively ruled out," continued the lawyer. "This leaves Friday as the last day they can hang you. But they can't hang you on Friday because by Thursday afternoon only two days would remain: Friday and Saturday. Since Saturday is not a possible day, the hanging would have to be on Friday. Your knowledge of that fact would violate the judge's decree again. So Friday is out. This leaves Thursday as the last possible day. But Thursday is out because if you're alive Wednesday afternoon, you'll know that Thursday is to be the day."

"I get it," said the prisoner, who was beginning to feel much better. "In exactly the same way I can rule out Wednesday, Tuesday, and Monday. That leaves only tomorrow. But they can't hang me tomorrow because I know it today!"

In brief, the judge's decree seems to be self-refuting. There is nothing logically contradictory in the two statements that make up his decree; nevertheless, it cannot be carried out in practice. That is how the paradox appeared to Donald John O'Connor, a philosopher at the University of Exeter, who was the first to discuss the paradox in print (*Mind,* July 1948). O'Connor's version of the paradox concerned a military commander who announced that there would be a Class A blackout during the following week. He then defined a Class A blackout as one that the participants could not know would take place until after 6 P.M. on the day it was to occur.

"It is easy to see," wrote O'Connor, "that it follows from the announcement of this definition that the exercise cannot take place at all." That is to say, it cannot take place without violating the definition. Similar views were expressed by the authors of the next two articles (L. Jonathan Cohen in *Mind* for January 1950, and Peter Alexander in *Mind* for October 1950), and even by George Gamow and Marvin Stern when they later included the paradox (in a man-to-be-hanged form) in their book *Puzzle Math* (New York: Viking, 1958).

Now, if this were all there was to the paradox, one could agree with O'Connor that it is "rather frivolous." But, as Scriven was the first to point out, it is by no means frivolous, and for a reason that completely escaped the first three authors. To make this clear, let us return to the man in the cell. He is convinced, by what appears to be unimpeachable logic, that he cannot be hanged without contradicting the conditions specified in his sentence. Then on Thursday morning, to his great surprise, the hangman arrives. Clearly he did not expect him. What is more surprising, the judge's decree is now seen to be perfectly correct. The sentence can be carried out exactly as stated. "I think this flavour of logic refuted by the world makes the paradox rather fascinating," writes Scriven. "The logician goes pathetically through the motions that have always worked the spell before, but somehow the monster, Reality, has missed the point and advances still."

In order to grasp more clearly the very real and profound linguistic difficulties involved here, it would be wise to restate the paradox in two other equivalent forms. By doing this we can eliminate various irrelevant factors that are often raised and that cloud the issue, such as the possibility of the judge's changing his mind, of the prisoner's dying before the hanging can take place, and so on.

The first variation of the paradox, taken from Scriven's article, can be called the paradox of the unexpected egg (see Figure 43.2).

Imagine that you have before you 10 boxes labeled from 1 to 10. While your back is turned, a friend conceals an egg in one of the boxes. You turn around. "I want you to open these boxes one at a time," he tells you, "in serial order. Inside one of them I guarantee that you will find an unexpected egg. By 'unexpected' I mean that you will not be able to deduce which box it is in before you open the box and see it."

Assuming that your friend is absolutely trustworthy in all his state-

Figure 43.2. The paradox of the unexpected egg

ments, can his prediction be fulfilled? Apparently not. He obviously will not put the egg in box 10, because after you have found the first nine boxes empty you will be able to deduce with certainty that the egg is in the only remaining box. This would contradict your friend's statement. Box 10 is out. Now consider the situation that would arise if he were so foolish as to put the egg in box 9. You find the first eight boxes empty. Only 9 and 10 remain. The egg cannot be in box 10. Ergo it must be in 9. You open 9. Sure enough, there it is. Clearly it is an *expected* egg, and so your friend is again proved wrong. Box 9 is out. But now you have started on your inexorable slide into unreality. Box 8 can be ruled out by precisely the same logical argument, and similarly boxes 7, 6, 5, 4, 3, 2, and 1. Confident that all 10 boxes are empty, you start to open them. What have we here in box 5? A totally unexpected egg! Your friend's prediction is fulfilled after all. Where did your reasoning go wrong?

To sharpen the paradox still more, we can consider it in a third form, one that can be called the paradox of the unexpected spade (see Figure 43.3). Imagine that you are sitting at a card table opposite a friend who shows you that he holds in his hand the 13 spades. He shuffles them, fans them with the faces toward him, and deals a single card face down on the table. You are asked to name slowly the 13 spades, starting with the ace and ending with the king. Each time you fail to name the card on the table he will say no. When you name the card correctly, he will say yes.

"I'll wager a thousand dollars against a dime," he says, "that you

Figure 43.3. The paradox of the unexpected spade

will not be able to deduce the name of this card before I respond with 'Yes.' "

Assuming that your friend will do his best not to lose his money, is it possible that he placed the king of spades on the table? Obviously not. After you have named the first 12 spades, only the king will remain. You will be able to deduce the card's identity with complete confidence. Can it be the queen? No, because after you have named the jack only the king and queen remain. It cannot be the king, so it must be the queen. Again, your correct deduction would win you $1,000. The same reasoning rules out all the remaining cards. Regardless of what card it is, you should be able to deduce its name in advance. The logic seems airtight. Yet it is equally obvious, as you stare at the back of the card, that you have not the foggiest notion which spade it is!

Even if the paradox is simplified by reducing it to two days, two boxes, two cards, something highly peculiar continues to trouble the situation. Suppose your friend holds only the ace and deuce of spades. It is true that you will be able to collect your bet if the card is the deuce. Once you have named the ace and it has been eliminated you will be able to say: "I deduce that it's the deuce." This deduction rests, of course, on the truth of the statement "The card before me is either the ace or the deuce of spaces." (It is assumed by everybody, in all three paradoxes, that the man *will* be hanged, that there *is* an egg in a box, that the cards *are* the cards designated.) This is as strong a deduction as mortal man can ever make about a fact of nature. You have, therefore, the strongest possible claim to the $1,000.

Suppose, however, your friend puts down the ace of spades. Cannot you deduce at the outset that the card is the ace? Surely he would not risk his $1,000 by putting down the deuce. Therefore it *must* be the ace. You state your conviction that it is. He says yes. Can you legitimately claim to have won the bet?

Curiously, you cannot, and here we touch on the heart of the mystery. Your previous deduction rested only on the premise that the card was either the ace or the deuce. The card is not the ace; therefore, it is the deuce. But now your deduction rests on the same premise as before plus an additional one, namely, on the assumption that your friend spoke truly; to say the same thing in pragmatic terms, on the assumption that he will do all he can to avoid paying you $1,000. But if it is possible for you to deduce that the card is the ace, he will lose his money just as surely as if he put down the deuce. Since he loses it either way, he has no rational basis for picking one card rather than the other. Once you realize this, your deduction that the card is the ace takes on an extremely shaky character. It is true that you would be wise to bet that it is the ace, because it probably is, but to win the bet you have to do more than that: you have to prove that you have deduced the card with iron logic. This you cannot do.

You are, in fact, caught up in a vicious circle of contradictions. First you assume that his prediction will be fulfilled. On this basis you deduce that the card on the table is the ace. But if it is the ace, his prediction is falsified. If his prediction cannot be trusted, you are left without a rational basis for deducting the name of the card. And if you cannot deduce the name of the card, his prediction will certainly be confirmed. Now you are right back where you started. The whole circle begins again. In this respect the situation is analogous to the vicious circularity involved in a famous card paradox first proposed by the English mathematician P.E.B. Jourdain in 1913 (see Figure 43.4). Since this sort of reasoning gets you no further than a dog gets in chasing its tail, you have no logical way of determining the name of the card on the table. Of course, you may *guess* correctly. Knowing your friend, you may decide that it is highly probable he put down the ace. But no self-

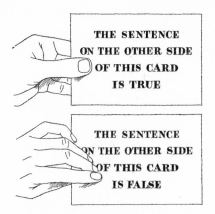

Figure 43.4. P.E.B. Jourdain's card paradox

THE SENTENCE
ON THE OTHER SIDE
OF THIS CARD
IS TRUE

THE SENTENCE
ON THE OTHER SIDE
OF THIS CARD
IS FALSE

respecting logician would agree that you have "deduced" the card with anything close to the logical certitude involved when you deduced that it was the deuce.

The flimsiness of your reasoning is perhaps seen more clearly if you return to the 10 boxes. At the start you "deduce" that the egg is in box 1, but box 1 is empty. You then "deduce" it to be in box 2, but box 2 is empty also. Then you "deduce" box 3, and so on. (It is almost as if the egg, just before you look into each box in which you are positive it must be, were cleverly transported by secret trap doors to a box with a higher number!) Finally you find the "expected" egg in box 8. Can you maintain that the egg is truly "expected" in the sense that your deduction is above reproach? Obviously you cannot, because your seven previous "deductions" were based on exactly the same line of reasoning, and each proved to be false. The plain fact is that the egg can be in any box, *including the last one.*

Even after having opened nine empty boxes, the question of whether you can "deduce" that there is an egg in the last box has no unambiguous answer. If you accept only the premise that one of the boxes contains an egg, then of course an egg in box 10 can be deduced. In that case, it is an expected egg and the assertion that it would not be is proved false. If you also assume that your friend spoke truly when he said the egg would be unexpected, then nothing can be deduced, for the first premise leads to an expected egg in box 10 and the second to an unexpected egg. Since nothing can be deduced, an egg in box 10 will be unexpected and both premises will be vindicated, but this vindication cannot come until the last box is opened and an egg is found there.

We can sum up this resolution of the paradox, in its hanging form, as follows. The judge speaks truly and the condemned man reasons falsely. The very first step in his chain of reasoning—that he cannot be hanged on the last day—is faulty. Even on the evening of the next-to-last day, as explained in the previous paragraph with reference to the egg in the last box—he has no basis for a deduction. This is the main point of Quine's 1953 paper. In Quine's closing words, the condemned man should reason: "We must distinguish four cases: first, that I shall be hanged tomorrow noon and I know it now (but I do not); second, that I shall be unhanged tomorrow noon and know it now (but I do not); third, that I shall be unhanged tomorrow noon and do not know it now; and fourth, that I shall be hanged tomorrow noon and do not know it now. The latter two alternatives are the open possibilities, and the last of all would fulfill the decree. Rather than charging the judge with self-contradiction, therefore, let me suspend judgment and hope for the best."

The Scottish mathematician Thomas H. O'Beirne, in an article with the somewhat paradoxical title "Can the Unexpected *Never* Happen?" (*The New Scientist,* May 25, 1961), has given what seems to me an excellent analysis of this paradox. As O'Beirne makes clear, the key to re-solving the paradox lies in recognizing that a statement about a future event can be known to be a true prediction by one person but not known to be true by another until after the event. It is easy to think of simple examples. Someone hands you a box and says: "Open it and you will find an egg inside." *He* knows that his prediction is sound, but *you* do not know it until you open the box.

The same is true in the paradox. The judge, the man who puts the egg in the box, the friend with the 13 spades—each knows that his prediction is sound. But the prediction cannot be used to support a chain of arguments that results eventually in discrediting the prediction itself. It is this roundabout self-reference that, like the sentence on the face of Jourdain's card, tosses the monkey wrench into all attempts to prove the prediction unsound.

We can reduce the paradox to its essence by taking a cue from Scriven. Suppose a man says to his wife: "My dear, I'm going to sur-prise you on your birthday tomorrow by giving you a completely un-expected gift. You have no way of guessing what it is. It is that gold bracelet you saw last week in Tiffany's window."

What is the poor wife to make of this? She knows her husband to be truthful. He always keeps his promises. But if he does give her the gold bracelet, it will not be a surprise. This would falsify his prediction. And if his prediction is unsound, what *can* she deduce? Perhaps he will keep his word about giving her the bracelet but violate his word that the gift will be unexpected. On the other hand, he may keep his word about the surprise but violate it about the bracelet and give her instead, say, a new vacuum cleaner. Because of the self-refuting character of her husband's statement, she has no rational basis for choosing between these alternatives; therefore, she has no rational basis for expecting the gold bracelet. It is easy to guess what happens. On her birthday she is surprised to receive a logically unexpected bracelet.

He knew all along that he could and would keep his word. *She* could not know this until after the event. A statement that yesterday appeared to be nonsense, that plunged her into an endless whirlpool of logical contradictions, has today suddenly been made perfectly true and noncontradictory by the appearance of the gold bracelet. Here in the starkest possible form is the queer verbal magic that gives to all the paradoxes we have discussed their bewildering, head-splitting charm.

Addendum

A great many trenchant and sometimes bewildering letters were received from readers offering their views on how the paradox of the unexpected hanging could be resolved. Several went on to expand their views in articles that are listed in the bibliography for this chapter.

Lennart Ekbom, a mathematics teacher at Östermalms College, in Stockholm, pinned down what may be the origin of the paradox. In 1943 or 1944, he wrote, the Swedish Broadcasting Company announced that a civil-defense exercise would be held the following week, and to test the efficiency of civil-defense units, no one would be able to predict, even on the morning of the day of the exercise, when it would take place. Ekbom realized that this involved a logical paradox, which he discussed with some students of mathematics and philosophy at Stockholm University. In 1947 one of these students visited Princeton, where he heard Kurt Gödel, the famous mathematician, mention a variant of the paradox. Ekbom adds that he originally believed the paradox to be older than the Swedish civil-defense announcement, but in view of Quine's statement that he first heard of the paradox in the early forties, perhaps this was its origin.

Papers on the unexpected hanging paradox continue to proliferate. I have greatly lengthened the chapter's bibliography to catch all the references in English that have come to my attention since 1969. It is amazing that philosophers continue to be bemused by this paradox and are still unable to agree on how best to resolve it.

The paradox has many variants. Douglas Hofstadter suggested that instead of an unexpected egg in a box there be an unexpected deadly snake. Roy Sorensen, in his 1982 paper, offered three versions in which a time sequence is not involved: one concerning what are called Moorean sentences, another about moves on a 3 × 3 game matrix, and a third based on a chain of sacrificial virgins.

Bibliography

The following bibliography has been updated by references taken from an extensive bibliography, much more complete than mine, that has been prepared by Timothy Chow. You can download his entire bibliography from his website: http://front.math.ucdavis.edu/search/author:Chow+and+Timothy.

P. Alexander, "Pragmatic Paradoxes," *Mind,* Vol. 59, October 1950, pp. 536–38.

A. K. Austin, "On the Unexpected Examination," *Mind,* Vol. 78, January 1969, p. 137.

A. K. Austin, "The Unexpected Examination," *Analysis,* Vol. 39, January 1979, pp. 63–64.

A. J. Ayer, "On a Supposed Antimony," *Mind,* Vol. 82, January 1973, pp. 125–26.

R. Binkley, "The Surprise Examination in Modal Logic," *Journal of Philosophy,* Vol. 65, 1968, pp. 127–36.

J. Bosch, "The Examination Paradox and Formal Prediction," *Logique et Analyse,* Vol. 59, September/December 1972.

L. Bovens, "The Backward Induction Argument for the Finite Iterated Prisoner's Dilemma and the Surprise Examination Paradox," *Analysis,* Vol. 57, 1997, pp. 179–86.

J. Cargile, "The Surprise Test Paradox," *Journal of Philosophy,* Vol. 64, September 1967, pp. 550–63.

J. Case, "Paradoxes Involving Conflicts of Interest," *American Mathematical Monthly,* Vol. 107, January 2000, pp. 33–43.

T. S. Chaplain, "Quine's Judge," *Philosophical Studies,* Vol. 29, May 1976, pp. 349–52.

J. M. Chapman and R. J. Butler, "On Quine's 'So-called Paradox,' " *Mind,* Vol. 74, July 1965, pp. 424–25.

T. Y. Chow, "The Surprise Examination or Unexpected Hanging Paradox," *American Mathematical Monthly,* Vol. 105, January 1998, pp. 41–51.

D. Clark, "How Unexpected is the Unexpected Hanging?" *Mathematics Magazine,* Vol. 67, February 1994, pp. 55–58.

R. Clark, "Pragmatic Paradox and Rationality," *Canadian Journal of Philosophy,* Vol. 24, 1994, pp. 229–42.

L. J. Cohen, "Mr. O'Connor's 'Pragmatic Paradoxes,' " *Mind,* Vol. 59, January 1950, pp. 85–87.

F. B. Ebersole, "The Definition of 'Pragmatic Paradox,' " *Mind,* Vol. 62, January 1953, pp. 80–85.

M. Edman, "The Prediction Paradox," *Theoria,* Vol. 40, 1974, pp. 166–75.

K. G. Ferguson, "Equivocation in the Surprise Examination Paradox," *Southern Journal of Philosophy,* Vol. 29, 1991, pp. 291–302.

F. B. Fitch, "A Goedelized Formulation of the Prediction Paradox," *American Philosophical Quarterly,* Vol. 1, 1964, pp. 161–64.

J. T. Fraser, "Note Relating to a Paradox of the Temporal Order," *The Voices of Time,* J. T. Fraser (ed.), New York: Braziller, 1966, pp. 524–26, 679.

J. S. Fulda, "The Paradox of the Surprise Test," *Mathematical Gazette,* Vol. 75, December 1991, pp. 419–23.

M. Gardner, "The Erasing of Philbert the Fudger," *Science Fiction Puzzle Tales,* New York: Clarkson Potter, 1981, Chapter 20.

M. Gardner, "Again, How's That Again?," *Riddles of the Sphinx,* New York: Mathematical Association of America, 1987, Chapter 27.

L. Goldstein, "Inescapable Surprises and Aquirable Intentions," *Analysis,* Vol. 53, 1993, pp. 93–99.

N. Hall, "How to Set a Surprise Exam," *Mind,* Vol. 108, 1999, pp. 647–703.

J. Y. Halpern and Y. Moses, "Taken by Surprise: The Paradox of the Surprise Test Revisited," *Journal of Philosophical Logic,* Vol. 15, 1986, pp. 281–304.

C. Harrison, "The Unexpected Examination in View of Kripne's Semantics for Modal Logic," *Philosophical Logic,* J. W. Davis et al. (eds.), Holland: Reidel, 1969.

J. M. Holtzman, "An Undecidable Aspect of the Unexpected Hanging Problem," *Philosophia,* Vol. 72, March 1987, pp. 195–98.

J. M. Holtzman, "A Note on Schrödinger's Cat and the Unexpected Examination Paradox," *British Journal of the Philosophy of Science,* Vol. 39, 1988, pp. 397–401.

C. Hudson, A. Solomon, and R. Walker, "A Further Examination of the 'Surprise Examination Paradox,' " *Eureka,* October 1969, pp. 23–24.

B. Jangeling and T. Koetsier, "A Reappraisal of the Hanging Paradox," *Philosophia,* Vol. 22, 1993, pp. 199–311.

C. Janoway, "Knowing About Surprises: A Supposed Antinomy Revisited," *Mind,* Vol. 98, 1989, pp. 391–409.

D. Kaplan and R. Montague, "A Paradox Regained," *Notre Dame Journal of Formal Logic,* Vol. 1, July 1960, pp. 79–90.

J. Kiefer and J. Ellison, "The Prediction Paradox Again," *Mind*, Vol. 74, July 1965, pp. 426–27.

P. L. Kirkham, "On Paradoxes and the Surprise Examination," *Philosophia*, Vol. 21, 1991, pp. 31–51.

I. Kvart, "The Paradox of the Surprise Examination," *Logique et Analyse*, Vol. 11, 1976, pp. 66–72.

A. Lyon, "The Prediction Paradox," *Mind*, Vol. 68, October 1959, pp. 510–17.

A. Margalit and M. Bar-Hillel, "Expecting the Unexpected," *Philosophia*, Vol. 18, October 1982, pp. 263–89.

J. McLelland, "Epistemic Logic and the Paradox of the Surprise Examination Paradox," *International Logic Review*, Vol. 3, January 1971, pp. 69–85.

J. McLelland and C. Chihara, "The Surprise Examination Paradox," *Journal of Philosophical Logic*, Vol. 4, February 1975, pp. 71–89.

B. Medlin, "The Unexpected Examination," *American Philosophical Quarterly*, Vol. 1, January 1964, pp. 1–7.

B. Meltzer, "The Third Possibility," *Mind*, Vol. 73, July 1964, pp. 430–33.

B. Meltzer and I. J. Good, "Two Forms of the Prediction Paradox," *British Journal for the Philosophy of Science*, Vol. 16, May 1965, pp. 50–51.

G. C. Nerlich, "Unexpected Examinations and Unprovable Statements," *Mind*, Vol. 70, October 1961, pp. 503–13.

T. H. O'Beirne, "Can the Unexpected *Never* Happen?," *The New Scientist*, Vol. 15, May 25, 1961, pp. 464–65; letters and replies, June 8, 1961, pp. 597–98.

D. J. O'Connor, "Pragmatic Paradoxes," *Mind*, Vol. 57, July 1948, pp. 358–59.

D. J. O'Connor, "Pragmatic Paradoxes and Fugitive Propositions," *Mind*, Vol. 60, October 1951, pp. 536–38.

D. Olin, "The Prediction Paradox Resolved," *Philosophical Studies*, Vol. 44, September 1983, pp. 225–33.

W. Poundstone, "The Unexpected Hanging," *Labyrinths of Reason*, New York: Doubleday, 1988, Chapter 6.

W. V. Quine, "On a So-called Paradox," *Mind*, Vol. 62, January 1953, pp. 65–67.

J. Schoenberg, "A Note on the Logical Fallacy in the Paradox of the Unexpected Examination," *Mind*, Vol. 75, January 1966, pp. 125–27.

M. Scriven, "Paradoxical Announcements," *Mind*, Vol. 60, July 1951, pp. 403–7.

S. C. Shapiro, "A Procedural Solution to the Unexpected Hanging and Sorites Paradoxes," *Mind*, Vol. 107, 1998, pp. 751–61.

R. A. Sharpe, "The Unexpected Examination," *Mind*, Vol. 74, April 1965, p. 255.

R. Shaw, "The Paradox of the Unexpected Examination," *Mind*, Vol. 67, July 1958, pp. 382–84.

B. H. Slater, "The Examiner Examined," *Analysis*, Vol. 35, December 1974, pp. 48–50.

J. W. Smith, "The Surprise Examination on the Paradox of the Heap," *Philosophical Papers*, Vol. 13, May 1984, pp. 43–56.

R. Smullyan, "Surprised?," *Forever Undecided*, New York: Knopf, 1987, Chapter 2.

R. A. Sorensen, "Recalcitrant Variations of the Prediction Paradox," *Australasian Journal of Philosophy*, Vol. 60, December 1982, pp. 355–62.

R. A. Sorensen, "Conditional Blindspots and the Knowledge Squeeze: A Solution to the Prediction Paradox," *Australasian Journal of Philosophy,* Vol. 62, June 1984, pp. 126–35.

R. A. Sorensen, "The Bottle Imp and the Prediction Paradox," *Philosophia,* Vol. 15, January 1986, pp. 421–24.

R. A. Sorensen, "A Strengthened Prediction Paradox," *Philosophical Quarterly,* Vol. 36, October 1986, pp. 504–13.

R. A. Sorensen, *Blindspots,* London: Oxford University Press, 1988, Chapters 7–9.

R. A. Sorenson, "The Earliest Unexpected Class Inspection," *Analysis,* Vol. 53, 1993, p. 252.

R. A. Sorenson, "Infinite 'Backward' Induction Arguments," *Pacific Philosophical Quarterly,* Vol. 80, 1999, pp. 278–83.

M. Stack, "The Surprise Examination Paradox," *Dialogue,* Vol. 26, June 1977.

I. Stewart, "Paradox Lost," *Manifold,* Autumn 1971, pp. 19–21.

R. Weintraub, "Practical Solutions to the Surprise Examination Paradox," *Ratio* (New Series), Vol. 8, 1995, pp. 161–69.

P. Weiss, "The Prediction Paradox," *Mind,* Vol. 61, April 1952, pp. 265–69.

P. Y. Windt, "The Liar and the Prediction Paradox," *American Philosophical Quarterly,* Vol. 10, January 1973, pp. 65–68.

D. R. Woodall, "The Paradox of the Surprise Examination," *Eureka,* No. 30, October 1967, pp. 31–32.

C. Wright and A. Sudbury, "The Paradox of the Unexpected Examination," *Australasian Journal of Philosophy,* May 1977, pp. 41–58.

J. A. Wright, "The Surprise Exam: Prediction on Last Day Uncertain," *Mind,* Vol. 76, January 1967, pp. 115–17.

Chapter 44

Newcomb's Paradox

> *A common opinion prevails that the juice has ages ago been pressed out of the free-will controversy, and that no new champion can do more than warm up stale arguments which every one has heard. This is a radical mistake. I know of no subject less worn out, or in which inventive genius has a better chance of breaking open new ground.* —WILLIAM JAMES

One of the perennial problems of philosophy is how to explain (or explain away) the nature of free will. If the concept is explicated within a framework of determinism, the will ceases to be free in any commonly understood sense, and it is hard to see how fatalism can be avoided. Che sarà, sarà. Why work hard for a better future for yourself or for others if what you do must always be what you do do? And how can you blame anyone for anything if he could not have done otherwise?

On the other hand, attempts to explicate will in a framework of indeterminism seem equally futile. If an action is not caused by the previous states of oneself and the world, it is hard to see how to keep the action from being haphazard. The notion that decisions are made by some kind of randomizer in the mind does not provide much support for what is meant by free will either.

Philosophers have never agreed on how to avoid the horns of this dilemma. Even within a particular school there have been sharp disagreements. William James and John Dewey, America's two leading pragmatists, are a case in point. Although Dewey was a valiant defender of democratic freedoms, his metaphysics regarded human behavior as completely determined by what James called the total "push of the past." Free will for Dewey was as illusory as it is in the psychology of B. F. Skinner. In contrast, James was a thoroughgoing indeterminist. He believed that minds had the power to inject genuine novelty into history—that not even God himself could know the future except partially. *"That,"* he wrote, "is what gives the palpitating reality to our moral life and makes it tingle . . . with so strange and elaborate an excitement."

A third approach, pursued in depth by Immanuel Kant, accepts both sides of the controversy as being equally true but incommensurable ways of viewing human behavior. For Kant the situation is something like that pictured in one of Piet Hein's "grooks":

> *A bit beyond perception's reach*
> *I sometimes believe I see*
> *That Life is two locked boxes, each*
> *Containing the other's key.*

Free will is neither fate nor chance. In some unfathomable way it partakes of both. Each is the key to the other. It is not a contradictory concept, like a square triangle, but a paradox that our experience forces on us and whose resolution transcends human thought. That was how Niels Bohr saw it. He found the situation similar to his "principle of complementarity" in quantum mechanics. It is a viewpoint that Einstein, a Spinozist, found distasteful, but many other physicists, J. Robert Oppenheimer for one, found enormously attractive.

What has free will to do with mathematical games? The answer is that in recent decades philosophers of science have been wrestling with a variety of queer "prediction paradoxes" related to the problem of will. Some of them are best regarded as a game situation. One draws a payoff matrix and tries to determine a player's best strategy, only to find oneself trapped in a maze of bewildering ambiguities about time and causality.

A marvelous example of such a paradox came to light in 1970 in the paper "Newcomb's Problem and Two Principles of Choice" by Robert Nozick, a philosopher at Harvard University. The paradox is so profound, so amusing, so mind-bending, with thinkers so evenly divided into warring camps, that it bids fair to produce a literature vaster than that dealing with the prediction paradox of the unexpected hanging.

Newcomb's paradox is named after its originator, William A. Newcomb, a theoretical physicist at the University of California's Lawrence Livermore Laboratory. (His great-grandfather was the brother of Simon Newcomb, the astronomer.) Newcomb thought of the problem in 1960 while meditating on a famous paradox of game theory called the prisoner's dilemma. A few years later Newcomb's problem reached Nozick by way of their mutual friend Martin David Kruskal, a Princeton University mathematician. "It is not clear that I am entitled to present this paper," Nozick writes. "It is a beautiful problem. I wish it were mine."

Although Nozick could not resolve it, he decided to write it up anyway. His paper appears in *Essays in Honor of Carl G. Hempel,* edited by Nicholas Rescher and published by Humanities Press in 1970. What follows is largely a paraphrase of Nozick's paper.

Two closed boxes, B1 and B2, are on a table. B1 contains $1,000. B2 contains either nothing or $1 million. You do not know which. You have an irrevocable choice between two actions:

1. Take what is in both boxes.
2. Take only what is in B2.

At some time before the test a superior Being has made a prediction about what you will decide. It is not necessary to assume determinism. You only need be persuaded that the Being's predictions are "almost certainly" correct. If you like, you can think of the Being as God, but the paradox is just as strong if you regard the Being as a superior intelligence from another planet or a supercomputer capable of probing your brain and making highly accurate predictions about your decisions. If the Being expects you to choose both boxes, he has left B2 empty. If he expects you to take only B2, he has put $1 million in it. (If he expects you to randomize your choice by, say, flipping a coin, he has left B2 empty.) In all cases B1 contains $1,000. You understand the situation fully, the Being knows you understand, you know that he knows, and so on.

What should you do? Clearly it is not to your advantage to flip a coin, so that you must decide on your own. The paradox lies in the disturbing fact that a strong argument can be made for either decision. Both arguments cannot be right. The problem is to explain why one is wrong.

Let us look first at the argument for taking only B2. You believe the Being is an excellent predictor. If you take both boxes, the Being almost certainly will have anticipated your action and have left B2 empty. You will get only the $1,000 in B1. On the other hand, if you take only B2, the Being, expecting that, almost certainly will have placed $1 million in it. Clearly it is to your advantage to take only B2.

Convincing? Yes, but the Being made his prediction, say a week ago, and then left. Either he put the $1 million in B2, or he did not. "If the money is already there, it will stay there whatever you choose. It is not going to disappear. If it is not already there, it is not going to suddenly appear if you choose only what is in the second box." It is assumed that no "backward causality" is operating; that is, your present actions cannot influence what the Being did last week. So why not take both boxes and get everything that is there? If B2 is filled, you get $1,001,000. If it is empty, you get at least $1,000. If you are so foolish as to take only B2, you know you cannot get more than $1 million, and there is even a slight possibility of getting nothing. Clearly it is to your advantage to take both boxes!

"I have put this problem to a large number of people, both friends and students in class," writes Nozick. "To almost everyone it is perfectly clear and obvious what should be done. The difficulty is that these people seem to divide almost evenly on the problem, with large numbers thinking that the opposing half is just being silly.

"Given two such compelling opposing arguments, it will not do to rest content with one's belief that one knows what to do. Nor will it do to just repeat one of the arguments, loudly and slowly. One must also disarm the opposing argument; explain away its force while showing it due respect."

Nozick sharpens the "pull" of the two arguments as follows. Suppose the experiment had been done many times before. In every case the Being predicted correctly. Those who took both boxes always got only $1,000; those who took only B2 got $1 million. You have no reason to suppose your case will be different. If a friend were observing the scene, it would be completely rational for him to bet, giving high odds, that if you take both boxes you will get only $1,000. Indeed, if there is a time delay after your choice of both boxes, you know it would be rational for you yourself to bet, offering high odds, that you will get only $1,000. Knowing this, would you not be a fool to take both boxes?

Alas, the other argument makes you out to be just as big a fool if you do not. Assume that B1 is transparent. You see the $1,000 inside. You cannot see into B2, but the far side is transparent and your friend is sitting opposite. He knows whether the box is empty or contains $1 million. Although he says nothing, you realize that, whatever the state of B2 is, he wants you to take both boxes. He wants you to because, regardless of the state of B2, you are sure to come out ahead by $1,000. Why not take advantage of the fact that the Being played first and cannot alter his move?

Nozick, an expert on decision theory, approaches the paradox by considering analogous game situations in which, as here, there is a conflict between two respected principles of choice: the "expected-utility principle" and the "dominance principle." To see how the principles apply, consider the payoff matrix for Newcomb's game (see Figure 44.1). The argument for taking only B2 derives from the principle that you should choose so as to maximize the expected utility (value to you) of the outcome. Game theory calculates the expected utility of each action by multiplying each of its mutually exclusive outcomes by the probability of the outcome, given the action. We have assumed that the Being predicts with near certainty, but let us be conservative and make the probability a mere .9. The expected utility of taking both boxes is

$$(.1 \times \$1,001,000) + (.9 \times \$1,000) = \$101,000.$$

The expected utility of taking only B2 is

$$(.9 \times \$1,000,000) + (.1 \times \$0) = \$900,000.$$

Guided by this principle, your best strategy is to take only the second box.

BEING

		MOVE 1 (PREDICTS YOU TAKE ONLY BOX 2)	MOVE 2 (PREDICTS YOU TAKE BOTH BOXES)
YOU	MOVE 1 (TAKE ONLY BOX 2)	$1,000,000	$0
	MOVE 2 (TAKE BOTH BOXES)	$1,001,000	$1,000

Figure 44.1. Payoff matrix for Newcomb's paradox

The dominance principle, however, is just as intuitively sound. Suppose the world divided into n different states. For each state k mutually exclusive actions are open to you. If in at least one state you are better off choosing a, and in all other states either a is the best choice or the choices are equal, then the dominance principle asserts that you should choose a. Look again at the payoff matrix. The states are the outcomes of the Being's two moves. Taking both boxes is strongly dominant. For each state it gives you $1,000 more than you would get by taking only the second box.

That is as far as we can go into Nozick's analysis, but interested readers should look it up for its mind-boggling conflict situations related to Newcomb's problem. Nozick finally arrives at the following tentative conclusions:

If you believe in absolute determinism and that the Being has in truth predicted your behavior with unswerving accuracy, you should "choose" (whatever that can mean!) to take only B2. For example, suppose the Being is God and you are a devout Calvinist, convinced that God knows every detail of your future. Or assume that the Being has a time-traveling device he can launch into the future and use to bring back a motion picture of what you did on that future occasion when you made your choice. Believing that, you should take only B2, firmly persuaded that your feeling of having made a genuine choice is sheer illusion.

Nozick reminds us, however, that Newcomb's paradox does *not* assume that the Being has perfect predictive power. If you believe that you possess a tiny bit of free will (or alternatively that the Being is sometimes wrong, say once in every 20 billion cases), then this may be one of the times the Being has erred. Your wisest decision is to take both boxes.

Nozick is not happy with this conclusion. "Could the difference between one in n and none in n, for arbitrarily large finite n, make this difference? And how exactly does the fact that the predictor is certain to have been correct dissolve the force of the dominance argument?" Both questions are left unanswered. Nozick hoped that publishing the problem might "call forth a solution which will enable me to stop returning, periodically, to it."

One such solution, "to restore [Nozick's] peace of mind," was attempted by Maya Bar-Hillel and Avishai Margalit, of Hebrew University in Jerusalem, in their paper "Newcomb's Paradox Revisited." They adopt the same game-theory approach taken by Nozick, but they come to an opposite conclusion. Even though the Being is not a perfect predictor, they recommend taking only the second box. You must, they argue, resign yourself to the fact that your best strategy is to behave as *if* the Being has made a correct prediction, even though you know there is a slight chance he has erred. You know he has played before you, but you cannot do better than to play as if he is going to play after you. "For you cannot outwit the Being except by knowing what he predicted, but you cannot know, or even meaningfully guess, at what he predicted before actually making your final choice."

It may seem to you, Bar-Hillel and Margalit write, that backward causality is operating—that somehow your choice makes the $1 million more likely to be in the second box—but this is pure flim-flam. You choose only B2 "because it is inductively known to correlate remarkably with the existence of this sum in the box, and though we do not assume a causal relationship, there is no better alternative strategy than to behave as if the relationship was, in fact, causal."

For those who argue for taking only B2 on the grounds that causality is independent of the direction of time—that your decision actually "causes" the second box to be either empty or filled with $1 million—Newcomb proposed the following variant of his paradox. Both boxes are transparent. B1 contains the usual $1,000. B2 contains a piece of paper with a fairly large integer written on it. You do not know whether the number is prime or composite. If it proves to be prime (you must not test it, of course, until after you have made your choice), then you get $1 million. The Being has chosen a prime number if he predicts you will take only B2 but has picked a composite number if he predicts you will take both boxes.

Obviously you cannot by an act of will make the large number change from prime to composite, or vice versa. The nature of the number is fixed for eternity. So why not take both boxes? If it is prime, you get $1,001,000. If it is not, you get at least $1,000. (Instead of a number, B2 could contain any statement of a decidable mathematical fact that you do not investigate until after your choice.)

It is easy to think of other variations. For example, there are 100 little boxes, each holding a $10 bill. If the Being expects you to take all of them, he has put nothing else in them. But if he expects you to take only one box—perhaps you pick it at random—he has added to that box a large diamond. There have been thousands of previous tests, half of them involving you as a player. Each time, with possibly a few exceptions, the player who took a single box got the diamond, and the player who took all the boxes got only the money. Acting pragmatically, on the basis of past experience, you should take only one box. But then how can you refute the logic of the argument that says you have everything to gain and nothing to lose if the next time you play you take all the boxes?

These variants add nothing essentially new. With reference to the original version, Nozick halfheartedly recommends taking both boxes. Bar-Hillel and Margalit strongly urge you to "join the millionaire's club" by taking only B2. That is also the view of Kruskal and Newcomb. But has either side really done more than just repeat its case "loudly and slowly"? Can it be that Newcomb's paradox validates free will by invalidating the possibility, in principle, of a predictor capable of guessing a person's choice between two equally rational actions with better than 50 percent accuracy?

Addendum

Scientific American received such a flood of letters about Newcomb's paradox that I asked Robert Nozick if he would be willing to look them over and write a guest column about them. To my delight, he agreed. I packed a large carton with the correspondence and took it along on a visit to Cambridge where I had the pleasure of lunching with Nozick and giving him the carton. You will find his trenchant discussion of the letters in my *Knotted Doughnuts and Other Mathematical Entertainments.* Nozick concluded that the paradox remains unsolved.

In 1974 Basic Books published Nozick's controversial defense of political libertarianism, *Anarchy, State, and Utopia.* It was followed by *Philosophical Explanations* and other books that have elevated Nozick into the ranks of major U.S. philosophers.

As my annotated bibliography indicates, philosophers are still far from agreement on how to resolve Newcomb's paradox. For whatever they are worth, here are my own tentative opinions.

My sympathies are with those who say the predictor cannot exist. Even if strict determinism in some sense holds for every event in the history of the universe, I believe that certain events are in principle unpredictable when predictions are allowed to interact causally with the event being predicted. We have here, I am persuaded, something analogous to the resolution of semantic paradoxes. Contradictions arise whenever a language is allowed to talk about the truth or falsity of its own statements, or when sets are allowed to be members of themselves. We can escape the semantic paradoxes by permitting talk about the truth of a sentence only in a metalanguage. "This sentence is false" simply is not a sentence. The notorious paradox of the barber who shaves every person and only those persons who do not shave themselves, and who himself belongs to the set of persons, is a barber who cannot exist. It is not logically inconsistent to suppose that the future is totally determined, whether or not an omniscient God exists, but as soon as we permit a superbeing to make predictions that interact with the event being predicted, we encounter contradictions that render the existence of such a superpredictor impossible.

Consider the simplest case. A superbeing knows that when you go to bed next Thursday you will take off your shoes. If the superbeing keeps this knowledge from you, there is no problem; but if the superbeing informs you of the prediction, you can falsify it easily by going to bed with your shoes on. I agree with those who say that Newcomb's problem in no way settles the question of whether the future is completely determined, but I do maintain that it brings us face to face with the eternal, and to me unanswerable, problem of defining what is meant by free choice.

Although I don't believe it, the state of the world a hundred years from now may be determined in every detail by the state of the world now. Innumerable future events obviously can be predicted with almost certain accuracy, but other events are the outcome of such complex causes that even if determinism is true it seems likely there is no possible way they could be predicted by any technique faster than allowing the universe itself to unroll to see what happens. (We leave aside the notion of a God outside of time who sees the past and future simultaneously, whatever that means.) All this is by the way. The main point is that when a prediction interacts with the predicted event, whether human wills are involved or not, logical contradictions can arise. A familiar example is the supercomputer asked to predict if a certain event will occur in the next three minutes. If the prediction is no, it turns on a green light. If yes, it turns on a red light. The computer is now asked to predict whether the green light will go on. By making the event part of the prediction, the computer is rendered logically impotent.

It is my view that Newcomb's predictor, even if accurate only 51 percent of the time, forces a logical contradiction that makes such a predictor, like Bertrand Russell's barber, impossible. We can avoid contradictions arising from two different "shoulds" (should you take one or two boxes?) by stating the contradiction as follows. One flawless argument implies that the best way to maximize your reward is to take only the closed box. Another flawless argument implies that the best way to maximize your reward is to take both boxes. Because the two conclusions are contradictory, the predictor cannot exist. Faced with a Newcomb decision, I would share the suspicions of Max Black and others that I was either the victim of a hoax or of a badly controlled experiment that had yielded false data about the predictor's accuracy. On this assumption, I would take both boxes.

But, you may ask, how would I decide if I made what I would regard as a counterfactual posit that the predictor was what it was claimed to be? I suppose if I could persuade myself that the predictor existed I might take only the closed box even though it would be logically irrational. But I cannot so persuade myself. It is as if someone asked me to put 91 eggs in 13 boxes, so each box held seven eggs, and then added that an experiment had proved that 91 is prime. On that assumption, one or more eggs would be left over. I would be given a million dollars for each leftover egg, and 10 cents if there were none. Unable to believe that 91 is a prime, I would proceed to put seven eggs in each box, take my 10 cents and not worry about having made a bad decision.

Bibliography

M. Bar-Hillel and A. Margalit, "Newcomb's Paradox Revisited," *The British Journal for the Philosophy of Science,* Vol. 23, 1972, pp. 295–304. Defends the pragmatic decision. "We hope to convince the reader to take just the one covered box, and join the millionaire's club!"

T. M. Benditt and D. J. Ross "Newcomb's 'Paradox,' " *The British Journal for the Philosophy of Science,* Vol. 27, 1976, pp. 161–64. Attack's Schlesinger's reasoning and concludes that "there is only one rational choice . . . namely choose box II alone."

M. Black, "Newcomb's Problem Demystified," *The Prevalence of Humbug and Other Essays,* Cornell University Press, 1983, Chapter 8. Argues that the existence of the predictor is so impossible that anyone faced with a Newcomb decision "would be well advised to suspect fraud, and to play safe by taking both boxes."

S. J. Brams, "Newcomb's Problem and the Prisoner's Dilemma," *The Journal of Conflict Resolution,* Vol. 19, 1975, pp. 596–619. Argues the close relationship of Newcomb's problem to the prisoner's dilemma. Applying metagame theory, Brams maintains that if you believe empirical evidence indicates the predictor's accuracy is greater than .5005, you should take only one box; if less than .5005, you should be indifferent as to your choice.

S. J. Brams, "A Paradox of Prediction," *Paradoxes in Politics,* Free Press, 1976, Chapter 8. Enlarges on the previous entry.

S. J. Brams "Newcomb's Problem," *Superior Beings,* Springer-Verlag, 1983, pp. 46–54. Summarizes his earlier views.

J. Broome, "An Economic Newcomb Problem," *Analysis,* Vol. 49, October 1989, pp. 220–22.

R. Campbell and L. Sowden (eds.), *Paradoxes of Rationality and Cooperation: Prisoner's Dilemma and Newcomb's Problem,* University of British Columbia Press, 1985.

J. Cargile, "Newcomb's Paradox," *The British Journal for the Philosophy of Science,* Vol. 26, 1975, pp. 234–39. Contends that there is "no basis for determining one course of action as the right one without additional information."

E. Eells, "Newcomb's Paradox and the Principle of Maximizing Conditional Expected Utility," Ph.D. dissertation, University of California, Berkeley, June 1980.

E. Eells, *Rational Decision and Causality,* Cambridge University Press, 1982.

E. Eells, "Newcomb's Many Solutions," *Theory and Decision,* Vol. 16, 1984, pp. 59–105.

E. Eells, "Levi's 'The Wrong Box,' " *The Journal of Philosophy,* Vol. 82, 1985, pp. 91–106.

A. Gibbard and W. Harper, "Counterfactuals and Two Kinds of Expected Utility," *Foundations and Applications of Decision Theory,* Reidel, 1978. Reprinted in *Ifs,* Reidel, 1981.

T. Horgan, "Counterfactuals and Newcomb's Problem," *The Journal of Philosophy,* Vol. 78, 1981, pp. 331–56.

N. Howard, *Paradoxes of Rationality: Theory of Metagames and Political Behavior,* MIT Press, 1971, pp. 168–84.

J. Leslie, "Ensuring Two Bird Deaths With One Throw," *Mind,* Vol. 100, January 1991, pp. 73–86.

D. Lewis, "Prisoner's Dilemma Is a Newcomb Problem," *Philosophy and Public Affairs,* Vol. 8, 1979, pp. 235–40.

I. Levi "Newcomb's Many Problems," *Theory and Decision,* Vol. 6, 1975, pp. 161–75.

I. Levi, "A Note on Newcombmania," *The Journal of Philosophy,* Vol. 79, 1982, pp. 337–42.

I. Levi "The Wrong Box," *The Journal of Philosophy,* Vol. 80, 1983, pp. 534–42.

D. Locke, "How to Make a Newcomb Choice," *Analysis,* January 1978, pp. 17–23. Takes the view that the situation is not sharply enough defined to make decision theory applicable. It is best to take both boxes,

confident that whether Box One is full or empty I cannot suffer from that choice, hopeful that I might at least demonstrate freedom of choice in the presence of a Newcomb Predictor, but resigned to receive a mere thousand where less rational mortals, who from the evidence of *Scientific American* (March 1974, p. 102) outnumber the rational by the order of 5 to 2, stand to gain a million. Perhaps it is precisely because I am a rational man that the Predictor is able to predict my choice, and leave Box One empty. Perhaps, in this respect at least, I would be better off were I less rational than I am. The penalties of philosophy are no less than its pleasures.

R. Nozick, "Newcomb's Problem and Two Principles of Choice," *Essays in Honor of Carl G. Hempel,* N. Rescher (ed.), Humanities Press, 1969.

W. Poundstone, *Labyrinths of Reason,* Doubleday, 1988, Chapter 12.

R. Richter, "Rationality Revisited," *The Australasian Journal of Philosophy,* Vol. 62, 1984, pp. 392–403.

G. Schlesinger, "The Unpredictability of Free Choices," *The British Journal for the Philosophy of Science,* Vol. 25, 1974, pp. 209–21. Argues that no amount of inductive evidence can support the belief that the predictor is capable of better than chance in its predictions and that the paradox brings out the fundamental unpredictability of human choices and the reality of free will. It is best to take both boxes.

G. Schlesinger, "Unpredictability: A Reply to Cargile and to Benditt and Ross," *The British Journal of the Philosophy of Science,* Vol. 27, 1976, pp. 267–74. The author defends his earlier position.

R. A. Sorensen, "Newcomb's Problem: Recalculations for the One-Boxer," *Theory and Decision,* Vol. 15, 1983, pp. 399–404.

R. A. Sorensen, "The Iterated Versions of Newcomb's Problem and the Prisoner's Dilemma," *Synthese,* Vol. 63, 1985, pp. 157–66.

R. A, Sorenson, *Blind Spots,* Oxford, 1988, pp. 371–77.

F. A. Wolf, "Newcomb's Paradox," *Taking the Quantum Leap,* Harper and Row, 1981. The funniest attempt at resolution yet, by a physicist deep into the paranormal and a defender of the solipsistic view that the world "out there" really isn't there until it is created by your observation. Since the money is not in the closed box until you perceive it, you have to put it there by choosing and opening the box. "It is your act of observation that resolves the paradox. Choosing both boxes creates box R empty. Choosing box R creates it one million dollars fuller. . . . your choices create the alternate possibilities as realities."

Chapter 45

Nothing

Nobody seems to know how to deal with it.
(He would, of course.) —P. L. Heath

Our topic is nothing. By definition nothing does not exist, but the concepts we have of it certainly exist as concepts. In mathematics, science, philosophy, and everyday life it turns out to be enormously useful to have words and symbols for such concepts.

The closest a mathematician can get to nothing is by way of the null (or empty) set. It is not the same thing as nothing because it has whatever kind of existence a set has, although it is unlike all other sets. It is the only set that has no members and the only set that is a subset of every other set. From a basket of three apples you can take one apple, two apples, three apples, or no apples. To an empty basket you can, if you like, add nothing.

The null set denotes, even though it doesn't denote anything. For example, it denotes such things as the set of all square circles, the set of all even primes other than 2, and the set of all readers of this book who are chimpanzees. In general it denotes the set of all x's that satisfy any statement about x that is false for all values of x. Anything you say about a member of the null set is true, because it lacks a single member for which a statement can be false.

The null set is symbolized by $\emptyset$. It must not be confused with 0, the symbol for zero. Zero is (usually) a number that denotes the number of members of $\emptyset$. The null set denotes nothing, but 0 denotes the number of members of such sets, for example, the set of apples in an empty basket. The set of these nonexisting apples is $\emptyset$, but the number of apples is 0.

A way to construct the counting numbers, discovered by the great German logician Gottlob Frege and rediscovered by Bertrand Russell, is to start with the null set and apply a few simple rules and axioms. Zero is defined as the cardinal number of elements in all sets that are

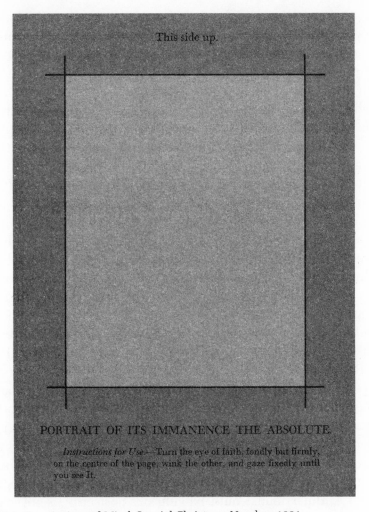

Figure 45.1. Frontispiece of *Mind,* Special Christmas Number, 1901

equivalent to (can be put in one-to-one correspondence with) the members of the null set. After creating 0, 1 is defined as the number of members in all sets equivalent to the set whose only member is 0. Two is the number of members in all sets equivalent to the set containing 0 and 1. Three is the number of members in all sets equivalent to the set containing 0, 1, 2, and so on. In general, an integer is the number of members in all sets equivalent to the set containing all previous numbers.

There are other ways of recursively constructing numbers by beginning with nothing, each with subtle advantages and disadvantages, in large part psychological. John von Neumann, for example, shortened Frege's procedure by one step. He preferred to define 0 as the null set, 1 as the set whose sole member is the null set, 2 as the set whose members are the null set and 1, and so on.

John Horton Conway then at the University of Cambridge hit on a remarkable new way to construct numbers that also starts with the null set. He first described his technique in a photocopied typescript of 13 pages, "All Numbers, Great and Small." It begins: "We wish to construct all numbers. Let us see how those who were good at constructing numbers have approached the problem in the past." It ends with 10 open questions, of which the last is: "Is the whole structure of any use?"

Conway explained his new system to Donald E. Knuth, a computer scientist at Stanford University, when they happened to meet at lunch one day in 1972. Knuth was immediately fascinated by its possibilities and its revolutionary content. In 1973 during a week of relaxation in Oslo, Knuth wrote an introduction to Conway's method in the form of a novelette. It was issued in paperback in 1974 by Addison-Wesley, which also publishes Knuth's well-known series titled *The Art of Computer Programming.* I believe it is the only time a major mathematical discovery has been published first in a work of fiction. A later book by Conway, *On Numbers and Games,* opens with an account of his number construction, then goes on to apply the theory to the construction and analysis of two-person games.

Knuth's novelette, *Surreal Numbers,* is subtitled *How Two Ex-Students Turned On to Pure Mathematics and Found Total Happiness.* The book's primary aim, Knuth explains in a postscript, is not so much to teach Conway's theory as "to teach how one might go about developing such a theory." He continues: "Therefore, as the two characters in this book gradually explore and build up Conway's number system, I have recorded their false starts and frustrations as well as their good ideas. I wanted to give a reasonably faithful portrayal of the important principles, techniques, joys, passions, and philosophy of mathematics, so I wrote the story as I was actually doing the research myself."

Knuth's two ex-mathematics students, Alice and Bill *(A and B),* have fled from the "system" to a haven on the coast of the Indian Ocean. There they unearth a half-buried black rock carved with ancient Hebrew writing. Bill, who knows Hebrew, manages to translate the opening sentence: "In the beginning everything was void, and J.H.W.H.

Conway began to create numbers." JHWH is a transliteration of how the ancient Hebrews wrote the name Jehovah. "Conway" also appears without vowels, but it was the most common English name Bill could think of that fitted the consonants.

Translation of the "Conway stone" continues: "Conway said, 'Let there be two rules which bring forth all numbers large and small. This shall be the first rule: Every number corresponds to two sets of previously created numbers, such that no member of the left set is greater than or equal to any member of the right set. And the second rule shall be this: One number is less than or equal to another number if and only if no member of the first number's left set is greater than or equal to the second number, and no member of the second number's right set is less than or equal to the first number.' And Conway examined these two rules he had made, and behold! they were very good."

The stone's text goes on to explain how on the zero day Conway created zero. He did it by placing the null set on the left and also on the right. In symbolic notation $0 = \{\varnothing \mid \varnothing\}$, where the vertical line divides the left and right sets. No member of the left $\varnothing$ is equal to or greater than a member of the right $\varnothing$ because $\varnothing$ *has* no members, so that Conway's first rule is satisfied. Applying the second rule, it is easy to show that 0 is less than or equal to 0.

On the next day, the stone reveals, Conway created the first two nonzero integers, 1 and -1. The method is simply to combine the null set with 0 in the two possible ways: $1 = \{0 \mid \varnothing\}$ and $-1 = \{\varnothing \mid 0\}$. It checks out. Minus 1 is less than but not equal to 0, and 0 is less than but not equal to 1. Now, of course, 1 and -1 and all subsequently created numbers can be plugged back into the left–right formula, and in this way all the integers are constructed. With 0 and 1 forming the left set and $\varnothing$ on the right, 2 is created. With 0, 1, and 2 on the left and $\varnothing$ on the right, 3 is created, and so on.

At this point readers might enjoy exploring a bit on their own. Jill C. Knuth's illustration for the front cover of *Surreal Numbers* shows some huge boulders shaped to symbolize $\{0 \mid 1\}$. What number does this define? And can the reader prove that $\{-1 \mid 1\} = 0$?

"Be fruitful and multiply," Conway tells the integers. By combining them, first into finite sets, then into infinite sets, the "copulation" of left–right sets continues, aided by no more than Conway's ridiculously simple rules. Out pour all the rest of the real numbers: first the integral fractions, then the irrationals. At the end of aleph-null days a big bang occurs and the universe springs into being. That, however, is not all. Taken to infinity, Conway's construction produces all of Georg Cantor's transfinite numbers, all infinitesimal numbers (they are reciprocals of infinite numbers), and infinite sets of queer new quantities such as the roots of transfinites and infinitesimals!

It is an astonishing feat of legerdemain. An empty hat rests on a table made of a few axioms of standard set theory. Conway waves two simple rules in the air, then reaches into almost nothing and pulls out an infinitely rich tapestry of numbers that form a real and closed field. Every real number is surrounded by a host of new numbers that lie closer to it than any other "real" value does. The system is truly "surreal."

"Man, that empty set sure gets around!" exclaims Bill. "I think I'll write a book called *Properties of the Empty Set.*" This notion that nothing has properties is, of course, commonplace in philosophy, science, and ordinary language. Lewis Carroll's Alice may think it nonsense when the March Hare offers her nonexistent wine or when the White King admires her ability to see nobody on the road and wonders why nobody did not arrive ahead of the March Hare because nobody goes faster than the hare. It is easy, however, to think of instances in which nothing actually does enter human experience in a positive way.

Consider holes. An old riddle asks how much dirt is in a rectangular hole of certain dimensions. Although the hole has all the properties of a rectangular parallelepiped (corners, edges, faces with areas, volume, and so on), the answer is that there is no dirt in the hole. The various holes of our body are certainly essential to our health, sensory awareness, and pleasure. In *Dorothy and the Wizard in Oz,* the braided man, who lives on Pyramid Mountain in the earth's interior, tells Dorothy how he got there. He had been a manufacturer of holes for Swiss cheese, doughnuts, buttons, porous plasters, and other things. One day he decided to store a vast quantity of adjustable postholes by placing them end to end in the ground, making a deep vertical shaft into which he accidentally tumbled.

The mathematical theory behind Sam Loyd's sliding-block puzzle (15 unit cubes inside a 4 × 4 box) is best explained by regarding the hole as a moving cube. It is analogous to what happens when a gold atom diffuses through lead. Bubbles of nothing in liquids, from the size of a molecule on up, can move around, rotate, collide, and rebound just like things. Negative currents are the result of free electrons jostling one another along a conductor, but holes caused by an absence of free electrons can do the same thing, producing a positive "hole current" that goes the other way.

Lao-tzu writes in Chapter 11 of *Tao Tê Ching:*

> *Thirty spokes share the wheel's hub;*
> *It is the center hole that makes it useful.*
> *Shape clay into a vessel;*
> *It is the space within that makes it useful.*
> *Cut doors and windows for a room;*
> *It is the holes which make it useful.*
> *Therefore profit comes from what is there;*
> *Usefulness from what is not there.*

Osborne Reynolds, a British engineer who died in 1912, invented an elaborate theory in which matter consists of microparticles of nothing moving through the ether the way bubbles move through liquids. His two books about the theory, *On an Inversion of Ideas as to the Structure of the Universe* and *The Sub-Mechanics of the Universe,* both published by the Cambridge University Press, were taken so seriously that W. W. Rouse Ball, writing in early editions of his *Mathematical Recreations and Essays,* called the theory "more plausible than the electron hypothesis."

Reynolds' inverted idea is less crazy than it sounds. P.A.M. Dirac, in his famous theory that predicted the existence of antiparticles, viewed the positron (antielectron) as a hole in a continuum of negative charge. When an electron and positron collide, the electron falls into the positron hole, causing both particles to vanish.

The old concept of a "stagnant ether" has been abandoned by physicists, but in its place is not nothing. The "new ether" consists of the metric field responsible for the basic forces of nature, perhaps also for the particles. John Archibald Wheeler proposes a substratum, called superspace, of infinitely many dimensions. Occasionally a portion of it twists in such a peculiar way that it explodes, creating a universe of three spatial dimensions, changing in time, with its own set of laws and within which the field gets tied into little knots that we call "matter." On the microlevel, quantum fluctuations give space a foamlike structure in which the microholes provide space with additional properties. There is still a difference between something and nothing, but it is purely geometrical and there is nothing behind the geometry.

Empty space is like a straight line of zero curvature. Bend the line, add little bumps that ripple back and forth, and you have a universe dancing with matter and energy. Outside the utmost fringes of our expanding cosmos are (perhaps) vast regions unpenetrated by light and gravity. Beyond those regions may be other universes. Shall we say that these empty regions contain nothing, or are they still saturated with a metric of zero curvature?

Greek and medieval thinkers argued about the difference between being and nonbeing, whether there is one world or many, whether a perfect vacuum can properly be said to "exist," whether God formed the world from pure nothing or first created a substratum of matter that was what St. Augustine called *prope nihil,* or close to nothing. Exactly the same questions were and are debated by philosophers and theologians of the East. When the god or gods of an Eastern religion created the world from a great Void, did they shape nothing or something that was almost nothing? The questions may seem quaint, but change the terminology a bit and they are equivalent to present controversies.

A special Christmas 1901 issue of the British philosophical magazine *Mind* consisted entirely of joke papers. Its frontispiece (reproduced here as Figure 45.1) was titled "Portrait of Its Immanence the Absolute." The Absolute was a favorite term of Hegelian philosophers.

"Turn the eye of faith," the instructions read, "fondly but firmly on the centre of the page, wink the other, and gaze fixedly until you see It."

The Absolute's picture was a totally blank rectangle, protected by covering tissue. Above the picture were the words, "This side up."

There are endless examples from the arts—some jokes, some not—of nothing admired as something. In 1951 Ad Reinhardt, a respected American abstractionist who died in 1967, began painting all-blue and all-red canvases. A few years later he moved to the ultimate—black. His all-black 5 × 5-foot pictures were exhibited in 1963 in leading galleries in New York, Paris, Los Angeles, and London. (See Figure 45.2) Although one critic called him a charlatan (Ralph F. Colin, "Fakes and Frauds in the Art World," *Art in America,* April 1963), more eminent critics (Hilton Kramer, *The Nation,* June 22, 1963, and Harold Rosenberg, *The New Yorker,* June 15, 1963) admired his black art. An "ultimate statement of esthetic purity," was how Kramer put it (*The New York Times,* October 17, 1976) in praising an exhibit of the black paintings at the Pace Gallery.

In 1965 Reinhardt had three simultaneous shows at top Manhattan galleries: one of all-blacks, one of all-reds, one of all-blues. Prices ranged from $1,500 to $12,000.

Since black is the absence of light, Reinhardt's black canvases come as close as possible to pictures of nothing, certainly much closer than the all-white canvases of Robert Rauschenberg and others. A *New Yorker* cartoon (September 23, 1944) by R. Taylor showed two ladies at an art exhibit, standing in front of an all-white canvas and reading from the catalogue: "During the Barcelona period he became enamored of the possibilities inherent in virgin space. With a courage born of the most profound respect for the enigma of the imponderable, he produced, at this time, a series of canvases in which there exists solely an expanse of pregnant white."

The Museum of Modern Art

Figure 45.2. Ad Reinhardt's *Abstract Painting, 1960–61* (Oil, 60″ × 60″)

I know of no piece of "minimal sculpture" that is reduced to the absolute minimum of nothing, though I expect to read any day now that a great museum has purchased such a work for many thousands of dollars. Henry Moore certainly exploited the aesthetics of holes (see Figure 45.3). In 1950 Ray Bradbury received the first annual award of The Elves', Gnomes', and Little Men's Science-Fiction Chowder and Marching Society at a meeting in San Francisco. The award was an invisible little man standing on the brass plate of a polished walnut pedestal. This was not entirely nothing, says my informant, Donald Baker Moore, because there were two black shoe prints on the brass plate to indicate that the little man was actually there.

There have been many plays in which principal characters say nothing. Has anyone ever produced a play or motion picture that consists, from beginning to end, of an empty stage or screen? Some of Andy Warhol's early films come close to it, and I wouldn't be surprised to learn that the limit was actually attained by some early avant-garde playwright.

John Cage's *4′33″* is a piano composition that calls for four minutes and thirty-three seconds of total silence as the player sits frozen on the piano stool. The duration of the silence is 273 seconds. This corresponds, Cage has explained, to −273 degrees centigrade, or absolute zero, the temperature at which all molecular motion quietly stops. I have not heard *4′33″* performed, but friends who have tell me it is Cage's finest composition.

There are many outstanding instances of nothing in print: Chapters 18 and 19 of the final volume of *Tristram Shandy,* for example. Elbert Hubbard's *Essay on Silence,* containing only blank pages, was bound in brown suede and gold-stamped. I recall as a boy seeing a similar book titled *What I Know about Women,* and a Protestant fundamentalist tract called *What Must You Do to Be Lost? Poème Collectif,* by Robert Filliou, issued in Belgium in 1968, consists of sixteen blank pages.

In 1972 the Honolulu Zoo distributed a definitive monograph called *Snakes of Hawaii: An authoritative, illustrated and complete guide to exotic species indigenous to the 50th State,* by V. Ralph Knight, Jr., B.S. A correspondent, Larry E. Morse, informs me that this entire monograph is reprinted (without credit) in *The Nothing Book.* This volume of blank pages was published in 1974, by Harmony House, in regular and deluxe editions. It sold so well that in 1975 an even more expensive ($5) deluxe edition was printed on fine French marble design paper and bound in leather. According to *The Village Voice* (December 30, 1974), Harmony House was threatened with legal action by a European author whose blank-paged book had been published a few years before *The Nothing Book.* He believed his copyright had been infringed, but nothing ever came of it.

Howard Lyons, a Toronto correspondent, points out that the null set has long been a favorite topic of song writers: "I ain't got nobody," "Nobody loves me," "I've got plenty of nothing," "Nobody lied when they said that I cried over you," "There ain't no sweet gal that's worth the salt of my tears," and hundreds of other lines.

Events can occur in which nothing is as startling as a thunderclap. An old joke tells of a man who slept in a lighthouse under a foghorn that boomed regularly every 10 minutes. One night at 3:20 A.M., when the mechanism failed, the man leaped out of bed shouting, "What was that?" As a prank all the members of a large orchestra once stopped playing suddenly in the middle of a strident symphony, causing the conductor to fall off the podium. One afternoon in a rural section of North Dakota, where the wind blew constantly, there was a sudden cessation of wind. All the chickens fell over. A Japanese correspondent tells me that the weather bureau in Japan issues a "no-wind warning" because an absence of wind can create damaging smog.

There are many examples that are not jokes. An absence of water can cause death. The loss of a loved one, of money, or of a reputation can push someone to suicide. The law recognizes innumerable occasions on which a failure to act is a crime. Grave consequences will follow when a man on a railroad track, in front of an approaching train and unable to decide whether to jump to the left or to the right, makes no decision. In the story "Silver Blaze," Sherlock Holmes based a famous deduction on the "curious incident" of a dog that "did nothing in the night-time."

Moments of escape from the omnipresent sound of canned music are becoming increasingly hard to obtain. Unlike cigar smoke, writes Edmund Morris in a fine essay, "Oases of Silence in a Desert of Din" (*The New York Times,* May 25, 1975), noise can't be fanned away. There is an old joke about a jukebox that offers, for a quarter, to provide three minutes of no music. Drive to the top of Pike's Peak, says Morris, "whose panorama of Colorado inspired Katharine Lee Bates to write 'America the Beautiful,' and your ears will be assailed by the twang and boom of four giant speakers—N, S, E, and W—spraying cowboy tunes into the crystal air." Even the Sistine Chapel is now wired for sound.

"At first," continues Morris, "there is something discomforting, almost frightening, about real silence. . . . You are startled by the apparent loudness of ordinary noises. . . . Gradually your ears become attuned to a delicate web of sounds, inaudible elsewhere, which George Eliot called 'that roar which lies on the other side of silence.' " Morris provides a list of a few Silent Places around the globe where one can escape not only from Muzak but from all the aural pollution that is the by-product of modern technology.

These are all examples of little pockets in which there is an absence of something. What about that monstrous dichotomy between all being—everything there is—and nothing? From the earliest times the most eminent thinkers have meditated on this ultimate split. It seems unlikely that the universe is going to vanish (although I myself once wrote a story, "Oom," about how God, weary of existing, abolished everything, including himself), but the fact that we ourselves will soon vanish is real enough. In medieval times the fear of death was mixed with a fear of eternal suffering, but since the fading of hell this fear has been replaced by what Sören Kierkegaard called an "anguish" or "dread" over the possibility of becoming nothing.

This brings us abruptly to what Paul Edwards has called the "super-ultimate question." "Why," asked Leibniz, Schelling, Schopenhauer, and a hundred other philosophers, "should something exist rather than nothing?"

Obviously it is a curious question, not like any other. Large numbers of people, perhaps the majority, live out their lives without ever considering it. If someone asks them the question, they may fail to understand it and believe the questioner is crazy. Among those who understand the question, there are varied responses. Thinkers of a mystical turn of mind, the late Martin Heidegger for instance, consider it the deepest, most fundamental of all metaphysical questions and look with contempt on all philosophers who are not equally disturbed by it. Those of a positivistic, pragmatic turn of mind consider it trivial. Since everyone agrees there is no way to answer it empirically or rationally, it is a question without cognitive content, as meaningless as asking if the number 2 is red or green. Indeed, a famous paper by Rudolf Carnap on the meaning of questions heaps scorn on a passage in which Heidegger pontificates about being and nothingness.

A third group of philosophers, including Milton K. Munitz, who wrote an entire book titled *The Mystery of Existence,* regards the question as being meaningful but insists that its significance lies solely in our inability to answer it. It may or may not have an answer, argues Munitz, but in any case the answer lies totally outside the limits of science and philosophy.

Whatever their metaphysics, those who have puzzled most over the superultimate question have left much eloquent testimony about those unexpected moments, fortunately short-lived, in which one is suddenly caught up in an overwhelming awareness of the utter mystery of why anything is. That is the terrifying emotion at the heart of Jean-Paul Sartre's great philosophical novel *Nausea*. Its red-haired protagonist, Antoine Roquentin, is haunted by the superultimate mystery. "A circle is not absurd," he reflects. "It is clearly explained by the rotation of a straight segment around one of its extremities. But neither does a circle exist." Things that do exist, such as stones and trees and himself, exist without any reason. They are just insanely *there*, bloated, obscene, gelatinous, unable not to exist. When the mood is on him, Roquentin calls it "the nausea." William James had earlier called it an "ontological wonder sickness." The monotonous days come and go, all cities look alike, nothing happens that means anything.

G. K. Chesterton is as good an example as any of the theist who, stunned by the absurdity of being, reacts in opposite fashion. Not that shifting to God the responsibility for the world's existence answers the superultimate question; far from it! One immediately wonders why God exists rather than nothing. But although none of the awe is lessened by hanging the universe on a transcendent peg, the shift can give rise to feelings of gratitude and hope that relieve the anxiety. Chesterton's existential novel *Manalive* is a splendid complement to Sartre's *Nausea*. Its protagonist, Innocent Smith, is so exhilarated by the privilege of existing that he goes about inventing whimsical ways of shocking himself into realizing that both he and the world are not nothing.

Let P. L. Heath, who had the first word in this article, also have the last. "If nothing whatsoever existed," he writes at the end of his article on nothing in *The Encyclopedia of Philosophy*, "there would be no problem and no answer, and the anxieties even of existential philosophers would be permanently laid to rest. Since they are not, there is evidently *nothing to worry about*. But that itself should be enough to keep an existentialist happy. Unless the solution be, as some have suspected, that it is not nothing that has been worrying them, but they who have been worrying it."

Addendum

When this chapter first appeared in *Scientific American* (February 1975), it prompted many delightful letters on aspects of the topic I had not known about or had failed to mention.

Hester Elliott was the first of several readers who were reminded, by my story of the lighthouse keeper, of what some New Yorkers used to call the "Bowery El phenomenon." After the old elevated on Third Avenue was torn down, police began receiving phone calls from people who lived near the El. They were waking at regular intervals during the night, hearing strange noises, and having strong feelings of foreboding. "The schedules of the absent trains," as Ms. Elliott put it, "reappeared in the form of patterned calls on the police blotters." This is discussed, she said, by Karl Pribram in his book *Languages of the Brain* as an example of how our brain, even during sleep, keeps scanning the flow of events in the light of past expectations. It is aroused by any sharp deviation from the accustomed pattern.

Psychologist Robert B. Glassman also referred in a letter to the El example, and gave others. The human brain, he wrote, has the happy facility of forgetting, of pushing out of consciousness whatever seems irrelevant at the moment. But the irrelevant background is still perceived subliminally, and changes in this background bring it back into consciousness. Russian psychologists, he said, have found that if a human or animal listens long enough to the repeated sound of the same tone, they soon learn to ignore it. But if the same tone is then sounded in a different way, even if sounded more *softly* or more *briefly,* there is instant arousal.

Vernon Rowland, a professor of psychology at Case Western Reserve, elaborated similar points. His letter, which follows, was printed in *Scientific American,* April 1975:

Sirs:

I enjoyed Martin Gardner's essay on "nothing." John Horton Conway's rule and Gardner's analysis of "nothing" are, like all human activity, expressions of the nervous system, the study of which helps in understanding the origins and evolution of "nothing."

The brain is marvelously tuned to detect change as well as constancies in the environment. Sharp change between constancies is a perceptually or intellectually recognizable boundary. "Nothing" is "knowable" with clarity only if it is well demarcated from the "non-nothing." Even if it is vaguely bounded, nothingness cannot be treated as an absolute. This is an example of the illogicality of absolutes, because "nothing" cannot be in awareness except as it is related to (contrasted with) non-nothing.

One can observe in the brains of perceiving animals, even animals as primitive as the frog, special neurons responding specifically to spatial

Logic and Philosophy

boundaries and to temporal boundaries. In the latter, for instance, neurons called "off" neurons, go into action when "something," say light, becomes "nothing" (darkness). "Nothing" is therefore positively signaled and is thereby endowed with existence. The late Polish neuropsychologist Jerzy Konorski pointed out the possibility that closing the eyes may activate off neurons, giving rise to "seeing" darkness and recognizing it as being different from not seeing at all.

I and others have used temporal nothingness as a food signal for cats by simply imposing 10 seconds of silence in an otherwise continuously clicking environment. Their brains show the learning of the significance of this silence in ways very similar to those for the inverse: 10 seconds of clicking presented on a continuous background of silence. "Nothing" and "something" can be treated in the same way as psychologists deal with other forms of figure-ground or stimulus-context reversal.

The nothingness of which we become aware by specific brain signals can be known only by discriminating it from other brain signals that reveal the boundaries and constancies of existing objects. This requires an act of attention. There is another form of "nothing" that is based on an attentional shift from one sense modality to another (as in the example of listening to music) or to a failure of the attentional mechanism. In certain forms of strokes the person "forgets" one part of his body and acts as if it simply does not exist, for example a man who shaves only one half of his face.

Animate systems obtain and conserve life-supporting energy by evolving mechanisms to offset or counter perturbations in their energy supply. Detecting absences ("nothings") in the energy domain had to be acquired early or survival could not have gone beyond the stage of actually living in the energy supply (protozoa in nutritious pools) rather than near it (animals that can leave the water and return).

If this pragmatic view of the biopsychological origins of "nothing" and "absence" is insufficient for trivializing the Leibnizian question ("Why should something exist rather than nothing?"), I would argue that the philosopher faces the necessity of showing that the statement "Nothing [in the absolute sense] exists" is not a self-contradiction.

The reference to my story "Oom" reminded Ms. Elliott of the following paragraph from Jorge Luis Borges' essay on John Donne's *Biathanatos* (a work which argues that Jesus committed suicide), in *Other Inquisitions, 1937–1952*:

As I reread this essay, I think of the tragic Philipp Batz, who is called Philipp Mainländer in the history of philosophy. Like me, he was an impassioned reader of Schopenhauer, under whose influence (and perhaps under the influence of the Gnostics) he imagined that we are fragments of a God who destroyed Himself at the beginning of time, because He did not wish to exist. Universal history is the obscure agony of those fragments. Mainländer was born in 1841; in 1876 he published his book *Philosophy of the Redemption.* That same year he killed himself.

Is Mainländer one of Borges' invented characters? No, he actually existed. You can read about him and his strange two-volume work in *The Encyclopedia of Philosophy,* Vol. 6, page 119.

Several readers informed me of the amusing controversy among graph theorists over whether the "null-graph" is useful. This is the graph that has no points or edges. The classic reference is a paper by Frank Harary and Ronald C. Read, "Is the Null-Graph a Pointless Concept?" (The paper was given at the Graphs and Combinatorial Conference at George Washington University in 1973 and appears in the conference lecture notes published by Springer-Verlag.)

"Note that it is not a question of whether the null-graph 'exists,' " the authors write. "It is simply a question of whether there is any point in it." The authors survey the literature, give pros and cons, and finally reach no conclusion. Figure 45.3, reproduced from their paper, shows what the null-graph looks like.

Figure 45.3. The null-graph

Wesley Salmon, the philosopher of science, sent a splendid ontological argument for the existence of the null set:

I have just finished reading, with much pleasure, your column on "nothing." It reminded me of a remark made by a brilliant young philosopher at the University of Toronto, Bas van Fraassen, who, in a lecture on philosophy of mathematics, asked why there might not be a sort of ontological proof for the existence of the null set. It would begin, "By the null set we understand that set than which none emptier can be conceived ..." Van Fraassen is editor in chief of the *Journal of Philosophical Logic.* I sent him the completion of the argument:

"The fool hath said in his heart that there is no null set. But if that were so, then the set of all such sets would be empty, and hence, it would be the null set. Q.E.D."

I still do not know why he did not publish this profound result.

Frederick Mosteller, a theoretical statistician at Harvard, made the following comments on the superultimate question:

Ever since I was about fourteen years old I have been severely bothered by this question, and by and large not willing to talk to other people about it because the first few times I tried I got rather unexpected responses, mainly rather negative putdowns. It shook me up when it first occurred to me, and has bothered me again and again. I could not understand why it wasn't in the newspapers once a week. I suppose, in a sense, all references to creation are a reflection of this same issue, but it is the simplicity of the question that seems to me so scary.

When I was older I tried it once or twice on physicists and again did not get much of a response—probably talked to the wrong ones. I did mention it to John Tukey once, and he offered a rather good remark. He said something like this: contemplating the question at this time doesn't seem to be producing much information—that is, we aren't making much progress with it—and so it is hard to spend time on it. Perhaps it is not yet a profitable question.

It seems so much more reasonable to me that there should be nothing than something that I have secretly concluded for myself that quite possibly physicists will ultimately prove that, were there a system containing nothing, it would automatically create a physical universe. (Of course, I know they can't quite do this.)

When my column on nothing was reprinted in my *Mathematical Magic Show* (1977) an editor at Knopf asked me if it was necessary to obtain MOMA's (Museum of Modern Art) permission to reprint Ad Reinhardt's all-black painting (Figure 45.2). I convinced her it was not necessary. And as I anticipated, somewhere along the production line I was asked for the missing art of Figure 45.3. This picture of the null graph, by the way, is reproduced (without credit to the artist) as Figure 257 in the *Dictionary of Mathematics,* edited by E. J. Borowski and J. M. Borwein (London: Collins, 1989).

Meditating on the recent flurry of interest in what is called the "anthropic principle" I suddenly realized that I could answer the superultimate question: Why is there something rather than nothing? Because if there wasn't anything we wouldn't be here to ask the question. I think this points up the essential absurdity of the weak anthropic principle. It's not wrong, but it contributes nothing significant to any philosophical or scientific question. The Danish poet Piet Hein, in one of his "grook" verses, says it this way:

> *The universe may*
> *Be as great as they say,*
> *But it wouldn't be missed*
> *If it didn't exist.*

Lakenan Barnes, an attorney in Missouri, reminded me that Joshua was the son of Nun (Joshua 1:1), that "love" in tennis means nothing, that the doughnut's hole is the "dough naught." He also passed along a quatrain of his that had appeared in the St. Louis *Post-Dispatch* (July 7, 1967):

> *In the world of math*
> *That Man has wrought,*
> *The greatest gain*
> *Was the thought of naught.*

Some readers were mystified by the chapter's epigraph. It is the second sentence of Heath's article on nothing in the *Encyclopedia of Philosophy.* Like Lewis Carroll, in the second Alice book, Heath is taking Nobody to be the name of a person. Here is the sentence in the context of Heath's playful opening paragraph:

NOTHING is an awe-inspiring yet essentially undigested concept, highly esteemed by writers of a mystical or existentialist tendency, but by most others regarded with anxiety, nausea, or panic. Nobody seems to know how to deal with it (he would, of course), and plain persons generally are reported to have little difficulty in saying, seeing, hearing, and doing nothing. Philosophers, however, have never felt easy on the matter. Ever since Parmenides laid it down that it is impossible to speak of what is not, broke his own rule in the act of stating it, and deduced himself into a world where all that ever happened was nothing, the impression has persisted that the narrow path between sense and nonsense on this subject is a difficult one to tread and that altogether the less said of it the better.

Will Shortz, *The New York Times* crossword puzzle editor, presented an April Fools' crossword in *Games* magazine, March/April 1979, page 31. The solution was to leave all the cells blank because "nothing" was the answer to each of the puzzle's 49 definitions of words.

One of Gary Larson's *Far Side* cartoons showed a man equipped for mountain climbing but who was about to descend into a huge hole in the ground. A reporter asks him why he is doing this. He replies, "Because it's not there."

Ray Smullyan on *Discover's* puzzle page (November 1996) gave this riddle which readers should have little difficulty answering: "What is it that is larger than the universe, the dead eat it, and if the living eat it, they die?"

Bibliography

R. M. Adams, *Nil: Episodes in the Literary Conquest of Void During the Nineteenth Century*, Oxford University Press, 1966.

J. H. Conway, *All Numbers Great and Small*, University of Calgary Mathematical Research Paper No. 149, February 1972.

J. H. Conway, *On Numbers and Games*, Academic Press, 1976.

P. Edwards, "Why," *The Encyclopedia of Philosophy*, MacmillanFree Press, 1967.

M. Gardner, "Oom," *The Journal of Science Fiction*, Fall 1951. Reprinted in M. Gardner, *The No-Sided Professor and Other Tales*, Prometheus, 1987.

M. Gardner, *The Annotated Snark*, Simon and Schuster, 1962.

M. Gardner, "Conway's Surreal Numbers," *Penrose Tiles to Trapdoor Ciphers*, W. H. Freeman, 1989, Chapter 4.

P. L. Heath, "Nothing," *The Encyclopedia of Philosophy*, MacmillanFree Press, 1967.

R. Kaplan, *The Nothing That Is: A Natural History of Zero*, Oxford University Press, 1999.

D. E. Knuth, *Surreal Numbers*, Addison-Wesley, 1974.

B. Rotman, *Signifying Nothing: The Semiotics of Zero*, Stanford, 1993.

C. Seife, *Zero: The Biography of a Dangerous Idea*, Viking, 2000.

I. Stewart, "Zero, Zilch, and Zip," *New Scientist*, April 25, 1998, pp. 41–44.

Chapter 46   *Everything*

*T*he topic of the previous chapter is "Nothing." I have nothing more to say about nothing, or about "something," since everything I know about something was said when I wrote about nothing. But "everything" is something altogether different.

Let us begin by noting the curious fact that some things, namely ourselves, are such complicated patterns of waves and particles that they are capable of wondering about everything. "What is man in nature?" asked Pascal. "A nothing in comparison with the infinite, an all in comparison with the nothing, a mean between nothing and everything."

In logic and set theory "things" are conveniently diagrammed with Venn circles. In Figure 46.1 the points inside circle *a* represent humans. The points inside circle *b* stand for feathered animals. The overlap, or intersection set, has been darkened to show that it has no members. It is none other than our old friend the empty set.

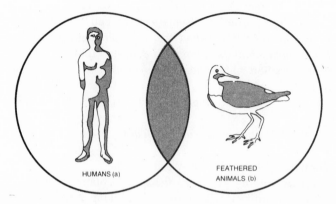

HUMANS (a)

FEATHERED ANIMALS (b)

Figure 46.1. A Venn diagram for "No humans have feathers"

So far, so clear. What about the points on the plane outside the two circles? Obviously they represent things that are not *a* and not *b,* not human and not feathered, but how far-ranging is this set? To clarify the question Augustus De Morgan invented the phrase "universe of discourse." It is the range of all the variables with which we are concerned. Sometimes it is explicitly defined, sometimes tacitly assumed, sometimes left fuzzy. In set theory it is made precise by defining what is called the universal set, or, for short, the universe. This is the set with a range that coincides with the universe of discourse. And that range can be whatever we want it to be.

With the Venn circles *a* and *b* we are perhaps concerned only with living things on the earth. If this is so, that is our universe. Suppose, however, we expand the universe by adding a third set, the set of all typewriters, and changing *b* to all feathered objects. As Figure 46.2 shows, all three intersection sets are empty. It is the same empty set, but the range of the null set has also been expanded. There is only one "nothing," but a hole in the ground is not the same as a hole in a piece of cheese. The complement of a set *k* is the set of all elements in the universal set that are not in *k*. It follows that the universe and the empty set are complements of each other.

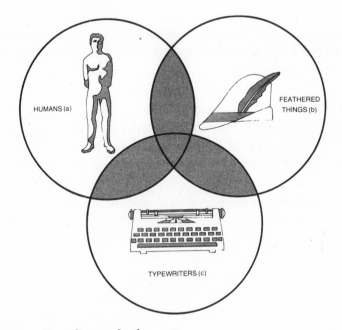

Figure 46.2. A Venn diagram for three sets

How far can we extend the universal set without losing our ability to reason about it? It depends on our concern. If we expand the universe of Figure 46.1 to include all concepts, the intersection set is no longer empty because it is easy to imagine a person growing feathers. The proofs of Euclid are valid only if the universe of discourse is confined to points in a Euclidean plane or in 3-space. If we reason that a dozen eggs can be equally divided only between one, two, three, four, six, or 12 people, we are reasoning about a universal set that ranges over the integers. John Venn (who invented the Venn diagram) likened the universe of discourse to our field of vision. It is what we are looking at. We ignore everything behind our head.

Nevertheless, we can extend the universe of discourse amazingly far. We certainly can include abstractions such as the number 2, pi, complex numbers, perfect geometric figures, even things we cannot visualize such as hypercubes and non-Euclidean spaces. We can include universals such as redness and cowness. We can include things from the past or in the future and things real or imaginary, and can still reason effectively about them. Every dinosaur had a mother. If it rains next week in Chicago, the old Water Tower will get wet. If Sherlock Holmes had actually fallen off that cliff at Reichenbach Falls, he would have been killed.

Suppose we extend our universe to include every entity that can be defined without logical contradiction. Every statement we can make about that universe, if it is not contradictory, is (in a sense) true. The contradictory objects and statements are not allowed to "exist" or be "true" for the simple reason that contradiction introduces meaninglessness. When a philosopher such as Leibniz talks about "all possible worlds," he means worlds that can be talked about. You can talk about a world in which humans and typewriters have feathers. You cannot say anything sensible about a square triangle or an odd integer that is a multiple of 2.

Is it possible to expand our universe of discourse to the ultimate and call it the set of all possible sets? No, this is a step we cannot take without contradiction. Georg Cantor proved that the cardinal number of any set (the number of its elements) is always lower than the cardinal number of the set of all its subsets. This is obvious for any finite set (if it has n elements, it must have 2^n subsets), but Cantor was able to show that it also applies to infinite sets. When we try to apply this theorem to everything, however, we get into deep trouble. The set of all sets must have the highest aleph (infinite number) for its cardinality; otherwise it would not be everything. On the other hand, it cannot have the highest aleph because the cardinality of its subsets is higher.

When Bertrand Russell first came across Cantor's proof that there is no highest aleph, and hence no "set of all sets," he did not believe it. He wrote in 1901 that Cantor had been "guilty of a very subtle fallacy, which I hope to explain in some future work," and that it was "obvious" there had to be a greatest aleph because "if everything has been taken, there is nothing left to add." When this essay was reprinted in *Mysticism and Logic* 16 years later, Russell added a footnote apologizing for his mistake. ("Obvious" is obviously a dangerous word to use in writing about everything.) It was Russell's meditation on his error that led him to discover his famous paradox about the set of all sets that are not members of themselves.

To sum up, when the mathematician tries to make the final jump from lots of things to everything, he finds he cannot make it. "Everything" is self-contradictory and therefore does not exist!

The fact that the set of all sets cannot be defined in standard (Zermelo-Fraenkel) set theory, however, does not inhibit philosophers and theologians from talking about everything, although their synonyms for it vary: being, *ens,* what is, existence, the absolute, God, reality, the Tao, Brahman, *dharma-kaya,* and so on. It must, of course, include everything that was, is, and will be, everything that can be imagined and everything totally beyond human comprehension. Nothing is also part of everything. When the universe gets this broad, it is difficult to think of anything meaningful (not contradictory) that does not in some sense exist. The logician Raymond Smullyan, in one of his several hundred marvelous unpublished essays, retells an incident he found in Oscar Mandel's book *Chi Po and the Sorcerer: A Chinese Tale for Children and Philosophers.* The sorcerer Bu Fu is giving a painting lesson to Chi Po. "No, no!" says Bu Fu. "You have merely painted what *is.* Anybody can paint what is! The real secret is to paint what isn't!" Chi Po, puzzled, replies: "But what is there that isn't?"

This is a good place to come down from the heights and consider a smaller, tidier universe, the universe of contemporary cosmology. Modern cosmology started with Einstein's model of a closed but unbounded universe. If there is sufficient mass in the cosmos, our 3-space curves back on itself like the surface of a sphere. (Indeed, it becomes the 3-space hypersurface of a 4-space hypersphere.) We now know that the universe is expanding from a primordial fireball, but there does not seem to be enough mass for it to be closed. The steady-state theory generated much discussion and stimulated much valuable scientific work, but it now seems to have been eliminated as a viable theory by such discoveries as that of the universal background radiation (which has no reasonable explanation except that it is radiation left over from the primordial fireball, or "big bang").

The large unanswered question is whether there is enough mass hidden somewhere in the cosmos (in black holes?) to halt the expansion and start the universe shrinking. If that is destined to happen, the contraction will become runaway collapse, and theorists see no way to prevent the universe from entering the "singularity" at the core of a black hole, that dreadful spot where matter is crushed out of existence and no known laws of physics apply. Will the universe disappear like the fabled Poof Bird, which flies backward in ever decreasing circles until—poof!—it vanishes into its own anus? Will everything go through the black hole to emerge from a white hole in some completely different spacetime? Or will it manage to avoid the singularity and give rise to another fireball? If reprocessing is possible, we have a model of an oscillating universe that periodically explodes, expands, contracts, and explodes again.

Among physicists who have been building models of the universe John Archibald Wheeler has gone further than anyone in the direction of everything. In Wheeler's wild vision our universe is one of an infinity of universes that can be regarded as embedded in a strange kind of space called superspace.

In order to understand (dimly) what Wheeler means by superspace let us start with a simplified universe consisting of a line segment occupied by two particles, one black and one gray (see Figure 46.3, *top*). The line is one-dimensional, but the particles move back and forth (we allow them to pass through each other) to create a space–time of two dimensions: one of space and one of time.

There are many ways to graph the life histories of the two particles. One way is to represent them as wavy lines, called world lines in relativity theory, on a two-dimensional space–time graph (see Figure 46.3). Where was the black particle at time k? Find k on the time axis, move horizontally to the black particle's world line, then move down to read off the particle's position on the space axis.

To see how beautifully the two world lines record the history of our

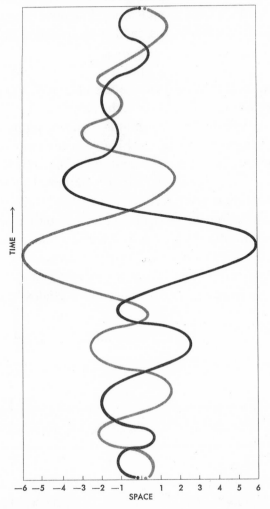

Figure 46.3. A space–time graph of a two-particle cosmos from birth to death

infant universe, cut a slot in a file card. The slot should be as long as the line segment and as wide as a particle. Place the card at the bottom of the graph where you can see the universe through it. Move the card upward slowly. Through the slot you will see a motion picture of the two particles. They are born at the center of their space, dance back and forth until they have expanded to the limits, and then dance back to the center, where they disappear into a black hole.

In kinematics it is sometimes useful to graph the changes of a system of particles as the motion of a single point in a higher space called con-

Everything

figuration space. Let us see how to do this with our two particles. Our configuration space again is two-dimensional, but now both coordinates are spatial. One coordinate is assigned to the black particle and the other, to the gray particle (see Figure 46.4). The positions of both particles can be represented by a single point called the configuration point. As the point moves, its coordinate values change on both axes. One axis locates one particle; the other axis, the other particle. The trajectory traced by the moving point corresponds to the changing pattern of the system of particles; conversely, the history of the system determines a unique trajectory. It is not a space–time graph. (Time enters later as an added parameter.) The line cannot form branches because that would split each particle in two. It may, however, intersect itself. If a system is periodic, the line will be a closed curve. To transform the graph into a spacetime graph we can, if we like, add a time coordinate and allow the point to trace a curve in three dimensions.

The technique generalizes to a system of N particles in a space with any number of dimensions. Suppose we have 100 particles in our lit-

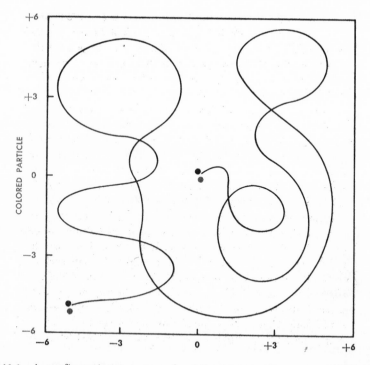

Figure 46.4. A configuration-space graph of the history of two particles in a one-dimensional universe

tle line-segment cosmos. Each particle has one degree of freedom, so our configuration point must move in a space of 100 dimensions. If our universe is a system of N particles on a plane, each particle has two degrees of freedom, so our configuration space must be a hyperspace of $2N$ dimensions. In 3-space a particle has three degrees of freedom, so the configuration space must have $3N$ dimensions. In general the hyperspace has an order equal to the total degrees of freedom in the system. Add another coordinate for time and the space becomes a space–time graph.

Unfortunately the position of a configuration point at any instant does not enable us to reconstruct the system's past or predict its future. Josiah Willard Gibbs, working on the thermodynamics of molecules, found a slightly more complicated space in which he could graph a system of molecules so that the record was completely deterministic. This is done by assigning six coordinates to each molecule: three to determine position and three to specify momentums. The movement of a single phase point in what Gibbs called a "phase space" of $6N$ dimensions will record the life history of N particles. Now, however, the position of the phase point provides enough information to reconstruct (in principle) the entire previous history of the system and to predict its future. As before, the trajectory cannot branch, but now it also cannot intersect itself. An intersection would mean that a state could be reached from two different states and could lead to two different states, but both possibilities are ruled out by the assumption that position and momentums (which include a vector direction) fully determine the next state. The curve may still loop, however, indicating that the system is periodic.

Our universe, with its non-Euclidean space–time and its quantum uncertainties, cannot be graphed in anything as simple as phase space, but Wheeler has found a way to do it in superspace. Like configuration space, superspace is timeless, but it has an infinity of dimensions. A single point in superspace has an infinite set of coordinates that specify completely the structure of our non-Euclidean 3-space: its size, the location of every particle, and the structure of every field (including the curvature of space itself) at every point. As the superpoint moves, its changing coordinate numbers describe how our universe changes, not failing to take into account the role of observers' frames of reference in relativity and the probability parameters of quantum mechanics. The motion of the superpoint gives the entire history of our universe.

At the same time (whatever that means!) that the present drama of our cosmos is being acted on the stage of superspace countless other superpoints, representing other 3-space universes, are going through their cycles. Superpoints close to one another describe universes that most resemble one another, like the parallel worlds that H. G. Wells introduced into science fiction with his *Men Like Gods.* These parallel universes, cut off from one another because they occupy different slices of superspace, are continually bursting into space–time through a singularity, flourishing for a moment of eternity, then vanishing back through a singularity into the pure and timeless "pregeometry" from whence they came.

Whenever such a cosmos explodes into being, random factors generate a specific combination of logically consistent (Leibniz called them compossible) particles, constants, and laws. The resulting structure has to be tuned exceedingly fine to allow life. Alter the fine-structure constant a trifle either way and a sun such as ours becomes impossible. Why are we here? Because random factors generated a cosmic structure that allowed us to evolve. An infinity of other universes, not so finely tuned, are living and dying without there being anyone in them capable of observing them.

These "meaningless" universes, meaningless because they contain no participator-observers, do not even "exist" except in the weak sense of being logically possible. Bishop Berkeley said that to exist is to be perceived, and Charles Sanders Peirce maintained that existence is a matter of degree. Taking cues from both philosophers, Wheeler argues that only when a universe develops a kind of self-reference, with the universe and its observers reinforcing one another, does it exist in a strong sense. "All the choir of heaven and furniture of earth have no substance without a mind" was how Berkeley put it.

As far as I can tell, Wheeler does not take Berkeley's final step: the grounding of material reality in God's perception. Indeed, the fact that a tree seems to exist in a strong sense, even when no one is looking at it, is the key to Berkeley's way of proving God's existence. Imagine a god experimenting with billions of cosmic models until he finds one that permits life. Would not these universes be "out there," observed by the deity? There would be no need for flimsy creatures like ourselves, observing and participating, to confer existence on these models.

Wheeler seems anxious to avoid this view. He argues that quantum mechanics requires participator-observers in the universe regardless of whether there is an outside observer. In one of his metaphors, a universe without internal observers is like a motor without electricity. The cosmos "runs" only when it is "guaranteed to produce somewhere, and for some little length of time in its history-to-be, life, consciousness, and observership." Internal observers and the universe are both essential to the existence of each other, even if the observers exist only in a potential sense. This raises unusual questions. How strongly does a universe exist before the first forms of life evolve? Does it exist in full strength from the moment of the big bang, or does its existence get stronger as life gets more complex? And how strong is the existence of a galaxy, far removed from the Milky Way, in which there may be no participator-observers? Does it exist only when it is observed by life in another galaxy? Or is the universe so interconnected that the observation of a minute portion of it supports the existence of all the rest?

There is a famous passage in which William James imagines a thousand beans flung onto a table. They fall randomly, but our eyes trace geometrical figures in the chaos. Existence, wrote James, may be no more than the order which our consciousness singles out of a disordered sea of random possibilities. This seems close to Wheeler's vision. Reality is not something out there, but a process in which our consciousness is an essential part. We are not what we are because the world is what it is but the other way around. The world is what it is because we are what we are.

When relativity theory first won the day, many scientists and philosophers with a religious turn of mind argued that the new theory supported such a view. The phenomena of nature, said James Jeans, are "determined by us and our experience rather than by a mechanical universe outside us and independent of us." The physical world, wrote Arthur Stanley Eddington, "is entirely abstract and without 'actuality' apart from its linkage to consciousness." Most physicists today would deny that relativity supports this brand of idealism. Einstein himself vigorously opposed it. The fact that measurements of length, time, and mass depend on the observer's frame of reference in no way dilutes the actuality of a space–time structure independent of all observers.

Nor is it diluted by quantum mechanics. What bearing does the statistical nature of quantum laws have on the independent existence of a structure to which those laws apply whenever it is observed? The fact that observations alter state functions of a system of particles does not entail that there is nothing "out there" to be altered. Einstein may have thought that quantum mechanics implies this curious reduction of physics to psychology, but there are not many quantum experts today who agree.

In any case, belief in an external world, independent of human existence but partly knowable by us, is certainly the simplest view and the one held today by the vast majority of scientists and philosophers. As I have suggested, to deny this commonsense attitude adds nothing of value to a theistic or pantheistic faith. Why adopt an eccentric terminology if there is no need for it?

But this is not the place for debating these age-old questions. Let me turn to a strange little book called *Eureka: A Prose Poem,* written by Edgar Allan Poe shortly before his death. Poe was convinced that it was his masterpiece. "What I have propounded will (in good time) revolutionize the world of Physical & Metaphysical Science," he wrote to a friend. "I say this calmly—but I say it." In another letter he wrote, "It is no use to reason with me *now;* I must die. I have no desire to live since I have done *Eureka.* I could accomplish nothing more." (I quote from excellent notes in *The Science Fiction of Edgar Allan Poe,* edited by Harold Beaver, Penguin Books, 1976.)

Poe wanted his publisher, George P. Putnam, to print 50,000 copies. Putnam advanced Poe $14 for his "pamphlet," and printed 500 copies. Reviews were mostly unfavorable. The book seems to have been taken seriously only in France, where it had been translated by Baudelaire. Now suddenly, in the light of current cosmological speculation, Poe's prose poem is seen to contain a vast vision that is essentially a theist's version of Wheeler's cosmology! As Beaver points out, the "I" in Poe's "Dreamland" has become the universe itself:

> By a route obscure and lonely,
> Haunted by ill angels only,
> Where an Eidolon, named NIGHT,
> On a black throne reigns upright,
> I have reached these lands but newly
> From an ultimate dim Thule—
> From a wild weird clime that lieth, sublime,
>   Out of SPACE—out of TIME.

A universe begins, said Poe, when God creates a "primordial particle" out of nothing. From it matter is "irradiated" spherically in all directions, in the form of an "inexpressibly great yet limited number of unimaginably yet not infinitely minute atoms." As the universe expands, gravity slowly gains the upper hand and the matter condenses to form stars and planets. Eventually gravity halts the expansion and the universe begins to contract until it returns again to nothingness. The final "globe of globes will instantaneously disappear" (how Poe would have exulted in today's black holes!) and the God of our universe will remain "all in all."

In Poe's vision each universe is being observed by its own deity, the way your eye watched the two particles dance in our created world of 1-space. But there are other deities whose eyes watch other universes. These universes are "unspeakably distant" from one another. No communication between them is possible. Each of them, said Poe, has "a new and perhaps totally different series of conditions." By introducing gods Poe implies that these conditions are not randomly selected. The fine-structure constant is what it is in our universe because our deity wanted it that way. In Poe's superspace the cyclical birth and death of an infinity of universes is a process that goes on "for ever, and for ever, and for ever; a novel Universe swelling into existence, and then subsiding into nothingness at every throb of the Heart Divine."

Did Poe mean by "Heart Divine" the God of our universe or a higher deity whose eye watches all the lesser gods from some abode in super-superspace? Behind Brahma the creator, goes Hindu mythology, is Brahman the inscrutable, so transcendent that all we can say about Brahman is *Neti neti* (not that, not that). And is Brahman being observed by a supersupersupereye? And can we posit a final order of superspace, with its Ultimate Eye, or is that ruled out by the contradiction in standard set theory of the concept of a greatest aleph?

This is the great question asked in the final stanza of the Hymn of Creation in the *Rig Veda*. The "He" of the stanza is the impersonal One who is above all gods:

> *Whether the world was made or was self-made,*
> *He knows with full assurance, He alone,*
> *Who in the highest heaven guards and watches;*
> *He knows indeed, but then, perhaps, He knows not!*

It is here that we seem to touch—or perhaps we are still infinitely far from touching—the hem of Everything. Let C. S. Lewis (I quote from Chapter 2 of his *Studies in Words*) make the final comment: " 'Everything' is a subject on which there is not much to be said."

Addendum

Charles Peirce once complained that universes are not as plentiful as blackberries, but today there are many top cosmologists who think otherwise. The notion that our universe is only one of an enormous number, perhaps an infinite number, of other universes is now a popular conjecture. Andrei Linde at Stanford University is the leading proponent of such a view. His "multiverse" should not be confused with the countless universes assumed in the many-worlds interpretation of quantum mechanics. In that theory the other universes are all part of a single metaverse that is sprouting countless branches as its monstrously complicated tree evolves. The many worlds of Linde and his sympathizers are modern versions of Poe's beautiful vision—universes unrelated to ours, each with its unique set of laws, particles, and physical constants.

There is no need to invoke intelligent design to explain why our universe is so finely tuned that it can evolve stars, planets, life, and such strange creatures as you and me. Naturally we could exist, as the anthropic principle maintains, only in one of the myriad of universes that had the necessary laws and constants. David Lewis and a few other eccentric philosophers plunge even deeper into fantasy. They argue that every logically possible universe—one free of contradictions—exists and is just as real as the one we are in! Leibniz considered what he called logically "compossible" worlds that are in the mind of God, but God chose only one, the best possible, to actually exist.

Defenders of Linde's multiverse, the many world's interpretation of quantum mechanics, and Lewis's logically possible worlds tend to be atheists or agnostics. The total ensemble of universes is all there is. No creator occupies a transcendent realm. On the other hand, cosmological and philosophical theists invoke Occam's razor. Is it not much simpler, they maintain, to posit a wholly other deity who, for reasons we cannot know, created the one universe in which we find ourselves? It is a universe destined either to last forever, or expand and die of the cold, or eventually to contract, as Poe imagined, to be annihilated in a Big Crunch.

Bibliography

D. Deutsch, *The Fabric of Reality: The Science of Parallel Universes and Its Implications,* Penguin, 1997.

J. Leslie, *Universes,* Oxford University Press, 1989.

D. Lewis, *On the Plurality of Worlds,* Blackwell, 1985.

A. Linde, *Particle Physics and Inflationary Cosmology,* Harvard Academic, 1990.

A. Linde, "The Self-Reproducing Inflationary Universe," *Scientific American,* November 1994, pp. 48–55.

A. N. Prior, "Existence," *The Encyclopedia of Philosophy,* MacmillanFree Press, 1967.

M. Rees, *Before the Beginning: Our Universe and Others,* Addison-Wesley, 1997.

L. Smolin, *The Life of the Cosmos,* Oxford University Press, 1997.

W. V. O. Quine "On What There Is," *From a Logical Point of View,* Harvard University Press, 1953; revised 1961.

XII
Miscellaneous

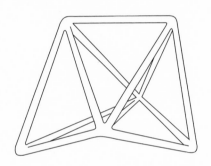

Chapter 47

Melody-Making Machines

*Mathematics and music! The most glaring
possible opposites of human thought! Yet
connected, mutually sustained!*

—HERMANN VON HELMHOLTZ,
Popular Scientific Lectures

*T*here is a trivial sense in which any work of art is a combination of a finite number of discrete elements. Not only that, the precise combination of the elements can be expressed by a sequence of digits or, if you will, by one enormous number.

Consider a poem. Assign distinct numbers to each letter of the alphabet, to each punctuation symbol, and so on. A certain digit, say zero, can be used to separate the numbers. It is obvious that one long string of digits can express the poem. If the books of a vast library contain every possible combination of words and punctuation marks, as they do in Jorge Luis Borges's famous story "The Library of Babel," then somewhere in the collection is every poem ever written or that can be written. Imagine those poems coded as digital sequences and indexed. If one had enough time, billions on billions of years, one could locate any specified great poem. Are there algorithms by which one could find a great poem not yet written?

Consider a painting. Rule the canvas into a matrix of minute cells. The precise color of each cell is easily coded by a number. Scanning the cells yields a chain of numbers that expresses the painting. Since numbers do not decay, a painting can be re-created as long as the number sequence is preserved. Future computers will be able to reproduce a painting more like the original than the original itself, since after a few decades the original will have physically deteriorated to some extent. If a vast art museum contains every combination of colored cells for matrixes not exceeding a certain size, somewhere in that monstrous museum will hang every picture ever painted or that can be painted. Are there algorithms by which a computer could search a list of the museum's code numbers and identify a sequence for a great painting not yet painted?

Consider a symphony. It is a fantastically complex blend of discreteness and continuity; a violin or a slide trombone can move up and down the scale continuously, but a piano cannot produce quarter-tones. We know, however, from Fourier analysis that the entire sound of a symphony, from beginning to end, can be represented by a single curve on an oscilloscope. "This curve," wrote Sir James Jeans in *Science and Music* (Dover, 1968), "*is* the symphony—neither more nor less, and the symphony will sound noble or tawdry, musical or harsh, refined or vulgar, according to the quality of this curve." On a long-playing record a symphony is actually represented by one long space curve.

Because curves can be coded to any desired precision by numbers, a symphony, like a painting or a poem, can be quantized and expressed by a number chain. A vast library recording all combinations of symphonic sounds, would contain every symphony ever written or that could be written. Are there algorithms by which a computer could scan the number sequences of such a library and pick out a great symphony not yet written?

Such procedures would, of course, be so stupendously complex that man may never come close to formulating them, but that is not the point. Do they exist in principle? Is it worthwhile to look for bits and pieces of them? Consider one of the humblest of such aesthetic tasks, the search for rules that govern the invention of a simple melody. Is there a procedure by which a person or a computer can compose a pleasing tune, using no more than a set of combinatorial rules?

If we restrict the tune to a finite length and a finite number of pure tones and rhythms, the number of possible melodies is finite. John Stuart Mill, in his autobiography, recalls that as a young man he was once "seriously tormented by the thought of the exhaustibility of musical compositions." Suppose our tune is made up of just 10 notes chosen from the set of eight notes in a single octave. The number of melodies is the same as the number of 10-letter words that can be formed with eight distinct letters, allowing duplications. It is $8^{10} = 1,073,741,824$, and this without even considering varying rhythms which create, in effect, varying melodies. Many of these tunes will be dull (mi, mi, mi, mi, mi, mi, mi, mi, mi, mi for instance), but some will be extremely pleasing. Are there rules by which a computer or a person could pick out the pleasing combinations?

Attempts to formulate such rules and embody them in a mechanical device for composing tunes have a curious history that began in 1650 when Athanasius Kircher, a German Jesuit, published in Rome his *Musurgia universalis sive ars magna consoni et dissoni* (see Figure 47.1). Kircher was an ardent disciple of Ramón Lull, the Spanish medieval mystic whose *Ars magna* derived from the crazy notion that significant new knowledge could be obtained in almost every field simply by exploring all combinations of a small number of basic elements. It was natural that Kircher, who later wrote a 500-page elaboration of Lull's "great art," would view musical composition as a combinatorial problem. In his music book, he describes a Lullian technique of creating polyphony by sliding columns alongside one another, as with Napier's bones, and reading off rows to obtain various permutations and combinations. Like all of Kircher's huge tomes, the book is a fantastic mix of valuable information and total nonsense, illustrated with elaborate engravings of vocal cords, bones in the ears of various animals, birds and their songs, musical instruments, mechanical details of music boxes, water-operated organ pipes with animated figures of animals and people, and hundreds of other curious things.

The Lullian device described by Kircher was actually built, circa 1670, for the diarist Samuel Pepys, who owned a copy of Kircher's music book and much admired it. The original machine, called "musarithmica mirifica," is in the Pepys Museum at Pepys's alma mater, Magdalene College, Cambridge.

During the early 18th century many German music scholars became interested in mechanical methods of composition. Lorenz Christoph Mizler wrote a book in 1739 describing a system that produced figured bass for baroque ensemble music. In 1757 Bach's pupil, Johann Philipp Kirnberger, published in Berlin his *Ever-ready Composer of Polonaises and Minuets,* using a die for randomizing certain choices. In 1783 another book by Kirnberger extended his methods to symphonies and other forms of music.

Toward the close of the 18th century the practice of generating melodies with the aid of tables and randomizers such as dice or teetotums became a popular pastime. Maximilian Stadler, an Austrian composer, published in 1779 a set of musical bars and tables for producing minuets and trios with the help of dice. At about the same time, a London music publisher, Welcker, issued a "tabular system whereby any person, without the least knowledge of music, may compose ten thousand minuets in the most pleasing and correct manner." Similar anony-

Figure 47.1. Frontispiece of Kircher's *Musurgia universalis* (1650). The artist was J. Paul Schor. Here is how the picture is described in *Athanasius Kircher: A Renaissance Man and the Quest for Lost Knowledge,* by Joscelyn Godwin (Thames and Hudson, Ltd., 1979): "The symbol of the Trinity sheds its rays on the nine choirs of angels, who sing a 36-part canon (by Romano Micheli), and thence on the earth. The terrestrial sphere is shown encircled by the Zodiac and surmounted by Musica, who holds Apollo's lyre and the pan-pipes of Marsyas. In the landscape are seen dancing mermaids and satyrs, a shepherd demonstrating an echo, and Pegasus, the winged horse of the Muses. On the left is Pythagoras, the legendary father of musical theory. He points with one hand to his famous theorem, and with the other to the blacksmiths whose hammers, ringing on the anvil, first led him to discover the relation of tone to weight. On the right is a muse (Polymnia?) with a bird-perched on her head—possibly one of the nine daughters of Pierus, who for their presumption in attempting to rival the Muses were turned into birds. These figures are surrounded respectively by antique and modern instruments."

mous works were falsely attributed to well-known composers such as C.P.E. Bach (son of Johann Sebastian Bach) and Joseph Haydn. The "Haydn" work, "Gioco filarmonico" ("Philharmonic Joke"), Naples, 1790, was discovered by the Glasgow mathematician Thomas H. O'Beirne to be a plagiarism. Its bars and tables are identical with Stadler's.

The most popular work explaining how a pair of dice can be used "to compose without the least knowledge of music" as many German waltzes as one pleases was first published in Amsterdam and in Berlin in 1792, a year after Mozart's death. The work was attributed to Mozart. Most Mozart scholars say it is spurious, although Mozart was fond of mathematical puzzles and did leave handwritten notes showing his interest in musical permutations. (The same pamphlet was issued in Bonn a year later, with a similar work, also attributed to Mozart, for dice composition of country dances. The contredanses pamphlet was reprinted in 1957 by Heuwekemeijer in Amsterdam.)

Mozart's *Musikalisches Würfelspiel*, as the waltz pamphlet is usually called, has been reprinted many times in many languages. In 1806 it appeared in London as *Mozart's Musical Game, Fitted in an Elegant Box, Showing by Easy System How to Compose an Unlimited Number of Waltzes, Rondos, Hornpipes and Reels.* In New York in 1941 the Hungarian composer and concert pianist Alexander Laszlo brought it out under the title *The Dice Composer*, orchestrating the music so that it could be played by chamber groups and orchestras. The system popped up again in West Germany in 1956 in a score published by B. Schott. Photocopies of Schott's charts and musical bars appear in the instruction booklet for *The Melody Dicer*, issued early in 1974 by Carousel Publishing Corporation, Brighton, MA. This boxed set also includes a pair of dice and blank sheets of music paper.

The "Mozart" system consists of a set of short measures numbered 1 through 176. The two dice are thrown 16 times. With the aid of a chart listing 11 numbers in each of eight columns, the first eight throws determine the first eight bars of the waltz. A second chart is used for the second eight throws that complete the 16-bar piece. The charts are constructed so that the waltz opens with the tonic or keynote, modulates to the dominant, then finds its way back to the tonic on its final note. Because all bars listed in the eighth column of each chart are alike, the 11 choices (sums 2 through 12 on the dice) are available for only 14 bars. This allows the system to produce 11^{14} waltzes, all with a distinct

Mozartean flavor. The number is so large that any waltz you generate with the dice and actually play is almost certainly a waltz never heard before. If you fail to preserve it, it will be a waltz that will probably never be heard again.

The first commercial recording of "Mozart" dice pieces was made by O'Beirne. Both the randomizing of the bars and the actual playing of the melodies was done by Solidac, a small and slow experimental computer designed and built between 1959 and 1964 by the Glasgow firm of Barr and Stroud, where O'Beirne was then chief mathematician. It was the first computer built in Scotland. O'Beirne programmed Solidac to play the pieces in clarinetlike tones, and a long-playing recording of selected waltzes and contredanses was issued by Barr and Stroud in 1967. (This recording is no longer available.) O'Beirne is the author of an excellent book on mathematical recreations, *Puzzles and Paradoxes* (Oxford University Press, 1965). He has been of invaluable help in the preparation of this account.

Other methods of producing tunes mechanically were invented in the early 19th century. Antonio Calegari, an Italian composer, used two dice for composing pieces for the pianoforte and harp. His book on the system was published in Venice in 1801, and later in a French translation. The *Melographicon,* an anonymous and undated book issued in London about 1805, is subtitled: "A new musical work, by which an interminable number of melodies may be produced, and young people who have a taste for poetry enabled to set their verses to music for the voice and pianoforte, without the necessity of a scientific knowledge of the art." The book has four parts, each providing music for poetry with a certain meter and rhyme scheme. Dice are not used. One simply selects any bar from group *A,* any from group *B,* and so on to the last letter of the alphabet for that section.

A photograph of a boxed dice game appears in Plate 42 of *The Oxford Companion to Music,* but without mention of date, inventor, or place of publication. Apparently, it uses 32 dice, their sides marked to indicate tones, intervals, chords, modulations, and so on. There also are ivory men whose purpose, the caption reads, is "difficult to fathom."

In 1822 a machine called the Kaleidacousticon was advertised in a Boston music magazine, *The Euterpiad.* By shuffling cards it could compose 214 million waltzes. The Componium, a pipe organ that played its own compositions, was invented by M. Winkel of Amsterdam and created a sensation when it was exhibited in Paris in 1824. Listeners could not believe that the machine actually constructed the melodies it played. Scientists from the French Academy investigated.

"When this instrument has received a varied theme," their report stated, "which the inventor has had time to fix by a process of his own, it decomposes the variations of itself, and reproduces their different parts in all the orders of possible permutation. . . . None of the airs which it varies lasts above a minute; could it be supposed that but one of these airs was played without interruption, yet, the principle of variability which it possesses, it might, without ever resuming precisely the same combination, continue to play . . . during so immense a series of ages that, though figures might be brought to express them, common language could not."

The report, endorsed by physicist Jean Baptiste Biot, appeared in a British musical journal (*The Harmonicon* 2, 1824, pp. 40–41). Winkel's machine inspired a Vienna inventor, Baron J. Giuliani, to build a similar device, the construction of which is given in detail on pages 198–200 of the same volume.

In 1865, a composing system called the Quadrille Melodist, invented by J. Clinton, was advertised in *The Euterpiad.* By shuffling a set of composing cards, a pianist at a quadrille party could "keep the evening's pleasure going by means of a modest provision of 428,000,000 quadrilles."

Joseph Schillinger, a Columbia University teacher who died in 1943, published his mathematical system of musical composition in a booklet, *Kaleidophone,* in 1940. George Gershwin is said to have used the system in writing *Porgy and Bess.* In 1940 Heitor Villa-Lobos, using the system, translated a silhouette of New York City's skyline into a piano composition (see Figure 47.2). *The Schillinger System of Musical Composition* is a two-volume work by L. Dowling and A. Shaw, published by Carl Fischer in 1941. A footnote on page 673 of Schillinger's eccentric opus *The Mathematical Basis of the Arts* (Philosophical Library, 1948) says that he left plans for music-composing machines, protected by patents, but nothing is said about their construction.

In the 1950s, information theory was applied to musical composition by J. R. Pierce and others. In a pioneering article "Information Theory and Melody," chemist Richard C. Pinkerton included a graph which he called the "banal tune-maker." By flipping a coin to determine paths

THE SKYLINE HAS ITS OWN MUSICAL PATTERN TRANSLATED
FROM SILHOUETTE TO MUSIC NOTES WITH THE HELP OF

THE SCHILLINGER SYSTEM OF MUSICAL COMPOSITION

Figure 47.2. New York skyline translated into music by Villa-Lobos, who used the Schillinger system

along the network, one can compose simple nursery tunes. Most of them are monotonous, but hardly more so, Pinkerton reminds us, than "A Tisket, a Tasket."

During the 1960s and early 1970s the proliferation of computers and the development of sophisticated electronic tone synthesizers opened a new era in machine composition of music. It is now possible to write computer programs that go far beyond the crude devices of earlier days. Suppose one wishes to compose a melody in imitation of one by Chopin. A computer analysis is made of all Chopin melodies so that the computer has in its memory a set of "transition probabilities." These give the probability that any set of one, two, three, or more notes in a Chopin melody is followed by any other note. Of course, one must also take into account the type of melody one wishes to compose, the rhythms, the position of each note within the melody, the overall pattern, and other things. In brief, the computer makes random choices within a specified general structure, but these choices are subject to rules and weighted by Chopin's transition preferences. The result is a

"Markoff chain" melody, undistinguished but nevertheless sounding curiously like Chopin. The computer can quickly dash off several hundred such pieces, from which the most pleasing may be selected.

There is now a rapidly growing literature on computer composition, not only of music in traditional styles but also music that takes full advantage of the computer's ability to synthesize weird sounds that resemble none of the sounds made by familiar instruments. Microtones, strange timbres, unbelievably complex rhythms, and harmonics are no problem. The computer is a universal musical instrument. In principle, it can produce any kind of sound the human ear is capable of hearing. Moreover, a computer can be programmed to play one of its own compositions at the same time it is composing it.

How can we sum up? Computers certainly can compose mediocre music, frigid and forgettable, even though the music has the flavor of a great composer. No one, however, has yet found an algorithm for producing even a simple melody that will be as pleasing to most people of a culture as one of their traditional popular songs. We simply do not know what magic takes place inside the brain of a composer when he creates a superior tune. We do not even know to what extent a tune's merit is bound up with cultural conditioning or even with hereditary traits. About all that can be said is that a good melody is a mixture of predictable patterns and elements of surprise. What the proportions are and how the mixture is achieved, however, still eludes everybody, including composers.

O'Beirne has called my attention to how closely some systems of musical composition resemble the buzz-phrase generator (see Figure 47.3). This is a give-away of Honeywell Incorporated. Pick at random any four-digit number, such as 8,751, then read off phrase 8 of module *A*, phrase 7 of module *B*, and so on. The result is a SIMP (Simplified Integrated Modular Prose) sentence. "Add a few more four-digit numbers," the instructions say, "to make a SIMP paragraph. After you have mastered the basic technique, you can realize the full potential of SIMP by arranging the modules in *DACB* order, *BACD* order, or *ADCB* order. In these advanced configurations, some additional commas may be required."

SIMP sounds very much like authentic technical prose, but on closer inspection one discovers that something is lacking. Computer-generated melodies are perhaps less inane, closer to the random abstract art of a kaleidoscope, but still something essential (nobody knows

SIMP TABLE A	SIMP TABLE B
1. In particular,	1. a large portion of the interface coordination communication
2. On the other hand,	2. a constant flow of effective information
3. However,	3. the characterization of specific criteria
4. Similarly,	4. initiation of critical subsystem development
5. As a resultant implication,	5. the fully integrated test program
6. In this regard,	6. the product configuration baseline
7. Based on integral subsystem considerations,	7. any associated supporting element
8. For example,	8. the incorporation of additional mission constraints
9. Thus,	9. the independent functional principle
0. In respect to specific goals,	0. a primary interrelationship between subsystem and/or subsystem technologies

Figure 47.3. Honeywell's buzz-phrase generator for writing Simplified Integrated Modular Prose (SIMP)

what) is missing. Indeed, a good simple tune is much harder to compose than an orchestral piece in the extreme avant-garde manner, so loaded with randomness and dissonance that one hesitates to say, as Mark Twain (or was it Bill Nye?) said of Wagner's music: It is better than it sounds.

When a computer generates a melody that becomes as popular as (think of the title of your favorite song), you will know that a colossal breakthrough has been made. Will it ever occur? If so, when? Experts disagree on the answers as much as they do on if and when a computer will write a great poem, paint a great picture, or play grand-master chess.

Addendum

Carousel, the company that brought out Mozart's *The Melody Dicer,* later issued a similar set called *The Scott Joplin Melody Dicer.* Using the same system of dice and cards, one can compose endless rags of the Scott Joplin variety.

In 1977 a curious 284-page book titled *The Directory of Tunes and Musical Themes,* by Denys Parsons, was published in England by the mathematician G. Spencer Brown. Parsons discovered that almost every melody can be identified by a ridiculously simple method. Put

SIMP TABLE C	SIMP TABLE D
1. must utilize and be functionally interwoven with	1. the sophisticated hardware
2. maximizes the probability of project success and minimizes the cost time required for	2. the anticipated fourth-generation equipment
3. adds explicit performance limits to	3. the subsystem compatibility and testing
4. necessitates that urgent consideration be applied to	4. the structural design, based on system engineering concepts
5. requires considerable systems analysis and trade off studies to arrive at	5. the preliminary qualification limit
6. is further compounded, when taking into account	6. the evolution of specifications over a given time period
7. presents extremely interesting challenges to	7. the philosophy of commonality and standardization
8. recognizes the importance of other systems and the necessity for	8. the greater fight-worthiness concept
9. effects a significant implementation of	9. any discrete configuration mode
0. adds overriding performance constraints to	

down an asterisk for the first note. If the second note is higher, write down *U* for "up." If lower, write *D* for "down." If the same, use *R* for "repeat." Continue with succeeding notes until you have a sequence of up to 16 letters. This is almost always sufficient to identify the melody. For example *UDDUUUU* is enough to key "White Christmas." The book lists alphabetically in two sections, popular and classical, some 15,000 different sequences followed by the title of the work, its composer, and the date.

The idea behind the buzz-phrase generator—random selection of words and phrases to create prose or poetry—is an old idea. *Rational Recreations,* a four-volume work by W. Hooper (the fourth edition was published in London in 1794), has a section in Volume 2 on how to use dice for composing Latin verse. The technique surely is much older. Similar randomizing is used in modern computer programs that generate "poems" and various kinds of imitation prose.

First-grade teachers, who call it "stringing," use the technique for teaching reading. Children are given a simple pattern sentence with blanks into which they insert words. *The New York Times Book Review*

(June 4, 1978), printed Randolph Hogan's list of buzzwords that enable anyone to write impressive literary criticism. (Someone should do a similar list, perhaps already has, for art critics.) *Mad Magazine* (October 1974) featured Frank Jacobs's 12 columns of buzzwords and phrases for writing impeachment newspaper stories. Tom Koch (*Mad Magazine*, March 1982) gave a similar technique for stand-up comics. Jacobs returned in *Mad* (September 1982) with another 12 columns of words and phrases for writing the lyrics of country-western songs.

Donald Knuth, after reading this chapter, told me that Glenn Miller also studied under Schillinger, who is mentioned in the book *The Glenn Miller Story*. Knuth also informed me that circa 1960 Fred Brooks, a computer scientist at the University of North Carolina, Chapel Hill, published some computer-generated hymn tunes based on data from Methodist hymnals.

The best imitations so far of great music are the computer compositions of David Cope, professor of music at the University of California, Santa Cruz. His program EMI (Experiments in Musical Intelligence), pronounced Emmy, has created very convincing music in the styles of Mozart, Bach, Beethoven, Brahms, Chopin, Scott Joplin, and others. Douglas Hofstadter has expressed amazement at how authentic the music sounds. After his "Mozart's 42nd Symphony" was produced by the college orchestra at Santa Cruz in 1997, Cope told a reporter: "There's no expert in the world who could, without knowing its source, say for certain that it's not Mozart. . . ."

Bibliography

D. Cope, *Compositions and Music Style*, Oxford University Press, 1991.

M. Gardner, "The Ars Magna of Ramón Lull," *Logic Machines and Diagrams*, McGraw-Hill, 1958; Revised edition University of Chicago Press, 1982.

J. Godwin, *Athanasius Kircher*, Thames and Hudson, 1979.

L. A. Hiller, Jr., "Computer Music," *Scientific American*, December 1959, pp. 109–120.

L. A. Hiller, Jr., and L. M. Isaacson, *Experimental Music*, McGraw-Hill, 1959.

R. Holmes, "Requiem for the Soul," *New Scientist*, August 9, 1997, pp. 23–27. On David Cope's music.

K. Jones, "Dicing with Mozart," *New Scientist*, December 14, 1991, pp. 26–29.

T. H. O'Beirne, "From Mozart to the Bagpipe with a Small Computer," *Bulletin of the Institute of Mathematics and its Applications* 7, January 1971, pp. 3–8.

C. A. Pickover, "Music Beyond Imagination, Part III," *Mazes of the Mind*, St. Martin's, 1992.

R. C. Pinkerton, "Information Theory and Melody," *Scientific American*, February 1956, pp. 77–86.

J. Reichardt (ed.), *Cybernetic Serendipity: The Computer and the Arts*, Frederick A. Praeger, 1969.

F. Schillinger, *Joseph Schillinger: A Memoir*, De Capo, 1976.

P. A. Scholes, "Composition Systems and Mechanisms," *The Oxford Companion to Music*, J. O. Ward (ed.), Oxford University Press, 1970.

H. Von Foerster and J. W. Beauchamp (eds.), *Music by Computers*, Wiley, 1969.

Chapter 48

Mathematical Zoo

*T*here has never been a zoo designed to display animals with features of special interest to recreational mathematicians, yet such a zoo could be both entertaining and instructive. It would be divided, as I visualize it, into two main wings, one for live animals, the other for pictures, replicas, and animated cartoons of imaginary creatures. Patrons of the "mathzoo" would be kept informed of new acquisitions by a newsletter called *ZOONOOZ* (with the permission of the Zoological Society of San Diego, which issues a periodical of that name), a title that is both palindromic and the same upside down.

A room of the live-animal wing would contain microscopes through which one could observe organisms too tiny to be seen otherwise. Consider the astonishing geometrical symmetries of radiolaria, the one-celled organisms that flourish in the sea. Their intricate silica skeletons are the nearest counterparts in the biological world to the patterns of snow crystals. In his *Monograph of the Challenger Radiolaria,* the German biologist Ernst Haeckel described thousands of radiolaria species that he discovered on the *Challenger* expedition of 1872–76. The book contains 140 plates of drawings that have never been excelled in displaying the geometric details of these intricate, beautiful forms.

Figure 48.1, which originally appeared in Haeckel's book, is of special interest to mathematicians. The first radiolarian is basically spherical, but its six clawlike extensions mark the corners of a regular octahedron. The second skeleton has the same solid at its center. The third is a regular icosahedron of 20 faces. The fifth is the 12-sided dodecahedron. Other plates in Haeckel's book show radiolaria that approximate cubical and tetrahedral forms.

It is well known that there are just five Platonic solids, three of which have faces that are equilateral triangles. Not so widely known is that

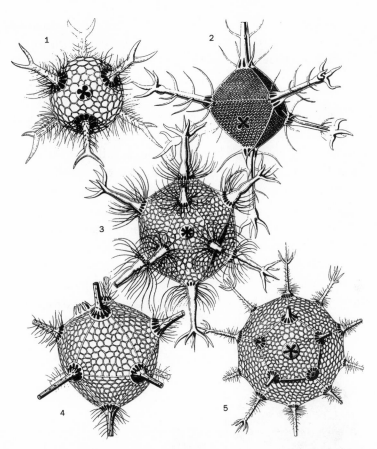

Figure 48.1. Radiolaria skeletons in Ernst Haeckel's *Monograph of the Challenger Radiolaria*

there are an infinite number of semi-regular solids also with sides that are equilateral triangles. They are called "deltahedra" because their faces resemble the Greek letter delta. Only eight deltahedra are convex: those with 4, 6, 8, 10, 12, 14, 16, and 20 faces. The missing 18-sided convex deltahedron is mysterious. One can almost prove it should exist, and it is not so easy to show why it cannot. It is hard to believe, but the proof that there are only eight convex deltahedra was not known until B. L. van der Waerden and Hans Freudenthal published it in 1947. If concavity is allowed, a deltahedron can have any number of faces of eight or greater.

The four-faced deltahedron is the regular tetrahedron, the simplest of the Platonic solids. The six-faced deltahedron consists of two tetrahedra sharing one face. Note the fourth radiolarian in Haeckel's picture.

It is a 10-faced deltahedron, or rather one that is inflated slightly toward a sphere. It may surprise you to learn that there are two topologically distinct eight-sided deltahedra. One is the familiar regular octahedron. Can you construct a model of the other one (it is not convex)?

Surfaces of radiolaria are often covered with what seems to be a network of regular hexagons. The regularity is particularly striking in *Aulonia hexagona,* shown in Figure 48.2. Such networks are called "regular maps" if each cell has the same number of edges and each vertex has the same number of edges joined to it. Imagine a regular tetrahedron, octahedron, or icosahedron inflated like a balloon but preserving its edges as lines on the resulting sphere. The tetrahedron will form a regular map of triangles with three edges at each vertex, the octahedron will form a map of triangles with four edges at each vertex, and the icosahedron will form a map of triangles with five edges at each vertex. Inflating a cube produces a regular map of four-sided cells with three edges at each vertex. Inflating a dodecahedron produces a regular map of pentagons with three edges at each vertex.

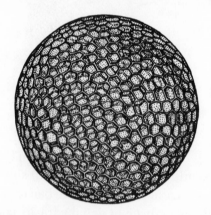

Figure 48.2. The radiolarian *Aulonia hexagona*

Aulonia hexagona raises an interesting question. Is it possible to cover a sphere with a regular map of hexagons, three edges at each vertex? Only the topological properties of the map concern us. The hexagons need not be regular or even convex. They may have any size or shape, and their edges may twist and curve any way you like provided they do not intersect themselves or one another and provided three of them meet at each vertex.

The answer is no, and it is not hard to prove impossibility with a famous formula that Leonhard Euler discovered for the skeletons of all simply connected (no "holes") polyhedra. The formula is $F + C - E = 2$, where the letters stand for faces, corners, and edges. Since all such polyhedra can be inflated to spheres, the formula applies also to maps on the sphere. In Chapter 13 of *Enjoyment of Mathematics,* by Hans Rademacher and Otto Toeplitz, you will find it explained how Euler's formula can be used in proving that no more than five regular maps can be drawn on a sphere and that therefore no more than five regular convex solids can exist. As a second problem, can you use Euler's formula to show that a regular map of hexagons is impossible on a sphere?

D'Arcy Wentworth Thompson, whose classic work *On Growth and Form* contains an excellent section on radiolaria, liked to tell about a biologist who claimed to have seen a spherical radiolarian covered with a perfect map of hexagons. But, said Thompson, Euler had proved this impossible. "That," replied the biologist, "proves the superiority of God over mathematics."

"Euler's proof happened to be correct," writes Warren S. McCulloch in an essay where I found this anecdote, "and the observations, inaccurate. Had both been right, far from proving God's superiority to logic, they would have impugned his wit by catching him in a contradiction." If you look carefully at the picture of *Aulonia hexagona* you will see cells with more or fewer than six sides.

Under electron microscopes in our zoo's micro room would be the many viruses that have been found to crystallize into macromolecules shaped like regular icosahedra: the measles virus, the herpes, the triola iridescent, and many others (see R. W. Horne's article cited in the bibliography). Viruses may also have dodecahedral shapes, but as far as I know this remains unsettled. Another modern discovery is that some viruses, such as the one that causes mumps, are helical. It had formerly been thought that helical structures were restricted to plants and to parts of animals: hair, the umbilical cord, the cochlea of the human ear, the DNA molecule, and so on. A section of our zoo would feature such spectacular helical structures as molluscan seashells, the twisted horns of certain sheep, goats, antelopes, and other mammals, and such curiosities as "devil's corkscrews"—the huge fossil burrows of extinct beavers.

In the macro world of fishes, reptiles, birds, insects, mammals, and human beings the most striking geometrical aspect of the body is its overall bilateral symmetry. It is easy to understand why this symmetry

Mathematical Zoo

evolved. On the earth surface gravity creates a marked difference between up and down, and locomotion creates a marked difference between front and back. But for any moving, upright creature the left and right sides of its surroundings—in the sea, on the land, or in the air—are fundamentally the same. Because an animal needs to see, hear, smell, and manipulate the world equally well on both sides, there is an obvious survival value in having nearly identical right and left sides.

Animals with bilateral symmetry are of no interest for our mathzoo—you can see them at any zoo—but it would be amusing to assemble an exhibit of the most outrageous violations of bilateral symmetry. For example, an aviary would feature the crossbill, a small red bird in the finch family that has its upper and lower beaks crossed in either of the two mirror-image ways. The bird uses its crossed bill for prying open evergreen cones in the same way a cook uses a plierlike device to pry off the lid of a jar or can. A medieval legend has it that the bill became twisted as the bird was trying vainly to pull the nails from the cross when Jesus was crucified; in the effort the bird's plumage became stained with blood. In the same aviary would be some wry-billed plovers from New Zealand. The entire bill of this funny bird is twisted to the right. The bill is used for turning over stones to find food. As you would expect, foraging wry-billed plovers search mainly on the right.

An aquarium in our mathzoo would exhibit similar instances of preposterous asymmetry among marine life: the male fiddler crab, for example, with its enormous left (or right) claw. Flatfish are even more grotesque examples. The young are bilaterally symmetric, but as they grow older one eye slowly migrates over the top of the head to the other side. The poor fish, looking like a face by Picasso, sinks to the bottom, where it lies in the ooze on its eyeless side. The eyes on top turn independently so that they can look in different directions at the same time.

Another tank would contain specimens of the hagfish. This absurd fish looks like an eel, has four hearts, teeth on its tongue, and reproduces by a technique that is still a mystery. When its single nostril is clogged, it sneezes. The hagfish is in our zoo because of its amazing ability to tie itself into an overhand knot of either handedness. By sliding the knot from tail to head it scrapes slime from its body. The knot trick is also used for getting leverage when the hagfish tears food from a large dead fish and also for escaping a predator's grasp (see David Jensen's article listed in the bibliography).

Knots are, of course, studied by mathematicians as a branch of topology. Another exhibit in our aquarium would be beakers filled with *Leucothrix mucor,* a marine bacterium shaped like a long filament. A magnifying glass in front of each beaker would help visitors see the flimsy filaments. They reproduce by tying themselves into knots—overhands, figure-eights, even more complicated knots—that get tighter and tighter until they pinch the filament into two or more parts (see Thomas D. Brock's paper listed in the bibliography). Do higher animals ever tie parts of themselves into knots? Fold your arms and think about it.

The most popular of our aquarium exhibits would probably be a tank containing specimens of *Anableps,* a small (eight-inch) Central American carp sometimes called the stargazer. It looks as if it has four eyes. Each of its two bulging eyes is divided into upper and lower parts by an opaque band. There is one lens but separate corneas and irises. This little BEM (bug-eyed monster) swims with the band at water level. The two upper "eyes" see above water while the two lower ones see below. The *Anableps* is in our zoo because of its asymmetric sex life. The young are born alive, which means that the male must fertilize the eggs inside the female. The female opening is on either the left side or the right. The male organ also is either on the left or the right. This makes it impossible for two fish of the same handedness to mate. Fortunately both males and females are equally left- or right-sexed, and so the species is in no danger of extinction.

In a larger tank one would hope to see some narwhals, although until now they have not survived in captivity. This small whale, from north-polar seas, has been called the sea unicorn because the male has a single "horn" that projects straight forward from its upper jaw and is about half the whale's body length. Both sexes are born with two small side-by-side teeth. The teeth stay small on the female, but the male's left tooth grows into an ivory tusk, straight as a javelin and seven to 10 feet long. This ridiculous tooth, the longest in the world, has a helical groove that spirals around it like a stripe on a barber pole. Nobody knows what function the tusk serves. It is not used for stabbing enemies or punching holes in ice, but during the mating season narwhals have been seen fencing with each other, so that its main purpose may be a role in sexual ritual (see John Tyler Bonner's article in the bibliography). Incidentally, the narwhal is also unusual in having a name starting with the letter *n.* It is easy to think of mammalian names beginning with any letter of the alphabet except *n.*

Among snakes, species that sidewind across the desert sands are mathematically interesting because of their highly asymmetric tracks: sets of parallel line segments that slant either right or left at angles of about 60 degrees from the line of travel. Many species of snakes are capable of sidewinding, notably the sidewinder itself, a small rattlesnake of Mexico and the U.S. Southwest, and the African desert viper. Exactly how sidewinding works is rather complicated, but you will find it clearly explained in Carl Gans's article.

The insect room of our mathzoo would certainly display the nests of bees and social wasps. They exhibit a hexagonal tessellation even more regular than the surfaces of radiolaria. A large literature, going back to ancient Greece and still growing, attempts to explain the factors that play a role in producing this pattern. D'Arcy Thompson, in his book cited earlier, has a good summary of this literature. In times before Darwin bees were usually regarded as being endowed by the Creator with the ability to design nests so that the cells use the least amount of wax to hold a maximum amount of honey. Even Darwin marveled at the bee's ability to construct a honeycomb, calling that ability "the most wonderful of known instincts," and "absolutely perfect in economizing labor and wax."

Actual honeycombs are not as perfect as early writers implied, and there are ways of tessellating space with polyhedral cells that allow an even greater economy of wax. Moreover, it seems likely that the honeycomb pattern is less the result of evolution finding a way to conserve wax than an accidental product of how bees use their bodies and the way they form dense clusters when they work. Surface tension in the semiliquid wax may also play a role. The matter is still far from settled. The best discussion I know is a paper by the Hungarian mathematician L. Fejes Tóth.

No actual animal propels itself across the ground by rolling like a disk or a sphere, but our insect room would be incomplete without an exhibit of a remarkable insect that transports its food by rolling near-perfect spheres. I refer to the dung beetle, the sacred scarab of ancient Egypt. These sometimes beautiful insects (in the Tropics they have bright metallic colors) use their flat, sharp-edged heads as shovels to dig a supply of fresh ordure that their legs then fashion into spheres. By pushing with its hind legs and walking backward the dung beetle will roll the little ball to its burrow where it will be consumed as food. No one has described the process with more literary skill and humor than the French entomologist, Jean Henri Fabre, in his essay on "The Sacred Beetle."

Our zoo's imaginary wing would lack the excitement of living creatures but would make up for it in wild fantasy. In Flaubert's *Temptation of St. Anthony,* for example, there is a beast called the Nasnas that is half of an animal bisected by its plane of symmetry. Jorge Luis Borges, in his delightful *Book of Imaginary Beings,* refers to an earlier invention of such a creature by the Arabs. L. Frank Baum's fantasy, *Dot and Tot of Merryland,* tells of a valley inhabited by wind-up animals. The toys are kept wound by a Mr. Split, whose left half is bright red and whose right half is white. He can unhook his two sides, each of which hops about on one leg so that he gets twice as much winding done. Conversing with a half of Mr. Split is difficult because Mr. Left Split speaks only the left halves of words and Mr. Right Split speaks only the right halves.

A variety of mythical "palindromic" beasts violate front and back asymmetry by having identical ends. Borges writes of the fabled *amphisbaena* (from the Greek for "go both ways"), a snake with a head at each end. Dante puts the snake in the seventh circle of Hell, and in Milton's *Paradise Lost* some of Satan's devils are turned into *amphisbaenas.* Alexander Pope writes in his *Dunciad:*

> Thus Amphisbaena (I have read)
> At either end assails;
> None knows which leads,
>   or which is led,
> For both Heads are but Tails.

The fable is not without foundation. There are actual snakes called amphisbaenas that crawl both ways and have such tiny eyes that it is hard to distinguish one end from the other. If a flatworm's head is cut off, another grows at the base of the severed head, so palindromic animals actually can exist. In Baum's *John Dough and the Cherub* one meets Duo, a dog with a head and forelegs at both ends (see Figure 48.3). The animal anticipates the Pushmi-Pullyu (it has a two-horned head at each end) that flourishes in the African jungle of Hugh Lofting's Dr. Dolittle books.

Rectangular parallelepipeds are never the parts of real animals, but in Baum's *Patchwork Girl of Oz* there is a block-headed, thick-skinned, dark blue creature called the Woozy (see Figure 48.4). The animal's head, body, legs, and tail are shaped like blocks. It is friendly as long

Figure 48.3. Duo, a palindromic dog

Figure 48.4. L. Frank Baum's Woozy

as no one says "Krizzle-kroo." This makes the Woozy so angry that its eyes dart fire. Nobody, least of all the Woozy, knows what Krizzle-kroo means, and that is what makes it so furious. Borges reminds us of the Gillygoo, a bird in the Paul Bunyan mythology, that nests on steep slopes and lays cubical eggs that will not roll down and break. Minnesota lumberjacks hard-boil them and use them for dice. In Stanley G. Weinbaum's story, "A Martian Odyssey," a species of nondescript animals on Mars excrete silica bricks that they use for building pyramidal dwellings.

Baum also imagined spherical creatures. The Roly-Rogues, in *Queen Zixi of Ix,* are round like a ball and attack enemies by rolling at them. In *John Dough and the Cherub,* one of the main characters is Para Bruin, a large rubber bear that likes to roll into a rubber ball and bounce around.

Borges, writing about animals in the form of spheres, tells us that Plato, in the *Laws,* conjectures that the earth, planets, and stars are alive. The notion that the earth is a living, breathing organism was later defended by such mystics as Giordano Bruno, Kepler, the German psychologist Gustav Theodor Fechner, and Rudolf Steiner (who broke away from theosophy to found his rival cult of anthroposophy). The same notion is basic to the plot of one of Conan Doyle's stories about Professor George Edward Challenger of *Lost World* fame. When Professor Challenger drills a deep hole through the earth's epidermis, in a story called "When the Earth Screamed," the planet howls with pain.

Rotating wheels and propellers are common mechanisms for transporting man-made vehicles across ground, and through the sea and the air, but until a few decades ago it was assumed that evolution had been unable to exploit rotational devices for propulsion. Biologists were amazed to discover that the flagella of bacteria actually spin like propellers (see the article by Howard C. Berg).

The imaginary wing of our zoo would display two of Baum's creatures that use the wheel for propulsion. In *Ozma of Oz* Dorothy has an unpleasant encounter with the Wheelers, a race of fierce, four-legged humanoids that have wheels instead of feet (see Figure 48.5). In *The Scarecrow of Oz* we read about the Ork, a huge bird with a propeller at the tip of its tail (see Figure 48.6). The propeller can spin both ways, enabling the bird to fly backward as well as forward.

I know of only two imaginary beasts that bend themselves into wheels and roll across the ground. From time to time, in most parts of the world, people have claimed to have seen "hoop snakes" that bite their tails to form a hoop and then go rolling across the terrain. Some snakes, such as the American milk snake, travel by gathering their body into large vertical loops and pushing forward so rapidly that they create an optical illusion of a rolling ring. These animals may be the origin of hoop-snake fables.

The Dutch artist M. C. Escher made several pictures featuring his curl-up, the beast shown in Figure 48.7. This unlikely animal moves slowly on six humanlike feet, but when it wants to go faster it curls up and rolls like a wheel.

Figure 48.5. A Wheeler

Figure 48.6. The Ork

Most animals, particularly the earthworm, may be thought of as being basically toroidal—a shape topologically equivalent to a doughnut. There must be many science-fiction animals shaped like toruses, but I can recall only the undulating silver ringfish, floating on the canals of Ray Bradbury's *Martian Chronicles,* that closes like an eye's iris around food particles.

Topologists know that any torus can be turned inside out through a hole in its surface. There is no parallel in earth zoology, but there is a spherical organism called volvox that actually does turn inside out through a hole. It is a strange freshwater-pond colony of hundreds of flagellated cells bound together in a spherical jellylike mass that rotates

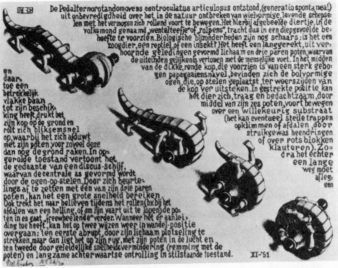

Collection Hoags Gemeente-museum—the Hague.

Figure 48.7. The curl-up, an animal imagined by M. C. Escher, can roll like a wheel when it wants to.

as it moves through the water. Volvox is one of those twilight things that can be called a green plant (because it obtains food by photosynthesis) or an animal (because it moves freely about). One is equally hard put to decide whether it is a colony or a single organism.

Young volvox colonies grow inside the mother sphere, but the cells have their flagella ends pointing inward. At the spot where each infant sphere is attached to the inside of the mother, there is a small hole in the infant sphere. When the infant reaches a certain size, it breaks away from the mother and turns inside out through the hole! Flagella quickly sprout at the ends of the cells that now point outward, and the newborn colony goes spinning about inside the mother. The mother eventually dies by splitting open and allowing her offspring to escape, one of the earliest examples on the evolutionary tree of nonaccidental death (see the article by John Tyler Bonner).

We could have considered volvox earlier, but I kept it for now to introduce the ta-ta, a mythical but much higher form of animal capable of turning inside out. It was invented by Sidney H. Sime, the British artist who so wondrously illustrated Lord Dunsany's fantasies. Sime drew and described the ta-ta in his only book, *Bogey Beasts,* a rare collection of original verses set to music:

The Ta-Ta

There is a cosy Kitchen
Inside his roomy head
Also a tiny bedroom
In which he goes to bed.

So when his walk is ended
And he no more would roam
Inside out he turns himself
To find himself at Home.

He cleared away his brain stuff
Got pots and pans galore!
Sofas, chairs, and tables,
And carpets for the floor.

He found his brains were useless,
As many others would
If they but tried to use them
A great unlikelihood.

He pays no rent, no taxes
No use has he for pelf
Infested not with servants
He plays with work himself.

And when his chores are ended
And he would walk about,
Outside in he turns himself
To get himself turned out.

Addendum

I was mistaken in saying that no animal propels itself across the ground by rolling like a disk or sphere. Brier Lielst, Philip Schultz, and a geologist with the appropriate name of Paul Pushcar, were among many who informed me of a *National Geographic* television special on March 6, 1978, about the Namib desert of Africa. It showed a small spider that lives in burrows in the sides of sand dunes. When attacked by a wasp, it extends its legs like the spokes of a wheel and escapes by rolling down the dune.

Peter G. Trei of Belgium sent me a copy of a note in *Journal of Mammalogy* (February 1975) by Richard R. Tenaza, an American zoologist.

Tenaza describes how on Siberut, an island west of Sumatra, he witnessed the technique by which pangolins, a species of scaly anteater, elude capture: they curl themselves into a tight ball and roll rapidly down a steep slope (see Figure 48.8). In fact, the name "pangolin" is from a Malay word meaning "to roll."

Figure 48.8. An uncurled pangolin

Roy L. Caldwell, a zoologist, wrote to me about an inch-long crustacean called *Nannosquilla decemspinosa,* found in the sands off the coast of Panama. "When exposed on dry land, the animal's short legs are not sufficiently strong to drag its long, slender body, so it flips over on its back, brings its tail up over on its head, and takes off rolling much like a tank-track. While it does not actually close the perimeter by grasping the tail in its mouth-parts, they are usually kept in close proximity. The path taken is usually a straight line, the animal can actually climb a five-degree slope, and it can make a speed of about six cm/sec."

Thomas H. Hay told me about a species of wood lice (also called slaters, sow-bugs, and pill-bugs) that, when alarmed, curl into a ball and roll away. Hay said his children call them roly-polys. While working underneath his car, with a trouble light beside him, he often finds the roly-polys "advancing in inexorable attack. I fantasize that they are Martian armored vehicles, released from a tiny spacecraft. Fortunately, they move so slowly that my work is finished before they constitute a threat."

Robert G. Rogers, in a letter that appeared in *Discover* (October 1983) had this interesting comment on an earlier *Discover* article, "Why Animals Run on Legs, Not on Wheels" (see the bibliography):

The concept of animals developing wheels for locomotion is not so far-fetched. A wheel with a diameter of one foot has a circumference a little over three feet. If it were mounted on a bone-bearing joint, with flexible veins and arteries, and a continuous series of circumferential pads (as on a dog's paw), the wheel could be wound back one turn by its internal muscles, then placed on the ground and rotated forward two full turns, traveling about six and a quarter feet. While one wheel (or pair of wheels in a four-legged animal) is driving the creature along the ground, the other would be lifted up and rotated back in preparation for its next turn at propulsion. At a speed of ten m.p.h., the creature would be traveling about 15 feet per second—not an impossible pace.

Matthew Hodgart, writing from England, reminded me that the human animal is capable of moving by repeated somersaults, cartwheels, forward and back flips, and that two persons can grab each other's feet and roll like a hoop. Hodgart quoted these lines from Andrew Marvell's poem "To His Coy Mistress":

> Let us roll all our strength, and all
> Our sweetness, up into one ball:
> And tear our pleasures with rough strife,
> Through the iron gates of Life.

"I don't quite know what's going on here," Hodgart adds.

Chandler Davis supplemented my list of imaginary creatures that roll by calling attention to such an animal in George MacDonald's fantasy *The Princess and Curdie*. Ian F. Rennie thought I should have mentioned the Wumpetty-Dumps, found in *The Log of the Ark*, by Kenneth Walker and Geoffrey Boumphrey.

Is the narwhal the only animal with a name starting with N? Garth Slade cited the numbat, a small marsupial that lives in Western Australia. An article in *Word Ways* (May 1973) gave two other examples: the nutria, a web-footed South American aquatic rodent (now also flourishing on the Gulf Coast and the coasts of the Pacific northwest), and the nilgai, an antelope in India that is commonly called a "blue bull" because of its bluish-gray color. It is curious that the best one can do with common names are the colloquial nag and nanny-goat.

Arthur C. Statter sent his reasons for thinking that the drawing by Haeckel which is reproduced in Figure 48.1 (I picked it up from D'Arcy Thompson's *On Growth and Form*), was one of many drawings that Haeckel deliberately faked. The forms shown, Statter says, simply do not exist. I have not tried to investigate this, and would welcome opinions from radiolaria experts.

Rufus P. Isaacs, commenting on the impossibility of tessellating a sphere with hexagons, sent a proof of a surprising theorem he discovered many years earlier. If a sphere is tessellated with hexagons and pentagons, there must be exactly 12 pentagons, no more and no fewer.

A soccerball is tessellated with 20 hexagonal "faces" and 12 that are pentagons. In 1989 chemists succeeded in creating the world's tiniest soccerball—a carbon molecule with 60 atoms at the vertices of a spherical structure exactly like that of a soccerball (see Figure 48.9). It is called a buckyball, or more technically, a buckminsterfullerene, after its resemblance to Buckminster Fuller's famous geodesic domes. It belongs to a class of highly symmetrical molecules called fullerenes.

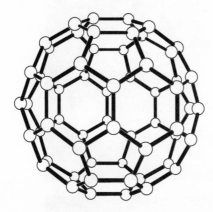

Figure 48.9. The "buckyball" molecule—the world's smallest soccerball

The buckyball is known to geometers as a truncated icosahedron because it can be constructed by slicing off the 12 corners of a regular icosahedron. It is still not clear what properties this third form of carbon (the other two are graphite and diamond) might have. Because of the molecule's near spherical shape, it might provide a marvelous lubricant. (See "Buckyball: The Magic Molecule," Edward Edelson, *Popular Science,* August 1991, p. 52ff.)

It also can be shown that if a sphere is tessellated with hexagons and triangles, there must be an even number of hexagons and exactly four triangles. These results suggest the following general question: What are the integral values of k such that the sphere can be tessellated with hexagons and exactly k polygons of side n? As far as I know, this question has not been completely answered, though many special cases have been proved.

Emerson Frost sent photographs of his paper models of the eight convex deltahedra, as well as many of the nonconvex forms. Figure 48.10 shows his model of the 18-sided deltahedron that is so close to being convex that William McGovern (see the bibliography) has dubbed it the

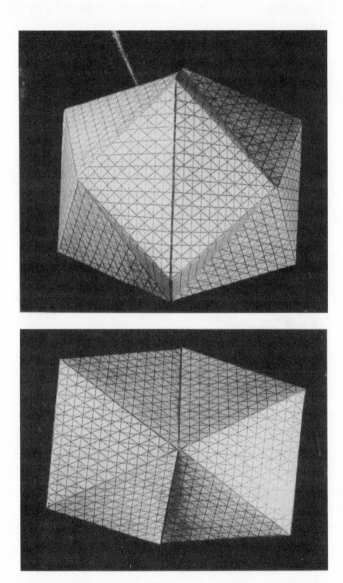

Figure 48.10. Two views of the "deceptahedron," an 18-sided deltahedron that is almost convex

"deceptahedron." It is a pity that teachers are not as familiar with the eight convex deltahedra as they are with the five Platonic solids because constructing models and proving the set unique by way of Euler's formula are splendid classroom challenges.

As early as 1993 more than 1,500 papers were published on buckyballs, and a thousand more have appeared since then. In 1996 the Nobel Prize in chemistry went to three discoverers of buckyballs: Robert Curl, Harold Kyoto, and Richard Smalley.

Answers

Figure 48.11 shows the answer to the first problem: an eight-sided deltahedron (all faces equilateral triangles) that is not a regular octahedron. The regular octahedron has four edges meeting at each corner. On this solid two corners are meeting spots for three edges, two for four edges, and two for five edges.

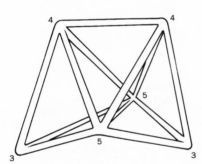

Figure 48.11. The "other" eight-sided deltahedron

The second problem was to use Euler's formula, $F + C - E = 2$, to show that no sphere can be covered with a "regular map" of hexagons, each vertex the meeting point of three edges. Assume such a map exists. Each hexagon has six edges and six corners. Therefore if the hexagons did not share corners and edges, there would be six times as many edges as faces. Each corner is shared, however, by three faces; therefore, the number of corners in such a map must be $6F/3$. Similarly, each edge is shared by two faces; therefore, the number of corners in such a map must be $6F/2$. Substituting these values in Euler's formula gives the equation $F + 6F/3 - 6F/2 = 2$, which simplifies to $F + 2F - 3F = 2$, or $0 = 2$. This contradiction proves the original assumption to be false.

What happens when the above argument is applied to the regular maps formed by the edges of the five Platonic solids? In each case we get a formula that gives F a unique value: 4, 8, and 20 for the tetrahedron, octahedron, and icosahedron respectively, 6 for the cube, and 12 for the dodecahedron. Since a regular polyhedron cannot have faces with more than six edges, we have proved that no more than five regular solids can exist.

Euler's formula also underlies an elementary proof that there are exactly eight convex deltahedra. See the paper by Beck, Bleicher, and Crowe cited in the bibliography.

Bibliography

J. T. Bonner, "Volvox: A Colony of Cells," *Scientific American*, May 1950, pp. 52–55.

J. T. Bonner, "The Horn of the Unicorn," *Scientific American*, March 1951, pp. 42–46.

J. L. Borges, *The Book of Imaginary Beings*, Dutton, 1969.

T. D. Brock, "Knots in Leucothrix Mucor," *Science*, Vol. 144, 1964, pp. 870–71.

C. Gans, "How Snakes Move," *Scientific American*, June 1970, pp. 82–96.

M. Gardner, *The New Ambidextrous Universe*, W. H. Freeman, 1989.

R. W. Horne, "The Structure of Viruses," *Scientific American*, January 1963, pp. 48–52.

D. Jensen, "The Hagfish," *Scientific American*, February 1966, pp. 82–90.

A. C. Neville, *Animal Asymmetry*, London: Edwin Arnold, 1976.

On The Bee Honeycomb

D. F. Siemens, Jr., "The Mathematics of the Honeycomb," *Mathematics Teacher*, April 1965, pp. 334–37.

L. Fejes Tóth, "What the Bees Know and What They Don't Know." *Bulletin of the American Mathematical Society*, Vol. 70, 1964, pp. 468–86.

On Animal Wheels

H. C. Berg, "How Bacteria Swim," *Scientific American*, August 1975, pp. 36–44.

J. Diamond, "Why Animals Run On Legs, Not On Wheels," *Discover*, September 1983, pp. 64–67.

L. James, "Bacteria's Motors Work in Forward, Reverse, and Twiddle," *Smithsonian*, September 1983, pp. 127–34.

W. Marston, "Roll with it: The Life and Times of a Somersaulting Crustacean," *The Sciences*, January/February 1994, p. 12.

On Deltahedra

A. Beck and D. Crowe, "Deltahedra," *Excursions into Mathematics*, Worth, 1969, pp. 21–26.

J. D. Bernal, "The Structure of Liquids," *Scientific American,* August 1960, pp. 124–30. The author maintains that molecules in liquids form deltahedral structures.

H. M. Cundy, "Deltahedra," *Mathematical Gazette,* Vol. 36, 1952, pp. 263–66.

W. E. McGowan, "A Recursive Approach to the Construction of the Deltahedra," *Mathematics Teacher,* Vol. 71, 1978, pp. 204–10.

C. W. Trigg, "An Infinite Class of Deltahedra," *Mathematics Magazine,* Vol. 51, 1978, pp. 55–58.

On Buckyballs

H. Aldersey-Williams, *The Most Beautiful Molecule: The Discovery of the Buckyball,* Wiley, 1995.

F. Chung and S. Sternberg, "Mathematics and the Buckyball," *American Scientist,* Vol. 81, January/February 1993, pp. 56–71.

R. Curl and R. Smalley, "Fullerines," *Scientific American,* October 1991, pp. 54–63.

R. Smalley, "Great Balls of Carbon," *The Sciences,* Vol. 31, March/April 1991, pp. 22–28.

C. Wu, "Buckyballs Can Come From Outer Space," *Science News,* Volume 157, March 25, 2000, p. 490.

Chapter 49   Gödel, Escher, Bach

This sentence no verb.

—Douglas R. Hofstadter,
*Gödel, Escher, Bach: an
Eternal Golden Braid*

Every few decades an unknown author brings out a book of such depth, clarity, range, wit, beauty, and originality that it is recognized at once as a major literary event. *Gödel, Escher, Bach: an Eternal Golden Braid,* a hefty (777 pages) volume published by Basic Books (1979), is such a work. The author (and the illustrator and typesetter) is Douglas R. Hofstadter, then a young computer scientist at Indiana University who is the son of the well-known physicist Robert Hofstadter.

What can Kurt Gödel, M. C. Escher, and Johann Sebastian Bach have in common? The answer is symbolized by the objects shown in the photograph that is Figure 49.1, and in the photograph on the book's jacket. In each photograph two wood blocks floating in space are illuminated so that their shadows on the three walls meeting at the corner of a room form the initials of the three surnames Gödel, Escher, and Bach. More precisely, the upper block casts *"GEB"* (Gödel, Escher, Bach) the heading of the book's first half, and the lower block casts *"EGB"* (Eternal Golden Braid) the heading of its second half. The letters *G, E,* and *B* may be thought of as the labels for three strands that are braided by repeatedly switching a pair of letters. Six steps are required to complete a cycle from *GEB* (through *EGB*) back to *GEB.*

Dr. Hofstadter (his Ph.D. is in physics from the University of Oregon) calls such a block a "trip-let," a shortened form of "three letters." The idea came to him, he explains, "in a flash." Intending to write a pamphlet about Gödel's theorem, his thoughts gradually expanded to include Bach and Escher until finally he realized that the works of these men were "only shadows cast in different directions by some central solid essence." He "tried to reconstruct the central object, and came up with this book."

Figure 49.1. *G, E,* and *B* cast as shadows by a pair of "trip-lets"

Hofstadter carved the trip-let blocks from redwood, using a band saw and an end mill. The basic idea behind this form is an elaboration of the classic puzzle that asks what solid shape will cast the shadows of a circle, a square, and a triangle. Can a trip-let be constructed for any set of three different letters? If the letters may be distorted sufficiently, the answer is yes, and so to make the problem interesting some restrictions must be imposed. To begin with, the letters (preferably uppercase) must all be conventionally shaped, and they must fit snugly into the three rectangles that are the orthogonal projections of a rectangular block. In addition the solid must be connected, that is, it must not fall apart into separate pieces. It is not easy to determine except by trial and error whether such a trip-let can be made for any three given letters. As it turns out, some trip-lets are not possible. This problem suggests exotic variations, for example *n*-tup-lets that project *n* letters; four-dimensional tup-lets that project solid trip-lets that in turn project plane shadows of letters; solids that project numbers, pictures, or words, and so on. (For this description of trip-lets I am indebted to Hofstadter's friend Scott Kim, who worked closely with him on many aspects of the book.)

What reality does Hofstadter see behind the work of his three giants? One aspect of that reality is the formal structure of mathematics: a structure that, as Gödel's famous undecidability proof shows, has infinitely many levels, none of which are capable of capturing all truth in one consistent system. Hofstadter puts it crisply: "Provability is a weaker notion than truth." In any formal system, rich enough to contain arithmetic, true statements can be made that cannot be proved within the system. To prove them one must jump to a richer system, in which again true statements can be made that cannot be proved, and so on. The process goes on forever.

Is the universe Gödelian in the sense that there is no end to the discovery of its laws? Perhaps. It may be that no matter how deeply science probes there will always be laws uncaptured by the theories, an endless sequence of wheels within wheels. Hofstadter argues eloquently for a kind of Platonism in which science, at any stage of its history, is like the shadow projections on the wall of Plato's cave. The ultimate reality is always out of reach. It is the Tao about which nothing can be said. "In a way," Hofstadter writes at the end of his preface, "this book is a statement of my religion."

For laymen I know of no better explanation than this book presents of what Gödel achieved and of the implications of his revolutionary discovery. That discovery concerns in particular recursion, self-reference, and endless regress, and Hofstadter finds those three themes vividly mirrored in the art of Escher, the most mathematical of graphic artists, and in the music of Bach, the most mathematical of the great composers. The book's own structure is as saturated with complex counterpoint as a Bach composition or James Joyce's *Ulysses.* The first half of the book serves as a prelude to the second, just as a Bach prelude introduces a fugue. Moreover, each chapter is preceded by a kind of prelude, which early in the book takes the form of a "Dialogue" between Achilles and the Tortoise. Other characters enter later: the Sloth, the Anteater, the Crab, and finally Alan Turing, Charles Babbage, and the author himself. Each Dialogue is patterned on a composition by Bach, and in several instances the mapping is strict. For example, if the composition has n voices, so does the corresponding Dialogue. If the composition has a theme that is turned upside down or played backward, so does the Dialogue. Each Dialogue states in a comic way, with incredible wordplay (puns, acrostics, acronyms, anagrams, and more), the themes that will be more soberly explored in the chapter that follows.

There are two main reasons for Achilles and the Tortoise having been chosen to lead off the Dialogues. First, they play the major roles in Zeno's paradox (the topic of the book's first Dialogue), in which Achilles must catch the Tortoise by escaping from an infinite regress. Second, they are the speakers in an equally ingenious but less familiar paradox devised by Lewis Carroll. In Carroll's paradox, which Hofstadter reprints as his second Dialogue, Achilles wishes to prove Z, a theorem of Euclid's, from premises A and B. The Tortoise, however, will not accept the theorem until Achilles postulates a rule of inference C, which explicitly states that Z follows from A and B. Achilles adds the rule to his proof, thinking the discussion is over. The Tortoise then, however, jumps to a higher level, demanding another rule of inference D, which states that Z follows from A, B, and C, and so it goes. The resulting endless regress seems to invalidate all reasoning in much the same way that Zeno's paradox seems to invalidate all motion. "Plenty of blank leaves, I see!" exclaims Carroll's Tortoise, glancing at Achilles' notebook. "We shall need them ALL!" The warrior shudders.

One of Hofstadter's Dialogues, "Contracrostipunctus," is an acrostic (complete with punctuation marks) asserting that if the words in it are taken backward, they provide a second-order acrostic spelling out "J. S. Bach." Another Dialogue, "Crab Canon," which is illustrated with an Escher periodic tessellation of crabs, is based on Bach's "Crab Canon" in his *Musical Offering*. As the Tortoise discusses Bach, his sentences are interspersed with those of Achilles, who is discussing Escher. The Tortoise and Achilles use the same sentences in reverse order. The Crab enters briefly at the crossing point to knot the halves of their discourse together, halves that interweave in time in the same way that the positive and negative crabs of Escher's tessellation interweave in space.

The initials A, T, and C (for Achilles, Tortoise, and Crab) correspond to the initials of adenine, thymine, and cytosine, three of the four nucleotides of DNA, the molecule with the extraordinary ability to replicate itself. Just as Achilles pairs with the Tortoise, so adenine pairs with thymine along the DNA double helix. Cytosine pairs with guanine. The fact that the initial G can be taken to stand for "gene" prompted Hofstadter to do a "little surgery on the Crab's speech" so that it would reflect this coincidence. The striking parallel between the tenets of mathematical logic and the "central dogma" of molecular biology is dramatized in a chart Hofstadter calls the "Central Dogmap."

The letter G also stands for Gödel's sentence: the sentence at the heart of his proof that asserts its own unprovability. To Hofstadter the sentence provides an example of what he calls a Strange Loop, exemplifying the self-reference that is one of the book's central themes. (A framework in which a Strange Loop can be realized is called a Tangled Hierarchy, and the letters of "sloth" turn out to stand for "Strange Loops, or Tangled Hierarchies.") Dozens of examples of Strange Loops are discussed, from Bach's endlessly rising canon (which modulates to higher and higher keys until it loops back to the original key) to the looping flow of water in Escher's *Waterfall* and the looping staircase of his *Ascending and Descending* (see Figure 17.5). One of the most amusing models of G is a record player X that self-destructs when a record titled "I Cannot Be Played on Record Player X" is played on it.

A particularly striking example of a two-step Strange Loop is Escher's drawing of two hands (see Figure 24.1), each one sketching the other. We who see the picture can escape the paradox by "jumping out of the system" to view it from a metalevel, just as we can escape the traditional paradoxes of logic by jumping into a metalanguage. We too, however, have Strange Loops, because the human mind has the ability to reflect on itself, that is, the firing of neurons creates thoughts about neurons. From a broader perspective the human brain is at a level of the universe where matter has acquired the awesome ability to contemplate itself.

By the end of *Gödel, Escher, Bach* Hofstadter has introduced his readers to modern mathematical logic, non-Euclidean geometries, computability theory, isomorphisms, Henkin sentences (which assert their own provability), Peano postulates (the pun on "piano" is not overlooked), Feynman diagrams for particles that travel backward in time, Fermat's last theorem (with a pun on "fermata"), transfinite numbers, Goldbach's conjecture (which is cleverly linked with Bach's *Goldberg Variations*), Turing machines, computer chess, computer music, computer languages (Terry Winograd, an expert on the computer simulation of natural language, appears in one Dialogue under the anagrammed name of Dr. Tony Earrwig), molecular biology, the "mind" of an anthill called Aunt Hillary, artificial intelligence, consciousness, free will, holism v. reductionism, and a kind of sentence philosophers call a counterfactual.

Counterfactuals are statements based on hypotheses that are contrary to fact, for example, "If Lewis Carroll were alive today, he would greatly enjoy Hofstadter's book." These statements pose difficult problems in the semantics of science, and there is now a great deal of literature about them. For Hofstadter they are instances of what he calls slipping, progressing from an event to something that is almost a copy of it. The Dialogue that precedes a chapter on counterfactuals and artificial intelligence concerns a Subjunc-TV set that enables an observer to get an "instant replay" of any event in a football game and see how the action would have looked if certain parameters were altered, that is, if the ball were spherical, if it were raining, if the game were on the moon, if it were played in four-dimensional space, and so on.

The book's discussion of artificial intelligence is also enormously stimulating. Does the human brain obey formal rules of logic? Hofstadter sees the brain as a Tangled Hierarchy: a multilevel system with an intricately interwoven and deep self-referential structure. It follows logical rules only on its molecular substrate, the "formal, hidden, hardware level" where it operates with eerie silence and efficiency. No computer, he believes, will ever do all a human brain can do until it somehow reproduces that hardware, but he has little patience with the celebrated argument of the Anglican philosopher J. R. Lucas that Gödel's work proves a human brain can think in ways that are in principle impossible for a computer.

Only a glimpse can be given here of the recreational aspects of this monstrously complicated book. In "The Magnificrab, Indeed" (a pun on Bach's *Magnificat* in D), the Dialogue that introduces a discussion of deep theorems of Alonzo Church, Turing, Alfred Tarski, and others, appears a whimsical Indian mathematician named Mr. Najunamar. Najunamar has proved three theorems: he can color a map of India with no fewer than 1,729 colors; he knows that every even prime is the sum of two odd numbers, and he has established that there is no solution to $a^n + b^n = c^n$ when n is zero. All three are indeed true.

Some readers will recognize 1,729 as the number of the taxi in which G. H. Hardy rode to visit the Indian mathematician Srinivasa Ramanujan ("Najunamar" spelled backward) in a British hospital. Hardy remarked to Ramanujan that 1,729 was a rather dull number. Ramanujan rejoined instantly that on the contrary it was the smallest positive integer that is the sum of two different pairs of cubes. Hardy then asked his friend if he knew the smallest such number for fourth powers. Ramanujan did not know the number, although he guessed that it would turn out to be fairly big. Hofstadter supplies the answer: 635,318,657, or $134^4 + 133^4$ or $158^4 + 59^4$. He also wonders if his readers can find the smallest number that can be expressed as the sum of two squares in two different ways, but he hides the answer. Can you determine it before I supply it in the answer section?

To explain the meaning of the term "formal system" Hofstadter opens his book with a simple example that uses only the symbols M, I, and U. These symbols can be arranged in strings called theorems according to the following rules:

1. If the last letter of a theorem is I, U can be added to the theorem.
2. To any theorem Mx, x can be added. (For example, MUM can be transformed into $MUMUM$, and MU can be transformed into MUU.)
3. If III is in a theorem, it can be replaced by U, but the converse operation is not acceptable. (For example, $MIII$ can be transformed into MU, and $UMIIIMU$ can be transformed into $UMUMU$.)
4. If UU is in a theorem, it can be dropped. (For example, UUU can be transformed into U, and $MUUUIII$ can be transformed into $MUIII$.)

There is only one "axiom" in the system: In forming theorems one must begin with MI. Every string that can be made by applying the rules, in any order, is a theorem of the system. Thus $MUIIU$ is a theorem because it can be generated from MI in six steps. If you play with the M, I, and U system, constructing theorems at random, you will soon discover that all theorems begin with M and that M can occur nowhere else.

Now for a puzzle: Is MU a theorem? I shall say no more about MU here except that it plays many other roles in the book, in particular serving as the first two letters of "Mumon," the name of a Zen monk who appears in a delightful chapter on Zen koans.

Even as simple a system as that of M, I, and U enables Hofstadter to introduce a profound question. If from all the possible strings in the system we subtract all the strings that are theorems, we are left with all the strings that are not theorems. Hence the "figure" (the set of theorems) and the "ground" between the theorems (the set of nontheorems) seem to carry equivalent information. Do they really? Is the system like an Escher tessellation in which the spaces between animals of one kind are animals of another kind, so that reproducing the shapes of either set automatically defines the other? (Or so that a black zebra with white stripes is the same as a white zebra with black stripes?) In this connection Hofstadter reproduces a remarkable tessellation by Kim in which the word "FIGURE" is periodically repeated in black so that the white ground between the black letters forms the same shapes (see Figure 49.2). The same concept is playfully illustrated in the Dialogue "Sonata for Unaccompanied Achilles" (modeled on Bach's sonatas for unaccompanied violin), in which we hear only Achilles' end of a tele-

From *Inversions* (W. H. Freeman & Co., 1989)

Figure 49.2. Scott Kim's FIGURE FIGURE figure

Gödel, Escher, Bach

phone conversation with the Tortoise about "figure" and "ground." From Achilles' half of the conversation we can reconstruct the Tortoise's.

Other figure-ground examples are provided by the counting numbers. For example, given all the primes, we can determine all the nonprimes simply by removing the primes from the set of positive integers. Is the same true of all formal systems? Can we always take all the theorems from the set of all possible statements in the system and find that what is left—the set of nontheorems—is another complementary formal system? An unexpected discovery of modern set theory is that this is not always the case. To put it more technically, there are recursively enumerable sets that are not recursive. Thus does Hofstadter lead his readers from trivial beginnings into some of the deepest areas of modern mathematics.

The book closes with the wild Dialogue "Six-Part Ricercar," which is simultaneously patterned after Bach's six-part ricercar and the story of how Bach came to write his *Musical Offering*. (A ricercar is a complicated kind of fugue.) In this Dialogue the computer pioneers Turing and Babbage improvise at the keyboard of a flexible computer called a "smart–stupid," which can be as smart or as stupid as the programmer wants. (The computer's name is a play on "pianoforte," which means "soft–loud.") Turing produces on his computer screen a simulation of Babbage. Babbage, however, is seen looking at the screen of his own smart–stupid, on which he has conjured up a simulation of Turing. Each man insists he is real and the other is no more than a program. An effort is made to resolve the debate by playing the Turing Game, which was proposed by Turing as a way to distinguish a human being from a computer program by asking shrewd questions. The conversation in this scene parodies the conversation Turing gives in his classic paper on the topic.

At this point Hofstadter himself walks into the scene and convinces Turing, Babbage, and all the others that they are creatures of his own imagination. He, however, is as unreal as any of the other characters of the Dialogue, because he too is imagined by the author. The situation resembles a painting by René Magritte titled *The Two Mysteries*, in which a small picture of a tobacco pipe is displayed with a caption that says (to translate from the French) "This is not a pipe." (see Figure 49.3.) Floating above the fake pipe is a presumably genuine larger pipe, but of course it too is painted on the canvas.

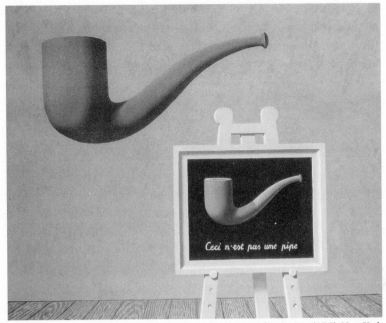

Figure 49.3. *The Two Mysteries,* by Magritte (1966)

And how real was Magritte? How real are Hofstadter, you, and I? Are we in turn no more than shadows on the rapidly flipping leaves of the book we call the universe? We are back to a Gödelian Platonism in which reality is infinitely layered. Who can say what reality really is? The book's final word, "RICERCAR," is a multilevel pun anticipated by a series of acronyms such as Turing's remark "Rigid Internal Codes Exclusively Rule Computers and Robots" as well as Bach's inscription "Regis Iustu Cantio et Reliqua Canonica Arte Resoluta" on a sheet of music he sent to Frederick the Great. Bach's six-part ricercar is from his *Musical Offering,* the story of which opens the book. In this way RICERCAR serves, in much the same way as "riverrun," the first word of *Finnegans Wake,* to twist the work into one gigantic self-referential loop.

One of the book's most amusing instances of a Strange Loop is in the Dialogue that introduces Chapter 16. The Crab speaks of browsing through a "crackpot" book on "metal-logic" titled *Copper, Silver, Gold: an Indestructible Metallic Alloy.* Hofstadter's annotated bibliography reveals that this book is by one Egbert B. Gebstadter (note the *EGB* of Egbert, the *GEB* of Gebstadter, and the *EBG* initials) and was published in

1979 by Acidic Books (Hofstadter's publisher being Basic Books). Here is Hofstadter's comment: "A formidable hodgepodge, turgid, and confused—yet remarkably similar to the present work. Contains some excellent examples of indirect self-reference. Of particular interest is a reference in its well-annotated bibliography to an isomorphic, but imaginary, book."

Addendum

Hofstadter's *GEB* took off like a rocket, staying long on the best seller lists, and winning the 1980 Pulitzer Prize for general nonfiction. Vintage Books paid $200,000 for paperback rights—the largest sum it had ever paid for nonfiction rights, and the largest sum Basic Books ever received for such a work.

Reviews in 1979 were lavish in their praise. Especially noteworthy were reviews by Brian Hayes (*The New York Times Book Review*, April 29), Walter Kerrick (*Village Voice*, November 19), and Edward Rothstein (*New York Review of Books*, December 6). Other reviews ran in *Commonweal, Technology Review, Psychology Today, American Scientist, Yale Review, American Scholar*, and *New Republic*.

In an amusing review in the *Journal of Recreational Mathematics* (14, 1981–82, pp. 52–54), Leon Bankoff observed that *GEB* has exactly 777 pages, and by using the cipher $A = 1$, $B = 2$, $C = 3$, and so on, one discovers that $G = E + B$.

Hofstadter became my successor in writing the Mathematical Games column in *Scientific American*, after he changed the department's name to Metamagical Themas, an anagram of its former title. His columns were reprinted in *Metamagical Themas: Questing For the Essence of Mind and Pattern* (Basic Books, 1985), a work of 852 pages. A few years earlier, Hofstadter and Daniel C. Dennett had edited a marvelous anthology, *The Mind's I: Fantasies and Reflections on Self and Soul* (Basic Books, 1981). At present Hofstadter is professor of cognitive science and computer science and technology at the Indiana University in Bloomington.

I must confess that I could never have written my review of *GEB* had I not had on hand a 33-page analysis of the book written by Scott Kim titled *Strange Loop Gazette*. Kim has since obtained his doctorate under Donald Knuth, in the computer science department of Stanford University. Kim's beautiful book *Inversions*, containing scores of names and phrases drawn in such a way that they magically remain the same when inverted or mirror-reflected (or turn into another word or phrase) has been reissued by W. H. Freeman. (A book of similar inversions by Hofstadter, titled *Ambigrammi,* was published in Italy in 1987 but has yet to have a U.S. edition.) Kim is now working on a book about how to use one's fingers to model such things as the skeleton of a cube or tetrahedron, or to entwine the fingers to produce such topological structures as a trefoil knot.

Answers

I showed how *MUIIU* could be generated from *MI* in six steps. Several readers lowered this to five, and one reader, Raymond Aaron, did it in four: *MI* to *MII* (rule 2), to *MIIII* (rule 2), to *MIIIIIIII* (rule 2), and finally to *MUIIU* (rule 3).

I did not give a proof that *MU* is not a theorem. Here is how Hofstadter handled it. Every theorem begins with *M,* which occurs nowhere else. The number of I's in a theorem is not a multiple of 3 because this is true for the axiom *MI,* and every permissible operation preserves this property. Therefore *MU,* whose number (zero) of I's is a multiple of 3, cannot be obtained by the permissible operations.

Several readers wrote programs for determining Ramanujan numbers that solve the Diophantine equation $A^n + B^n = C^n + D^n$. When n is 3, William J. Butler, Jr.'s program found 4,724 solutions for values less than 10^{10}, of which the largest is

$$1,956^3 + 1,360^3 = 2,088^3 + 964^3.$$

Of the 4,724 solutions, 26 are triples, the smallest being

$$414^3 + 255^3 = 423^3 + 228^3 = 436^3 + 167^3.$$

The number of primitive solutions (no common factor of the four numbers) is infinite, but the number of triples, Butler conjectures, could be finite because their density declines rapidly as the numbers grow in size.

Hofstadter disclosed in a letter that the phrase "formidable hodge-podge" in his joke review of *GEB* (quoted in my final paragraph) was taken from a reviewer's comment when Indiana University considered publishing the book. (The book was also rejected, incidentally, by an editor then at W. H. Freeman.) Because W. V. Quine, the Harvard philosopher, was one of the two reviewers, the chances are fifty percent that the phrase was Quine's. "Turgid and confused" (the phrase also appears on p. 3 of *GEB*) is from a comment on Bach's style by one of his pupils. In *GEB*'s second printing Hofstadter added the following to his hoax review: "Professor Gebstadter's Shandean digressions include some excellent examples of indirect self-reference." The first four words, with the change of name, are from Brian Hayes's *The New York Times* review of *GEB*.

Answers

The first problem was to find the smallest positive integer that can be expressed as the sum of two squares in two different ways. The number is 50, which equals $5^2 + 5^2$ or $1^2 + 7^2$. If zero squares are allowed, however, the number is 25, which equals $5^2 + 0^2$ or $3^2 + 4^2$. If the two squares must be nonzero and different, the solution is 65, which equals $8^2 + 1^2$ or $7^2 + 4^2$.

The second problem was to determine whether or not *MU* is a theorem in the *M*, *I*, and *U* formal system. A simple proof of why *MU* is not a theorem can be found on pages 260 and 261 of *Gödel, Escher, Bach: an Eternal Golden Braid*.

Bibliography

J. Gleick, "Exploring the Labyrinth of the Human Mind," *The New York Times Magazine*, August 21, 1983. A cover story on Douglas Hofstadter.

G. Johnson, "The Copycat Project," *Machinery of the Mind: Inside the New Science of Artificial Intelligence*, Times Books, 1986, Chapter 15. The chapter is about Hofstadter's approach to artificial intelligence.

Reviews of Godel, Escher, Bach

H. Gardner, "Strange Loops of the Mind," *Psychology Today*, March 1980, pp. 72–86.

B. Hayes, "This Head is False," *The New York Times Book Review*, April 29, 1978, pp. 13, 18.

E. Rothstein, "The Dream of Mind and Machines," *New York Review of Books*, December 6, 1979, pp. 34–39.

Other books by Douglas Hofstadter

The Mind's I: Fantasies and Reflections on Self and Soul. With Daniel C. Dennett, Basic, 1981.

Metamagical Themas, Basic, 1985. A collection of Hofstadter's *Scientific American* columns.

Ambigrams, Italy: Hopeful Monster, 1987.

Fluid Concepts and Creative Analogies: Computer Models of the Fundamental Mechanisms of Thought, Basic, 1995.

Le Ton Beau de Marot: The Spark and Sparkle of Creative Translation, Basic, 1997.

Chapter 50

Six Sensational Discoveries

$\boldsymbol{A}$s a public service, I shall comment briefly on six major discoveries of 1974 that for one reason or another were inadequately reported to both the scientific community and the public at large. The most sensational of that year's discoveries in pure mathematics was surely the finding of a counterexample to the notorious four-color-map conjecture. That theorem is that four colors are both necessary and sufficient for coloring all planar maps so that no two regions with a common boundary are the same color. It is easy to construct maps that require only four colors, and topologists long ago proved that five colors are enough to color any map. Closing the gap, however, had eluded the greatest minds in mathematics. Most mathematicians have believed that the four-color theorem is true and that eventually it would be established. A few suggested it might be Gödel-undecidable. H.S.M. Coxeter, a geometer at the University of Toronto, stood almost alone in believing that the conjecture is false.

Coxeter's insight was vindicated. In November 1974 William McGregor, a graph theorist of Wappingers Falls, NY., constructed a map of 110 regions that cannot be colored with fewer than five colors (see Figure 50.1). McGregor's technical report appeared in 1978 in the *Journal of Combinatorial Theory,* Series B.

In number theory the most exciting discovery of 1974 was that when the transcendental number e is raised to the power of π times $\sqrt{163}$, the result is an integer. The Indian mathematician Srinivasa Ramanujan had conjectured that e to the power of $\pi\sqrt{163}$ is integral in a note in the *Quarterly Journal of Pure and Applied Mathematics* (vol. 45, 1913–1914, p. 350). Working by hand, he found the value to be 262,537,412,640,768,743.999,999,999,999,. . . . The calculations were tedious, and he was unable to verify the next decimal digit. Modern

Figure 50.1. The four-color-map theorem is exploded.

computers extended the 9's much farther; indeed, a French program of 1972 went as far as two million 9's. Unfortunately, no one was able to prove that the sequence of 9's continues forever (which, of course, would make the number integral) or whether the number is irrational or an integral fraction.

In May 1974 John Brillo of the University of Arizona found an ingenious way of applying Euler's constant to the calculation and managed to prove that the number exactly equals 262,537,412,640,768,744. How the prime number 163 manages to convert the expression to an integer is not yet fully understood.

There were rumors late in 1974 that π would soon be calculated to six million decimal places. This may seem impressive to laymen, but it is a mere computer hiccup compared with the achievement of a special-purpose chess-playing computer built in 1973 by the Artificial Intelligence Laboratory at the Massachusetts Institute of Technology. Richard Pinkleaf, who designed the computer with the help of ex-world-chess-champion Mikhail Botvinnik of the former U.S.S.R., calls his machine MacHic because it so often plays as if it were intoxicated.

Unlike most chess-playing programs, MacHic is a learning machine that profits from mistakes, keeping a record of all games in its memory and thus steadily improving. Early in 1974 Pinkleaf started MacHic playing against itself, taking both sides and completing a game on an average of every 1.5 seconds. The machine ran steadily for about seven months.

At the end of the run, MacHic announced an extraordinary result. It had established, with a high degree of probability, that pawn to king's rook 4 is a win for White. This was quite unexpected because such an opening move has traditionally been regarded as poor. MacHic could not, of course, make an exhaustive analysis of all possible replies. In constructing a "game tree" for the opening, however, MacHic extended every branch of the tree to a position that any chess master would unhesitatingly judge to be so hopeless for Black that Black should at once resign.

Pinkleaf has been under enormous pressure from world chess leaders to destroy MacHic and suppress all records of its analysis. The Russians are particularly concerned. I am told by one reliable source that a meeting between Kissinger and Brezhnev will take place in June, at which the impact on world chess of MacHic's discovery will be discussed.

Bobby Fischer reportedly said that he had developed an impregnable defense against P-KR4 at the age of 11. He has offered to play it against MacHic, provided that arrangements can be made for the computer to play silently and provided that he (Fischer) is guaranteed a win-or-lose payment of $25 million.

The reaction of chess grand masters to MacHic's discovery was mild compared with the shock waves generated among leading physicists by the discovery that the special theory of relativity contains a logical flaw. The crucial "thought experiment" is easily described. Imagine a meterstick traveling through space like a rocket, on a straight line colinear with the stick. A plate with a circular hole one meter in diameter is parallel to the stick's path and moving perpendicularly to it (see Figure 50.2). We idealize the experiment by assuming that both the plate and the meterstick have zero thickness. The two objects are on a precise collision course. At the same instant, the center of the meter stick and the center of the hole will coincide.

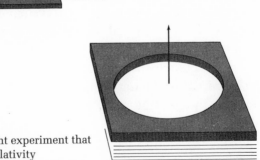

Figure 50.2. A thought experiment that disproves special relativity

Assume that the plate is the fixed inertial frame of reference and the meterstick is moving so fast that it is Lorentz-contracted by a factor of 10. In this inertial frame the stick has a length of 10 centimeters. As a result it will pass easily through the hole in the rapidly rising plate. (The speed of the rising plate is immaterial.)

Now consider the situation from the standpoint of the meterstick's inertial frame. The plate is moving in the opposite horizontal direction, and so now it is the hole that is Lorentz-contracted, along its diameter parallel to the stick, to 10 centimeters. There is no way the 10 centimeter by 1 meter elliptical hole can move up past the meterstick without a collision. The two situations are not equivalent, and thus a fundamental assumption of special relativity is violated.

Physicists have long realized that the general theory of relativity is weakly confirmed, but the special theory had been confirmed in so many ways that its sudden collapse came as a great surprise. Humbert Pringle, the British physicist who discovered the fatal *Gedankenexperiment,* reported it in a short note in *Reviews of Modern Physics,* but the full impact of all of this has not yet reached the general public.

When facsimiles of two lost notebooks of Leonardo da Vinci's were published in 1974 by McGraw-Hill, they were widely reviewed. The public learned of many hitherto unknown inventions made by Leonardo: a system of ball bearings surrounding a conical pivot (thought to have been first devised by Sperry Gyroscope in the 1920s), a worm screw credited to an 18th-century clockmaker, and dozens of other devices, including a bicycle with a chain drive.

In view of the publicity given the McGraw-Hill volumes, it is hard to understand why the media failed to report in December 1974 on the discovery of a drawing that had been missing from the first notebook.

This notebook, known as *Codex Madrid I* (it had been found 10 years earlier in the National Library in Madrid), is a systematic treatise of 382 pages on theoretical and applied mechanics (see Ladislao Reti, "Leonardo on Bearings and Gears," *Scientific American,* February 1971). There had been much speculation on the nature of the missing page. Augusto Macaroni of the Catholic University of Milan observed that the sketch was in a section on hydraulic devices, and he speculated that it dealt with some type of flushing mechanism.

The missing page was found shortly before Christmas by Ramón Paz y Bicuspid, head of the manuscript division of the Madrid Library. It was Bicuspid who had originally found the two lost notebooks. The missing page had been torn from the manuscript and inserted in a 15th-century treatise on the Renaissance art of perfume making. Figure 50.3 reproduces a photocopy of the original drawing. As the reader can see at once, Professor Macaroni was on target. The drawing establishes Leonardo as the first inventor of the valve flush toilet.

It had long been known that Leonardo had invented a folding toilet seat and had proposed a water closet with continuously running water

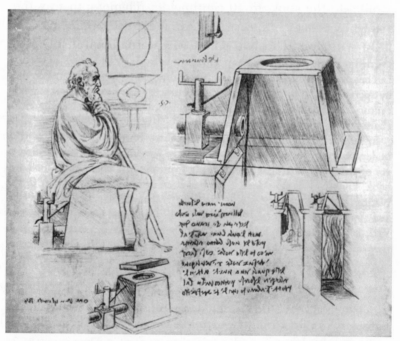

courtesy N.Y. Public Library

Figure 50.3. Lost Leonardo da Vinci drawing

in channels inside walls, a ventilating shaft to the roof, and suspended weights to make sure the entrance door closed. Until now, however, the first valve flush toilet has always been credited to Sir John Harington, a godson of Queen Elizabeth. Harington described it amusingly in his book *The Metamorphosis of Ajax* (1596) a cloacal satire that got him banished from the court. Although his "Ajax" actually was built at Kelston near Bath, it was not until 200 years later that it came into general use.

The first English patent for a valve flush toilet was granted in 1775 to Alexander Cummings, a watchmaker. Modern mechanisms, in which a ball float and automatic cutoff stopper limit the amount of water released with each flush, date from the early 19th-century patents of Thomas Crapper, a British manufacturer of plumbing fixtures who died in 1910. (See L. Wright, R. Paul, and K. Paul, *Clean and Decent: The Fascinating History of the Bathroom and Water Closet,* 1960 and W. Reyburn, *Flushed with Pride: The Story of Thomas Crapper,* Prentice-Hall, 1971.)

Although hundreds of books on parapsychology spewed forth from reputable publishing houses in 1974, not one reported the most sensational psi discovery of the century: a simple motor that runs on psi energy. It was constructed in 1973 by Robert Ripoff, the noted Prague parapsychologist and founder of the International Institute for the Investigation of Mammalian Auras. When Henrietta Birdbrain, an American expert on Kirlian photography, visited Prague in early 1974, Dr. Ripoff taught her how to make his psychic motor. Ms. Birdbrain demonstrated the device many times in her lectures, but as far as I am aware the only published report on it appeared in the Boston monthly newspaper *East West Journal* (May 1974, p. 21).

Readers are urged to construct and test a model of the motor. The first step is to cut a three-by-seven-inch rectangle from a good grade of bond paper. Make a tiny slot in the paper at the spot shown (see Figure 50.4). The slot must be three-eighths inch long and exactly in the center of the strip, one-eighth inch from the top edge. Bend the paper into a cylinder, overlapping the ends five-sixteenths inch, and glue the ends together. Cut a second slot in the center of the overlap, directly opposite the preceding one. It must be the same size and the same distance from the top.

From a file card or a piece of pasteboard of similar weight, cut a strip three-eighths inch by three inches. Insert a fine, sharp-pointed needle

STEP 1

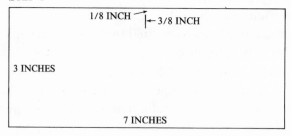

1/8 INCH ⌐ ← 3/8 INCH

3 INCHES

7 INCHES

STEP 2   **STEP 3**

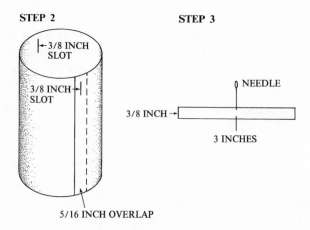

|← 3/8 INCH
SLOT

3/8 INCH →|
SLOT

5/16 INCH OVERLAP

○ NEEDLE

3/8 INCH →

3 INCHES

STEP 4   **STEP 5**

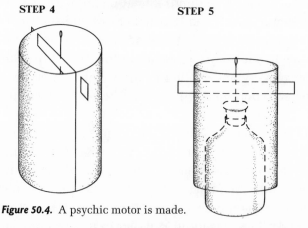

Figure 50.4. A psychic motor is made.

twice through the center of the strip, as shown in step 3. The point of the needle should be no more than one-fourth inch below the bottom edge of the strip. Push the ends of the strip through the cylinder's two slots, as shown in step 4, taking care not to bend the strip. The final step is to balance the needle on top of a narrow bottle at least four inches high (step 5). It is essential that the top of the bottle be either glass (preferable) or very hard smooth plastic.

Adjust the strip in the slots until the cylinder hangs perfectly straight, its side the same distance from the bottle all around. With scissors snip the ends of the strip so that each end projects one-fourth inch on each side.

Place the little motor on a copy of the Bible or the *I Ching*, with the book's spine running due north and south. Sit in front of the motor, facing north. Hold either hand, cupped as shown in Figure 50.5, as close to the cylinder as you can without actually touching it. You must be in a quiet room, where the air is still. Make your mind blanker than usual and focus your mental energy on the motor. Strongly will it to rotate either clockwise or counterclockwise. Be patient. It is normally at least one full minute before the psi energy from your aura takes effect. When it does, the cylinder will start to rotate slowly.

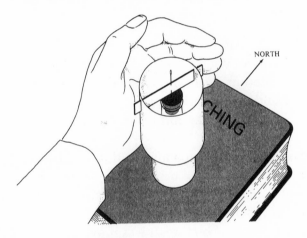

Figure 50.5. How to apply psi energy to the psychic motor.

Some people, of course, have a stronger psi field than others. A lot depends on your mental state. At times the motor refuses to turn. At other times it begins to turn almost as soon as you start concentrating. Experiments show that for most people it is easier to make the motor rotate counterclockwise with psi energy from your right hand and clockwise with the energy from your left hand. At times negative psi takes over, and the motor turns in the direction opposite to the direction being willed. As Dr. J. B. Rhine has taught us, psi effects are elusive, skittish, and unpredictable.

The motor underwent extensive investigation at numerous parapsychology laboratories around the world. Russian experts were convinced the energy that turns the motor is the same as the psychokinetic energy that enabled the Israeli psychic Uri Geller to bend silverware, the Russian "sensitive" Ninel Kulagina to levitate table-tennis balls, and the Brooklyn psychic Dean Kraft to make pieces of candy leap out of bowls and pens crawl across rugs. When Kulagina held both hands near the motor, the cylinder flew straight up in the air for several meters. A book on the Ripoff rotor (as it is called in Prague), with papers by 12 of the world's leading parapsychologists, was edited by Ms. Birdbrain.

James Randi, the magician, contends that by using trickery, he can make the motor spin rapidly in either direction. Of course, that does not explain why the motor operates so efficiently for thousands of people who know nothing about conjuring.

Addendum

The foregoing chapter, when it ran in the April 1975 issue of *Scientific American,* was intended as an April Fools' joke. It was so crammed with preposterous ideas and outlandish names that I never dreamed anyone would take it seriously, yet it produced more than a thousand letters from readers who did not recognize the column as a hoax.

The map was designed by correspondent William McGregor (his real name), who gave me permission to print it. Hundreds of readers sent me copies of the map, colored with four colors. Some said they had worked on it for days before they found a way to do it. The four-color map theorem is no longer a conjecture. It was proved in 1976 by Wolfgang Haken and Kenneth Appel, with the aid of a long-running computer program. (See "The Solution of the Four-Color-Map Problem," *Scientific American,* October 1977, pp. 108–21.) Whether a simple, elegant proof not requiring a computer will ever be found, is still an open question.

When Norman K. Roth published an article, "Map Coloring," in *Mathematics Teacher* (December 1975), many readers informed him that *Scientific American* had published a map disproving the four-color theorem. A letter from Roth in the following May issue pointed out that my column was "an apparently successful April Fools' article."

In 1977 the *Vancouver Sun* reported a British mathematician's claim (it turned out to be invalid) that he had proved the four-color map theorem. On January 17, 1977, the newspaper ran a letter from a lady in Port Moody, which said in part:

> To set the record straight, I would like to bring to your notice the fact that the theorem has already been disproved by William McGregor, a graph theorist . . . in November 1975. He constructed a map of 110 regions that cannot be colored with less than five colors. . . .

Artificial Intelligence (Vol. 10, 1978, p. 116) reported that a computer program had managed to color McGregor's map with four colors, with only two backtracks in the tree search. Apparently the author did not realize the map was a joke.

The following letter, signed by "Ivan Guffvanoff III," who claimed to be a mathematician at the University of Wisconsin, was a bit frightening to the staff of *Scientific American,* until they realized that it, too, was a joke:

> This is to inform you that my lawyer will soon be contacting you for a damage case of $25 million.
>
> In the mathematics section of your April 1975 issue, Martin Gardner wrote that the four-color problem had been solved. I have been working on this problem for 25 years. I had prepared a paper to be submitted to the *American Mathematical Monthly.* The paper was over 300 pages in length. In it I had proved that the answer to the four-color problem was no and that it would take five colors instead of four. Upon reading Gardner's article that someone else would publish the solution before I could, I destroyed my paper. Last week I read in *Time* magazine that Gardner's article was a farce. I did not read Gardner's entire article, only the part on the four-color problem, so I was not aware of the farce. Now that I have destroyed my article, it will not be possible to reproduce all 300 pages, since the work has extended over such a long time. I therefore believe that damages are due me.
>
> I believe that Gardner's article was the most unprofessional article I have ever seen in your's or any other journal. This kind of activity is below the dignity of what I thought your magazine stood for. I am not only suing you but I am cancelling my membership, and I will ask all my friends to cancel theirs.

In Italy the noted mathematician Beniamino Segre published a serious research note (*Rendiconti* 59, 1975, pp. 411–12) in which he reproduced McGregor's map, showing how it could be four-colored. "It is shown the falsehood," his summary reads, "of a presumed counterexample for the four-color conjecture."

Manifold, a journal published by mathematics students at the University of Warwick, ran the following lines in its Autumn 1975 issue. They are to be sung to the tune of "Oh Mr. Porter, what shall I do?"

> *"Oh Mr. Gardner,*
> *What have you done?*
> *You've started up a rumour*
> *You should never have begun!*
> *A four-colour hoax can't*
> *Be undone so quick . . .*
> *Oh Mr. Gardner, what*
> *A bloody silly trick!"*

When e is raised to the power of the product of π and the square root of 163, the result is the 18-digit number I gave, minus .000,000,000,000,75. . . . John Brillo, to whom I attributed this hoax, is a play on the name of the distinguished number theorist John Brillhart. The reference to Ramanujan's paper is legitimate. In it the Indian mathematician discusses a family of remarkable near-integer numbers to which this one belongs, but of course he knew that none were integral. Indeed, as many readers pointed out, it is not hard to prove that they are transcendental.

The value of the number to 39 significant decimal digits was given by D. H. Lehmer in *Mathematical Tables and Aids to Computation* (Vol. 1, January 1943, pp. 30–31). The digit following the run of 9's is 2. See also "What Is the Most Amazing Approximate Integer in the Universe?" by I. J. Good, in the *Pi Mu Epsilon Journal* (Vol. 5, Fall 1972, pp. 314–15).

The description of Richard Pinkleaf's chess-playing program, MacHic, is a play on the chess program MacHack, written by Richard Greenblatt of the Massachusetts Institute of Technology. The relativity paradox that I hung on Humbert Pringle (a play on the name of Herbert Dingle, a British physicist who maintained that relativity theory is disproved by the famous twin paradox) is well known. It appears as a problem on page 99 of the paperback edition of *Spacetime Physics,* by Edwin F. Taylor and John A. Wheeler (W. H. Freeman and Company, 1966), and the solution is given on page 25 of the answer section. The paradox is discussed at greater length by George Gamow in *Mr. Tomp-*

kins in Wonderland (Macmillan, 1947), W. Rindler in *American Journal of Physics,* (Vol. 29, 1961, p. 365 ff.), R. Shaw (*ibid.,* Vol. 30, 1962, p. 72 ff.), and P. T. Landsberg in *The Mathematical Gazette,* (Vol. 47, 1964, p. 197 ff).

A stationary outside observer will see the meterstick just make it through the hole. If the plate and the stick have thickness, the stick must, of course, be a trifle shorter than the hole to prevent an end from catching. To an observer on the plate the stick will appear Lorentz-contracted, but it will also appear rotated, so that it seems to approach the hole on a slant. The stick's *back* end actually seems to go through the hole before its front end, so that it gets through with the same clearance as before. To an observer on the stick the plate will appear Lorentz-contracted, its hole becoming elliptical, but the plate also appears to be rotated. In this case the hole first goes over the *front* end of the slanted stick, again with the same close fit. "Contractions" and "rotations" are ways of speaking in a Euclidean language. In a four-dimensional, non-Euclidean language of space–time the objects retain their shapes and orientations. Having at one time written a book on relativity, I was abashed to receive more than 100 letters from physicists pointing out the stupid "blunder" I had made.

Those who enjoyed explaining the paradox may wish to consider how to escape from the following variant. Assume that the meterstick is sliding at high speed along the surface of an enormous flat plate of metal toward a hole slightly larger than the stick. We idealize the thought experiment by assuming that there is no friction and that the stick and the plate are extremely thin. When the stick is over the hole, gravity (or some other force) pulls it down and through. For an observer on the stick the sheet slides under it, and the hole is Lorentz-contracted enough to prevent the stick from dropping through. In this case the stick and the plate cannot rotate relative to each other. How does the stick get through? (Please, no letters! I know the answer.)

The Leonardo da Vinci drawing was done by Anthony Ravielli, a graphic artist well known for his superb illustrations in books on sports, science, and mathematics. It was an earlier version of the sketch that suggested to me the idea of a hoax column. Many years ago a friend of Ravielli's had jokingly made a bet with a writer that Leonardo had invented the first valve flush toilet. The friend persuaded Ravielli to do a Leonardo drawing in brown ink on faded paper. It was smuggled into the New York Public Library, stamped with a catalogue file number, and placed in an official library envelope. Confronted with this evidence, the writer paid off the bet.

Augusto Macaroni is a play on Augusto Marinoni, a da Vinci specialist at the Catholic University of Milan, and Ramón Paz y Bicuspid is a play on Ramón Paz y Remolar, the man who actually found the two missing da Vinci notebooks. My data on the history of the water closet are accurate, including the reference to Thomas Crapper. The book by Wallace Reyburn *Flushed with Pride: The Story of Thomas Crapper* does exist.

For many years I assumed that Reyburn's book was the funniest plumbing hoax since H. L. Mencken wrote his fake history of the bathtub. I thought this for two reasons: (1) The book implies that the slang words "crap" and "crapper" derived from Mr. Crapper's name, but "crap" and "crapping case" are both listed in *The Slang Dictionary,* published in London in 1873. (2) Reyburn wrote a later book titled *Bust-up: The Uplifting Tale of Otto Titzling and the Development of the Bra.* It turns out, though, that both Thomas Crapper and Otto Titzling were real people, and neither of Reyburn's books is entirely a hoax.

The Ripoff Rotor is a modification of a psychic motor described in Hugo Gernsback's lurid magazine *Science and Invention* (November 1923, p. 651). Prizes were awarded in March 1924 to readers who gave the best explanations of why the cylinder turned. The motion can be caused by any of three forces: slight air currents in the room, convection currents produced by heat from the hand, and currents from breathing. The three forces combine in unpredictable ways. If a person who believes he or she has psychokinetic powers is willing the motor to turn, it may turn in the direction willed, or it may go the other way.

Nandor Fodor, in his *Encyclopedia of Psychic Science* (Citadel, 1966), under the heading "Fluid Motor," credits the paper-cylinder device to one Count de Tromelin, but he doesn't say who the Count was or when he invented the motor.

I had no individual in mind when I mentioned Ms. Henrietta Birdbrain, but there *is* an *East West* newspaper in Boston, and the reader who bothers to check the issue cited, will find a sober report by Stanley Krippner on a psychic motor that was demonstrated to him in Prague by Robert Pavalita. (On Pavalita, see S. Ostrander and L. Schroeder, *Psychic Discoveries behind the Iron Curtain,* Prentice-Hall, 1970, Chapter 28.)

Hereward Carrington, in *The Story of Psychic Science* (1931, p. 138), mentions half a dozen early French psychic motors. Opposite page 138 is a photograph of Paul Joire's "sthenometer," a straw balanced on a needle and under a bell jar. Incidently, Dr. Joseph B. Rhine for years tried to find evidence that a person could turn an arrow delicately balanced on a needle in a vacuum, but without success. His negative results were never published. Surely one of the great scandals of parapsychology is its claim that the mind can affect heavy objects like falling dice—and if Uri Geller is taken seriously, can bend spoons and keys—but as yet no one has been able, under controlled conditions, to rotate a featherweight pointer.

After my hoax column appeared, several psychic motors went on sale. The most colorful was a cardboard device called the Mind Machine, sold by Unicorn Products in Santa Fe. "As with the Ouiji Board," the instructions said, "results are not always immediate, and often vary from time to time. . . . Yet for countless people the phenomenon is a reality. At parties, skeptics have been known to chop the base apart, expecting to find a hidden motor. How does the Mind Machine work? Theories abound—but no one really knows." A similar cardboard device called a "psionic generator," marketed by Monarch Manufacturing Company, in Roseville, MI, was advertised in *Fate* (July 1975, p. 91).

Professor L. E. van der Tweel, at the University of Amsterdam, sent the photograph reproduced in Figure 50.6. It shows the noted neurophysiologist and philosopher Warren McCulloch demonstrating a paper psychic motor to several colleagues. The picture was taken in 1953.

Some readers who believed in psychokinesis took the Ripoff Rotor seriously, but most readers did not. Mark J. Hagmann reported his discovery that the motor's rotation relative to the room was an inverse function of the contents of the liquor bottle he used to support the needle. The rotation's speed increased as the level of booze went down.

I've selected the following four letters as the funniest:

Dear Mr. Gardner:

As a citizen deeply concerned about the energy crisis and full public disclosure, I express my gratitude for your release of the psi-engine data. I took drawings of a larger model, coupled to an electric generator, to Ralph Nadir and obtained his tentative endorsement that, used as a power generator, the Ripoff system would have a minimum environmental impact.

Figure 50.6. Warren McCulloch demonstrating a psychic motor

Since any mass, once set in motion, will remain in motion unless force is applied to stop it, the Ripoff generator need only be started and once started will continue to produce electricity indefinitely. The prospect that this system holds the promise of eliminating those dirty power plants and unsightly solar energy collectors is indeed exciting.

I then contacted my old friend Dr. P. D. Feckless whose work in the psi-energy field over the past 20 years speaks for itself. P. D. is now a senior investigator at Fulsome Psi-Tech and editor of their journal "Fulsome Dynamic Systems" (F.D.S.).

To my delight, I found that he was not only familiar with the Ripoff engine but also has succeeded in making improvements resulting in a tenfold increase in power output. Just as there is a magnetic field associated with the flow of electricity, he found that there is a psi-magnetic analogue associated with the flow of psi-force. It extends at an angle of 45° to the path of the psi-force.

In the Ripoff engine, if the strip through which the needle is inserted is twisted in the fashion of a propeller to a blade angle of 45°, it can then take full advantage of the psi-magnetic vector.

Dr. Feckless must also be credited with the application of his own discovery, the psi-temperature, to the Ripoff machine. Each person radiates energy at an optimum psi-temperature, but calculating this involves improper fractions as well as surds. Therefore, he suggests that an average be used which is room temperature plus 100°. To operate with maximum power output, the Ripoff machine must "feel" this temperature. This is accomplished by filling the bottle with water heated to 100° above room temperature—but in no case above boiling.

The machine, thus adjusted and filled, will work well with only a watchful look from the experimenter. He does not need to use his hands.

Another finding of great importance was his measurement of aura power from various types of printed material. Using a torsion balance wire tipped with a psi-magnetically oriented crystal of mezzanine sulphate, he found that purely scientific material was more potent than the religious material used by Ripoff. Four volumes of Tom Swift or the equivalent weight in AAAS Science magazines will do nicely. Greater power levels can be sustained if these are also warmed to the psi-temperature.

Finally, in a personal (telepathic) communication with Dr. Ripoff, Dr. Feckless learned that the northern orientation favored by Dr. Ripoff was empirically determined. Analysis showed that one of Europe's greatest intellectual centers was due north of the Ripoff lab. Considering a great focus of intellectual power to be the critical variable, the most favorable orientation for U.S. experiments would therefore be toward Washington D.C.

I hope these findings will be helpful and will move us closer to the era of psi-power.

Sincerely yours,
George N. Chatham

Dear Mr. Gardner:

Your psi energy motor. I've got it! I know how it works. It really does work! And I know how it works. Even Eureka!

But first let me tell you, in a scholarly fashion, how I came to this unexpected revelation of the psi energy motor's operational mechanism through your lucid instructions in the April *Scientific American.* Let me also reassure you that, although I know several people who teach psychology, I know very little of the subject myself, other than infrequent casual reading about their feeling bumps on other people's heads to tell fortunes. The psychologists I do know, during our Friday-afternoon beer drinking sessions, do not talk of parapsychology; here in Seattle we call parasols *umbrellas.* The rain, you know. As for my knowledge of energy sources, I know almost nothing of physics, aside from an occasional Ex-Lax, as needed.

The experiment, though. I was going to tell you about the experiment. I came home on a wet Friday afternoon to find my wife out on some do-gooding mission so I picked up the April issue of *Scientific American* and chanced on your startling article. I immediately went to work building the motor you described. I used the recommended good grade of bond paper. It was stationery from the college where I teach political science. I cut the three by seven inch piece from the center, lest the letterhead portion might influence the psychic energy. The cardboard portion of the assembly was easily cut from a student grade-change request card on my desk. Finding the needle was more difficult, since my wife was out and I had to hunt down where she hides her sewing equipment. I found it. I found the needles, too, scattered in the bottom of her thread-filled case. I can report that the motor runs with a slight smear of blood on the paper and a band-aid on the finger.

With the psychic energy motor assembled on my desk, I set out in search of the rest of the equipment necessary to make it operate. There was no hard, smooth, plastic bottle in the house. A glass catsup bottle could not function because of its metal cap. A water glass was too great in diameter. At that point I spied a half-full scotch bottle in the bar. It had a long enough neck to allow the motor to hang perfectly. But it was half full. I poured that into the too-big water glass. I do not have a copy of *I Ching* around the house, and I found the family Bible had so many photographs, pressed flowers, decades-old dance programs, baby footprints, and grade-school report-cards that the family Bible could not have provided a flat surface on which to set the bottle.

North and south, for the alignment of the book, was not difficult to ascertain because the house faces west and the desk in the study is along that wall. But a book. I first tried the required text in an introductory American government course. The results were inconclusive; nothing but turbulent action and indecision from the psi motor. I then tried a tract from the John Birch Society, and the motor rotated to the right. Wondering if that had any political significance, I tried an anthology of the works of Lenin, and sure enough, it turned to the left. I then placed the motor assembly on Merkin Swiver's *Collected Speeches of President Ford,* and noted that the motor revolved in a quite mediocre fashion, again to the right. It did not turn at all with a football helmet over it. Barry Goldwater's *Conscience of a Conservative* made the motor spin so rapidly I could feel the draft. When placed on N. Comium's biography of Henry M. "Stoop" Jackson, the motor rose slowly accelerating marvelously toward the north on what I assumed was a great-circle course. I then tried a copy of M. P. Chment's long study of ex-President Nixon. The motor kept falling off the bottle to the South, toward California, and kept rolling into hiding under the desk. At last, I was beginning to gain insight into the motive power of the psychic energy motor. But by now the glass into which I had poured the scotch was empty, and I had to go open a new bottle.

On my return to the study, I placed the refilled glass on the desk and I pried the phlebitic motor out from under the desk. Then, finding a monograph on the functioning of Congress, I placed the bottle and the psi motor on that legislative tome. You will not believe it. I did only after I had closed one eye to focus more carefully. The motor was turning both directions—right and left—at the same time and was doing so in a most indecisive fashion.

At that instant, in a flash of intuition, I learned the secret of the motive power of the psychic motor: hot air.

Soon thereafter, however, other psychokinetic manifestations occurred. The floor rose up rapidly, and with a thump, hit the side of the chair in which I was sitting. From that position, most curiously, the books on their shelves did not fall from their obviously unaccustomed position on the ceiling, even though the room, too, was rotating.

Some time later my wife returned. I attempted to show her how the psychic energy motor would work, but it was damaged beyond repair by the floor rising to hit it during those psychokinetic events. The nearly finished glass of scotch, quite unaccountably, still maintained its position on the desk.

Yours in learning,
Charles W. Harrington

Dear Sirs:

This evening I was sitting in my garage, which is $\sqrt{163}$ meters long and has a door at each end, waiting for my friend Wade Wykert to drive his psychic energy powered automobile through it at his usual speed of 0.995 times the velocity of light. His automobile is also $\sqrt{163}$ meters long, but it is Lorentz-contracted by a factor of 10, so I have no difficulty in closing the entrance door before I have to open the exit door to let him out of the garage. My friend, however, thinks that my garage is Lorentz-contracted to one-tenth the length of his automobile and that, as he passes through both doors have to be open at the same time. So far, I have managed to open and close the doors at appropriate times so that his belief in the special theory of relativity has not been destroyed.

As he passed by he threw out to me a copy of the April 1975 issue of *Scientific American* open to Martin Gardner's Mathematical Games Section. I hoped that he had not read the part about Humbert Pringle's *Gedanken-experiment* disproving special relativity. At best this knowledge would make him nervous next time he drove through, and might result in the splintering of my garage doors.

I noticed that Wade had succeeded in coloring William McGregor's 110-region map with only four colors, thus unvindicating H. S. M. Coxeter's foursighted insight. A handwritten note indicated a desire to have his achievement publicized before 1978, when McGregor's report is scheduled to appear.

I then read with great interest the rest of Martin Gardner's article on inadequately reported major discoveries of 1974, noticing that I was sitting in an attitude rather similar to that of Leonardo (the figure in the drawing must surely be Leonardo himself) in the reproduction on page 127.

I was surprised that Gardner was not aware of the analysis made by the chess-playing computer built at the editorial offices of *Ms.* and named MacHaec. (One of the editors is a Latin scholar as well as a feminist.) MacHaec's analysis indicates that the pawn to king's rook 4 opening leads to a sure win for black. Clearly a match between MacHic and MacHaec is indicated, and there have been rumors that Bobby Fischer might put up the $25 million prize money.

Martin Gardner's point that such discoveries as these are frequently not brought to the public notice is well taken. There is, however, a satisfactory alternative for the authors of such world-shaking communications. They should be submitted to either the *International Journal of Hypercritical Obfuscation* or the *Archives of Information Pollution.* I am the editor of both these journals and I will guarantee that articles submitted will be dealt with expeditiously.

Sincerely yours,
John L. Howarth
Professor of Physics

Dear Sir,

Having completed reading your article in the April issue of *Scientific American* I feel that it is my duty to inform you that the reason the psi motor runs was discovered approximately one year ago. You see there is an aura of glacomoridine which can envelop the person's hands.

According to the papers I have received over the past five years, it seems that some humans produce more of it than others. It now appears certain that glacomoridine is a hormone. However, there was much discussion on whether or not it was produced internally or not. This problem was cleared by R.E. Coleman et al. After many arduous months of searching for this elusive organ, they discovered what appeared to be a satellite on the pituitary gland. Analyzation proved that this was what they were looking for. Because of its white appearance it was decided to name it the glacier gland.

But this doesn't really explain how the psi motor is able to work. It remained for Richard Douglas of Washington and Jefferson College to provide a mechanism which would solve this problem. It appears that glacomoridine enters the blood stream after being secreted by the glacier gland. (There would seem to be a conscious control of the gland by the individual.) After being circulated it is then collected in some way (still unknown) in the sweat glands of the hands. From there it is released along with any perspiration. However, instead of diffusing into air the physical properties of glacomoridine enable it to remain close to the hand. The glacomoridine begins to "flow" distally. As can be imagined this movement would cause the psi motor to move. Your term psi motor can still be used, because studies have shown that the level of glacomoridine actually increases when thoughts are directed to moving the motor. More concrete evidence of the increase is demonstrated by observing an EEG taken during the tests. It was shown that both the alpha and delta waves were dominant.

One problem which wasn't adequately answered until a month ago dealt with why some people were able to rotate the psi motor in the reverse direction. The answer to this is also credited to Richard Douglas. A recessive gene, which when homozygous, is responsible for the flow of glacomoridine in the opposite direction than is normal.

Sincerely yours,
William Hughes

And now for the sad story of Martin von Strasser Caidin, prolific author and pilot who died of cancer at age 69 in 1997. He wrote some 140 books, most of them nonfiction, about such topics as aviation, space flight, and UFOs, but more than 30 were science-fiction novels. His novel *Cyborg* was the basis of the TV series *The Six Million Dollar Man* and its spinoff *The Bionic Woman*. *Marooned* became a movie with Gregory Peck and Gene Hackman as stranded astronauts. Other novels include *The Messiah Stone, Exit Earth, Prison Ship, Three Corners to Nowhere, Zoboa,* and *Ghosts in the Air.*

Why do I mention Caidin here? Because in his declining years he convinced himself that his mind could rotate psychic motors. His home contained a "target room" filled with little paper, cardboard, and foil "umbrellas" suspended on needles and toothpicks. The room had no ventilation. Caidin would sit for hours in an adjoining room, staring through a window and trying to control the movements of the little rotors with his brain. The devices behaved erratically; sometimes not moving, sometimes rotating one way, sometimes the other way. The motions were surely caused by heat convection currents, or perhaps by air moving under doors, but Caidin believed he had proved the reality of a mysterious force totally unknown to science, a force responsible for what he called telekinesis. He was convinced that his discovery would lift him into the ranks of Tesla and Einstein! He couldn't understand why physicists ignored him. Here are a few references:

M. Caidin, "Fiction That Ain't," *New Destinies, Summer 1988, pp. 211–22.*

M. Caidin, "Telekinesis Demonstrated," *Fate,* January 1994, pp. 52–68. Includes photos of Caidin's "rotors."

W. Cox, "Mind Moving Matter," *Florida Today,* January 8, 1989, pp.1D–2D.

W. Robkin, "Martin Caidin," *Starlog,* October 1985, pp. 63–65.

In concluding this long addendum, let me pass along some sage advice supplied by my brother Jim. Every responsible science writer should constantly keep in mind the following four words: "Accuracy above all."

Selected Titles by the Author on Mathematics

*T*he "Mathematical Games" columns in this book are arranged by subject. The entire collection of "Mathematical Games" columns that appeared in *Scientific American* has been brought together by Martin Gardner in chronological order in the series of books listed below. In 2001 all of them were in print or being reissued. If you enjoyed the selection here, you will want to look at these books by Martin Gardner as well.

Martin Gardner's "Mathematical Games" books:

The Scientific American *Book of Mathematical Puzzles and Games,* Simon and Schuster, 1959 (later retitled *The First* Scientific American *Book of Mathematical Puzzles and Games*), republished as *Hexaflexagons and Other Mathematical Diversions: The* Scientific American *Book of Mathematical Puzzles and Games,* University of Chicago Press, 1988.

The Second Scientific American *Book of Mathematical Puzzles & Diversions,* Simon and Schuster, 1961, republished by University of Chicago Press, 1987.

New Mathematical Diversions from Scientific American, Simon and Schuster, 1966; revised edition, *New Mathematical Diversions,* The Mathematical Association of America, 1995.

The Unexpected Hanging and Other Mathematical Diversions, Simon and Schuster, 1969, republished by University of Chicago Press, 1991.

The Numerology of Dr. Matrix, Simon and Schuster, 1967; expanded and republished as *The Incredible Dr. Matrix,* Scribner, 1979. See below.

Martin Gardner's Sixth Book of Mathematical Games from Scientific American, W. H. Freeman and Company, 1971, revision in press by The Mathematical Association of America, 2001.

Mathematical Carnival, Knopf, 1975, revised with a foreword by John H. Conway, The Mathematical Association of America, 1992.

Mathematical Magic Show, Knopf, 1977, revised with a foreword by Ronald L. Graham, The Mathematical Association of America, 1990.

Mathematical Circus, Knopf, 1979, revised with a foreword by Donald E. Knuth, The Mathematical Association of America, 1992.

The Incredible Dr. Matrix, Scribner, 1979 (see *The Numerology of Dr. Matrix,* above), revised and expanded as *The Magic Numbers of Dr. Matrix,* Prometheus Books, 1985.

Wheels, Life, and Other Mathematical Amusements, W. H. Freeman and Company, 1983.

Knotted Doughnuts and Other Mathematical Entertainments, W. H. Freeman and Company, 1986.

Time Travel and Other Mathematical Bewilderments, W. H. Freeman and Company, 1988.

Penrose Tiles to Trapdoor Ciphers, W. H. Freeman and Company, 1989.

Fractal Music, Hypercards, and More . . . : Mathematical Recreations from Scientific American, W. H. Freeman and Company, 1992.

Last Recreations: Hydras, Eggs, and Other Mathematical Mystifications, Copernicus Books, Springer Verlag, 1997.

Index

Page numbers in *italics* refer to illustrations.

Index

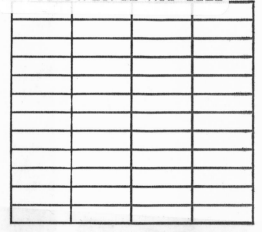